Inductive Databases and
Constraint-Based Data Mining

Sašo Džeroski • Bart Goethals • Pance Panov
Editors

Inductive Databases and Constraint-Based Data Mining

 Springer

Editors
Sašo Džeroski
Jožef Stefan Institute
Dept. of Knowledge Technologies
Jamova cesta 39
SI-1000 Ljubljana
Slovenia
Saso.Dzeroski@ijs.si

Panče Panov
Jožef Stefan Institute
Dept. of Knowledge Technologies
Jamova cesta 39
SI-1000 Ljubljana
Slovenia
Pance.Panov@ijs.si

Bart Goethals
University of Antwerp
Mathematics and Computer Science Dept.
Middelheimlaan 1
B-2020 Antwerpen
Belgium
Bart.Goethals@ua.ac.be

ISBN 978-1-4899-8217-9   ISBN 978-1-4419-7738-0 (eBook)
DOI 10.1007/978-1-4419-7738-0
Springer New York Dordrecht Heidelberg London

Springer is part of Springer Science+Business Media (www.springer.com)

Preface

This book is about inductive databases and constraint-based data mining, emerging research topics lying at the intersection of data mining and database research. The aim of the book as to provide an overview of the state-of- the art in this novel and exciting research area. Of special interest are the recent methods for constraint-based mining of global models for prediction and clustering, the unification of pattern mining approaches through constraint programming, the clarification of the relationship between mining local patterns and global models, and the proposed integrative frameworks and approaches for inducive databases. On the application side, applications to practically relevant problems from bioinformatics are presented.

Inductive databases (IDBs) represent a database view on data mining and knowledge discovery. IDBs contain not only data, but also generalizations (patterns and models) valid in the data. In an IDB, ordinary queries can be used to access and manipulate data, while inductive queries can be used to generate (mine), manipulate, and apply patterns and models. In the IDB framework, patterns and models become "first-class citizens" and KDD becomes an extended querying process in which both the data and the patterns/models that hold in the data are queried.

The IDB framework is appealing as a general framework for data mining, because it employs declarative queries instead of ad-hoc procedural constructs. As declarative queries are often formulated using constraints, inductive querying is closely related to constraint-based data mining. The IDB framework is also appealing for data mining applications, as it supports the entire KDD process, i.e., nontrivial multi-step KDD scenarios, rather than just individual data mining operations.

The interconnected ideas of inductive databases and constraint-based mining have the potential to radically change the theory and practice of data mining and knowledge discovery. The book provides a broad and unifying perspective on the field of data mining in general and inductive databases in particular. The 18 chapters in this state-of-the-art survey volume were selected to present a broad overview of the latest results in the field.

Unique content presented in the book includes constraint-based mining of global models for prediction and clustering, including predictive models for structured outputs and methods for bi-clustering; integration of mining local (frequent) patterns and global models (for prediction and clustering); constraint-based mining through constraint programming; integrative IDB approaches at the system and framework level; and applications to relevant problems that attract strong interest in the bioinformatics area. We hope that the volume will increase in relevance with time, as we witness the increasing trends to store patterns and models (produced by humans or learned from data) in addition to data, as well as retrieve, manipulate, and combine them with data.

This book contains sixteen chapters presenting recent research on the topics of inductive databases and queries, as well as constraint-based data, conducted within the project IQ (Inductive Queries for mining patterns and models), funded by the EU under contract number IST-2004-516169. It also contains two chapters on related topics by researchers coming from outside the project (Siebes and Puspitaningrum; Wicker et al.)

This book is divided into four parts. The first part describes the foundations of and frameworks for inductive databases and constraint-based data mining. The second part presents a variety of techniques for constraint-based data mining or inductive querying. The third part presents integration approaches to inductive databases. Finally, the fourth part is devoted to applications of inductive querying and constraint-based mining techniques in the area of bioinformatics.

The first, introductory, part of the book contains four chapters. Džeroski first introduces the topics of inductive databases and constraint-based data mining and gives a brief overview of the area, with a focus on the recent developments within the IQ project. Panov et al. then present a deep ontology of data mining. Blockeel et al. next present a practical comparative study of existing data-mining/inductive query languages. Finally, De Raedt et al. are concerned with mining under composite constraints, i.e., answering inductive queries that are Boolean combinations of primitive constraints.

The second part contains six chapters presenting constraint-based mining techniques. Besson et al. present a unified view on itemset mining under constraints within the context of constraint programming. Bringmann et al. then present a number of techniques for integrating the mining of (frequent) patterns and classification models. Struyf and Džeroski next discuss constrained induction of predictive clustering trees. Bingham then gives an overview of techniques for finding segmentations of sequences, some of these being able to handle constraints. Cerf et al. discuss constrained mining of cross-graph cliques in dynamic networks. Finally, De Raedt et al. introduce ProbLog, a probabilistic relational formalism, and discuss inductive querying in this formalism.

The third part contains four chapters discussing integration approaches to inductive databases. In the Mining Views approach (Blockeel et al.), the user can query the collection of all possible patterns as if they were stored in traditional relational tables. Wicker et al. present SINDBAD, a prototype of an inductive database system that aims to support the complete knowledge discovery process. Siebes and Puspitaningrum discuss the integration of inductive and ordinary queries (relational algebra). Finally, Vanschoren and Blockeel present experiment databases.

The fourth part of the book, contains four chapters dealing with applications in the area of bioinformatics (and chemoinformatics). Vens et al. describe the use of predictive clustering trees for predicting gene function. Slavkov and Džeroski describe several applications of predictive clustering trees for the analysis of gene expression data. Rigotti et al. describe how to use mining of frequent patterns on strings to discover putative transcription factor binding sites in gene promoter sequences. Finally, King et al. discuss a very ambitious application scenario for inductive querying in the context of a robot scientist for drug design.

The content of the book is described in more detail in the last two sections of the introductory chapter by Džeroski.

We would like to conclude with a word of thanks to those that helped bring this volume to life: This includes (but is not limited to) the contributing authors, the referees who reviewed the contributions, the members of the IQ project and the various funding agencies. A more complete listing of acknowledgements is given in the Acknowledgements section of the book.

September 2010   Sašo Džeroski
                                                                                                                 Bart Goethals
                                                                                                                 Panče Panov

Acknowledgements

Heartfelt thanks to all the people and institutions that made this volume possible and helped bring it to life.

First and foremost, we would like to thank the contributing authors. They did a great job, some of them at short notice. Also, most of them showed extraordinary patience with the editors.

We would then like to thank the reviewers of the contributed chapters, whose names are listed in a separate section. Each chapter was reviewed by at least two (on average three) referees. The comments they provided greatly helped in improving the quality of the contributions.

Most of the research presented in this volume was conducted within the project IQ (Inductive Queries for mining patterns and models). We would like to thank everybody that contributed to the success of the project: This includes the members of the project, both the contributing authors and the broader research teams at each of the six participating institutions, the project reviewers and the EU officials handling the project. The IQ project was funded by the European Comission of the EU within FP6-IST, FET branch, under contract number FP6-IST-2004-516169.

In addition, we want to acknowledge the following funding agencies:

- Sašo Džeroski is currently supported by the Slovenian Research Agency (through the research program *Knowledge Technologies* under grant P2-0103 and the research projects *Advanced machine learning methods for automated modelling of dynamic systems* under grant J2-0734 and *Data Mining for Integrative Data Analysis in Systems Biology* under grant J2-2285) and the European Commission (through the FP7 project PHAGOSYS *Systems biology of phagosome formation and maturation - modulation by intracellular pathogens* under grant number HEALTH-F4-2008-223451). He is also supported by the Centre of Excellence for Integrated Approaches in Chemistry and Biology of Proteins (operation no. OP13.1.1.2.02.0005 financed by the European Regional Development Fund (85%) and the Slovenian Ministry of Higher Education, Science and Technology (15%)), as well as the Jozef Stefan International Postgraduate School in Ljubljana.

- Bart Goethals wishes to acknowledge the support of FWO-Flanders through the project "Foundations for inductive databases".
- Panče Panov is supported by the Slovenian Research Agency through the re-search projects *Advanced machine learning methods for automated modelling of dynamic systems* (under grant J2-0734) and *Data Mining for Integrative Data Analysis in Systems Biology* (under grant J2-2285).

Finally, many thanks to our Springer editors, Jennifer Maurer and Melissa Fearon, for all the support and encouragement.

September 2010   Sašo Džeroski
                                                                  Bart Goethals
                                                                  Panče Panov

List of Reviewers

Hendrik Blockeel	Katholieke Universiteit Leuven, Belgium
Marko Bohanec	Jožef Stefan Institute, Slovenia
Jean-Francois Boulicaut	University of Lyon, INSA Lyon, France
Mario Boley	University of Bonn and Fraunhofer IAIS, Germany
Toon Calders	Eindhoven Technical University, Netherlands
Vineet Chaoji	Yahoo! Labs, Bangalore, India
Amanda Clare	Aberystwyth University, United Kingdom
James Cussens	University of York, United Kingdom
Tomaž Curk	University of Ljubljana, Ljubljana, Slovenia
Ian Davidson	University of California - Davis, USA
Luc Dehaspe	Katholieke Universiteit Leuven, Belgium
Luc De Raedt	Katholieke Universiteit Leuven, Belgium
Jeroen De Knijf	University of Antwerp, Belgium
Tijl De Bie	University of Bristol, United Kingdom
Sašo Džeroski	Jožef Stefan Institute, Slovenia
Elisa Fromont	University of Jean Monnet, France
Gemma C. Garriga	University of Paris VI, France
Christophe Giraud-Carrier	Brigham Young University, USA
Jiawei Han	University of Illinois at Urbana-Champaign, USA
Hannes Heikinheimo	Aalto Universit, Finland
Cristoph Hema	In Silico Toxicology, Switzerland
Andreas Karwath	Albert-Ludwigs-Universitat, Germany
Jörg-Uwe Kietz	University of Zurich, Switzerland
Arno Knobbe	University of Leiden, Netherlands
Petra Kralj Novak	Jožef Stefan Institute, Slovenia
Stefan Kramer	Technische Universität München, Germany
Rosa Meo	University of Torino, Italy
Pauli Miettinen	Max-Planck-Institut für Informatik, Germany
Siegfried Nijssen	Katholieke Universiteit Leuven, Belgium
Markus Ojala	Aalto University, Finland
Themis Palpanas	University of Trento, Italy

Contents

Part I
Introduction

Chapter 1
Inductive Databases and Constraint-based Data Mining: Introduction and Overview

Sašo Džeroski

Abstract We briefly introduce the notion of an inductive database, explain its relation to constraint-based data mining, and illustrate it on an example. We then discuss constraints and constraint-based data mining in more detail, followed by a discussion on knowledge discovery scenarios. We further give an overview of recent developments in the area, focussing on those made within the IQ project, that gave rise to most of the chapters included in this volume. We finally outline the structure of the book and summarize the chapters, following the structure of the book.

1.1 Inductive Databases

Inductive databases (IDBs, Imielinski and Mannila 1996, De Raedt 2002a) are an emerging research area at the intersection of data mining and databases. Inductive databases contain both data and patterns (in the broader sense, which includes frequent patterns, predictive models, and other forms of generalizations). IDBs embody a database perspective on knowledge discovery, where knowledge discovery processes become query sessions. KDD thus becomes an extended querying process (Imielinski and Mannila 1996) in which both the data and the patterns that hold (are valid) in the data are queried.

Roughly speaking, an *inductive database* instance contains: (1) Data (e.g., a relational database, a deductive database), (2) Patterns (e.g., itemsets, episodes, subgraphs, substrings, ...), and (3) Models (e.g., classification trees, regression trees, regression equations, Bayesian networks, mixture models, ...). The difference between patterns (such as frequent itemsets) and models (such as regression trees) is that patterns are *local* (they typically describe properties of a subset of the data),

Sašo Džeroski
Jožef Stefan Institute, Jamova cesta 39, 1000 Ljubljana, Slovenia
e-mail: saso.dzeroski@ijs.si

3

whereas models are *global* (they characterize the entire data set). Patterns are typically used for descriptive purposes and models for predictive ones.

A *query language* for an inductive database is an extension of a database query
language that allows us to: (1) select, manipulate and query data in the database as in
current DBMSs, (2) select, manipulate and query "interesting" patterns and models
(e.g., patterns that satisfy *constraints* w.r.t. frequency, generality, etc. or models that
satisfy *constraints* w.r.t. accuracy, size, etc.), and (3) *match* patterns or models with
data, e.g., select the data in which some patterns hold, or predict a property of the
data with a model.

To clarify what is meant by the terms inductive database and inductive query, we
illustrate them by an example from the area of bio-/chemo-informatics.

1.1.1 Inductive Databases and Queries: An Example

To provide an intuition of what an inductive query language has to offer, consider the
task of discovering a model that predicts whether chemical compounds are toxic or
not. In this context, the data part of the IDB will consist of one or more sets of compounds. In our illustration below, there are two sets: the *active* (toxic) and the *inactive* (non-toxic) compounds. Assume, furthermore, that for each of the compounds,
the two dimensional (i.e., graph) structure of their molecules is represented within
the database, together with a number of attributes that are related to the outcome of
the toxicity tests. The database query language of the IDB will allow the user (say
a predictive toxicology scientist) to retrieve information about the compounds (i.e.,
their structure and properties). The inductive query language will allow the scientist
to generate, manipulate and apply patterns and models of interest.

As a first step towards building a predictive model, the scientist may want
to find local patterns (in the form of compound substructures or molecular fragments), that are "interesting", i.e., satisfy certain constraints. An example induc-
tive query may be written as follows: $F = \{\tau | (\tau \in AZT) \wedge (freq(\tau, Active) \geq$
15%$) \wedge (freq(\tau, Inactive) \leq 5\%)\}$. This should be read as: "Find all molecular
fragments that appear in the compound AZT (which is a drug for AIDS), occur
frequently in the active compounds ($\geq 15\%$ of them) and occur infrequently in the
inactive ones ($\leq 5\%$ of them)."

Once an interesting set of patterns has been identified, they can be used as descriptors (attributes) for building a model (e.g., a decision tree that predicts activity).
A data table can be created by first constructing one feature/column for each pattern,
then one example/row for each data item. The entry at a given column and row has
value "true" if the corresponding pattern (e.g., fragment) appears in the corresponding data item (e.g., molecule). The table could be created using a traditional query
in a database query language, combined with IDB matching primitives.

Suppose we have created a table with columns corresponding to the molecular
fragments *F* returned by the query above and rows corresponding to compounds
in *Active*∪*Inactive*, and we want to build a global model (decision tree) that dis-

tinguishes between active and inactive compounds. The toxicologist may want to constrain the decision tree induction process, e.g., requiring that the decision tree contains at most k leaves, that certain attributes are used before others in the tree, that the internal tests split the nodes in (more or less) proportional subsets, etc. She may also want to impose constraints on the accuracy of the induced tree.

Note that in the above scenario, a sequence of queries is used. This requires that the *closure property* be satisfied: the result of an inductive query on an IDB instance should again be an IDB instance. Through supporting the processing of sequences of inductive queries, IDBs would support the entire KDD process, rather than individual data mining steps.

1.1.2 Inductive Queries and Constraints

In inductive databases (Imielinski and Mannila 1996), patterns become "first-class citizens" and can be stored and manipulated just like data in ordinary databases. Ordinary queries can be used to access and manipulate data, while inductive queries (IQs) can be used to generate (mine), manipulate, and apply patterns. KDD thus becomes an extended querying process in which both the data and the patterns that hold (are valid) in the data are queried. In IDBs, the traditional KDD process model where steps like pre-processing, data cleaning, and model construction follow each other in succession, is replaced by a simpler model in which all operations (pre-processing, mining, post-processing) are queries to an IDB and can be interleaved in many different ways.

Given an IDB that contains data and patterns (or other types of generalizations, such as models), several different types of queries can be posed. Data retrieval queries use only the data and their results are also data: no pattern is involved in the query. In IDBs, we can also have cross-over queries that combine patterns and data in order to obtain new data, e.g., apply a predictive model to a dataset to obtain predictions for a target property. In processing patterns, the patterns are queried without access to the data: this is what is usually done in the post-processing stages of data mining. Inductive (data mining) queries use the data and their results are patterns (generalizations): new patterns are generated from the data: this corresponds to the traditional data mining step.

A general statement of the problem of data mining (Mannila and Toivonen 1997) involves the specification of a language of patterns (generalizations) and a set of constraints that a pattern has to satisfy. The constraints can be language constraints and evaluation constraints: The first only concern the pattern itself, while the second concern the validity of the pattern with respect to a given database. Constraints thus play a central role in data mining and constraint-based data mining (CBDM) is now a recognized research topic (Bayardo 2002). The use of constraints enables more efficient induction and focusses the search for patterns on patterns likely to be of interest to the end user.

In the context of IDBs, inductive queries consist of constraints. Inductive queries can involve language constraints (e.g., find association rules with item A in the head) and evaluation constraints, which define the validity of a pattern on a given dataset (e.g., find all item sets with support above a threshold or find the 10 association rules with highest confidence).

Different types of data and patterns have been considered in data mining, including frequent itemsets, episodes, Datalog queries, and graphs. Designing inductive databases for these types of patterns involves the design of inductive query languages and solvers for the queries in these languages, i.e., CBDM algorithms. Of central importance is the issue of defining the primitive constraints that can be applied for the chosen data and pattern types, that can be used to compose inductive queries. For each pattern domain (type of data, type of pattern, and primitive constraints), a specific solver is designed, following the philosophy of constraint logic programming (De Raedt 2002b).

1.1.3 The Promise of Inductive Databases

While knowledge discovery in databases (KDD) and data mining have enjoyed great popularity and success over the last two decades, there is a distinct lack of a generally accepted framework for data mining (Fayyad et al. 2003). In particular, no framework exists that can elegantly handle simultaneously the mining of complex/structured data, the mining of complex (e.g., relational) patterns and use of domain knowledge, and support the KDD process as a whole, three of the most challenging/important research topics in data mining (Yang and Wu 2006).

The IDB framework is an appealing approach towards developing a generally accepted framework/theory for data mining, as it employs declarative queries instead of ad-hoc procedural constructs: Namely, in CBDM, the conditions/constraints that a pattern has to satisfy (to be considered valid/interesting) are stated explicitly and are under direct control of the user/data miner. The IDB framework holds the promise of facilitating the formulation of an "algebra" for data mining, along the lines of Codd's relational algebra for databases (Calders et al. 2006b, Johnson et al. 2000).

Different types of structured data have been considered in CBDM. Besides itemsets, onther types of frequent/local patterns have been mined under constraints, e.g., on strings, sequences of events (episodes), trees, graphs and even in a first-order logic context (patterns in probabilistic relational databases). More recently, constraint-based approaches to structured prediction have been considered, where models (such as tree-based models) for predicting hierarchies of classes or sequences / time series are induced under constraints.

Different types of local patterns and global models have been considered as well, such as rule-based predictive models and tree-based clustering models. When learning in a relational setup, background / domain knowledge is naturally taken into account. Also, the constraints provided by the user in CBDM can be viewed as a form of domain knowledge that focuses the search for patterns / model towards interesting and useful ones.

The IDB framework is also appealing for data mining applications, as it supports the entire KDD process (Boulicaut et al. 1999). In inductive query languages, the results of one (inductive) query can be used as input for another. Nontrivial multi-step KDD scenarios can be thus supported in IDBs, rather than just single data mining operations.

1.2 Constraint-based Data Mining

"Knowledge discovery in databases (KDD) is the non-trivial process of identifying valid, novel, potentially useful, and ultimately understandable patterns in data", state Fayyad et al. (1996). According to this definition, data mining (DM) is the central step in the KDD process concerned with applying computational techniques (i.e., data mining algorithms implemented as computer programs) to actually find patterns that are valid in the data. In constraint-based data mining (CBDM), a pattern/model is valid if it satisfies a set of constraints.

The basic concepts/entities of data mining include data, data mining tasks, and generalizations (e.g., patterns and models). The validity of a generalization on a given set of data is related to the data mining task considered. Below we briefly discuss the basic entities of data mining and the task of CBDM.

1.2.1 Basic Data Mining Entities

Data. A data mining algorithm takes as input a set of data. An individual datum in the data set has its own structure, e.g., consists of values for several attributes, which may be of different types or take values from different ranges. We assume all data items are of the same type (and share the same structure).

More generally, we are given a data type T and a set of data D of this type. It is of crucial importance to be able to deal with structured data, as these are attracting an ever increasing amount of attention within data mining. The data type T can thus be an arbitrarily complex data type, composed from a set of basic/primitive types (such as Boolean and Real) by using type constructors (such as Tuple, Set or Sequence).

Generalizations. We will use the term generalization to denote the output of different data mining tasks, such as pattern mining, predictive modeling and clustering. Generalizations will thus include probability distributions, patterns (in the sense of frequent patterns), predictive models and clusterings. All of these are defined on a given type of data, except for predictive models, which are defined on a pair of data types. Note that we allow arbitrary (arbitrarily complex) data types. The typical case in data mining considers a data type $T = \text{Tuple}(T_1, \ldots, k)$, where each of $T_1, \ldots, T_k$ is Boolean, Discrete or Real.

We will discuss briefly here local patterns and global models (predictive models and clusterings). Note that both are envisaged as first-class citizens of inductive databases. More detailed discussions of all types of generalizations are given by Panov et al. (2010/this volume) and Džeroski (2007).

A *pattern P on type T* is a Boolean function on objects of type T: A pattern on type T is true or false on an object of type T. We restrict the term pattern here to the sense that it is most commonly used, i.e., in the sense of frequent pattern mining. A *predictive model M for types* T_d, $\mathbf{T}_c$ is a function that takes an object of type T_d (description) and returns one of type $\mathbf{T}_c$ (class/target). We allow both T_d and T_c to be arbitrarily complex data types, with classification and regression as special cases (when T_c has nominal, respectively numeric values). A *clustering C on a set of objects S of type T* is a function from S to $\{1, \dots, k\}$, where k is the number of clusters (with $k \le |S|$). It partitions a set of objects into subsets called clusters by mapping each object to a cluster identifier.

Data Mining Tasks. In essence, the task of data mining is to produce a generalization from a given set of data. A plethora of data mining tasks has been considered so far in the literature, with four covering the majority of data mining research: approximating the (joint) probability distribution, clustering, learning predictive models, and finding valid (frequent) patterns. We will focus here on the last two of these.

In *learning a predictive model*, we are given a dataset consisting of example input/output pairs (d, c), where each d is of type T_d and each c is of type T_c. We want to find a model m (mapping from T_d to T_c), for which the observed and predicted outputs, i.e., c and $\hat{c} = m(d)$, match closely. In *pattern discovery*, the task is to find all local patterns from a given pattern language (class) that satisfy the required conditions. A prototypical instantiation of this task is the task of finding frequent itemsets (sets of items, such as $\{bread, butter\}$), which occur frequently (in a sufficiently high proportion) in a given set of transactions (market baskets) (Aggrawal et al 1993). In *clustering*, we are given a set of examples (object descriptions), and the task is to partition these examples into subsets, called clusters. The notion of a distance (or conversely, similarity) is crucial here: The goal of clustering is to achieve high similarity between objects within a cluster (intra-cluster similarity) and low similarity between objects from different clusters (inter-cluster similarity).

1.2.2 The Task(s) of (Constraint-Based) Data Mining

Having set the scene, we can now attempt to formulate a very general version of the problem addressed by data mining. We are given a dataset D, consisting of objects of type T. We are also given a data mining task, such as learning a predictive model or pattern discovery. We are further given C_G a family/class of generalizations (patterns/models), such as decision trees, from which to find solutions to the data mining task at hand. Finally, a set of constraints C is given, concerning both the syntax (form) and semantics (validity) that the generalizations have to satisfy.

The problem addressed by constraint-based data mining (CBDM) is to find a set of generalizations G from C_G that satisfy the constraints in C: A desired cardinality on the solution set is usually specified.

In the above formulation, all of data mining is really constraint-based. We argue that the 'classical' formulations of and approaches to data mining tasks, such as clustering and predictive modelling, are a special case of the above formulation. A major difference between the 'classical' data mining paradigm and the 'modern' constraint-based one is that the former typically considers only one quality metric, e.g., minimizes predictive error or intra-cluster variance, and produces only one solution (predictive model or clustering).

A related difference concerns the fact that most of the 'classical' approaches to data mining are heuristic and do not give any guarantees regarding the solutions. For example, a decision tree generated by a learning algorithm is typically not guaranteed to be the smallest or most accurate tree for the given dataset. On the other hand, CBDM approaches have typically been concerned with the development of so-called 'optimal solvers', i.e., data mining algorithms that return the complete set of solutions that satisfy a given set of constraints or the k best solutions (e.g., the k itemsets with highest correlation to a given target).

1.3 Types of Constraints

Constraints in CBDM are propositions/statements about generalizations (e.g., patterns or models). In the most basic setting, the propositions are either true or false (Boolean valued): If true, the generalization satisfies the constraint. In CBDM, we are seeking generalizations that satisfy a given set of constraints.

Many types of constraints are currently used in CBDM, which can be divided along several dimensions. Along the first dimension, we distinguish between primitive and composite constraints. Along the second dimension, we distinguish between language and evaluation constraints. Along the third dimension, we have Boolean (or hard) constraints, soft constraints and optimization constraints. In this section, we discuss these dimensions in some detail.

1.3.1 Primitive and Composite Constraints

Recall that constraints in CBDM are propositions on generalizations. Some of these propositions are atomic in nature (and are not decomposable into simpler propositions). In mining frequent itemsets, the constraints "item *bread* must be contained in the itemsets of interest" and "itemsets of interest should have a frequency higher than 10" are atomic or primitive constraints.

Primitive constraints can be combined by using boolean operators, i.e., negation, conjunction and disjunction. The resulting constraints are called composite constraints. The properties of the composite constrains (such as monotonicity/anti-monotonicity discussed below) depend on the properties of the primitive constraints and the operators used to combine them.

1.3.2 Language and Evaluation Constraints

Constraints typically refer to either the form / syntax of generalizations or their semantics / validity with respect to the data. In the first case, they are called language constraints, and in the second evaluation constraints. Below we discuss primitive language and evaluation constraints. Note that these can be used to form composite language constraints, composite evaluation constraints, and composite constraints that mix language and evaluation primitives.

Language constraints concern the syntax / representation of a pattern/model, i.e., refer only to its form. We can check whether they are satisfied or not without accessing the data that we have been given as a part of the data mining task. If we are in the context of inductive databases and queries, post-processing queries on patterns / models are composed of language constraints.

A commonly used type of language constraints is that of subsumption constraints. For example, in the context of mining frequent itemsets, we might be interested only in itemsets where a specific item, e.g., *beer* occurs (that is itemsets that subsume *beer*). Or, in the context of learning predictive models, we may be interested only in decision trees that have a specific attribute in the root node.

Another type of language constraints involves (cost) functions on patterns / models. An example of these is the size of a decision tree: We can look for decision trees of at most ten nodes. Another example would be the cost of an itemset (market basket), in the context where each item has a price. The cost functions as discussed here are mappings from the representation of a pattern/model to non-negative reals: Boolean (hard) language constraints put thresholds on the values of these functions.

Evaluation constraints concern the semantics of patterns / models, in particular as applied to a given set of data. Evaluation constraints typically involve evaluation functions, comparing them to constant thresholds. Evaluation functions measure the validity of patterns/models on a given set of data.

Evaluation functions take as input a pattern or a model and return a real value as output. The set of data is an additional input to the evaluation functions. For example, the frequency of a pattern on a given dataset is an evaluation function, as is the classification error of a predictive model. Evaluation constraints typically compare the value of an evaluation function to a constant threshold, e.g., minimum support or maximum error.

Somewhat atypical evaluation constraints are used in clustering. Must-link constraints specify that two objects x, y in a dataset should be assigned to the same cluster by the clustering C, i.e., $C(x) = C(y)$, while cannot-link constraints specify that x, y should be assigned to different clusters $C(x) \neq C(y)$. These constraints do not concern the overall quality of a clustering, but still concern its semantics.

Constraints on the pattern / model may also involve some general property of the pattern / model, which does not depend on the specific dataset considered. For example, we may only consider predictive models that are convex or symmetric or monotonic in certain variables. These properties are usually defined over the entire domain of the model, i.e., the corresponding data type, but may be checked for the specific dataset at hand.

1.3.3 Hard, Soft and Optimization Constraints

Hard constraints in CBDM are Boolean functions on patterns / models. This means that a constraints is either satisfied or not satisfied. The fact that constraints actually define what patterns are valid or interesting in data mining, and that interestingness is not a dichotomy (Bistarelli and Bonchi 2005), has lead to the introduction of so-called soft constraints.

Soft constraints do not dismiss a pattern for violating a constraint; rather, the pattern incurring a penalty for violating a constraint. In the cases where we typically consider a larger number of binary constraints, such as must-link and cannot-link constraints in constrained clustering (Wagstaff and Cardie 2000), a fixed penalty may be assigned for violating each constraint. In case we are dealing with evaluation constraints that compare an evaluation function to a threshold, the penalty incurred by violating the constraint may depend on how badly the constraint is violated. For example, if we have a size threshold of five, and the actual size is six, a smaller penalty would be incurred as compared to the case where the actual size is twenty.

In the hard constraint setting, a pattern/model is either a solution or not. In the soft constraint setting, all patterns/models are solutions to a different degree. Patterns with lower penalty satisfy the constraints better (to a higher degree), and patterns that satisfy the constraint(s) completely get zero penalty. In the soft-constraint version of CBDM, we look for patterns with minimum penalty.

Optimization constraints allow us to ask for (a fixed-size set of) patterns/models that have a maximal/minimal value for a given cost or evaluation function. Example queries with such constraints could ask for the k most frequent itemsets or the top k correlated patterns. We might also ask for the most accurate decision tree of size five, or the smallest decision tree with classification accuracy of at least 90%.

In this context, optima for the cost/evaluation function at hand are searched for over the entire class of patterns/models considered, in the case the optimization constraint is the only one given. But, as illustrated above, optimization constraints often appear in conjunction with (language or evaluation) Boolean constraints. In this case, optima are searched for over the patterns/models that satisfy the given Boolean constraints.

1.4 Functions Used in Constraints

This section discusses the functions used to compose constraints in CBDM. Language constraints use language cost functions, while evaluation constraints use evaluation functions. We conclude this section by discussing monotonicity, an important property of such functions, and closedness, an important property of patterns.

1.4.1 Language Cost Functions

The cost functions that are used in language constraints concern the representation of generalizations (patterns/models/...). Most often, these functions are related to the size/complexity of the representation. They are different for different classes of generalizations, e.g., for itemsets, mixture models of Gaussians, linear models or decision trees. For itemsets, the size is the cardinality of the itemset, i.e., the number of items in it. For decision trees, it can be the total number of nodes, the number of leaves or the depth of the tree. For linear models, it can be the number of variables (with non-zero coefficients) included in the model.

More general versions of cost functions involve costs of the individual language elements, such as items or attributes, and sum/aggregate these over all elements appearing in the pattern/model. These are motivated by practical considerations, e.g., costs for items in an itemset and total cost of a market basket. In the context of predictive models, e.g., attribute-value decision trees, it makes sense to talk about prediction cost, defined as the total cost of all attributes used by the model. For example, in medical applications where the attributes correspond to expensive lab tests, it might be useful to upper-bound the prediction cost of a decision tree.

Language constraints as commonly used in CBDM involve thresholds on the values of cost functions (e.g., find a decision tree of size at most ten leaves). They are typically combined with evaluation constraints, be it threshold or optimization (e.g., find a tree of size at most 10 with classification error of at most 10% or find a tree of size at most 10 and the smallest classification error). Also, optimization constraints may involve the language-related cost functions, e.g., find the smallest decision tree with classification error lower than 10%.

In the 'classical' formulations of and approaches to data mining tasks, scoring functions often combine evaluation functions and language cost functions. The typical score function is a linear combination of the two, i.e., $Score(G, D) = w_E \times Evaluation(G.function, D) + w_L \times LanguageCost(G.data)$, where G is the generalization (pattern/model) scored and D is the underlying dataset. For predictive modelling, this can translate to $Score = w_E \times Error + w_S \times Size$.

1.4.2 Evaluation Functions

The evaluation functions used in evaluation constraints are tightly coupled with the data mining task at hand. If we are solving a predictive modelling problem, the evaluation function used will most likely concern predictive error. If we are solving a frequent pattern mining problem, the evaluation function used will definitely concern the frequency of the patterns.

For the task of **pattern discovery**, with the discovery of frequent patterns as the prototypical instantiation, the primary evaluation function is frequency. Recall that patterns are Boolean functions, assigning a value of true or false to a data item. For a dataset D, the frequency of a pattern p is $f(p,D) = |\{e|e \in D, p(e) = true\}|$.

For **predictive models**, predictive error is the function typically used in constraints. The error function used crucially depends on the type of the target predicted. For a discrete target (classification), misclassification error/cost can be used; for a continuous target (regression), mean absolute error can be used.

In general, for a target of type T_c, we need a distance (or cost) function d_c on objects of type T_c to define the notion of predictive error. For a given model m and a dataset D, the average predictive error of the model is defined as $1/|D| \times \sum_{e=(a,t) \in D} d_c(t,m(a))$. For each example $e = (a,t)$ in the dataset, which consists of a descriptive (attribute) part a and target (class) part t, the prediction of the model $m(a)$ is obtained and its distance to the true class value t is calculated. Analogously, the notion of mean squared error would be defined as $1/|D| \times \sum_{e=(a,t) \in D} d_c^2(t,m(a))$.

The notion of cost-sensitive prediction has been recently gaining increasing amounts of attention in the data mining community. In this setting, the errors incurred by predicting x instead of y and predicting y instead of x, are typically not the same. The corresponding misprediction (analogous to misclassification) cost function is thus not symmetric, i.e., is not a distance. The notion of average misprediction cost can be defined as above, with the distance $d(x,y)$ replaced by a cost function $c(x,y)$.

Similar evaluation functions can be defined for **probabilistic predictive modeling**, a subtask of predictive modeling. For the data mining task of **clustering**, the quality of a clustering is typically evaluated with intra-cluster variance (ICV) in partition-based clustering. For density-based clustering, a variant of the task of **estimating the probability distribution**, scoring functions for distributions / densities are used, typically based on likelihood or log-likelihood (Hand et al. 2001).

1.4.3 Monotonicity and Closedness

The notion of monotonicity of an evaluation (or cost) function on a class of generalizations is often considered in CBDM. In mathematics, a function $f(x)$ is monotonic (monotonically increasing) if $\forall x,y : x < y \rightarrow f(x) \le f(y)$, i.e., the function preserves the $<$ order. If the function reverses the order, i.e., $\forall x,y : x < y \rightarrow f(x) \ge f(y)$, we call it monotonically decreasing.

In data mining, in addition to the order on Real numbers, we also have a generality order on the class of generalizations. The latter is typically induced by a refinement operator. We say that $g_1 \leq_{ref} g_2$ if g_2 can be obtained from g_1 through a sequence of refinements (and thus g_1 is more general than g_2): we will refer to this order as the refinement order.

An evaluation (or cost) function is called monotonic if it preserves the refinement order or anti-monotonic if it reverses it. More precisely, an evaluation function f is called monotonic if $\forall g_1, g_2 : g_1 \leq_{ref} g_2 \to f(g_1) \leq f(g_2)$ and anti-monotonic (or monotonically decreasing) if $\forall g_1, g_2 : g_1 \leq_{ref} g_2 \to f(g_1) \geq f(g_2)$.

Note that the above notions are defined for both evaluation functions / constraints and for language cost functions / constraints. In this context, the frequency of itemsets is anti-monotonic (it decreases monotonically with the refinement order). The total cost of an itemset and the total prediction cost of a decision tree, on the other hand, are monotonic.

In the CBDM literature (Boulicaut and Jeudy 2005), the refinement order considered is typically the subset relation on itemsets ($\leq_{ref}$ is identical to $\subseteq$). A constraint C (taken as a Boolean function) is considered monotonic if $i_1 \leq_{ref} i_2 \wedge C(i_1)$ implies $C(i_2)$. A maximum frequency constraint of the form $freq(i) \leq \theta$, where θ is a constant, is monotonic. Similarly, minimum frequency/support constraints of the form $freq(i) \geq \theta$, the ones most commonly considered in data mining, are anti-monotonic. A disjunction or a conjunction of anti-monotonic constraints is an anti-monotonic constraint. The negation of a monotonic constraint is anti-monotonic and vice versa.

The notions of monotonicity and anti-monotonicity are important because they allow for the design of efficient CBDM algorithms. Anti-monotonicity means that when a pattern does not satisfy a constraint C, then none of its refinements can satisfy C. It thus becomes possible to prune huge parts of the search space which can not contain interesting patterns. This has been studied within the learning as search framework (Mitchell, 1982) and the generic levelwise algorithm from (Mannila and Toivonen, 1997) has inspired many algorithmic developments.

Finally, let us mention the notion of closedness. A pattern (generalization) is closed, with respect to a given refinement operator $\leq_{ref}$ and evaluation function f, if refining the pattern in any way decreases the value of the evaluation function. More precisely, x is closed if $\forall y, x \leq_{ref} y : f(y) < f(x)$. This notion has primarily been considered in the context of mining frequent itemsets, where a refinement adds an item to an itemset and the evaluation function is frequency. There it plays an important role in condensed representations (Calders et al. 2005). However, it can be defined analogously for other types of patterns, as indicated above.

1.5 KDD Scenarios

Real-life applications of data mining typically require interactive sessions and involve the formulation of a complex sequence of inter-related inductive queries (including data mining operations), which we will call a KDD scenario (Boulicaut et al. 1999). Some of the inductive queries would generate or manipulate patterns, others would apply these patterns to a given dataset to form a new dataset, still others would use the new dataset to to build a predictive model. The ability to formulate and execute such sequences of queries crucially depends on the ability to use the output of one query as the input to another (i.e., on compositionality and closure).

KDD scenarios can be described at different levels of detail and precision and can serve multiple purposes. At the lowest level of detail, the specific data mining algorithms used and and their exact parameter settings employed would be included, as well as the specific data analyzed. Moving towards higher levels of abstraction, details can be gradually omitted, e.g., first the parameter setting of the algorithm, then the actual algorithm may be omitted but the class of generalizations produced by it can be kept, and finally the class of generalizations can be left out (but the data mining task kept).

At the most detailed level of description, KDD scenarios can serve to document the exact sequence of data mining operations undertaken by a human analyst on a specific task. This would facilitate, for example, the repetition of the entire sequence of analyses after an erroneous data entry has been corrected in the source data. At this level of detail, the scenario is a sequence of inductive queries in a formal (data mining) query language.

At higher levels of abstraction, the scenarios would enable the re-use of already performed analyses, e.g., on a new dataset of the same type. To abstract from a sequence of inductive queries in a query language, we might move from the specification of an actual dataset to a specification of the underlying data type and further to data types that are higher in a taxonomy/hierarchy of data types. Having taxonomies of data types, data mining tasks, generalizations and data mining algorithms would greatly facilitate the description of scenarios at higher abstraction levels: the abstraction can proceed along each of the respective ontologies.

We would like to argue that the explicit storage and manipulation of scenarios (e.g., by reducing/increasing the level of detail) would greatly facilitate their re-use. This in turn can increase the efficiency of the KDD process as a whole by reducing human effort in complex knowledge discovery processes. Thus, a major bottleneck in applying KDD in practice would be alleviated.

1.6 A Brief Review of Literature Resources

The notions of inductive databases and queries were introduced by Imielinski and Mannila (1996). The notion of constraint-based data mining (CBDM) appears in the data mining literature for the first time towards the end of the 20th century (Han et al. 1999). A special issue of the SIGKDD Explorations bulletin devoted to constraints in data mining was edited by Bayardo (2002).

A wide variety of research on IDBs and queries, as well as CBDM, was conducted within two EU-funded projects. The first (contract number FP5-IST 26469) took place from 2001 to 2004 and was titled cInQ (consortium on discovering knowledge with Inductive Queries). The second (contract number FP6-IST 516169) took place from 2005 to 2008 and was titled IQ (Inductive Queries for mining patterns and models).

A series of five workshops titled *Knowledge Discovery in Inductive Databases (KDID)* took place in the period of 2002 to 2006, each time in conjunction with the European Conference on Machine Learning and European Conference on Principles and Practice of Knowledge Discovery in Databases (ECML/PKDD).

- R. Meo, M. Klemettinen (Eds) *Proceedings International Workshop on Knowledge Discovery in Inductive Databases (KDID'02)*, Helsinki
- J-F. Boulicaut, S. Džeroski (Eds) *Proc 2nd Intl Wshp KDID'03*, Cavtat
- B. Goethals, A. Siebes (Eds) *Proc 3rd Intl Wshp KDID'04*, Pisa
- F. Bonchi, J-F. Boulicaut (Eds.) *Proc 4th Intl Wshp KDID'05*, Porto
- J. Struyf, S. Džeroski (Eds.) *Proc 5th Intl Wshp KDID'06*, Berlin

This was followed by a workshop titled *International Workshop on Constraint-based mining and learning (CMILE'07)* organized by S. Nijssen and L. De Raedt at ECML/PKDD'07 in Warsaw, Poland.

Revised and extended versions of the papers presented at the last three KDID workshops were published in edited volumes within the Springer LNCS series:

- B. Goethals, A. Siebes (Eds). *Knowledge Discovery in Inductive Databases 3rd Int. Workshop (KDID'04) Revised Selected and Invited Papers.* Springer LNCS 3377, 2005.
- F. Bonchi, J-F. Boulicaut (Eds.) *Knowledge Discovery in Inductive Databases 4th Int. Workshop (KDID'05) Revised Selected and Invited Papers.* Springer LNCS Volume 3933, 2006.
- S. Džeroski and J. Struyf (Eds.) *Knowledge Discovery in Inductive Databases 5th Int. Workshop (KDID'05) Revised Selected and Invited Papers.* Springer LNCS Volume 4747, 2007.

Two edited volumes resulted from the cInQ project.

- R. Meo, P-L. Lanzi, M. Klemettinen (Eds) *Database Support for Data Mining Applications - Discovering Knowledge with Inductive Queries.* Springer- LNCS 2682, 2004.
- J-F. Boulicaut, L. De Raedt, and H. Mannila (Eds) *Constraint-based mining and inductive databases.* Springer- LNCS 3848, 2005.

The first contains among others revised versions of KDID'02 papers. The second contains an overview of the major results of the cInQ project and related research outside the project.

The most recent collection on the topic of CBDM is devoted to constrained clustering.

- S. Basu, I. Davidson, K. Wagstaff (Eds.) *Clustering with Constraints.* CRC Press, 2008.

The above review lists the major collections of works on the topic. Otherwise, papers on IDBs/queries and CBDM regularly appear at major data mining conferences (such as *ACM SIGKDD International Conference on Knowledge Discovery and Data Mining (KDD)*, *European Conference on Machine Learning and European Conference on Principles and Practice of Knowledge Discovery in Databases (ECML/PKDD)*, and *SIAM International Conference on Data Mining (SDM)*) and journals (such as *Data Mining and Knowledge Discovery*). Overview articles on topics such as CBDM and data mining query languages appear in reference works on data mining (such as the *Data Mining and Knowledge Discovery Handbook*, edited by O. Z. Maimon and L. Rokach).

1.7 The IQ (Inductive Queries for Mining Patterns and Models) Project

Most of the research presented in this volume was conducted within the project IQ (Inductive Queries for mining patterns and models). In this section, we first discuss the background of the IQ project, then present its structure and organization. Finally, we give an overview of the major results of the project.

1.7.1 Background (The cInQ project)

Research on inductive databases and constraint-based data mining was first conducted in an EU-funded project by the cInQ consortium (consortium on discovering knowledge with Inductive Queries), funded within FP5-IST under contract number 26469, which took place from 2001 to 2004. The project involved the following institutions: Institut National des Sciences Appliquées (INSA), Lyon (France, coordinator: Jean-Francois Boulicaut), Universitá degli Studi di Torino (Italy, Rosa Meo and Marco Botta), the Politecnico di Milano (Italy, Pier-Luca Lanzi and Stefano Ceri), the Albert-Ludwigs- Universitaet Freiburg (Germany, Luc De Raedt), the Nokia Research Center in Helsinki (Finland, Mika Klemettinen and Heikki Mannila), and the Jozef Stefan Institute in Ljubljana (Slovenia, Sašo Džeroski).

A more detailed overview of the results of the cInQ project is given by Boulicaut et al. (2005). The major contributions of the project, however, can be briefly summarized as follows:

- An important theoretical framework was introduced for local/frequent pattern mining (e.g., itemsets, strings) under constraints (see, e.g., De Raedt 2002a), in which arbitrary boolean combinations of monotonic and anti-monotonic primitives can be used to specify the patterns of interest.
- Major progress was achieved in the area of condensed representations that compress/condense sets of solutions to inductive queries (see, e.g., Boulicaut et al.

2003) enabling one to mine dense and/or highly correlated transactional data sets, such as WWW usage data or boolean gene expression data, that could not be mined before.

- For frequent itemsets and association rules, cInQ studied the incorporation of inductive queries in query languages such as SQL and XQuery, also addressing the problems of inductive query evaluation and optimization in this context (Meo et al. 2003).
- The various approaches to mining sets of (frequent) patterns were successfully used in real-life applications from the field of bio- and chemo-informatics, most notably for finding frequent molecular fragments (Kramer et al. 2001) and in gene expression data (Becquet et al. 2002).

However, many limitations of IDBs/queries and CBDM remained to be addressed at the end of the cInQ project. Most existing approaches to inductive querying and CBDM focused on mining local patterns for a specific type of data (such as itemsets) and a specific set of constraints (based on frequency-related primitives). Inductive querying of global models, such as mining predictive models or clusterings under constraints remained largely unexplored. Although some integration of frequent pattern mining into database query languages was attempted, most inductive querying/CBDM systems worked in isolation and were not integrated with other data mining tools. No support was available for interactive querying sessions that involve the formulation of a complex sequence of inter-related inductive queries, where, e.g., some of the queries generate local patterns and other use these local patterns to build global models. As such support is needed in real-life applications, applications of IDBs/queries and CBDM to practically important problems remained limited.

1.7.2 IQ Project Consortium and Structure

The IQ project set out to address the challenges to IDBs/queries and CBDM remaining at the end of the cInQ project, as described above. The project, funded within FP6-IST under contract number 516169, whose full title was *Inductive Queries for mining patterns and models*, took place from 2005 to 2008. The IQ consortium evolved from the cInQ consortium. Its composition was as follows: Jozef Stefan Institute, Ljubljana, Slovenia (overall project coordinator: Sašo Džeroski), Albert-Ludwigs-Universitaet Freiburg, Germany and Katholieke Universiteit Leuven, Belgium (principal investigator Luc De Raedt), Institut National des Sciences Appliquées (INSA), Lyon, France (Jean-Francois Boulicaut), University of Wales Aberystwyth, United Kingdom (Ross King), University of Helsinki / Helsinki Institute for Information Technology, Finland (Heikki Mannila), and University of Antwerp, Belgium (Bart Goethals).

The overall goal of the IQ project was to develop a sound theoretical understanding of inductive querying that would enable us to develop effective inductive database systems and to apply them on significant real-life applications. To realize this aim, the IQ consortium made major developments of the required theory, representations and primitives for local pattern and global model mining, and integrated these into inductive querying systems, inductive database systems and query languages, and general frameworks for data mining. Based on these advances, it developed a number of significant show-case applications of inductive querying in the area of bioinformatics.

The project was divided into five inter-related workpackages. **Applications** in bio- and chemo-informatics were considered, and in particular drug design, gene expression data analysis, gene function prediction and genome segmentation. These were a strong motivating factor for all the other developments, most notably providing insight into the **KDD Scenarios**, i.e., sequences of (inductive) queries, that need to be supported. The execution of the scenarios was to be supported by **Inductive Querying Systems**, designed to answer inductive queries for specific pattern domains. For the different pattern domains, **Database and Integration Issues** were studied as well, including the integration of different pattern domains, integration with databases, scalability to large databases, and condensed representations. The results that go beyond those of individual pattern domains, solvers and applications contribute to a generalized overall **Theory of Inductive Querying**.

1.7.3 Major Results of the IQ project

In sum, the IQ project has made major progress in several directions. In the first instance, these include further developments in constraint-based mining of frequent patterns, as well as advances in mining global models (predictive models and clusterings) under constraints. At another level, approaches for mining frequent patterns have been integrated with the mining of predictive models (classification) and clusterings (bi-clustering or co-clustering) under constraints. In the quest for integration, inductive query languages, inductive database systems and frameworks for data mining in general have been developed. Finally, applications in bioinformatics which use the abovementioned advances have been developed.

Advances in mining frequent patterns have been made along several dimensions, including the generalization of the notion of closed patterns. First, the one-dimensional (closed sets) and two-dimensional (formal concepts) cases have been lifted to the case of n-dimensional binary data (Cerf et al. 2008; 2010/this volume). Second, the notion of closed patterns (and the related notion of condensed representations) have been extended to the case of multi-relational data (Garriga et al. 2007). Third, and possibly most important, a unified view on itemset mining under constraints has been formulated (De Raedt et al. 2008; Besson et al. 2010/this volume) where a highly declarative approach is taken. Most of the constraints used in itemset mining can be reformulated as sets or reified summation constraints, for which efficient solvers exist in constraint programming. This means that, once the constraints have been appropriately formulated, there is no need for special purpose CBDM algorithms.

Additional contributions in mining frequent patterns include the mining of patterns in structured data, fault-tolerant approaches for mining frequent patterns and randomization approaches for evaluating the results of frequent pattern mining. New approaches have been developed for mining frequent substrings in strings (cf. Rigotti et al. 2010/this volume), frequent paths, trees, and graphs in graphs (cf., e.g., Bringman et al. 2006; 2010/this volume), and frequent multi-relational patterns in a probabilistic extension of Prolog named ProbLog (cf. De Raedt et al. 2010/this volume). Fault-tolerant approaches have been developed to mining bi-sets or formal concepts (cf. Besson et al. 2010/this volume), as well as string patterns (cf. Rigotti et al. 2010/this volume): The latter has been used to to discover putative transcription factor binding sites in gene promoter sequences. A general approach to the evaluation of data mining results, including those of mining frequent patterns, has been developed: The approach is based on swap randomization (Gionis et al. 2006).

Advances in mining global models for prediction and clustering have been made along two major directions. The first direction is based on predictive clustering, which unifies prediction and clustering, and can be used to build predictive models for structured targets (tuples, hierarchies, time series). Constraints related to prediction (such as maximum error bounds), as well as clustering (such as must-link and cannot link constraints), can be addressed in predictive clustering trees (Struyf and Džeroski 2010/this volume). Due to its capability of predicting structured outputs, this approach has been successfully used for applications such as gene function prediction (Vens et al. 2010/this volume) and gene expression data analysis (Slavkov and Džeroski 2010/this volume).

The second direction is based on integrated mining of (frequent) local patterns and global models (for prediction and clustering). For prediction, the techniques developed range from selecting relevant patterns from a previously mined set for propositionalization of the data, over inducing patternbased rule sets, to integrating pattern mining and model construction (Bringmann et al. 2010/this volume). For clustering, approaches have been developed for constrained clustering by using local patterns as features for a clustering process, computing co-clusters by post-processing collections of local patterns, and using local patterns to characterize given co-clusters (cf., e.g., Pensa et al. 2008).

Finally, algorithms have also been developed for constrained prediction and clustering that do not belong to the above two paradigms. These include algorithms for constrained induction of polynomial equations for multi-target prediction (Pečkov et al. 2007). A large body of work has been devoted to developing methods for the segmentation of sequences, which can be viewed as a form of constrained clustering (Bingham 2010/ this volume), where the constraints relate the segments to each other and make the end result more interpretable for the human eye, and/or make the computational task simpler. The major application area for segmentation methods has been the segmentation of genomic sequences.

Advances in integration approaches have been made concerning inductive query languages, inductive database systems and frameworks for data mining based on the notions of IDBs and queries, as well as CBDM. Several inductive query languages have been proposed within the project, such as IQL (Nijssen and De Raedt

2007), which is an extension of the tuple relational calculus with functions, a typing system and various primitives for data mining. IQL is expressive enough to support the formulation of non trivial KDD scenarios, e.g., the formal definition of a typical feature construction phase based on frequent pattern mining followed by a decision tree induction phase.

An example of an inductive database system coming out of the IQ project is embodied within the MiningViews approach (Calders et al. 2006a; Blockeel et al. 2010/this volume). This approach uses the SQL query language to access data, patterns (such as frequent itemsets) and models (such as decision trees): The patterns/models are stored in a set of relational tables, called mining views, which virtually represent the complete output of the respective data mining tasks. In reality, the mining views are empty and the database system finds the required tuples only when they are queried by the user, by extracting constraints from the SQL queries accessing the mining views and calling an appropriate CBDM algorithm.

A special purpose type of inductive database are experiment databases (Vanschoren and Blockeel 2010/this volume): These are databases designed to collect the details of data mining (machine learning) experiments, which run different data mining algorithms on different datasets and tasks, and their results. Like all IDBs, experiment databases store the results of data mining: They store information on datasets, learners, and models resulting from running those learners on those datasets: The datasets, learners and models are described in terms of predefined properties, rather than being stored in their entirety. A typical IDB stores one datasets and the generalizations derived from them (complete patterns/model), while experiment databases store summary information on experiments concerning multiple datasets. Inductive queries on experiment databases analyze the descriptions of datasets and models, as well as experimental results, in order to find possible relationships between them: In this context, meta-learning is well-supported.

Several proposals of frameworks for data mining were considered within the project, such as the data mining algebra of Calders et al. (2006b). Among these, the general framework for data mining proposed by Džeroski (2007) defines precisely and formally the basic concepts (entities) in data mining, which are used to frame this chapter. The framework has also served as the basis for developing OntoDM, an ontology of data mining (Panov and Džeroski 2010/this volume): While a number of data mining ontologies have appeared recently, the unique advantages of OntoDM include the facts that (a) it is deep, (b) it follows best practices from ontology design and engineering (e.g., small number of relations, alignment with top-level ontologies), and (c) it covers structured data, different data mining tasks, and IDB/CBDM concepts, all of which are orthogonal dimensions that can be combined in many ways.

On the **theory** front, the most important contributions (selected from the above) are as follows. Concerning frequent patterns, they include the extensions of the notion of closed patterns to the case of n-dimensional binary data and multi-relational data and the unified view on itemset mining under constraints in a constraint programming setting. Concerning global models, they include advances in predictive clustering, which unifies prediction and clustering and can be used for structured prediction, as well as advances in integrated mining of (frequent) local patterns and global models (for prediction and clustering). Finally, oncerning integration, they include the MiningViews approach and the general framework/ontology for data mining.

On the **applications** front, the tasks of drug design, gene expression data analysis, gene function prediction, and genome segmentation were considered. In drug design, the more specific task of QSAR (quantitative structure-activity relationships) modeling was addressed: The topic is treated by King et al. (2010/this volume). Several applications in gene expression data analysis are discussed by Slavkov and Džeroski (2010/this volume). In addition, human SAGE gene expression data have been analyzed (Blachon et al. 2007), where frequent patterns are found first (in a fault-tolerant manner), clustered next, and the resulting clusters (called also quasi-synexpression groups) are then explored by domain experts, making it possible to formulate very relevant biological hypotheses.

Gene function prediction was addressed for several organisms, a variety of datasets, and two annotation schemes (including the Gene Ontology): This application area is discussed by Vens et al. (2010/this volume). Finally, in the context of genome segmentation, the more specific task of detecting isochore boundaries has been addressed (Haiminen and Mannila 2007): Simplified, isochores are large-scale structures on genomes that are visible in microscope images and correspond well (but not perfectly) with GC rich areas of the genome. This problem has been adressed by techniques such as constrained sequence segmentation (Bingham 2010/this volume).

More information on the IQ project and its results can be found at the **project website** http://iq.ijs.si.

1.8 What's in this Book

This book contains eighteen chapters presenting recent research on the topic of IDBs/queries and CBDM. Most of the chapters (sixteen) describe research conducted within the EU project IQ (Inductive Queries for mining patterns and models), as described above. The book also contains two chapters on related topics by researchers the project (Siebes and Puspitaningrum; Wicker et al.)

The book is divided into four parts. The first part, containing this chapter, is introductory. The second part presents a variety of techniques for constraint-based data mining or inductive querying. The third part presents integration approaches to inductive databases. Finally, the fourth part is devoted to applications of inductive querying and constraint-based mining techniques in the area of bio- and chemo-informatics.

1.8.1 Introduction

The first part contains four chapters. This introductory chapter is followed by the chapter of Panov and Džeroski that briefly presents a general framework for data mining and focusses on a detailed presentation of a deep ontology of data mining. The ontology includes well formalized basic concepts of data mining through which more advanced notions (such as constraint-based data mining) can be also described. Unlike other existing data mining ontologies, this ontology is much broader (covering mining of structured data) and follows best practices in ontology engineering.

An important component of inductive databases are inductive query languages, also known as data mining query languages. Blockeel et al. present a practical comparative study of existing data mining query languages on prototypical tasks of itemset and association rule mining. The last chapter in this part (by De Raedt et al.) is concerned with mining under composite constraints, i.e., answering inductive queries that are arbitrary Boolean combinations of monotonic and anti-monotonic constraints.

1.8.2 Constraint-based Data Mining: Selected Techniques

The second part contains six chapters presenting constraint-based mining techniques. The first chapter in this part by Besson et al. presents a unified view on itemset mining under constraints within the context of constraint programming. This is of great importance, as many approaches exist to constraint-based mining of frequent itemsets, typically designing different mining algorithms to handle different types of constraints.

Bringmann et al. then present a number of techniques for integrating the mining of (frequent) patterns and classification models. The techniques span the entire range from approaches that select relevant patterns from a previously mined set for propositionalization of the data, over inducing patternbased rule sets, to algorithms that integrate pattern mining and model construction. Struyf and Džeroski next discuss constrained induction of predictive clustering trees, which includes aspects of both constrained clustering and constrained predictive modeling.

The three chapters in the second half of this part concern constraint-based mining of structured data. Bingham first gives an overview of techniques for finding segmentations of sequences, some of these being able to handle constraints. Cerf et al. discuss constrained mining of cross-graph cliques in dynamic networks. Finally, De Raedt et al. introduce ProbLog, a probabilistic relational formalism, and discuss inductive querying in this formalism.

1.8.3 Inductive Databases: Integration Approaches

The third part contains four chapters discussing integration approaches to inductive databases. These include two inductive querying languages and systems, namely Mining Views and SINDBAD. They are followed by the presentations of an approach to solving composite inductive queries and experiment databases that store the results of data mining experiments.

In the Mining Views approach (Blockeel et al.), the user can query the collection of all possible patterns as if they were stored in traditional relational tables: This can be done for itemset mining, association rule discovery and decision tree learning. Wicker et al. present SINDBAD, a prototype of an inductive database system that aims to support the complete knowledge discovery process, and currently supports basic preprocessing and data mining operations that can be combined arbitrarily.

Siebes and Puspitaningrum discuss the integration of inductive and ordinary queries (relational algebra). They first try to lift relational algebra operators to inductive queries on frequent itemsets. They then use a model learned on a given database to improve the efficiency of learning a model on a subset of the database resulting from an ordinary query. Finally, Vanschoren and Blockeel present experiment databases: These are (inductive) databases that log and organize all the details of one's machine learning experiments, providing a full and fair account of the conducted research.

1.8.4 Applications

The fourth part of the book, devoted to applications, contains four chapters. All of them deal with applications in the area of bioinformatics (and chemoinformatics). The first two describe applications of predictive clustering trees, which allow for predicting structured outputs.

Vens et al. describe the use of predictive clustering trees for predicting gene function. This is a problem of hierarchical multi-label classification, where each gene can have multiple functions, with the functions organized into a hierarchy (such as the Gene Ontology). Slavkov and Džeroski describe several applications of predictive clustering trees for the analysis of gene expression data, which include the prediction of the clinical picture of the patient (multiple parameters) and constrained clustering of gene expression profiles.

In the next chapter, Rigotti et al. describe how to use mining of frequent patterns on strings to discover putative transcription factor binding sites in gene promoter sequences. Finally, King et al. discuss a very ambitious application scenario for inductive querying in the context of a robot scientist for drug design. To select new experiments to conduct, the robot scientist would use inductive queries to build structure-activity models predicting the activity of yet unassessed chemicals from data collected through experiments already performed.

Acknowledgements This work was carried out within the project IQ (Inductive Queries for Mining Patterns and Models), funded by the European Comission of the EU within FP6-IST, FET branch, under contract number 516169. Thanks are due to the members of the project for contributing the material surveyed in this chapter and presented in this book.

For a complete list of agencies, grants and institutions currently supporting the Sašo Džeroski, please consult the Acknowledgements chapter of this volume.

References

1. R. Agrawal, T. Imielinski, and A. Swami (1993). Mining association rules between sets of items in large databases. In *Proc. ACM SIGMOD Conf. on Management of Data*, pages 207–216. ACM Press, New York.
2. R. Bayardo, guest editor (2002). Constraints in data mining. Special issue of *SIGKDD Explorations*, 4(1).
3. C. Becquet, S. Blachon, B. Jeudy, J-F. Boulicaut, and O. Gandrillon (2002). Strong-association-rule mining for large-scale gene-expression data analysis: a case study on human SAGE data. *Genome Biology*, 3(12):research0067.
4. S. Bistarelli and F. Bonchi (2005). Interestingness is not a Dichotomy: Introducing Softness in Constrained Pattern Mining. In *Proc. 9th European Conf. on Principles and Practice of Knowledge Discovery in Databases*, pages 22–33. Springer, Berlin.
5. S. Blachon, R. G. Pensa, J. Besson, C. Robardet, J.-F. Boulicaut, and O. Gandrillon (2007). Clustering formal concepts to discover biologically relevant knowledge from gene expression data. *In Silico Biology*, 7(4-5): 467-483.
6. J-F. Boulicaut, A. Bykowski, C. Rigotti (2003). Free-sets: a condensed representation of boolean data for the approximation of frequency queries. *Data Mining and Knowledge Discovery*, 7(1):5–22.
7. J.-F. Boulicaut, L. De Raedt, and H. Mannila, editors (2005). *Constraint-Based Mining and Inductive Databases*. Springer, Berlin.
8. J-F. Boulicaut and B. Jeudy (2005). Constraint-based data mining. In O. Maimon and L. Rokach, editors, *The Data Mining and Knowledge Discovery Handbook*, pages 399–416. Springer, Berlin.
9. J.-F. Boulicaut, M. Klemettinen, and H. Mannila (1999). Modeling KDD processes within the inductive database framework. In *Proc. 1st Intl. Conf. on Data Warehousing and Knowledge Discovery*, pages 293–302. Springer, Berlin.
10. B. Bringmann, A. Zimmermann, L. De Raedt, and S. Nijssen (2006) Don't be afraid of simpler patterns. In *Proc 10th European Conf. on Principles and Practice of Knowledge Discovery in Databases*, pages 55–66. Springer, Berlin.
11. T. Calders, B. Goethals and A.B. Prado (2006a). Integrating pattern mining in relational databases. In *Proc. 10th European Conf. on Principles and Practice of Knowledge Discovery in Databases*, pages 454–461. Springer, Berlin.
12. T. Calders, L.V.S. Lakshmanan, R.T. Ng and J. Paredaens (2006b). Expressive power of an algebra for data mining. *ACM Transactions on Database Systems*, 31(4): 1169–1214.
13. T. Calders, C. Rigotti and J.-F. Boulicaut (2005). A survey on condensed representations for frequent sets. In J.-F. Boulicaut, L. De Raedt, and H. Mannila, eds., *Constraint-Based Mining and Inductive Databases*, pages 64–80. Springer, Berlin.
14. L. Cerf, J. Besson, C. Robardet, and J-F. Boulicaut (2008). Data-Peeler: Constraint-based closed pattern mining in n-ary relations. In *Proc. 8th SIAM Intl. Conf. on Data Mining*, pages 37– 48. SIAM, Philadelphia, PA,
15. L. De Raedt (2002a). A perspective on inductive databases. *SIGKDD Explorations*, 4(2): 69–77.

16. L. De Raedt (2002b). Data mining as constraint logic programming. In A.C. Kakas and F. Sadri, editors, *Computational Logic: Logic Programming and Beyond – Essays in Honour of Robert A. Kowalski, Part II*, pages 113–125. Springer, Berlin.

17. L. De Raedt, T. Guns, and S. Nijssen (2008). Constraint programming for itemset mining. In *Proc. 14th ACM SIGKDD Intl. Conf. on Knowledge Discovery and Data Mining*, pages 204–212. ACM Press, New York.

18. S. Džeroski (2007). Towards a general framework for data mining. In *5th Intl. Wshp. on Knowledge Discovery in Inductive Databases: Revised Selected and Invited Papers*, pages 259–300. Springer, Berlin.

19. U. Fayyad, G. Piatetsky-Shapiro and P. Smyth (1996). From data mining to knowledge discovery: An overview. In U. Fayyad, G. Piatetsky-Shapiro, P. Smyth and R. Uthurusamy, editors, *Advances in Knowledge Discovery and Data Mining*, pages 495–515. MIT Press, Cambridge, MA.

20. U. Fayyad, G. Piatetsky-Shapiro, and R. Uthurusamy (2003). Summary from the KDD-2003 panel – "Data Mining: The Next 10 Years". *SIGKDD Explorations*, 5(2):191–196.

21. G. C. Garriga, R. Khardon, and L. De Raedt (2007). On mining closed sets in multirelational data. In *In Proc. 20th Intl. Joint Conf. on Artificial Intelligence*, pages 804–809. AAAI Press, Menlo Park, CA.

22. A. Gionis, H. Mannila, T. Mielikainen, and P. Tsaparas (2006). Assessing data mining results via swap randomization. In *Proc. 12th ACM SIGKDD Intl. Conf. on Knowledge Discovery and Data Mining*, pages 167–176. ACM Press, New York.

23. N. Haiminen and H. Mannila (2007). Discovering isochores by least-squares optimal segmentation. *Gene*, 394(1-2):53–60.

24. J. Han, L.V.S. Lakshmanan, R.T. Ng (1999). Constraint-Based Multidimensional Data Mining. *IEEE Computer*, 32(8):46-50.

25. D.J. Hand, H. Mannila, and P. Smyth (2001). *Principles of Data Mining*. MIT Press, Cambridge, MA.

26. T. Imielinski and H. Mannila. A database perspective on knowledge discovery. *Communications of the ACM*, 39(11):58–64, 1996.

27. T. Johnson, L.V. Lakshmanan and R. Ng (2000). The 3W model and algebra for unified data mining. In *Proc. of the Intl. Conf. on Very Large Data Bases*, pages 21–32. Morgan Kaufmann, San Francisco, CA.

28. S. Kramer, L. De Raedt, C. Helma (2001). Molecular feature mining in HIV data. In *Proc. 7th ACM SIGKDD Intl. Conf. on Knowledge Discovery and Data Mining*, pages 136–143. ACM Press, New York.

29. H. Mannila and H. Toivonen. Levelwise search and borders of theories in knowledge discovery. *Data Mining and Knowledge Discovery*, 1(3):241–258, 1997.

30. R. Meo (2003) Optimization of a language for data mining. In *Proc. 18th ACM Symposium on Applied Computing*, pages 437–444. ACM Press, New York.

31. T.M. Mitchell (1982). Generalization as search. *Artificial Intelligence*, 18(2): 203–226.

32. S. Nijssen and L. De Raedt. IQL: a proposal for an inductive query language. In *5th Intl. Wshp. on Knowledge Discovery in Inductive Databases: Revised Selected and Invited Papers*, pages 189–207. Springer, Berlin.

33. A. Pečkov, S. Džeroski, and L. Todorovski (2007). Multi-target polynomial regression with constraints. In *Proc. Intl. Wshp. on Constrained-Based Mining and Learning*, pages 61–72. ECML/PKDD, Warsaw.

34. R.G. Pensa, C. Robardet, and J-F. Boulicaut (2008). Constraint-driven co-clustering of 0/1 data. In S. Basu, I. Davidson, and K. Wagstaff, editors, *Constrained Clustering: Advances in Algorithms, Theory and Applications*, pages 145–170. Chapman & Hall/CRC Press, Boca Raton, FL.

35. K. Wagstaff and C. Cardie (2000). Clustering with instance-level constraints. In *Proc. 17th Intl. Conf. on Machine Learning*, pages 1103–1110. Morgan Kaufmann, San Francisco, CA.

36. Q. Yang and X. Wu (2006). 10 Challenging problems in data mining research. *International Journal of Information Technology & Decision Making*, 5(4): 597–604.

Chapter 2
Representing Entities in the OntoDM Data Mining Ontology

Pance Panov, Larisa N. Soldatova, and Sašo Džeroski

Abstract Motivated by the need for unification of the domain of data mining and the demand for formalized representation of outcomes of data mining investigations, we address the task of constructing an ontology of data mining. Our heavy-weight ontology, named OntoDM, is based on a recently proposed general framework for data mining. It represent entites such as data, data mining tasks and algorithms, and generalizations (resulting from the latter), and allows us to cover much of the diversity in data mining research, including recently developed approaches to mining structured data and constraint-based data mining. OntoDM is compliant to best practices in ontology engineering, and can consequently be linked to other domain ontologies: It thus represents a major step towards an ontology of data mining investigations.

2.1 Introduction

Traditionally, ontology has been defined as the philosophical study of what exists: the study of kinds of entities in reality, and the relationships that these entities bear to one another [41]. In recent years, the use of the term ontology has become prominent in the area of computer science research and the application of computer science methods in management of scientific and other kinds of information. In this sense, the term ontology has the meaning of a standardized terminological framework in terms of which the information is organized.

Pance Panov · Sašo Džeroski
Jožef Stefan Institute, Jamova cesta 39, 1000 Ljubljana, Slovenia
e-mail: (Pance.Panov, Saso.Dzeroski)@ijs.si

Larisa N. Soldatova
Aberystwyth University, Penglais, Aberystwyth, SY23 3DB, Wales, UK
e-mail: lss@aber.ac.uk

The ontological problem in general is focused on adopting a set of basic categories of objects, determining what (kinds of) entities fall within each of these categories of objects, and determining what relationships hold within and among different categories in the ontology. The ontological problem for computer science is identical to many of the problems in philosophical ontology: The success of constructing such an ontology is thus achievable by applying methods, insights and theories of philosophical ontology. Constructing an ontology then means designing a representational artifact that is intended to represent the universals and relations amongst universals that exist, either in a given domain of reality (e.g the domain of data mining research) or across such domains.

The engineering of ontologies is still a relatively new research field and some of the steps in ontology design remain manual and more of an art than craft. Recently, there has been significant progress in automatic ontology learning [31], applications of text mining [7], and ontology mapping [29]. However, the construction of a high quality ontology with the use of automatic and even semi-automatic techniques still requires manual definition of the key upper level entities of the domain of interest. Good practices in ontology development include following an upper level ontology as a template, the use of formally defined relations between the entities, and not allowing multiple inheritances [44].

In the domain of data mining and knowledge discovery, researchers have tried to construct ontologies describing data mining entities. These ontologies are developed to solve specific problems, primarily the task of automatic planning of data mining workflows [2, 49, 24, 11, 22, 26]. Some of the developments are concerned with describing data mining services on the GRID [8, 5].

The currently proposed ontologies of data mining are not based on upper level categories nor do they have use a predefined set of relations based on an upper level ontology. Most of the semantic representations for data mining proposed so far are based on so called light-weight ontologies [33]. Light-weight ontologies are often shallow, and without rigid relations between the defined entities. However, they are relatively easy to develop by (semi)automatic methods and they still greatly facilitate several applications. The reason these ontologies are more frequently developed then heavy-weight ontologies is that the development of the latter is more difficult and time consuming. In contrast to many other domains, data mining requires elaborate inference over its entities, and hence requires rigid heavy-weight ontologies, in order to improve the Knowledge Discovery in Databases (KDD) process and provide support for the development of new data mining approaches and techniques.

While KDD and data mining have enjoyed great popularity and success in recent years, there is a distinct lack of a generally accepted framework that would cover and unify the data mining domain. The present lack of such a framework is perceived as an obstacle to the further development of the field. In [52], Yang and Wu collected the opinions of a number of outstanding data mining researchers about the most challenging problems in data mining research. Among the ten topics considered most important and worthy of further research, the development of an unifying framework for data mining is listed first. One step towards developing a general framework for data mining is constructing an ontology of data mining.

In this chapter, we present our proposal for an ontology of data mining (DM) named OntoDM [35, 36]. Our ontology design takes into consideration the best practices in ontology engineering. We use an upper level ontology Basic Formal Ontology (BFO)[1] to define the upper level classes. We also use the OBO Relational Ontology (RO)[2] and other related ontologies for representing scientific investigations, to define the semantics of the relationships between the data mining entities, and provide is-a completeness and single is-a inheritance for all DM entities.

The OntoDM ontology is based on a recent proposal for a general framework for data mining [13]. We have developed our ontology in the most general fashion in order to be able to represent complex data mining entities. These are becoming more and more popular in research areas such as mining structured data and constraint-based mining.

The rest of the chapter is structured as follows. In Section 2.2, we present the ontology design principles and we put the ontology in context of other ontologies for representing scientific investigations. Section 2.3 presents the ontology upper level structure, the ontological relations employed, and the division of OntoDM into logical modules. In the following section (Section 2.4) we present the basic entities in data mining, following the basic principles from the proposal of a general framework for data mining. In Section 2.5, we describe how we represent the data mining entities in all three modules of the ontology. We conclude the chapter with a critical overview of related work (Section 2.6), discussion and conclusions (Section 2.7).

2.2 Design Principles for the OntoDM ontology

2.2.1 Motivation

The motivation for developing an ontology of data mining is multi-fold. First, the area of data mining is developing rapidly and one of the most challenging problems deals with developing a general framework for mining of structured data and constraint-based data mining. By developing an ontology of data mining we are taking one step toward solving this problem. The ontology would formalize the basic entities (e.g., dataset and data mining algorithm in data mining) and define the relations between the entities. After the basic entities are identified and logically defined, we can build upon them and define more complex entities (e.g., constraints, constraint-based data mining task, data mining query, data mining scenario and data mining experiment).

---

[1] BFO: http://www.ifomis.org/bfo

[2] RO: http://www.obofoundry.org/ro/

Second, there exist several proposals for ontologies of data mining, but the majority of them are light-weight, aimed at covering a particular use-case in data mining, are of a limited scope, and highly use-case dependent. Most of the developments are with the aim of automatic planning of data mining workflows [2, 49, 50, 24, 22, 26, 11, 12]. Some of the developments are aimed at describing of data mining services on the GRID [8, 5]. Data mining is a domain that needs a heavy-weight ontology with a broader scope, where much attention is paid to the precise meaning of each entity, semantically rigorous relations between entities and compliance to an upper level ontology, and compatibility with ontologies for the domains of application (e.g., biology, environmental sciences).

Finally, an ontology of data mining should define what is the minimum information required for the description of a data mining investigation. Biology is leading the way in developing standards for recording and representation of scientific data and biological investigations [16] (e.g., already more than 50 journals require compliance of the reporting in papers results of microarray experiments to the Minimum Information About a Microarray Experiment - MIAME standard [14]). The researchers in the domain of data mining should follow this good practice and the ontology of data mining should support the development of standards for performing and recording of data mining investigations.

To summarize, the major goal of our ontology is to provide a structured vocabulary of entities sufficient for the description of the scientific domain of data mining. In order to achieve this goal the ontology should:

- represent the fundamental data mining entities;
- allow support for representing entities for mining structured data at all levels: the entities representing propositional (single table) data mining should be a special case (subclass) of a more general framework of mining structured data;
- be extensible, i.e., support representing complex data mining entities using fundamental data mining entities;
- use an upper level ontology and formally defined relations based on upper-level classes in order to provide connections to other domain ontologies and provide reasoning capabilities across domains;
- reuse classes and relations from other ontologies representing scientific investigations and outcomes of research and
- support the representation of data mining investigations.

2.2.2 OntoDM design principles

The OntoDM ontology design takes into consideration the best practices in ontology engineering. We use the upper level ontology BFO (Basic Formal Ontology)[3] to define the upper level classes, We use the OBO Relational Ontology (RO)[4] and an

---

[3] BFO: http://www.ifomis.org/bfo
[4] RO: http://www.obofoundry.org/ro/

extended set of RO relations to define the semantics of the relationships between the data mining entities: in this way, we achieve is-a completeness and single is-a inheritance for all data mining entities.

OntoDM aims to follow the OBO Foundry principles[5] in ontology engineering that are widely accepted in the biomedical domain. The main OBO Foundry principles state that "the ontology is open and available to be used by all", "is in a common formal language", "includes textual definition of all terms", "uses relations which are unambiguously defined", "is orthogonal to other OBO ontologies" and "follows a naming convention" [39]. In this way, OntoDM is built on a sound theoretical foundation and will be compliant with other (e.g., biological) domain ontologies. Our ontology will be compatible with other formalisms, and thus widely available for sharing and reuse of already formalized knowledge.

OntoDM is "in a common formal language": it is expressed in OWL-DL, a de-facto standard for representing ontologies. OntoDM is being developed using the Protege[6] ontology editor. It consists of three main components: classes, relations (a hierarchical structure of is-a relations and relations other than is-a), and instances.

2.2.3 Ontologies for representing scientific investigations

Concerning the relationship to other ontologies, we note here that there exist several formalisms for describing scientific investigations and outcomes of research. Below we review five proposals that are relevant for describing data mining investigations: the Basic Formal Ontology (BFO) as an upper level ontology, the Ontology for Biomedical Investigations (OBI)[7], the Information Artifact Ontology (IAO) [8], the Ontology of Scientific Experiments (EXPO) [45] and its extension LABORS [28] ,and the Ontology of Experiment Actions (EXACT) [43]. In the design of the OntoDM ontology, we reuse and further extend their structure and use their philosophy to identify and organize the OntoDM entities in an *is-a* class hierarchy, folowing the MIREOT (The Minimum Information to Reference an External Ontology Term) principle [10].

Basic Formal Ontology - BFO. The philosophy of BFO [20] overlaps in some parts with the philosophy of other upper level ontologies, such as DOLCE (Descriptive Ontology for Linguistic and Cognitive Engineering) [19] and SUMO (Suggested Upper Merged Ontology)[34]. However, BFO is narrowly focused on the task of providing a genuine upper ontology which can be used in support of domain ontologies developed for scientific research, as for example in biomedicine. It is included within the framework of the OBO Foundry.

---

[5] OBO Foundry: http://ontoworld.org/wiki/OBO_foundry

[6] Protege: http://protege.stanford.edu

[7] OBI: http://purl.obolibrary.org/obo/obi

[8] IAO:http://code.google.com/p/information-artifact-ontology/

BFO recognizes a basic distinction between two kinds of entities: substantial entities or continuants and processual entities or occurrents. Continuants, represent entities that endure through time, while maintaining their identity. Occurents represent entities that happen, unfold and develop in time. The characteristic feature of occurents, or processual entities, is that they are extended both in space and time.

Ontology of biomedical investigations - OBI. The OBI ontology aims to provide a standard for the representation of biological and biomedical investigations. The OBI Consortium is developing a set of universal terms that are applicable across various biological and technological domains and domain specific terms relevant only to a given domain. The ontology supports consistent annotation of biomedical investigations regardless of the particular field of the study [6]. OBI defines an investigation as a process with several parts, including planning an overall study design, executing the designed study, and documenting the results.

The OBI ontology employs rigid logic and semantics as it uses an upper level ontology BFO and the RO relations to define the top classes and a set of relations. OBI defines occurrences (processes) and continuants (materials, instruments, qualities, roles, functions) relevant to biomedical domains. The Data Transformation Branch is an OBI branch with the scope of identifying and representing entities and relations to describe processes which produce output data given some input data, and the work done by this branch is directly relevant to the OntoDM ontology.

OBI is fully compliant with the existing formalisms in biomedical domains. OBI is an OBO Foundry candidate [15]. The OBO Foundry requires all member ontologies to follow the same design principles, the same set of relations, the same upper ontology, and to define a single class only once within OBO to facilitate integration and automatic reasoning.

Information Artifact Ontology - IAO. Due to the limitations of BFO in dealing with information, an Information Artifact Ontology (IAO) has been recently proposed as a spin-off of the OBI project. The IAO ontology aims to be a mid-level ontology, dealing with information content entities (e.g., documents, file formats, specifications), processes that consume or produce information content entities (e.g., writing, documenting, measuring), material bearers of information (e.g., books, journals) and relations in which one of the relata is an information content entity (e.g., is-about, denotes, cites). IAO is currently available only in a draft version, but we have included the most stable and relevant classes into OntoDM.

Ontology of experiments - EXPO and LABORS. The formal definition of experiments for analysis, annotation and sharing of results is a fundamental part of scientific practice. A generic ontology of experiments EXPO [45] tries to define the principal entities for representation of scientific investigations. EXPO defines types of investigations: *EXPO:computational investigation*, *EXPO:physical investigation* and their principal components: *EXPO:investigator*, *EXPO:method*, *EXPO:result*, *EXPO:conclusion*.

The EXPO ontology is of a general value in describing experiments from various areas of research. This was demonstrated with the use of the ontology for the description of high-energy physics and phylogenetics investigations. The ontology uses a subset of SUMO as top classes, and a minimized set of relations in order to provide compliance with the existing formalisms.

The LABORS ontology is an extension of EXPO for the description of automated investigations (the Robot Scientist Project [9]). LABORS defines research units, such as investigation, study, test, trial and replicate: These are required for the description of complex multilayered investigations carried out by a robot [28].

Ontology of experiment actions - EXACT The ontology of experiment actions (EXACT) [43] aims to provide a structured vocabulary of terms for the description of protocols in biomedical domains. The main contribution of this ontology is the formalization of biological laboratory protocols in order to enable repeatability and reuse of already published experiment protocols. This ontology and the COW (Combining Ontologies with Workflows) software tool were used as a use case to formalize laboratory protocols in the form of workflows [30].

2.3 OntoDM Structure and Implementation

The upper level structure of the OntoDM ontology is mapped and aligned closely to the structure of the OBI ontology, a state-of-the-art ontology for describing biomedical investigations. In order to describe informational entities, the OBI ontology uses classes from the IAO ontology. A design decision was made to include relevant classes from IAO into OntoDM for the same purpose. As both the OBI and IAO ontologies to use BFO as a top ontology, we decided use BFO top level classes to represent entities which exist in the real world. In addition, we follow the design philosophy of EXPO/LABORS to represent mathematical entities.

The OntoDM ontology aims at interoperability among the ontologies: It thus includes formally defined ontological relations, based on upper level ontology classes, in order to achieve the desired level of expressiveness and interoperability. The set of relations is composed of relations from the relational ontology (RO) [42], a relation from the EXACT ontology [43], and relations from IAO and OBI. All of the relations used are formally defined on an instance and class level.

In the remainder of this section, we present an overview of the upper level classes, and the relations used in OntoDM, and then discuss how design decisions on the structure of the ontology allow us to establish a modular ontology for representing the domain of data mining. The modular structure of the ontology is a necessity in order to represent different aspects of the data mining and knowledge discovery process and to facilitate the different needs of the potential users of the ontology.

---

[9] http://www.aber.ac.uk/compsci/Research/bio/robotsci/

2.3.1 Upper level is-a hierarchy

In Figure 2.1, we present the upper level OntoDM class hierarchy. Bellow we give more details on the meaning of each upper level class. The upper level classes are further extended in the OntoDM ontology.

Continuants. An entity that exists in full at any time in which it exists at all, persists through time while maintaining its identity, and has no temporal parts in the BFO ontology is called a *BFO:continuant* (e.g., a person, a heart). A *BFO:dependent continuant* is a continuant that is either dependent on one or other independent continuant bearers or inheres in or is borne by other entities. Dependent continuants in BFO can be generically dependend or specifically dependent. A *BFO:generically dependent continuant* is a dependent continuant where every instance of A requires some instance of B, but which instance of B serves can change from time to time (e.g., a certain PDF file that exists in different and in several hard drives). For a *BFO:specifically dependent continuant*, every instance of A requires some specific instance of B which must always be the same (e.g., the role of being a doctor, the function of the heart in the body etc.).

The *IAO:information content entity* (ICE) was recently introduced into IAO (motivated by the need of OBI) and denotes all entities that are generically dependent on some artifact and stand in relation of aboutness (*is-about*) to some entity. Examples of ICE include data, narrative objects, graphs etc. The introduction of ICE enables the representation of different ways that information relates to the world, sufficient for representing scientific investigations (and in case of OBI, specifically biomedical investigations).

A *BFO: Realizable entity* (RE) is a specifically dependent continuant and includes all entities that can be executed (manifested, actualized, realized) in concrete occurrences (e.g., processes). RE are entities whose instances contain

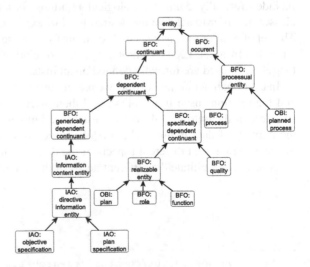

Fig. 2.1 OntoDM top-level class hirearchy (*is-a* hirearchy). The rectangle objects in the figure represent ontology classes. The *is-a* relation is represented with a directed labeled arrow.

periods of actualization, when they are manifested through processes in which their bearers participate. Examples of RE are plans, roles, functions and dispositions.

An *IAO:directive informational entity*[10] (DIC) is an information content entity that concerns a realizable entity. DICs are information content entities whose concretizations indicate to their bearer how to realize them in a process. Examples of DICs are: objective specification, plan specification, action specification, etc. An *IAO:objective specification* describes an intended process endpoint. An *IAO:plan specification* includes parts such as: objective specification, action specifications and conditional specifications. When concretized, it is executed in a process in which the bearer tries to achieve the objectives, in part by taking the actions specified.

Occurents. An entity that has temporal parts and that happens, unfolds or develops through time in the BFO ontology is called an *BFO:occurent* (e.g., the life of an organism). A *BFO:processual entity* is an occurrent that exists in time by occurring or happening, has temporal parts and always involves and depends on some entity. A *BFO: process* is a processual entity that is a maximally connected spatiotemporal whole and has beginnings and endings (e.g., the process of sleeping).

An *OBI:planned process* is a processual entity that realizes a *OBI:plan* which is the concretization of a *IAO:plan specification* in order to achieve the objectives *IAO:objective specification*. Process entities have as participants continuants , and participants can be also active and in that case they are called agents.

2.3.2 Ontological relations

Relations are the most essential part of a well designed ontology. It is thus crucial that the relations are logically defined. At every point of ontology development, from the initial conceptualization, through the construction, to its use, all the relations introduced should not change their meaning. The consistent use of rigorous definitions of formal relations is a major step toward enabling the achievement of interoperability among ontologies in the support of automated reasoning across data derived from multiple domains. The full set of relations used in the OntoDM ontology is presented in Table 2.1. Below we give a brief overview of their formal meaning.

Fundamental relations. The fundamental relations *is-a* and *has-part* are used to express subsumption and part-whole relationships between entities. The relation *has-instance* is a relation that connects a class with an instance of that class. The fundamental relations are formally defined in the Relational Ontology [42], both at class and instance level.

Information entity relations. We included a primitive relation from IAO (*is-about*) that relates an information artifact to an entity. In this ontology we reuse

---

[10] A directive information entity, before the OBI RC1 version, was named informational entity about a realizable.

Table 2.1 Relations in OntoDM. The relations are presented with the name of the relation, the origin of the relation, the domain and range of use and the inverse relation (where defined)

Relation	Origin	Domain	Range	Inverse relation
is-a	RO	entity	entity	sub-class-of
has-part	RO	entity	entity	part-of
has-instance	RO	entity	instance	instance-of
has-participant	RO	BFO:occurent	BFO:continuant	participates-in
has-agent				agent-of
is-about	IAO	IAO:information entity	BFO:entity	
has-information	EXACT	agent of a process	IAO:information content entity	
has-specified input	OBI	BFO:processual entity	BFO:dependent continuant	is-specified input-of
has-specified output				is-specified output-of
inheres-in	OBI	BFO:dependent continuant	BFO:continuant	bearer-of
is-concretization-of	OBI	BFO:specifically dependent continuant	BFO:generically dependent continuant	is-concretized-as
realizes	OBI	BFO:process	BFO:realizable entity	is-realized-by
achieves-planned-objective	OBI	OBI:planned process	IAO:objective specification	objective-achieved-by

the relation *has-information* defined in the EXACT ontology [43] to relate an agent of a process to a certain portion of information (information entity) that is essential for participating in the process.

Process relations. The relations *has-participant* and *has-agent* (both defined in RO) express the relationship between a process and participants in a process, that can be passive or active (in case of agents). The relations *has-specified-input* and *has-specified-output* have been recently introduced into the OBI ontology and are candidate relations for RO. These relations are specializations of the relation *has-participant*, and are used for relating a process with special types of participants, inputs and outputs of the process. We made a design decision to include them in OntoDM in order to increase the expressiveness and interoperability with the OBI ontology.

Role and quality relations. The relation between a dependent continuant and an entity is expressed via the relation *inheres-in* (defined in the OBI ontology and candidate for inclusion into RO). This relation links qualities, roles, functions, dispositions and other dependent continuants to their bearers. It is a super-relation of the relations *role-of* and *quality-of*.

Relations between information entities, realizable entities and processes. The relation *is-concretization-of* (introduced by the IAO ontology) expresses the relationship between a generically dependent continuant (GDC) and a specifically dependent continuant (SCD). In the OBI ontology, this relation is defined in the following way: "A GDC may inhere in more than one entity. It does so by virtue of the fact that there is, for each entity that it inheres, a specifically dependent 'concretization' of the GDC that is specifically dependent".

The relation *realizes* is used to express the relation between a process and a function (realizable entity), where the unfolding of the process requires execution of a function (execution of the realizable entity). The relation *achieves-planned-objective* links a planned process with its planned objectives. The planned process realizes a plan which is a concretization of a plan specification, which has as a part an objective specification. The objectives listed in the objective specification are met at the end of the planned process. Both relations were introduced by the OBI ontology.

2.3.3 Modularity: Specification, implementation, application

In Figure 2.3.3, we present three modules of the ontology capable of describing three different aspects of data mining. The first module, named "specification", is aimed to contain and represent the informational entities in data mining. Examples of such entities are: data mining task, algorithm specification, dataset description, generalization specification etc. The second module, named "implementation", is aimed to describe concrete implementations of algorithms, implementations of components of algorithms, such as distance functions and generalizations produced by the mining process. The third module, named "application", aims at describing the data mining process and the participants of the process in the context of data mining scenarios. Example of processual entities are: the application of an algorithm

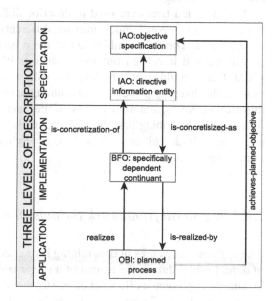

Fig. 2.2 Three levels of description: specification, implementation and application. The rectangle objects in the figure represent ontology classes. The ontological relations are represented with directed labeled arrows. The relations that do not have an attached label are *is-a* relations.

implementation (execution of an algorithm) and the application of a predictive model on new data etc.

The modules are inter connected with the previously introduced relations. In that fashion, a specification *is-concretizied-as* an implementation. Next, an implementation *is-realized-by* an application. Finally, an application *achieves-planned-objective* specification.

It is necessary to have all three aspects represented separately in the ontology as they have distinctly different nature. This will facilitate different usage of the ontology. For example, the specification aspect can be used to reason about components of data mining algorithms; the implementation aspect can be used for search over implementations of data mining algorithms and to compare various implementations and the application aspect can be used for constructing data mining scenarios and workflows, definition of participants of workflows and its parts.

2.4 Identification of Data Mining Entities

One of the fist major steps in domain ontology construction is the identification of domain terms. In the case of OntoDM, we performed the identification following the principles from a proposal for general framework for data mining [13]. This enables us to have a general ontology of data mining, that can cover different aspects of the data mining domain and allow easy extensions of the ontology with new entities in a principled way. From the framework proposal, we identified a set of basic terms of data mining that are used to construct the basic data mining entities that form the core of our ontology.

The identified terms are used to describe different dimensions of data mining. These are all orthogonal dimensions and different combinations among these should be facilitated. Through combination of these basic terms and other support terms already defined in related ontologies such as BFO, IAO, OBI, EXPO/LABORS, EXACT one should be able to describe, with principled extensions of the ontology, most of the diversity present in data mining approaches today. In the remainder of this section, we present an overview of thegeneral framework for (structured) data mining [13], describing first the basic principles of the framework, followed by an overview of basic enities such as data, generalizations, data mining task and data mining algorithms.

2.4.1 A general framework for data mining: Basic principles

One of the main features of data mining is its concern with *analyzing different types of data*. Besides data in the format of a single table, which is most commonly used in data mining, complex (in most cases structured) data are receiving and increasing amount of interest. These include data in the form of sequences and graphs, but also text, images, video, and multi-media data. Much of the current research in data mining is about mining such complex data, e.g., text mining, link mining, mining social network data, web mining, multi-media data mining. A major challenge is to *treat the mining of different types of structured data in a uniform fashion.*

Many different data mining tasks have been considered so far within the field of data mining. By far the most common is the task of predictive modeling, which includes classification and regression. Mining frequent patterns is the next most popular, with the focus shifting from mining frequent itemsets to mining frequent patterns in complex data. Clustering, which has strong roots in the statistical community, is also commonly encountered in data mining, with distance-based and density-based clustering as the two prevailing forms. A variety of other tasks has been considered, such as change and deviation detection and others, but it is not clear whether these are of fundamental nature or can be defined by composing some of the tasks listed above. The task of a general framework for data mining would be to *define the fundamental (basic) data mining tasks* and allow definition of more complex tasks by combining the fundamental ones.

Finally, *different types of generalizations (patterns/models)* may be used for the same data mining task. This is most obvious for predictive modelling, where a variety of methods/approaches exist, ranging from rules and trees, through support vector machines, to probabilistic models (such as Naive Bayes or Bayesian networks for classification). The different types of models are interpreted in different ways, and different algorithms may exist for building the same kind of model (cf. the plethora of algorithms for building decision trees).

2.4.2 Data

Data is the most basic data mining entity. A data mining algorithm takes as input a set of data (*dataset*). An individual datum (*data example*) in the dataset has its own structure, e.g., consists of values for several attributes, which may be of different types or take values from different ranges. We typically assume that all data examples are homogeneous (of the same type) and share the same structure.

More generally, we are given a *data type T* and a set of data *D* of this type. It is important to notice, though, that a *set of basic/primitive types* is typically taken as a starting point, and more *complex data types* are built by using *type constructors*. It is of crucial importance to be able to deal with structured data, as these are attracting an increasing attention within data mining.

Assume we are given a set of primitive data types, such as *Boolean* or *Real*. Other primitive data types might include *Discrete(S)*, where *S* is a finite set of identifiers, or Integer. In addition, we are given some type constructors, such as *Tuple* and *Set*, that can be used to construct more complex data types from existing ones. For example, *Tuple(Boolean, Real)* denotes a data type where each datum consists of a pair of a *Boolean* value and a real number, while *Set(Tuple(Boolean, Real))* denotes a data type where each datum is a set of such pairs.

Other type constructors might include $Sequence(T)$, which denotes a sequence of objects of type T, or $LabeledGraph(VL, EL)$, which denotes a graph where vertex labels are of type VL and edge labels are of type EL. With these, we can easily represent the complex data types that are of practical interest. For example, DNA sequences would be of type $Sequence(Discrete(\{A, C, G, T\}))$, while molecules would be labeled graphs with vertices representing atoms and edges representing bonds between atoms: atoms would be labeled with the type of element (e.g., nitrogen, oxygen) and edges would be labeled with the type of bond (e.g., single, double, triple).

2.4.3 Generalizations

Generalization is a broad term that denotes the output of different data mining tasks, such as pattern mining, predictive modeling and clustering. Generalizations include probability distributions, patterns (in the sense of frequent patterns) and global models (predictive models and clusterings). All of these are defined on a given type of data, except for predictive models, which are defined on a pair of data types.

Generalizations inherently have a dual nature. They can be treated as *data structures* and as such represented, stored and manipulated. On the other hand, they are *functions* that take as input data points and map them to probabilities (in the case of probability distributions), class predictions (in the case of predictive models), cluster assignments (in the case of clusterings), or Booleans (in the case of local patterns).

The remainder of this sub-section, we first list the fundamental types of generalizations in data mining, then describe classes of generalizations (that refer to the data structure nature) and finally we describe interpreters of generalizations (that refer to the function nature).

Fundamental types of generalizations. Fundamental types of generalizations include: probability distributions, patterns, predictive models and clusterings.

A *probability distribution D on type T* is a mapping from objects of type T to non-negative Reals, i.e., has the signature $d :: T \to R^{0+}$. For uncountably infinite types, probability densities are used instead. The sum of all probabilities (the integral of the probability densities) over T is constrained to amount to one.

A *pattern P on type T* is a Boolean function on objects of type T, i.e., has the signature $p :: T \to bool$. A pattern on type T is true or false on an object of type T. A pattern is defined as a statement (expression) in a given language, that describes (relationships among) the facts in (a subset of) the data [17].

A *predictive model M for types T_d, T_c* is a function that takes an object of type T_d and returns one of type T_c, i.e., has the signature $m :: T_d \to T_c$. Most often, predictive modelling is concerned with classification, where T_c would be Boolean (for binary classification), Discrete(S) (for multi-class classification), or regression, where T_c

would be Real. In our case, we allow both T_d (description) and T_c (class/target) to be arbitrarily complex data types.

A *clustering C on a set of objects S of type T* is a function from S to $\{1,\ldots,k\}$, where k is the number of clusters, which has to obey $k \leq |S|$. Unlike all the previously listed types of patterns, a clustering is not necessarily a total function on T, but rather a partial function defined only on objects from S. Overlapping and soft clusterings, where an element can (partially) belong to more that one cluster have the signature $T \rightarrow (\{1,\ldots,k\} \rightarrow R^{0+})$. In hierarchical clustering, in addition to the function C, we get a hierarchy on top of the set $1,\ldots,k$.

In predictive clustering, C is a total function on T. In addition, we have $T=(T_d,T_c)$ and we have a predictive model associated with each cluster through a mapping M $:: \{1,\ldots,k\} \rightarrow (T_d \rightarrow T_c)$. Performing the function composition of M and C, i.e., applying first C and then M, we get a predictive model on T.

Classes of Generalizations. Many different kinds of generalizations have been considered in the data mining literature. Classification rules, decision trees and linear models are just a few examples. We will refer to these as generalization classes.

A *class of generalizations C_G* on a set on a datatype T is a set of generalizations on T expressed in a language L_G. For each specific type of generalization we can define a specific generalization class. The languages L_G refer to the data part of the generalizations. They essentially define data types for representing the generalizations. For example, a *class of models C_M* on types T_d, T_c is a set of models M on types T_d, T_c, expressed in a language L_M.

Interpreters. There is usually a unique mapping from the data part of a generalization to the function part. This takes the data part of a generalization as input, and returns the corresponding function as an output. This mapping can be realized through a so-called interpreter. The interpreter is crucial for the semantics of a class of generalzations: a class of generalizations is only completely defined when the corresponding interpreter is defined (e.g., interpreter for models I_M is part of the definition of the class C_M).

For illustration, given a data type T, an example E of type T, and a pattern P of type $p :: T \rightarrow bool$, an interpreter I returns the result of applying P to E, i.e., $I(P,E) = P(E)$. The signature of the interpreter is $i :: p \rightarrow T \rightarrow bool$. If we apply the interpreter to a pattern and an example, we obtain a Boolean value.

2.4.4 Data mining task

In essence, the *task of data mining* is to produce a generalization from a given set of data. Here we will focus on four fundamental tasks, according to the generalizations produced: estimating the (joint) probability distribution, learning predictive models, clustering and finding valid (frequent) patterns.

Estimating the (Joint) Probability Distribution. Probably the most general data mining task [21] is the task of estimating the (joint) probability distribution D over type T from a set of data examples or a sample drawn from that distribution.

Learning a Predictive Model. In this task, we are given a dataset that consists of examples of the form (d, c), where each d is of type T_d and each c is of type T_c. We will refer to d as the description and c as the class or target. To learn a predictive model means to find a mapping from the description to the target, $m :: T_d \rightarrow T_c$, that fits the data closely. This means that the observed target values and the target values predicted by the model, i.e., c and $\hat{c} = m(d)$, have to match closely.

Clustering The task of clustering in general is concerned with grouping objects into classes of similar objects [25]. Given a set of examples (object descriptions), the task of clustering is to partition these examples into subsets, called clusters. The goal of clustering is to achieve high similarity between objects within individual clusters (intra-cluster similarity) and low similarity between objects that belong to different clusters (inter-cluster similarity).

Pattern Discovery. In contrast to the previous three tasks, where the goal is to build a single global model describing the entire set of data given as input, the task of pattern discovery is to find all *local patterns* from a given *pattern language* that satisfy the required conditions. A prototypical instantiation of this task is the task of finding frequent itemsets (sets of items, such as $\{bread, butter\}$), which are often found together in a transaction (e.g., a market basket) [1].

2.4.5 Data mining algorithms

A *data mining algorithm* is an algorithm (implemented in a computer program), designed to solve a data mining task. It takes as input a dataset of examples of a given datatype and produces as output a generalization (from a given class) on the given datatype. A data mining algorithm can typically handle examples of a limited set (class) of datatypes: For example, a rule learning algorithm might handle only tuples of Boolean attributes and a boolean class.

Just as we have classes of datatypes, classes of generalizations and data mining tasks, we have classes of data mining algorithms. The latter are directly related to the input and output of the algorithm, but can depend also on the specifics of the algorithm, such as the basic components of the algorithm (e.g., heuristic function, search method). For example, for the class of decision tree building algorithms, we can have two subclasses corresponding to top-down induction and beam-search (cf. Chapter 7) of this volume).

As stated earlier in this chapter, a very desirable property of a data mining framework is to treat the mining of different types of structured data in a uniform fashion. In this context, data mining algorithms should be able to handle as broad classes of datatypes at the input as possible. We will refer to algorithms that can handle arbitrary types of structured data at the input as generic. Generic data mining algorithms would typically have as parameters some of their components, e.g., a heuristic function in decision tree induction or a distance in distance-based clustering.

The general framework for data mining proposed by Džeroski [13] discusses several types of data mining algorithms and components thereof. The basic components include distances, features, kernels and generality/refinement operators. The framework proposes that the components of data mining should be treated as first-class citizens in inductive databases, much like generalizations (including patterns and models). We follow this approach and represent the entities corresponding to algorithm components in OntoDM: We thus give a brief overview thereof below.

Distances. The major components of distance-based algorithms are distance and prototype functions. A *distance function d* for type T is a mapping from pairs of objects of type T to non-negative reals: $d :: T \times T \rightarrow R^{0+}$. Distances are of crucial importance for clustering and predictive modelling. In clusters, we want to minimize the distance between objects in a cluster. In predictive modelling, we need to compare the true value of the target to the predicted one, for any given example. This is typically done by finding their distance.

A *prototype* is a representative of all the objects in a given set S. In the context of a given distance d, this is the object o that has the lowest average square distance to all of the objects in S. A *prototype function p* for objects of type T, takes as input a set S of objects of type T, and returns an object of type T, i.e., the prototype: $p :: Set(T) \rightarrow T$.

It is quite easy to formulate *generic distance-based algorithms* for data mining, which have the distance as a parameter. For example, hierarchical agglomerative clustering only makes use of the distances between the objects clustered and distances between sets of such objects. For a predictive problem of type $T_i \rightarrow T_j$, the nearest neighbor method applies as long as we have a distance on T_i.

To make a prediction for a new instance, the distance between the (descriptive part of) new instance and the training instances is calculated. The target part is copied from the nearest training instance and returned as a prediction.

To use the *k*-nearest neighbor algorithm (*k*-NN), we also need a prototype function on the target data type: the prediction returned is the prototype of the target parts of the k nearest (in the description space) instances. In the 1-NN case, we do not need this prototype function, as the prediction is simply copied from the nearest neighbor.

Features and feature based representation. Most of data mining algorithms use a feature based representation. Defining an appropriate set of features for a data mining problem at hand is still much of an art. However, it is also a step of key importance for the successful use of data mining.

Suppose d is a datum (structured object) of type T. Note that d can be, e.g., an image represented by an array of real numbers, or a recording of speech, represented by a sequence of real numbers. A *feature f* of objects of type T is a mapping from objects of type T to a primitive data type (Boolean, Discrete or Real) and $f(d)$ refers to the value of the feature for the specific object d.

There are at least three ways to identify features for a given object d of type T. First, the feature may have been directly observed and thus be a part of the representation of d. The other two ways are related to background knowledge concerning the structure of the object or concerning domain knowledge.

Kernels and Kernel Based Algorithms. Technically, a *kernel* k corresponds to the inner product in some feature space. The computational attractiveness of kernel methods[40] (KM) comes from the fact that quite often a closed form of these feature space inner products exists. The kernel can then be calculated directly, thus performing the feature transformation only implicitly without ever computing the coordinates of the data in the 'feature space'. This is called the *kernel trick*.

KMs in general can be used to address different tasks of data mining, such as clustering, classification, and regression, for general types of data, such as sequences, text documents, sets of points, vectors, images, etc. KMs (implicitly) map the data from its original representation into a high dimensional feature space, where each coordinate corresponds to one feature of the data items, transforming the data into a set of points in a Euclidean / linear space. Linear analysis methods are then applied (such as separating two classes by a hyperplane), but since the mapping can be nonlinear, nonlinear concepts can effectively be captured.

At the conceptual level, kernels elegantly relate to both features and distances. At the practical level, kernel functions have been introduced for different types of data, such as vectors, text, and images, including structured data, such as sequences and graphs [18]. There are also many algorithms capable of operating with kernels, and the most well known of which are SVMs (Support Vector Machines).

Refinement Orders and Search of Generalization Space. The notion of generality is a key notion in data mining, in particular for the task of pattern discovery. To find generalizations valid in the data, data mining algorithms search the *space of generalizations* defined by the class of generalizations considered, possibly additionally restricted by constraints. To make the search efficient, the space of generalizations is typically ordered by a *generality or subsumption relation*.

The generality relation typically refers to the function part of a generalization. The corresponding notion for the data part is that of *refinement*. A typical example of a refinement relation is the subset relation on the space of itemsets. This relation is a partial order on itemsets and structures itemsets into a lattice structure, which is typically explored during the search for, e.g., frequent itemsets. The refinement relation is typically the closure of a refinement operator, which performs minimal refinements (e.g., adds one item to an itemset).

The prototypical algorithm for mining frequent patterns starts its search with the empty pattern (set/sequence/graph), which is always frequent. It then proceeds level-wise, considering at each level the refinements of the patterns from the previous level and testing their frequencies. Only frequent patterns are kept for further refinement as no refinement of an infrequent pattern can be frequent.

2.4.6 OntoDM modeling issues

The identification of domain terms is just the first step in the construction of a domain ontology. Next, there is a need to revise the terms in the sense of ontology design principles and form ontological entities. In this phase, one has to form ontological classes, represent them with their unique characteristics (called properties), relate them to other classes using ontological relations, and place them adequately in the is-a hierarchy of classes.

An identified term is not always automatically mapped to an ontological class. Often a manual adjustment by an ontology engineer is required. For example, the term "data mining algorithm" can be used in three conceptually different aspects, which should be modeled separately in the ontology.

The first aspect is a specification of the algorithm. Here an algorithm would be described with the specification of the inputs and outputs, types of generalizations produced, data mining tasks it solves, the components of algorthms, parameters that can be tuned, the publication where the algorithm has been published etc. The second aspect is a concrete implementation of the algorithm. Here we have concrete executable version of the algorithm, and several different implementations can exist based on the same specification. Finally, a third aspect is the application of an algorithm implementation to a concrete dataset and the production of an output generalization. Here we deal with the data mining process, where essential entities are the participants in the process, the sub-processes, and how the sub-processes are connected between each other (which sub-process preceeds the other) etc.

The same can be exemplified with other entities in the ontology. Let us take, for example, a predictive model. The first aspect of a predictive model is its specification. Here we describe general characteristics of the model, what tasks they are produced from, model structure, parameters of the model structure, the language in which they are expressed (e.g., language of decision trees). The second aspect is a concrete (instantiated) model which is the result of execution of an algorithm implementation (a process) on a dataset. Here the instantiated model has a link to the dataset that produced it, the process that produced it, the quality measure instantiations on the data from which the model was produced etc. The final aspect is the execution of the model on new data, which is itself a process with the goal prediction. The inputs of the process are the model and the new data; the outputs are the predictions and the evaluation measures calculated.

Another important aspect in modeling the terms into an ontology is the treatment of the roles of entities. When modeling, one should define an entity with its purest properties that would allow us to differentiate it from other entities. But to do this, one has to abstract the entity from different contexts where the entity can appear. Modeling of realizations of an entity in different contexts should be done via roles of entities [33]. A typical example of a role in data mining is an operator. An operator is a role of an implementation of a data mining algorithm in the context of data mining workflows.

2.5 Representing Data Mining Enitities in OntoDM

In this section, we report how the data mining entities discussed above are represented in the OntoDM ontology. Furthermore, we give an overview and examples of classes, relations and instances from the specification, implementation and application module of the ontology. In addition, we provide a discussion of the advantages of the chosen ontology design patterns.

2.5.1 Specification entities in OntoDM

One of the main goals of the OntoDM ontology is to represent entities for structured data mining. Our design decisions allow us to treat the traditional single-table data mining as a special case of structured data mining. Furthermore, the goal is to keep the design as general as possible, in order to allow easy extensions covering further new developments in the domain of data mining.

The specification module of OntoDM contains specification entities (classes and instances) for the domain of data mining. Examples of entities are datatype, dataset, generalization specification, data mining task specification and data mining algorithm specification. The specification classes are extensions of the *information content entity* class.

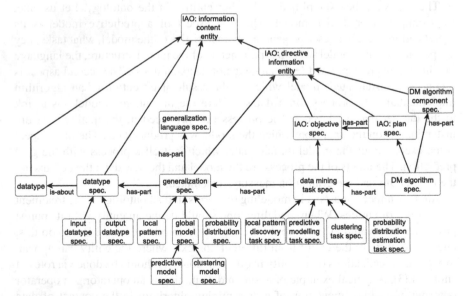

Fig. 2.3 Taxonomy and part-whole relations between basic data mining specification classes in OntoDM. The rectangle objects in the figure represent ontology classes. The ontological relations are represented with directed labeled arrows. The relations that do not have an attached label are *is-a* relations.

In Figure 2.3, we present the *is-a* hierarchy and the part-whole relations between the basic data mining entity classes of the specification module. The most fundamental specification class in OntoDM is the *datatype* (See Section 2.4.2 for more details). Next, we have the *datatype spec.* related to the datatype through the *is-about* relation. The *datatype spec.* has two subclasses at the first level: *input datatype spec.* and *output datatype spec.*. They are used to differentiate between input and output datatypes in the formal representation of generalizations.

A *generalization spec.* has as its parts a *datatype spec.* and *generalization language spec.*. It is further sub-classed at the first level with the following classes: *local pattern spec.*, *global model spec.* and *probability distribution spec.*. Having a *generalization language spec.* as a part of gives us the opportunity to further develop the taxonomy of generalizations by forming classes of generalizations (as discussed in Section 2.4.3).

Next, we have the *data mining task spec.* which is directly related to the types of generalizations via a *has-part* relation. This class is a subclass of *IAO:objective specification*. It is further sub-classed with the basic data mining tasks (See Section 2.4.4): *local pattern discovery task, predictive modeling task, clustering task* and *probability distribution estimation task*.

Finnaly, a *data mining algorithm spec.* has as its parts a *data mining task spec.* and *data mining algorithm component spec.* (See Section 2.4.5). A *data mining algorithm spec.* is a sub-class of *IAO:plan specification* and this is aligned with the IAO and OBI ontology structure, that is a *IAO:plan specification* has as its part *IAO:objective specification*.

The main advantage of having such a structure of classes (conected via *has-part* chains) is the ability to use the transitivity property of the *has-part* relation. For example, when we have an instance of *data mining algorithm spec.*, we can use reasoning to extract the data mining task, which is an objective of the algorithm, the type of generalization the algorithm gives at its output and the datatype specification on the input and output side.

In the remaining of this subsection we will discuss in more detail the *datatype* entity and the representation of structured datatypes and example of instances of structured datatypes.

Datatype. Figure 2.4 depicts the representation of datatypes in OntoDM. A *datatype* can be a *primitive datatype* or a *structured datatype* (See Figure 2.4c). According to [32], a primitive datatype is "a datatype whose values are regarded fundamental - not subject to any reduction". Primitive types can be non-ordered (e.g., *discrete datatype*) and ordered (e.g., *inst:real datatype, inst:integer datatype*). Furthermore, ordered datatypes can also be ordinal (e.g., *inst:boolean datatype*).

A *structured datatype* (or aggregated datatype in [32]) is "one whose values are made up of a number, in general more than one, of component values, each of which is a value of another datatype". A *structured datatype* has two parts: *datatype component spec.* and *aggregate datatype constructor spec.*. The *datatype component spec.* specifies the components of the structured datatype and *aggregate datatype constructor spec.* specifies the datatype constructor used to compose the structure.

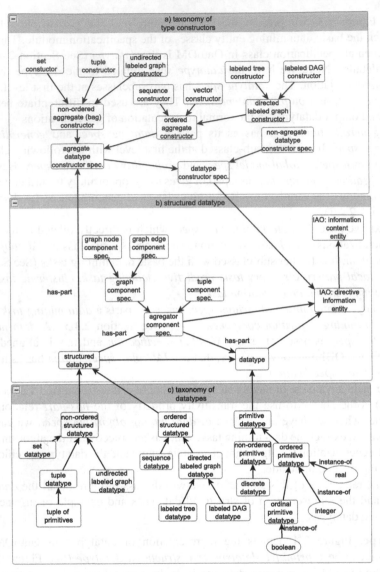

Fig. 2.4 Datatype specification in OntoDM: a) taxonomy of type constructors; b) structured datatype entity; c) taxonomy of datatypes. The rectangle objects in the figure represent ontology classes. The oval objects represent ontology instances. The ontological relations are represented with directed labeled arrows. The relations that don't have an attached label are *is-a* relations.

Providing an adequate and complete taxonomy of datatypes is a very challenging task. In the implementation of the OntoDM ontology, we decided to follow the guidelines from [32] to represent the *datatype* entity and construct a taxonomy of datatypes applicable to the domain of data mining. The construction is done in a general fashion that allows extensions in case new datatypes appear and are not covered so far. The taxonomy of datatypes is given in Figure 2.4c.

Datatype constructor. The taxonomy of the stuctured datatypes is based on the taxonomy of datatype constructors (See Figure 2.4a). A *datatype constructor spec.* can be non-aggregate or aggregate. A *non-aggregate datatype constr.* is defined in [32] as "datatypes that are produced from other datatypes by the methods familiar from languages that include them" (e.g., pointers, procedures, choices). An *aggregate datatype constr.* defines the aggregate that is used to combine the component datatypes. The aggregate type constructors classes can be further extended using different properties (e.g., ordering of components, homogeneity, how the components are distinguished - tagging or keying etc).

In this ontology, we distinguish between non-ordered and ordered aggregates. Non-ordered aggregate constructors (or bags) include: sets, tuples (or records) and undirected labeled graph. A *set constr.* is a constructor that does not allow duplicates. A *tuple constr.* is an aggregate where each component can be tagged.

A *sequence constr.* is the simplest ordered aggregate with a strict and unique ordering of the components. A *vector constr. is-a sequence constr.*, where components are indexed and the ordering of the index induces the ordering of the components. In a similar way, using properties of aggregates, we can define other aggregates like *directed labeled graph constr.* and its subclasses *labeled tree constr.* and *labeled DAG constr.*. All defined aggregates can be further sub-classed using constraints such as homogeneity, size (number of components), etc.

How do we define an instance of a structured datatype? Having the representation of a datatype and datatype constructor we can represent arbitrary datatype instances. In Fig.2.5a, we show how to represent *inst:tuple(boolean,real)*. It is an instance of the *tuple of primitive datatypes* class. *inst:tuple(boolean,real)* has two primitive datatype components (boolean and real) and a two element tuple

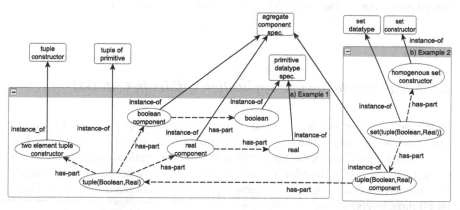

Fig. 2.5 An example of the structured datatype instances: a) The tuple(boolean,real) instance; b) The set{tuple(boolean,real)} instance. Dashed lines represent instance level relations and full lines represent class level relations.

constructor. In Fig.2.5b, we show how we can construct a more complex structured datatype using previously defined instances. *inst:set{tuple(boolean,real)}* has one component datatype (tuple(boolean,real)) and a homogeneous set constructor.

Dataset. Once we have representation of datatypes, we can represent datasets. A *dataset* is a *IAO:information content entity* and has as part *data example*. A *dataset spec.* is an information entity about a *dataset*, connected via the *is-about* relation. It has as its part a datatype specification, allowing us to have a classification of datasets using only datatype as a classification criteria.

This class can be further sub-classed with *unlabeled dataset spec.* class that has only input datatype specification as its part. We can further extend it with a special cases of unlabeled datasets: *unlabeled propositional dataset spec.* class, where the input specification is a tuple of primitives and *transactional dataset spec.* class where the input specification is a set of discrete. A *labeled dataset spec.* is a specialization of *unlabeled dataset spec.* class, where we have additionally defined output datatype specification.

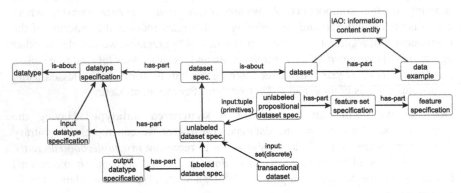

Fig. 2.6 The dataset entity in OntoDM. The ontological relations are represented with directed labeled arrows. The relations that don't have an attached label are *is-a* relations.

2.5.2 Implementation and application entities in OntoDM

In the previous subsection, we gave an overview of the specification module of the OntoDM ontology. The specification entities are connected via relations to their "counter part" entities in the implementation and application modules. In this subsection, we briefly describe the two modules and give an illustrative example how the three modules are interconnected, presenting example instances of classes on all three levels.

Implementation entities. Entities in the implementation module include implementations of algorithms, functions, instantiations of predictive models

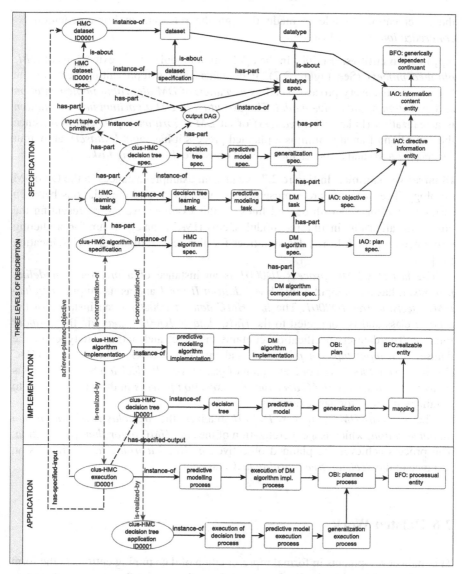

Fig. 2.7 An example of the connection between the three modules in OntoDM ontology: specification, implementation and application. The example shows defined instances of classes on all three levels. The rectangle boxes represent ontology classes. The oval boxes represent instances of classes. Dashed lines represent instance level relations, while the full lines represent class level relations. Relations that are not labeled are *is-a* relations.

resulting from the application of a data mining algorithm implementation on a concrete dataset. All classes are extensions of *BFO:realizable entity* (see Figure 2.7). A *data mining algorithm implementation* is an extension of *OBI:plan*. A *generalization* is an extension of the *mapping* class. The connection with the specification module is made through the relation: an implementation *is-concretization-of* specification.

Application entities. Entities in the application module are all extensions of *OBI: planned process* (See Figure 2.7). It contains entities representing parts of the knowledge discovery process, such as *execution of DM algorithm implementation* and *execution of predictive model*. The *execution of DM algorithm implementation* is a realization (linked with *realizes*) of an *DM algorithm implementation*. Since the execution of an algorithm is a planned process it has input (*dataset*), an output (*generalization*) and achieves the planned objective *data mining task*.

Illustrative example. In Figure 2.7, we present example instances in the OntoDM ontology. The instances are represented as oval objects and the relations between instances are marked with dashed lines. In this example, we are representing the clus-HMC algorithm in all three modules. clus-HMC in an algorithm for predicting structured outputs: it learns decision trees for hierarchical multi-label classification [48].

The *inst:clus-HMC process ID0001* is an instance of a *predictive modeling process*. It has as its input a *inst:HMC dataset ID0001* and as its output *inst:clus-HMC decision tree ID0001*. The *inst:HMC dataset ID0001* is an instance of the *dataset* class and is connected to the *HMC dataset ID0001 spec.* via the *is-about* relation. The dataset specification contains the input and output datatypes of the dataset (*inst:input tuple of primitives* and *inst:output DAG*). The *inst:clus-HMC decision tree ID0001* is a concretization of *inst:clus-HMC decision tree spec.* and is realized by a *inst:clus-HMC decision tree execution process* in the case we want to obtain predictions for new examples.

The *inst:clus-HMC process ID0001* realizes the *inst:clus-HMC algorithm implementation*, which is a concretization of the *clus-HMC algorithm specification*. The process achieves the planned objective *inst:HMC learning task*, which is an instance of the *decision tree learning task* class.

2.6 Related Work

The main developments in formal representation of data mining entities in the form of ontologies take place in the domain of data mining workflow construction, data mining services, and describing data mining resources on the GRID. Other research in ontologies for data mining include formal representations of machine learning experiments in context of experiment databases. Finally, there is an increasing interest in extracting data mining entities from the data mining literature. In the remainder of this section, we briefly summarize the contributions in all these domains.

Data mining workflows. A prototype of an Intelligent Discovery Assistant (IDA) has been proposed [2], which provides users with systematic enumerations of valid sequences of data mining operators (called data mining processes). Effective rankings of the processes by different criteria are also provided in order to facilitate the choice of data mining processes to execute or solve a concrete data mining task. This automated system takes an advantage of an explicit ontology of data mining operators (algorithms). A light-weight ontology is used that contains only a hierarchy of data mining operators divided into three main classes: preprocessing operators, induction algorithms and post processing operators. The leaves of the hierarchy are the actual operators. The ontology does not contain information about the internal structure of the operators and the taxonomy is produced only according to the role that the operator has in the knowledge discovery process.

Building upon this work has been proposed [24] in a proposal of an intelligent data mining assistant that combines planning and meta-learning for automatic design of data mining workflows. A knowledge driven planner relies on a knowledge discovery ontology [2], to determine the valid set of operators for each step in the workflow. A probabilistic meta-learner is proposed for selecting the most appropriate operators by using relational similarity measures and kernel functions.

The problem of semi-automatic design of workflows for complex knowledge discovery tasks has also been addressed by Žakova et al. [49, 50]. The idea is to automatically propose workflows for the given type of inputs and required outputs of the discovery process. This is done by formalizing the notions of a knowledge type and data mining algorithm in the form of an ontology (named KD ontology). The planning algorithm accepts task descriptions expressed using the vocabulary of the ontology.

Kietz et al. [26, 27] present a data mining ontology for workflow planning. The ontology is designed to contain all the information necessary to support a 3rd generation KDD Support System. This includes the objects manipulated by the system, the meta data needed, the operators (i.e., algorithms) used and a goal description. The vocabulary of the ontology is used further for Hierarchical Task Network planning (HTN).

Hilario et al. [22] present their vision of a data mining ontology designed to support meta-learning for algorithm and model selection in the context of data mining workflow optimization. The ontology (named DMOP) is viewed as the repository of the intelligent assistant's data mining expertise, containing representations of data mining tasks, algorithms and models.

Diamantini and Potena [11] introduce a semantic based, service oriented framework for tools sharing and reuse, in order to give support for the semantic enrichment through semantic annotation of KDD (Knowledge Discovery in Databases) tools and deployment of tools as web services. For describing the domain, they propose an ontology named KDDONTO [12] which is developed having in mind the central role of a KDD algorithm and their composition (similar to the work presented in [2, 49]).

GRID. In the context of GRID programming, Cannataro and Comito [8] propose a design and implementation of an ontology of data mining. The motivation for building the ontology comes from the context of the author's work in Knowledge

GRID [9]. The main goals of the ontology are to allow the semantic search of data mining software and other data mining resources and to assist the user by suggesting the software to use on the basis of the user's requirements and needs. The proposed DAMON (DAta Mining ONtology) ontology is built through a characterization of available data mining software.

Brezany et al. [5] introduce an ontology-based framework for automated construction of complex interactive data mining workflows as a means of improving productivity of GRID-enabled data systems. For this purpose they develop a data mining ontology which is based on concepts from industry standards such as: the predictive model mark-up language (PMML) [11], WEKA [51] and the Java data mining API [23].

Experiment databases. As data mining and machine learning are experimental sciences, insight into the performance of a particular algorithm is obtained by implementing it and studying how it behaves on different datasets. Blockeel and Vanschoren [3, 4] (also Vanschoren and Blockeel in this volume) propose an experimental methodology based on experiment database in order to allow repeatability of experiments and generalizability of experimental results in machine learning.

Vanschoren et al. [46] propose an XML based language (named ExpML) for describing classification and regression experiments. In this process, the authors identified the main entities for formalizing a representation of machine learning experiments and implemented it in an ontology (named Exposé) [47]. This ontology is based on the same design principles as the OntoDM ontology, presented in this chapter, and further uses and extends some of the OntoDM classes.

Identification of entities from literature. Peng et al. [37] survey a large collection of data mining and knowledge discovery literature in order to identify and classify the data mining entities into high-level categories using grounded theory approach and validating the classification using document clustering. As a result of the study the authors have identified eight main areas of data mining and knowledge discovery: data mining tasks, learning methods and tasks, mining complex data, foundations of data mining, data mining software and systems, high-performance and distributed data mining, data mining applications and data mining process and project.

2.7 Conclusion

In this chapter, we have presented the OntoDM ontology for data mining, based on a recent proposal for a general framework of data mining. OntoDM is developed as a heavy-weight ontology of the data mining, starting from first principles as laid out by the framework, and including a significant amount of detail on basic data

---

[11] http://www.dmg.org/

mining entities. Entities represented in OntoDM include data (datatypes, datasets), data mining tasks (e.g., predictive modeling, clustering), data mining algorithms and their components, and generalizations (e.g., patterns and models output by data mining algorithms).

OntoDM is very general and allows us to represent much of the diversity in data mining research, including recently developed approaches. For example, OntoDM covers the area of mining structured data, including both the mining of frequent patterns from structured data and the prediction of structured outputs. Also, entities from the area of constraint-based data mining and inductive databases are included, such as evaluation functions, constraints, and data mining scenarios.

In the design of OntoDM, we have followed best practices in ontology engineering. We reuse upper-level ontology categories and well-defined ontological relations accepted widely in other ontologies for representing scientific investigations.Using these design principles we can link the OntoDM ontology to other domain ontologies (e.g., ontologies developed under the OBO Foundry) and provides reasoning capabilities across domains. The ontology is divided into three logical modules (specification, implementation, application).

Consequently, OntoDM can be used to support a broad variety of tasks. For example, it can be used to search for different implementations of an algorithm, to support the composition of data mining algorithms from reusable components, as well as the construction of data mining scenarios and workflows. It can also be used for representing and annotating data mining investigations.

We are currently working on the further development of several aspects of the ontology, such as the taxonomies of generalizations, tasks and algorithms. Some of these will require further development and extension of the general framework for data mining that we have used a starting point (concerning, e.g., the more precise representation of DM algorithm components). Next, we plan to populate the ontology with specific instances of the present classes. Furthermore, we plan to connect the OntoDM ontology with ontologies of application domains (e.g., The Ontology for Drug Discovery Investigations [38]) by developing application specific use cases. Finally, applying the OntoDM design principles on the development of ontologies for other areas of computer science, is one of the most important long term objectives of our research.

Availability. The OntoDM ontology is available at: `http://kt.ijs.si/ pance_panov/OntoDM/`

Acknowledgements Part of the research presented in this chapter was conducted within the project IQ (*Inductive Queries for mining patterns and models*) funded by the European Commission of the EU under contract number FP6-IST 516169. Panče Panov and Sašo Džeroski are currently supported by the Slovenian Research Agency through the research projects *Advanced machine learning methods for automated modelling of dynamic systems* (under grant J2-0734) and *Data Mining for Integrative Data Analysis in Systems Biology* (under grant J2-2285). For a complete list of agencies, grants and institutions currently supporting Sašo Džeroski, please consult the Acknowledgements chapter of this volume.

References

1. R. Agrawal, T. Imielinski, and A. N. Swami. Mining association rules between sets of items in large databases. In *Proc. ACM SIGMOD Intl. Conf. on Management of Data*, pages 207–216. ACM Press, 1993.
2. A. Bernstein, F. Provost, and S. Hill. Toward intelligent assistance for a data mining process: An ontology-based approach for cost-sensitive classification. *IEEE Transactions on Knowledge and Data Engineering*, 17(4):503–518, 2005.
3. H. Blockeel. Experiment databases: A novel methodology for experimental research. In *Proc. 4th Intl. Wshp. on Knowledge Discovery in Inductive Databases*, LNCS 3933:72–85. Springer, 2006.
4. H. Blockeel and J. Vanschoren. Experiment databases: Towards an improved experimental methodology in machine learning. In *Proc. 11th European Conf. on Principles and Practices of Knowledge Discovery in Databases*, LNCS 4702:6–17. Springer, 2007.
5. P. Brezany, I. Janciak, and A. M. Tjoa. Ontology-Based Construction of Grid Data Mining Workflows. In H.O. Nigro, S. Gonzales Cisaro and D. Xodo, editors, *Data Mining with Ontologies: Implementations, Findings and Frameworks*, pages 182-210, IGI Global, 2007.
6. R. R. Brinkman, M. Courtot, D. Derom, J. M. Fostel, Y. He, P. Lord, J. Malone, H. Parkinson, B. Peters, P. Rocca-Serra, A. Ruttenberg, S-A. A. Sansone, L. N. Soldatova, C. J. Stoeckert, J. A. Turner, J. Zheng, and OBI consortium. Modeling biomedical experimental processes with OBI. *Journal of Biomedical Semantics*, 1(Suppl 1):S7+, 2010.
7. P. Buitelaar and P. Cimiano, editors. *Ontology Learning and Population: Bridging the Gap between Text and Knowledge*. IOS Press, 2008.
8. M. Cannataro and C. Comito. A data mining ontology for grid programming. In *Proc. 1st Intl. Wshop. on Semantics in Peer-to-Peer and Grid Computing*, pages 113–134. IWWWC, 2003.
9. M. Cannataro and D. Talia. The knowledge GRID. *Communications of the ACM*, 46(1):89–93, 2003.
10. M. Courtot, F. Gibson, A. L. Lister, R. R. Brinkman J. Malone, D. Schober, and A. Ruttenberg. MIREOT: The Minimum Information to Reference an External Ontology Term. In *Proc. Intl. Conf. on Biomedical Ontology*, 2009.
11. C. Diamantini and D. Potena. Semantic annotation and services for KDD tools sharing and reuse. In *Proc. IEEE International Conference on Data Mining Workshops*, pages 761–770, IEEE Computer Society, 2008.
12. C. Diamantini, D. Potena, and E. Storti. KDDONTO: An ontology for discovery and composition of KDD algorithms. In *Proc. 2nd Intl. Wshp. on Third Generation Data Mining: Towards Service-Oriented Knowledge Discovery*, pages 13–25. ECML/PKDD 2009.
13. S. Džeroski. Towards a general framework for data mining. In *Proc. 5th Intl. Wshp. on Knowledge Discovery in Inductive Databases*, LNCS 4747:259–300, Springer, 2007
14. A. Brazma et al. Minimum information about a microarray experiment (MIAME) - toward standards for microarray data. *Nature Genetics*, 29(4):365–371, 2001.
15. B. Smith et al. The OBO foundry: coordinated evolution of ontologies to support biomedical data integration. *Nature Biotechnology*, 25(11):1251–1255, 2007.
16. C.F. Taylor et al. Promoting coherent minimum reporting guidelines for biological and biomedical investigations: the MIBBI project. *Nature Biotechnology*, 26(8):889–896, 2008.
17. W. J. Frawley, G. Piatetsky-Shapiro, and C. J. Matheus. Knowledge discovery in databases: An overview. In G. Piatetsky-Shapiro and W. J. Frawley, editors. *Knowledge Discovery in Databases*, pages 1–30. AAAI/MIT Press, 1991.
18. T. Gaertner. A survey of kernels for structured data. *SIGKDD Explorations*, 2003.
19. A. Gangemi, N. Guarino, C. Masolo, A. Oltramari, and L. Schneider. Sweetening ontologies with DOLCE. In *Proc. 13th Intl. Conf. on Knowledge Engineering and Knowledge Management, Ontologies and the Semantic Web*, LNCS 2473:166-181, Springer, 2002.
20. P. Grenon and B. Smith. SNAP and PAN: Towards dynamic spatial ontology. *Spatial Cognition & Computation*, 4(1):69 – 104, 2004.
21. D. J. Hand, P. Smyth, and H. Mannila. *Principles of Data Mining*. MIT Press, 2001.

22. M. Hilario, A. Kalousis, P. Nguyen, and A. Woznica. A data mining ontology for algorithm selection and Meta-Mining. In *Proc. 2nd Intl. Wshp. on Third Generation Data Mining: Towards Service-Oriented Knowledge Discovery*, pages 76–88. ECML/PKDD, 2009.
23. M. F. Hornick, E. Marcadé, and S. Venkayala. *Java Data Mining: Strategy, Standard, and Practice*. Morgan Kaufmann, 2006.
24. A. Kalousis, A. Bernstein, and M. Hilario. Meta-learning with kernels and similarity functions for planning of data mining workflows. In *Proc. 2nd Intl. Wshp. on Planning to Learn*, pages 23–28. ICML/COLT/UAI, 2008.
25. L. Kaufman and P.J. Rousseeuw. *Finding Groups in Data: An Introduction to Cluster Analysis*. Wiley Interscience, 1990.
26. J. Kietz, F. Serban, A. Bernstein, and S. Fischer. Towards cooperative planning of data mining workflows. In *Proc. 2nd Intl. Wshp. on Third Generation Data Mining: Towards Service-Oriented Knowledge Discovery*, pages 1–13. ECML/PKDD, 2009.
27. J-U. Kietz, A. Bernstein F. Serban, and S. Fischer. Data mining workflow templates for intelligent discovery assistance and Auto-Experimentation. In *Proc. 2nd Intl. Wshop. Third Generation Data Mining: Towards Service-Oriented Knowledge Discovery*, pages 1–12. ECML/PKDD, 2010.
28. R.D. King, J. Rowland, S. G. Oliver, M. Young, W. Aubrey, E. Byrne, M. Liakata, M. Markham, P. Pir, L. N. Soldatova, A. Sparkes, K.E. Whelan, and A. Clare. The Automation of Science. *Science*, 324(5923):85–89, 2009.
29. A. Lister, Ph. Lord, M. Pocock, and A. Wipat. Annotation of SBML models through rule-based semantic integration. *Journal of Biomedical Semantics*, 1(Suppl 1):S3, 2010
30. A. Maccagnan, M. Riva, E. Feltrin, B. Simionati, T. Vardanega, G. Valle, and N. Cannata. Combining ontologies and workflows to design formal protocols for biological laboratories. *Automated Experimentation*, 2:3, 2010.
31. E. Malaia. *Engineering Ontology: Domain Acquisition Methodology and Pactice*. VDM Verlag, 2009.
32. B. Meek. A taxonomy of datatypes. *SIGPLAN Notes*, 29(9):159–167, 1994.
33. R. Mizoguchi. Tutorial on ontological engineering - part 3: Advanced course of ontological engineering. *New Generation Computing*, 22(2):193–220, 2004.
34. I. Niles and A. Pease. Towards a standard upper ontology. In *Proc. Intl. Conf. Formal Ontology in Information Systems*, pages 2–9. ACM Press, 2001.
35. P. Panov, S. Džeroski, and L. N. Soldatova. OntoDM: An ontology of data mining. In *Proc. IEEE International Conference on Data Mining Workshops*, pages 752–760. IEEE Computer Society, 2008.
36. P. Panov, L. N. Soldatova, and S. Džeroski. Towards an ontology of data mining investigations. In *Proc. 12th Intl. Conf. on Discovery Science*, LNCS 5808:257–271. Springer, 2009.
37. Y. Peng, G. Kou, Y. Shi, and Z. Chen. A descriptive framework for the field of data mining and knowledge discovery. *International Journal of Information Technology and Decision Making*, 7(4):639–682, 2008.
38. D. Qi, R. King, G. R. Bickerton A. Hopkins, and L. Soldatova. An ontology for description of drug discovery investigations. *Journal of Integrative Bioinformatics*, 7(3):126, 2010.
39. D. Schober, W. Kusnierczyk, S. E Lewis, and J. Lomax. Towards naming conventions for use in controlled vocabulary and ontology engineering. In *Proc. BioOntologies SIG*, pages 29–32. ISMB, 2007.
40. J. Shawe-Taylor and N. Cristianini. *Kernel Methods for Pattern Analysis*. Cambridge University Press, 2004.
41. B. Smith. Ontology. In Luciano Floridi, editor, *Blackwell Guide to the Philosophy of Computing and Information*, pages 155–166. Oxford Blackwell, 2003.
42. B. Smith, W. Ceusters, B. Klagges, J. Kohler, A. Kumar, J. Lomax, C. Mungall, F. Neuhaus, A. L. Rector, and C. Rosse. Relations in biomedical ontologies. *Genome Biology*, 6:R46, 2005.
43. L. N. Soldatova, W. Aubrey, R. D. King, and A. Clare. The EXACT description of biomedical protocols. *Bioinformatics*, 24(13):i295-i303, 2008.

44. L. N. Soldatova and R. D. King. Are the current ontologies in biology good ontologies? *Nature Biotechnology*, 23(9):1095–1098, 2005.
45. L. N. Soldatova and R. D. King. An ontology of scientific experiments. *Journal of the Royal Society Interface*, 3(11):795–803, 2006.
46. J. Vanschoren, H. Blockeel, B. Pfahringer, and G. Holmes. Experiment databases: Creating a new platform for meta-learning research. In *Proc. 2nd Intl. Wshp. on Planning to Learn*, pages 10–15. ICML/COLT/UAI, 2008.
47. J. Vanschoren and L. Soldatova. Exposé: An ontology for data mining experiments. In *Proc. 3rd Intl. Wshp. on Third Generation Data Mining: Towards Service-oriented Knowledge Discovery*, pages 31–44. ECML/PKDD, 2010.
48. C. Vens, J. Struyf, L. Schietgat, S. Džeroski, and H. Blockeel. Decision trees for hierarchical multi-label classification. *Machine Learning*, 73(2):185–214, 2008.
49. M. Žáková, P. Kremen, F. Zelezny, and N. Lavrač. Planning to learn with a knowledge discovery ontology. In *Proc. 2nd Intl. Wshop. Planning to Learn*, pages 29–34. ICML/COLT/UAI, 2008.
50. M. Žáková, V. Podpecan, F. Železný, and N. Lavrač. Advancing data mining workflow construction: A framework and cases using the orange toolkit. In V. Podpečan, N. Lavrač, J.N. Kok, and J. de Bruin, editors, *Proc. 2nd Intl. Wshop. Third Generation Data Mining: Towards Service-Oriented Knowledge Discovery*, pages 39–52. ECML/PKDD 2009.
51. I. H. Witten and E. Frank. *Data Mining: Practical Machine Learning Tools and Techniques*. 2nd ed., Morgan Kaufmann, 2005.
52. Q. Yang and X. Wu. 10 challenging problems in data mining research. *International Journal of Information Technology and Decision Making*, 5(4):597–604, 2006.

Chapter 3
A Practical Comparative Study Of Data Mining Query Languages

Hendrik Blockeel, Toon Calders, Élisa Fromont, Bart Goethals, Adriana Prado, and Céline Robardet

Abstract An important motivation for the development of inductive databases and query languages for data mining is that such an approach will increase the flexibility with which data mining can be performed. By integrating data mining more closely into a database querying framework, separate steps such as data preprocessing, data mining, and postprocessing of the results, can all be handled using one query language. In this chapter, we compare six existing data mining query languages, all extensions of the standard relational query language SQL, from this point of view: how flexible are they with respect to the tasks they can be used for, and how easily can those tasks be performed? We verify whether and how these languages can be used to perform four prototypical data mining tasks in the domain of itemset and association rule mining, and summarize their stronger and weaker points. Besides offering a comparative evaluation of different data mining query languages, this chapter also provides a motivation for a following chapter, where a deeper integration of data mining into databases is proposed, one that does not rely on the development of a new query language, but where the structure of the database itself is extended.

Hendrik Blockeel
Katholieke Universiteit Leuven, Belgium and Leiden Institute of Advanced Computer Science, Universiteit Leiden, The Netherlands e-mail: hendrik.blockeel@cs.kuleuven.be

Toon Calders
Technische Universiteit Eindhoven, The Netherlands e-mail: t.calders@tue.nl

Élisa Fromont · Adriana Prado
Université de Lyon (Université Jean Monnet), CNRS, Laboratoire Hubert Curien, UMR5516, F-42023 Saint-Etienne, France
e-mail: {elisa.fromont, adriana.bechara.prado}@univ-st-etienne.fr

Bart Goethals
Universiteit Antwerpen, Belgium e-mail: bart.goethals@ua.ac.be

Céline Robardet
Université de Lyon, INSA-Lyon, CNRS, LIRIS, UMR5205, F-69621, France
e-mail: celine.robardet@insa-lyon.fr

3.1 Introduction

An important motivation for the development of inductive databases and query languages for data mining is that such an approach will increase the flexibility with which data mining can be performed. By integrating data mining more closely into a database querying framework, separate steps such as data preprocessing, data mining, and postprocessing of the results, can all be handled using one query language. It is usually assumed that standard query languages such as SQL will not suffice for this; and indeed, SQL offers no functionality for, for instance, the discovery of frequent itemsets. Therefore, multiple researchers have proposed to develop new query languages, or extend existing languages, so that they offer true data mining facilities. Several concrete proposals have been implemented and evaluated.

In this chapter, we consider four prototypical data mining tasks, and six existing data mining query languages, and we evaluate how easily the tasks can be performed using these languages. The six languages we evaluate are the following: MSQL [8], MINE RULE operator [11], SIQL [17], SPQL [2], and DMX [16]. All six are based on extending SQL and have special constructs to deal with itemsets and/or association rules.

The four tasks with which the expressivity of the languages will be tested can all be situated in the association rule mining domain. The tasks are "typical" data mining tasks, in the sense that they are natural tasks in certain contexts, and that they have not been chosen with a particular data mining query language in mind. The four tasks are: discretizing a numerical attribute, mining itemsets with a specific *area* constraint, and two association rule mining tasks in which different constraints are imposed on the rules to be discovered. It turns out that the existing languages have significant limitations with respect to the tasks considered.

Many of the shortcomings of the six languages are not of a fundamental nature and can easily be overcome by adding additional elements to the query languages. Yet, when extending a query language, however, there is always the question of how much it should be extended. One can identify particular data mining problems and then extend the language so that these problems can be handled; but whenever a new type of data mining task is identified, a further extension may be necessary, unless one can somehow guarantee that a language is expressive enough to handle any kind of data mining problem.

While this chapter offers a comparative evaluation of different data mining query languages, this comparison is not the main goal; it is meant mostly as an illustration of the limitations that current languages have, and as a motivation for Chapter 11, where the idea of creating a special-purpose query language for data mining is abandoned, and the inductive database principle is implemented by changing the structure of the database itself, adding "virtual data mining views" to it (which can be queried using standard SQL), rather than by extending the query language.

We dedicate the next section to the description of the chosen data mining tasks. In Section 3.3, we introduce the data mining query languages and describe how they can be used for performing these tasks. Next, in Section 3.4, we summarize the positive and negative points of the languages, with respect to the accomplishment of the given tasks.

3.2 Data Mining Tasks

Inspired by [3], we verify whether and how four prototypical data mining tasks can be accomplished using a number of existing data mining query languages. To enable a fair comparison between them, we study here data mining tasks which mainly involve itemsets and association rules, as these patterns can be computed by almost all of the surveyed proposals. We also focus on intra-tuple patterns (i.e., patterns that relate values of different attributes of the same tuple), even though some of the languages considered can also handle inter-tuple patterns (which relate values of attributes of different tuples that are somehow connected) [3]. As the precise structure of the patterns that can be found typically also differs between query languages, we will for each task describe precisely what kind of pattern we are interested in (i.e., impose specific constraints that the patterns should satisfy).

For ease of presentation, we will assume that the data table Playtennis2 in Figure 3.1 forms the source data to be mined. The data mining tasks that we will discuss are the following:

- **Discretization task**: Discretize attribute *Temperature* into 3 intervals. The discretized attribute should be used in the subsequent tasks.
- **Area task**: Find all intra-tuple itemsets with relative support of at least 20%, size of at least 2, and area, that is, absolute support × size, of at least 10. The area of an itemset corresponds to the size of the tile that is formed by the items in the itemset in the transactions that support it. The mining of large tiles; i.e., itemsets with a high area is useful in constructing small summaries of the database [4].

Fig. 3.1 The data table Playtennis2.

Playtennis2

Day	Outlook	Temperature	Humidity	Wind	Play
D1	Sunny	85	High	Weak	No
D2	Sunny	80	High	Strong	No
D3	Overcast	83	High	Weak	Yes
D4	Rain	70	High	Weak	Yes
D5	Rain	68	Normal	Weak	Yes
D6	Rain	65	Normal	Strong	No
D7	Overcast	64	Normal	Strong	Yes
D8	Sunny	72	High	Weak	No
D9	Sunny	69	Normal	Weak	Yes
D10	Rain	75	Normal	Weak	Yes
D11	Sunny	75	Normal	Strong	Yes
D12	Overcast	72	High	Strong	Yes
D13	Overcast	81	Normal	Weak	Yes
D14	Rain	71	High	Strong	No

- **Right hand side task**: Find all intra-tuple association rules with relative support of at least 20%, confidence of at most 80%, size of at most 3, and a singleton right hand side.
- **Lift task**: Find, *from the result of the right hand side task*, rules with attribute *Play* as the right hand side that have a lift greater than 1.

While these tasks are only a very small sample from all imaginable tasks, they form a reasonably representative and informative sample. Discretization is a very commonly used preprocessing step. The discovery of itemsets and association rules are common data mining tasks, and the constraints considered here (upper/lower bounds on support, confidence, size, area, lift) are commonly used in many application domains. The fourth task is interesting in particular because it involves what we could call *incremental mining*: after obtaining results using a data mining process, one may want to refine those results, or mine the results themselves. This is one of the main motivating characteristics of inductive databases: the closure property implies that the results of a mining operation can be stored in the inductive database, and can be queried further with the same language used to perform the original mining operation.

3.3 Comparison of Data Mining Query Languages

We include six data mining query languages in our comparison : DMQL [6, 7], MSQL [8], SQL extended with the MINE RULE operator [11], SIQL [17], SPQL [2], and DMX, the extended version of SQL that is included in Microsoft SQL server 2005 [16]. As all these languages are quite different, we postpone a detailed discussion of each language until right before the discussion of how it can be used to perform the four tasks.

3.3.1 DMQL

The language DMQL (Data Mining Query Language) [6, 7] is an SQL-like data mining query language designed for mining several kinds of rules in relational databases, such as classification and association rules. It has been integrated into DBMiner [5], which is a system for data mining in relational databases and data warehouses.

As for association rules, a rule in this language is a relation between the values of two sets of predicates evaluated on the database. The predicates are of the form P(X, y), where P is a predicate that takes the name of an attribute of the source relation, X is a variable, and y is a value in the domain of this attribute. As an example, the association rule "if outlook is sunny and humidity is high, you should not play tennis" is represented in this language by

$$\text{Outlook}(X, \text{`Sunny'}) \wedge \text{Humidity}(X, \text{`High'}) \Rightarrow \text{Play}(X, \text{`No'}),$$

where X is a variable representing the tuples in the source relation that satisfy the rule.

DMQL also gives the user the ability to define a *meta-pattern* (template), which restricts the structure of the rules to be extracted. For example, the meta-pattern

$$P(X: \text{Playtennis2}, y) \wedge Q(X, w) \Rightarrow \text{Play}(X, z)$$

restricts the structure of the association rules to rules having only the attribute *Play* in the right hand side, and any 2 attributes in the left hand side. In addition to the meta-pattern resource, DMQL has also primitives to specify concept hierarchies on attributes. These can be used so as to extract generalized association rules [15] as well as for discretization of numerical attributes.

Next, we present how the tasks described above are executed in DMQL.

Discretization task. In DMQL, a discretization of a numerical attribute can be defined by a concept hierarchy as follows [7]:

```
1. define hierarchy temp_h for Temperature
   on Playtennis2 as
2. level1: {60..69} < level0:all
3. level1: {70..79} < level0:all
4. level1: {80..89} < level0:all
```

By convention, the most general concept, *all*, is placed at the root of the hierarchy, that is, at level 0. The notation ".." implicitly specifies all values within the given range. After constructing such hierarchy, it can be used in subsequent mining queries as we show later on.

Area task. As DMQL was specially designed for extracting rules from databases, the area task cannot be executed in this language.

Right hand side task. The following DMQL query is how intra-tuple association rules with a certain minimum support and minimum confidence are extracted in this language (note that we are not considering the constraint on the maximum size of the rules nor on their right hand sides yet):

```
1. use database DB
2. use hierarchy temp_h for attribute Temperature
3. in relevance to Outlook, Temperature,
                 Humidity, Wind, Play
4. mine associations as MyRules
5. from Playtennis2
6. group by Day
7. with support threshold = 0.20
8. with confidence threshold = 0.80
```

The language constructs allow to specify the relevant set of data (lines 3 and 5), the hierarchies to be assigned to a specific attribute (line 2, for the attribute

Temperature), the desired output, that is, the kind of knowledge to be discovered (line 4), and the constraints minimum support and minimum confidence (lines 7 and 8, respectively), as required by the current task.

DMQL is able to extract both intra- and inter-tuple association rules. For the extraction of intra-tuple rules (as requested by the current task), the group-by clause in line 6 guarantees that each group in the source data coincides with a unique tuple.

Concerning the remaining constraints, although it is possible to constrain the size of the right hand side of the rules using meta-patterns (as shown earlier), we are not aware of how meta-patterns can be used to constrain the maximum size of the rules, as also needed by the current task. An alternative solution to obtain the requested rules is to write two DMQL queries as the one above, the first using the meta-pattern:

$$P(X: \text{Playtennis2}, y) \wedge Q(X, w) \Rightarrow V(X, z)(\text{rules with size 3})$$

and the second, using the meta-pattern:

$$P(X: \text{Playtennis2}, y) \Rightarrow V(X, z)(\text{rules with size 2}).$$

A meta-pattern can be used in the query above by simply adding the clause "**matching** <meta-pattern>" between lines 4 and 5. We therefore conclude that the right hand side task can be performed in DMQL.

Lift task. In the system DBMiner [5], the mining results are presented to the user and an iterative refinement of these results is possible only through graphical tools. In fact, it is not clear in the literature whether nor how (with respect to the attributes) the rules are stored into the database. For this reason, we assume here that the mining results cannot be further queried and, consequently, the lift task cannot be accomplished in this language.

3.3.2 MSQL

The language MSQL [8] is an SQL-like data mining query language that focuses only on mining intra-tuple association rules in relational databases. According to the authors, the main intuition behind the language design has been to allow the representation and manipulation of rule components (left and right hand sides), which, being sets, are not easily representable in standard SQL [8].

In MSQL, an association rule is a propositional rule defined over *descriptors*. A descriptor is an expression of the form $(A_i = a_{ij})$, where A_i is an attribute in the database and a_{ij} belongs to the domain of A_i. A *conjunctset* is defined as a set containing an arbitrary number of descriptors, such that no two descriptors are formed using the same attribute. Given this, an association rule in this language is of the form $A \Rightarrow B$, where A is a conjunctset and B a single descriptor. The rule "if outlook is sunny and humidity is high, you should not play tennis" is therefore represented in MSQL as

$$(Outlook = \text{'Sunny'}) \land (Humidity = \text{'High'}) \Rightarrow (Play = \text{'No'}).$$

MSQL offers operators for extracting and querying association rules: these are called *GetRules* and *SelectRules*, respectively. Besides, it also provides an encode operator for discretization of numerical values.

In the following, we show how the given tasks are executed in MSQL.

Discretization task. MSQL offers an encode operator that effectively creates ranges of values, and assigns integers (an encoded value) to those ranges. The following MSQL statement creates a discretization for the attribute *Temperature*, as required.

```
1. create encoding temp_encoding
        on Playtennis2.Temperature as
2. begin
3. (60,69,1), (70,79,2), (80,89,3), 0
4. end
```

For every set (x,y,z) given in line 3, MSQL assigns the integer z to the range of values from x to y. In this example, the integer 0 is assigned to occasional values not included in any of the specified ranges (see end of line 3). The created encoding, called "temp_encoding", can be used in subsequent mining queries as we show below.

Area task. Similarly to DMQL, MSQL cannot perform the area task, as it was specially proposed for association rule mining.

Right hand side task. As described earlier, MSQL is able to extract only intra-tuple association rules, which, in turn, are defined as having a singleton right hand side. The current task can be completely performed by the operator *GetRules*, as follows.

```
1. GetRules(Playtennis2)
2. into MyRules
3. where length <= 2
4. and support >= 0.20
5. and confidence >= 0.80
6. using temp_encoding for Temperature
```

In line 1, the source data is defined between parentheses. Constraints on the rules to be extracted are posed in the where-clause: here, we constrain the size (length) of the left hand side of the rules, referred to in MSQL as body (line 3), their minimum support (line 4), and their minimum confidence (line 5). Finally, the using-clause allows the user to discretize numerical attributes on the fly. In line 6, we specify that the encoding called "temp_encoding" should be applied to the attribute *Temperature*.

MSQL also allows the user to store the resultant rules in a persistent rule base, although the format in which the rules are stored is opaque to the user. This storage is possible by adding the into-clause to the MSQL query, as in line 2. In this example, the name of the rule base is called "MyRules".

Lift task. As previously mentioned, MSQL offers an operator for querying mining results, which is called *SelectRules*. For example, the following MSQL query retrieves all rules with attribute *Play* as the right hand side, referred to in MSQL as consequent, from the rule base MyRules:

```
1. SelectRules(MyRules)
2. where Consequent is {(Play=*)}
```

The operator *SelectRules* retrieves the rules previously stored in the rule base MyRules (given in parentheses in line 1) that fulfill the constraints specified in the where-clause (line 2). These can only be constraints posed on the length of the rules, the format of the consequent (as in the query above), the format of the body, support or confidence of the rules. The constraint on lift, required by the current task, cannot be expressed in this language, which means that the lift task cannot be completely performed in MSQL.

3.3.3 MINE RULE

Another example is the operator MINE RULE [11] designed as an extension of SQL, which was also proposed for association rule mining discovery in relational databases. An interesting aspect of this work is that the operational semantics of the proposed operator is also presented in [11], by means of an extended relational algebra. Additionally, in [12], the same authors specified how the operator MINE RULE can be implemented on top of an SQL server.

As an example, consider the MINE RULE query given below:

```
1. Mine Rule MyRules as
2. select distinct 1..1 Outlook, Humidity as body,
                   1..1 Play as head,
                   support, confidence
3. from Playtennis2
4. group by Day
5. extracting rules with support: 0.20,
                        confidence: 0.80
```

This query extracts rules from the source table Playtennis2, as defined in line 3. The execution of this query creates a relational table called "MyRules", specified in line 1, where each tuple is an association rule. The select clause in line 2 defines the structure of the rules to be extracted: the body has schema {*Outlook*, *Humidity*}, while the head has schema {*Play*}. The notation "1..1" specifies the minimum and maximum number of schema instantiations in the body and head of the extracted rules, which is referred to as their cardinalities.

The select-clause also defines the schema of the relational table being created, which is limited to the body, head, support and confidence of the rules, the last 2 being optional. In the example query above, the schema of table MyRules consists of all these attributes.

Similar to DMQL, the operator MINE RULE is able to extract both inter- and intra-tuple association rules. For the extraction of intra-tuple rules, the group-by clause in line 4 assures that each group in the source data coincides with a unique tuple. Finally, in line 5, the minimum support and minimum confidence are specified.

Next, we show how the given tasks are performed with MINE RULE.

Discretization task We assume here that the MINE RULE operator has been integrated into a database system based on SQL (as discussed in [12]). Given this, although MINE RULE does not provide any specific operator for discretization of numerical values, the discretization required by this task can be performed by, e.g., the SQL CASE expression below. Such an expression is available in a variety of database systems, e.g., PostgreSQL[1], Oracle[2], Microsoft SQL Server[3] and MySQL[4].

```
1. create table MyPlaytennis as
2. select Day, Outlook,
3. case
   when Temperature between 60 and 69 then '[60,69]'
   when Temperature between 70 and 79 then '[70,79]'
   when Temperature between 80 and 89 then '[80,89]'
   end as Temperature,
4. Humidity, Wind, Play
5. from Playtennis2
```

The query above creates a table called "MyPlaytennis". It is in fact a copy of table Playtennis2 (see line 5), except that the attribute *Temperature* is now discretized into 3 intervals: [60,69],[70,79], and [80,89] (see line 3).

Area task. Similarly to MSQL, MINE RULE was specially developed for association rule discovery. Therefore, the area task cannot be performed with MINE RULE.

Right hand side task. Rules extracted with a MINE RULE query have only the body and head schemas specified in the query. For example, all rules extracted with the example MINE RULE query above have the body with schema {*Outlook,Humidity*} and head with schema {*Play*}. To perform this task, which asks for all rules of size of at most 3 and a singleton right hand side, we would need to write as many MINE RULE queries as are the possible combinations of disjoint body and head schemas. On the other hand, since MINE RULE is also capable of mining inter-tuple association rules, in particular single-dimensional association rules[5] [3], an alternative solution to obtain these rules is to firstly pre-process table MyPlaytennis into a new table, by breaking down each tuple *t* in MyPlaytennis

---

[1] http://www.postgresl.org/

[2] http://www.oracle.com/index.html/

[3] http://www.microsoft.com/sqlserver/2008/en/us/default.aspx/

[4] http://www.mysql.com/

[5] Single-dimensional association rules are rules that contain multiple occurrences of the same attribute, although over different values.

Fig. 3.2 Table MyPlaytennisTrans: the pre-processed MyPlaytennis data table created before using the MINE RULE operator.

MyPlaytennisTrans

Day	Condition
D1	"Outlook=Sunny"
D1	"Temperature=[80,89]"
D1	"Humidity=High"
D1	"Wind=Weak"
D1	"Play=No"
D2	"Outlook=Sunny"
...	...

into 5 tuples, each tuple representing one attribute-value pair in t (except the primary key). A sample of the new table, called "MyPlaytennisTrans", is depicted in Figure 3.2.

After the pre-processing step, the right hand side task can now be accomplished with the following query:

```
1. Mine Rule MyRules as
2. select distinct 1..2 Condition as body,
                   1..1 Condition as head,
                   support, confidence
3. from MyPlaytennisTrans
4. group by Day
5. extracting rules with support: 0.20,
                        confidence: 0.80
```

Here, the body and head of the extracted rules are built from the domain of the attribute *Condition* (attribute of table MyPlaytennisTrans). The body has cardinality of at most 2, while head has cardinality 1, as requested by this task.

Figure 3.3 shows the resulting table MyRules and illustrates how an association rule, e.g., "if outlook is sunny and humidity is high, you should not play tennis" is represented in this table. [6]

MyRules

body	head	support	confidence
{"Outlook=Sunny","Humidity=High"}	{"Play=No"}	0.21	1
...	...	...	...

Fig. 3.3 The table MyRules created by a MINE RULE query.

---

[6] For ease of presentation, we adopted here the same representation as in [11]. In [12] the authors suggest that the body and the head itemsets of the generated rules are stored in dedicated tables and referred to within the rule base table, in this case the table MyRules, by using foreign keys.

Lift task. For the execution of the previous task, the table MyRules, containing the extracted rules, was created. Note, however, that the table MyRules contains only the body, head, support and confidence of the rules. Indeed, to the best of our knowledge (see [3, 9, 11, 12]), the supports of the body and head of the rules are not stored in the database for being further queried. As a result, measures of interest, such as lift, cannot be computed from the mining results without looking again at the source data.

Although there is no conceptual restriction in MINE RULE that impedes the execution of this task, we assume here that the lift task cannot be performed using the operator MINE RULE, based on its description given in [3, 9, 11, 12].

3.3.4 SIQL

The system prototype SINDBAD (Structured Inductive Database Development), developed by Wicker et al. [17], provides an extension of SQL called SIQL (Structured Inductive Query Language). SIQL offers new operators for several data mining tasks, such as itemset mining, classification and clustering, and also for preprocessing, such as discretization and feature selection.
In the following, we present how the given tasks are performed in SIQL.

Discretization task. In SIQL, this task can be executed with the following query:

```
1. configure discretization numofintervals = 3
2. create table MyTable as
3. discretize Temperature in Playtennis2
```

In this language, all available operators should actually be seen as functions that transform tables into new tables. The query above, for example, produces a table called "MyTable"(see line 2), which is a copy of table Playtennis2, except that the attribute *Temperature* is now discretized according to the parameter previously configured in line 1. In this example, we discretize the attribute *Temperature* into 3 intervals, as requested by the discretization task.

Area task. For the frequent itemset mining task, SIQL allows the user to specify only the minimum support constraint, as follows:

```
1. configure apriori minSupport = 0.20
2. create table MySets as
3. frequent itemsets in MyTable
```

This query produces the table MySets (line 2) that contains the Boolean representation of the intra-tuple frequent itemsets found in table MyTable (line 3), which was previously created for the discretization task.[7]

---

[7] Before the mining can start, table MyTable needs to be encoded in a binary format such that each row represents a tuple with as many Boolean attributes as are the possible attribute-value pairs.

Items	
item_id	item
1	⟨Outlook, Sunny⟩
2	⟨Humidity, High⟩
...	...

Itemsets	
itemset_id	item_id
1	1
1	2
...	...

Supports	
itemset_id	support
1	3
...	...

Fig. 3.4 Materialization of SPQL queries.

Observe that, in this language, the attention is not focused on the use of constraints: the minimum support constraint is not posed within the query itself; it needs to be configured beforehand with the use of the so-called configure-clause (line 1). The minimum support constraint is therefore more closely related to a function parameter than to a constraint itself. Additionally, the number of such parameters is limited to the number foreseen at the time of implementation. For example, the constraints on size and area are not possible to be expressed in SIQL. We conclude, therefore, that the area task cannot be executed in this language.

Right hand side task. Although SIQL offers operators for several different mining tasks, there is no operator for association rule mining. This means that this task cannot be executed in SIQL.

Lift task. Due to the reason given above, the lift task is not applicable here.

3.3.5 SPQL

Another extension of SQL has been proposed by Bonchi et al. [2]. The language is called SPQL (Simple Pattern Query Language) and was specially designed for frequent itemset mining. The system called ConQueSt has also been developed, which is equipped with SPQL and a user-friendly interface.

The language SPQL supports a very large set of constraints of different types, such as anti-monotone [10], monotone [10], succinct [13], convertible [14], and soft constraints [1]. Additionally, it provides an operator for discretization of numerical values. Another interesting functionality of SPQL is that the result of the queries is stored into the database. The storage creates 3 different tables, as depicted in Figure 3.4. The figure also shows how the itemset (*Outlook* = 'Sunny' ∧ *Humidity* = 'High') is stored in these tables.

Below, we illustrate how the given tasks are executed in SPQL.

Discretization task. In SPQL, this task can be performed as below:

```
1. discretize Temperature as MyTemperature
2. from Playtennis2
3. in 3 equal width bins
4. smoothing by bin boundaries
```

In this example, we discretize the attribute *Temperature* into 3 intervals (bins), as requested by this task, with the same length (line 3), and we also want the bin boundaries to be stored as text (line 4) in a new attribute called "MyTemperature" (specified in line 1).

Area task. In SPQL, the user is allowed to constrain the support and the size of the frequent itemsets to be mined, as follows:

```
1. mine patterns with supp >= 3
2. select *
3. from Playtennis2
4. transaction Day
5. item Outlook, MyTemperature, Humidity, Wind, Play
6. constrained by length >= 2
```

The language allows the user to select the source data (lines from 2 to 5), the minimum absolute support, which is compulsory to be defined at the beginning of the query (line 1), and a conjunction of constraints, which is always posed at the end of the query. In this example, only the constraint on the size (length) of the itemsets is posed (line 6).

SPQL is able to extract both inter- and intra-tuple itemsets. For the extraction of intra-tuple itemsets, line 4 guarantees that each group in the source data corresponds to a unique tuple, while line 5 lists the attributes for exploration.

The constraint on the minimum area (absolute support × size), however, is apparently not possible to be expressed in this language, since the property support of an itemset cannot be referred to anywhere else but at the beginning of the query. Besides, it is not clear in [2] whether formulas such as support × length can be part of the conjunctions of constraints that are specified at the end of the queries. On the other hand, note that a post-processing query on the tables presented above, would be an alternative to complete this task. Contrary to SIQL, SPQL also stores the support of the extracted itemsets, which are crucial to compute their area. We therefore conclude that the area task can be accomplished by SQPL, provided that a post-processing query is executed.

Right hand side task. SPQL was specially designed for itemset mining. Consequently, the right hand side task cannot be performed in this language.

Lift task. Given the reason above, the lift task is not applicable.

3.3.6 DMX

Microsoft SQL server 2005 [16] provides an integrated environment for creating and working with data mining models. It consists of a large set of data mining algorithms for, e.g., association rule discovery, decision tree learning, and clustering. In order to create and manipulate the so-called data mining models, it offers an extended version of SQL, called DMX (Microsoft's Data Mining extensions). DMX is composed of data definition language (DDL) statements, data manipulation language (DML) statements, functions and operators.

In the following, we show how the given tasks can be performed by this language. DMX is able to extract both inter- and intra-tuple patterns. We focus here on the kind of patterns asked by the given tasks.

Discretization task. In DMX, the discretization of numerical values and creation of a data mining model itself can be done synchronously. This is shown below for the accomplishment of the right hand side task.

Area task. In DMX, frequent itemsets cannot be extracted independently from association rules. Nevertheless, the itemsets computed beforehand to form association rules can also be queried after mining such rules. Thus, to compute this task, the following steps are necessary. Firstly, a so-called association model has to be created as follows:

```
1. create mining model MyRules
2. (Day text, Outlook text,
      Temperature text discretized(Equal_Areas,3),
      Humidity text, Wind text, Play text)
3. using microsoft_association_rules
                       (minimum_support = 0.20,
                        minimum_probability = 0.80,
                        minimum_itemset_size = 2)
```

The above DMX query creates a model called "MyRules" that uses the values of the attributes defined in line 2 to generate association rules. The rules are extracted by the algorithm "Microsoft Association Rules", having as parameters minimum_support, minimum_itemset_size, as required by the current task, and also minimum_probability (the same as confidence, which is set to speed up computation only, as we are just interested in the itemsets). In addition, the user can specify which attributes he or she wants to have discretized, as in line 2. In this example, we specify that the values of the attribute *Temperature* should be discretized into 3 intervals, as demanded by the discretization task.

Having created the model, it needs to be trained through the insertion of tuples, as if it was an ordinary table:

```
1. insert into MyRules
2. (Day, Outlook, Temperature, Humidity, Wind, Play)
3. select Day, Outlook, Temperature,
           Humidity, Wind, Play
4. from Playtennis2
```

When training the model, we explicitly say from where the values of its associated attributes come (lines 3 and 4).

After training the model, it is necessary to query its content in order to visualize the computed itemsets. The content of an association model is stored in the database as shown in Figure 3.5 [16]. It consists of 3 levels. The first level has a single node, which represents the model itself. The second level represents the frequent itemsets computed to form the association rules. Each node represents one frequent itemset along with its characteristics, such as its corresponding support. The last level, in turn, represents the association rules. The parent of a rule node is the itemset that represents the left hand side of the rule. The right hand side, which is always a singleton, is kept in the corresponding rule node.

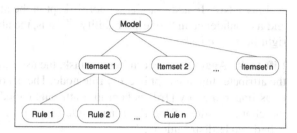

Fig. 3.5 The content of an association model in Microsoft SQL server 2005.

Each node keeps an attribute called "node_type", which defines the type of the node. For example, itemset nodes have node_type equal to 7, while rule nodes have node_type equal to 8. In addition to the attribute node_type, a text description of the itemset is kept in an attribute called "node_description", and its support in "node_support".[8] The description is a list of the items, displayed as a comma-separated text string, as in 'Outlook=Sunny, Humidity=High'. In order to query all itemsets in the model MyRules, along with their corresponding supports, the following DMX query is necessary:

```
1. select node_description, node_support
2. from MyRules.Content
3. where node_type = 7
```

In line 1, we select the text description and support of the rules. In line 2, we specify the model from which the content is to be queried. Finally, as we are only interested in the itemsets, we filter the nodes by their node types in line 3.

Note, however, that the current task asks for itemsets with size of at least 2 and area of at least 10. Therefore, a more complex DMX query is needed. As there is apparently no attribute in an itemset node keeping the size of the corresponding itemset, one needs to compute their sizes by processing the description of the itemset in the attribute node_description. By doing this, the area of the itemsets can also be computed. We assume therefore that this task can only be performed in DMX after the execution of a post-processing query.

Right hand side task. This task can be completely performed in DMX by following the same steps of the last task. Firstly, one needs to create a mining model similar to the one created above, except that here the parameter minimum_itemset_size is replaced with maximum_itemset_size, which is 3 for this task.

After training the model, we query for association rules as below:

---

[8] The specification of those attributes was found at http://technet.microsoft.com.

```
1. select node_description, node_support,
          node_probability
2. from MyRules.Content
3. where node_type = 8
```

For rules, the attribute node_description contains the left hand side and the right hand side of the rule, separated by an arrow, as in 'Outlook=Sunny, Humidity=High → Play = No'. In addition, its support is kept in the attribute called "node_support" and its confidence in "node_probability".[9] Thus, the above DMX query executes the right hand side task.

Lift task. Again, for performing this task, the user has to be aware of the names of the attributes that are kept in every rule node. The lift of an association rule (referred to as importance by [16]) is kept in a attribute called "msolap_node_score", while the characteristics of the right hand side of a rule can be found at a nested table called "node_distribution". [9]

The following DMX query performs the lift task:

```
1. select node_description, node_support,
          node_probability,
          (select attribute_name
          from node_distribution) as a
2. from MyRules.Content
3. where node_type=8
4.   and a.attribute_name = 'Play'
5.   and msolap_node_score >= 1
```

Here, we select from the content of the model MyRules only rules having attribute *Play* as the right hand side (line 4) that have a lift greater than 1 (line 5), just as required by this task. Thus, we can conclude that the lift task can be accomplished by DMX.

3.4 Summary of the Results

We now summarize the results achieved by the proposals presented in this chapter with respect to the accomplishment of the four given tasks. Table 3.1 shows, for each of the proposals, the performed tasks.

Discretization Task. Observe that the discretization task could be executed by all the proposals, although MINE RULE does not offer a specific operator for discretization. This shows that considerable attention is dedicated to pre-processing operations.

---

[9] The specification of those attributes was found at http://technet.microsoft.com.

Table 3.1 Results of each proposal for each task. The symbol $\sqrt{}\star$ means that the task was executed only after a pre- or post-processing query.

	Proposals					
	DMQL	MSQL	MINE RULE	SIQL	SPQL	DMX
Discretization task	√	√	√	√	√	√
Area task					√*	√*
Right hand side task	√	√	√*			√
Lift task				N/A	N/A	√

Area and Right Hand Side Tasks. From the results of the area and right hand side tasks, two main points can be concluded: firstly, the languages are not flexible enough to specify the kinds of patterns a user may be interested in. For example, MSQL and MINE RULE are entirely dedicated to the extraction of association rules, while SPQL was specially designed for frequent itemset mining. Concerning MINE RULE, the right hand side task could only be executed with a relatively high number of queries or after a pre-processing query (which was the approach we took). As for DMX, although it is able to perform the area task, we observe that there is not a clear separation between rules and itemsets.

The second point is that little attention is given to the flexibility of ad hoc constraints. For example, the constraint on area, which was required by the area task, could not be expressed in any of the proposals that can perform itemset mining. In fact, SPQL and DMX could only accomplish this task after the execution of a post-processing query. Note that the flexibility of these proposals is actually limited to the type of constraints foreseen by their developers; a new type of constraint in a mining operation which was not foreseen at the time of implementation will not be available for the user. In the particular cases of SIQL and DMX, a constraint is more closely related to a function parameter than to a constraint itself.

Lift Task. As for the lift task, we observed that little support is given to post-processing of mining results. Concerning DMQL, we are not aware of whether it considers the closure principle, that is, whether the results can be further queried, as opposed to the other data mining languages.

As for MSQL, although it gives the user the ability to store the mining rules in a rule base, the data space is totally opaque to the user. In other words, the rules can only be queried with the use of the operator *SelectRules*, and with a limit set of available constraints. In the case of MINE RULE, as opposed to MSQL, results are in fact stored in ordinary relational tables, but the format in which they are stored, with respect to the attributes, is not flexible enough. This restricts the number of possible constraints the user can express when querying those results.

Finally, observe that DMX is the only proposal that is able to perform the lift task. On the other hand, the models (and their properties) are stored in a very complex way with this language, making the access and browsing of mining results less intuitive.

3.5 Conclusions

Even though most of the limitations of the languages can be solved by minor extensions to the languages, the need to extend the languages itself is considered a drawback. In summary, we identify the following list of drawbacks noticed in at least one of the proposals surveyed in this chapter:

- There is little attention to the closure principle; the output of a mining operation cannot or only very difficultly be used as the input of another operation. While the closure principle is very important for the expressiveness of SQL, its data mining extensions mostly lack this advantage.
- The flexibility to specify different kinds of patterns and ad-hoc constraints is poor. If the user wants to express a constraint that was not explicitly foreseen by the developer of the system, he or she will have to do so with a post-processing query, if possible at all.
- The support for post-processing mining results is often poor due to a counter-intuitive way of representing mining results. Data mining results are often offered as static objects that can only be browsed or in a way that does not allow for easy post-processing.

In Chapter 11, we describe a new inductive database system which is based on the so-called virtual mining views framework. In addition, we show the advantages it has in comparison with the proposals described here.

Acknowledgements This work has been partially supported by the projects IQ (IST-FET FP6-516169) 2005/8, GOA 2003/8 "Inductive Knowledge bases", FWO "Foundations for inductive databases", and BINGO2 (ANR-07-MDCO 014-02). When this research was performed, Hendrik Blockeel was a post-doctoral fellow of the Research Foundation - Flanders (FWO-Vlaanderen), Élisa Fromont was working at the Katholieke Universiteit Leuven, and Adriana Prado was working at the University of Antwerp.

References

1. Bistarelli, S., Bonchi, F.: Interestingness is not a dichotomy: Introducing softness in constrained pattern mining. In: Proc. PKDD, pp. 22–33 (2005)
2. Bonchi, F., Giannotti, F., Lucchese, C., Orlando, S., Perego, R., Trasarti, R.: A constraint-based querying system for exploratory pattern discovery information systems. Information System (2008). Accepted for publication
3. Botta, M., Boulicaut, J.F., Masson, C., Meo, R.: Query languages supporting descriptive rule mining: A comparative study. In: Database Support for Data Mining Applications, pp. 24–51 (2004)
4. Geerts, F., Goethals, B., Mielikäinen, T.: Tiling databases. In: Discovery Science, pp. 77–122 (2004)
5. Han, J., Chiang, J.Y., Chee, S., Chen, J., Chen, Q., Cheng, S., Gong, W., Kamber, M., Koperski, K., Liu, G., Lu, Y., Stefanovic, N., Winstone, L., Xia, B.B., Zaiane, O.R., Zhang, S., Zhu, H.: Dbminer: a system for data mining in relational databases and data warehouses. In: Proc. CASCON, pp. 8–12 (1997)

6. Han, J., Fu, Y., Wang, W., Koperski, K., Zaiane, O.: DMQL: A data mining query language for relational databases. In: ACM SIGMOD Workshop DMKD (1996)
7. Han, J., Kamber, M.: Data Mining - Concepts and Techniques, 1st ed. Morgan Kaufmann (2000)
8. Imielinski, T., Virmani, A.: Msql: A query language for database mining. Data Mining Knowledge Discovery 3(4), 373–408 (1999)
9. Jeudy, B., Boulicaut, J.F.: Constraint-based discovery and inductive queries: Application to association rule mining. In: Proc. ESF Exploratory Workshop on Pattern Detection and Discovery in Data Mining, pp. 110–124 (2002)
10. Mannila, H., Toivonen, H.: Levelwise search and borders of theories in knowledge discovery. Data Mining and Knowledge Discovery 1(3), 241–258 (1997)
11. Meo, R., Psaila, G., Ceri, S.: An extension to sql for mining association rules. Data Mining and Knowledge Discovery 2(2), 195–224 (1998)
12. Meo, R., Psaila, G., Ceri, S.: A tightly-coupled architecture for data mining. In: Proc. IEEE ICDE, pp. 316–323 (1998)
13. Ng, R., Lakshmanan, L.V.S., Han, J., Pang, A.: Exploratory mining and pruning optimizations of constrained associations rules. In: Proc. ACM SIGMOD, pp. 13–24 (1998)
14. Pei, J., Han, J., Lakshmanan, L.V.S.: Mining frequent itemsets with convertible constraints. In: Proc. IEEE ICDE, pp. 433–442 (2001)
15. Srikant, R., Agrawal, R.: Mining generalized association rules. Future Generation Computer Systems 13(2–3), 161–180 (1997)
16. Tang, Z.H., MacLennan, J.: Data Mining with SQL Server 2005. John Wiley & Sons (2005)
17. Wicker, J., Richter, L., Kessler, K., Kramer, S.: Sinbad and siql: An inductive databse and query language in the relational model. In: Proc. ECML-PKDD, pp. 690–694 (2008)

Chapter 4
A Theory of Inductive Query Answering

Luc De Raedt, Manfred Jaeger, Sau Dan Lee, and Heikki Mannila

Abstract We introduce the Boolean inductive query evaluation problem, which is concerned with answering inductive queries that are arbitrary Boolean expressions over monotonic and anti-monotonic predicates. Boolean inductive queries can be used to address many problems in data mining and machine learning, such as local pattern mining and concept-learning, and actually provides a unifying view on many machine learning and data mining tasks. Secondly, we develop a decomposition theory for inductive query evaluation in which a Boolean query Q is reformulated into k sub-queries $Q_i = Q_A \wedge Q_M$ that are the conjunction of a monotonic and an anti-monotonic predicate. The solution to each sub-query can be represented using a version space. We investigate how the number of version spaces k needed to answer the query can be minimized and define this as the dimension of the solution space and query. Thirdly, we generalize the notion of version spaces to cover Boolean queries, so that the solution sets form a closed Boolean-algebraic space under the usual set operations. The effects of these set operations on the dimension of the involved queries are studied.

Luc De Raedt
Department of Computer Science, Katholieke Universiteit Leuven
e-mail: luc.deraedt@cs.kuleuven.be

Manfred Jaeger
Department of Computer Science, Aalborg Universitet
e-mail: jaeger@cs.aau.dk

Sau Dan Lee
Department of Computer Science, University of Hong Kong
e-mail: sdlee@cs.hku.hk

Heikki Mannila
Department of Information and Computer Science, Helsinki University of Technology
e-mail: mannila@cs.helsinki.fi

4.1 Introduction

Many data mining and learning problems address the problem of finding a set of patterns, concepts or rules that satisfy a set of constraints. Formally, this can be described as the task of finding the set of patterns or concepts $Th(Q, \mathcal{D}, \mathcal{L})$ $= \{\varphi \in \mathcal{L} \mid Q(\varphi, \mathcal{D})\}$ i.e., those patterns and concepts φ satisfying query Q on a data set $\mathcal{D}$. Here $\mathcal{L}$ is the language in which the patterns or concepts are expressed, and Q is a predicate or constraint that determines whether a pattern or concept φ is a solution to the data mining task or not [20]. This framework allows us to view the predicate or the constraint Q as an *inductive query* [7]. It is then the task of machine learning or data mining system to efficiently generate the answers to the query. This view of mining and learning as a declarative querying process is also appealing as the basis for a theory of mining and learning. Such a theory would be analogous to traditional database querying in the sense that one could study properties of different pattern languages $\mathcal{L}$, different types of queries (and query languages), as well as different types of data. Such a theory could also serve as a sound basis for developing algorithms that solve inductive queries.

It is precisely such a theory that we introduce in this chapter. More specifically, we study inductive queries that are Boolean expressions over monotonic and anti-monotonic predicates. An example query could ask for molecular fragments that have frequency at least 30 percent in the active molecules or frequency at most 5 percent in the inactive ones [15]. This type of Boolean inductive query is amongst the most general type of inductive query that has been considered so far in the data mining and the machine learning literature. Indeed, most approaches to constraint based data mining use either single constraints (such as minimum frequency), e.g., [2], a conjunction of monotonic constraints, e.g., [24, 10], or a conjunction of monotonic and anti-monotonic constraints, e.g., [8, 15]. However, [9] has studied a specific type of Boolean constraints in the context of association rules and itemsets. It should also be noted that even these simpler types of queries have proven to be useful across several applications, which in turn explains the popularity of constraint based mining in the literature. Inductive querying also allows one to address the typical kind of concept-learning problems that have been studied within computational learning theory [14] including the use of queries for concept-learning [3]. Indeed, from this perspective, there will be a constraint with regard to every positive and negative instance (or alternatively some constraints at the level of the overall dataset), and also the answers to queries to oracle (membership, equivalence, etc.) can be formulated as constraints.

Our theory of Boolean inductive queries is first of all concerned with characterizing the solution space $Th(Q, \mathcal{D}, \mathcal{L})$ using notions of convex sets (or version spaces [12, 13, 22]) and border representations [20]. This type of representations have a long history in the fields of machine learning [12, 13, 22] and data mining [20, 5]. Indeed, within the field of data mining it has been realized that the space of solutions w.r.t. a monotone constraint is completely characterized by its set (or border) of maximally specific elements [20, 5]. This property is also exploited by some effective data mining tools, such as Bayardo's MaxMiner [5], which output this border set. Border sets have an even longer history in the field of machine learning, where Mitchell recognized as early as 1977 that the space of solutions to a concept-learning task could be represented by two borders, the so-called S and G-set (where S represents the set of maximally specific elements in the solution space and G the set of maximally general ones). These data mining and machine learning viewpoints on border sets have been unified by [8, 15], who introduced the level-wise version space algorithm that computes the S and G set w.r.t. a conjunction of monotonic and anti-monotonic constraints.

In the present chapter, we build on these results to develop a decomposition approach to solving arbitrary Boolean queries over monotonic and anti-monotonic predicates. More specifically, we investigate how to decompose arbitrary queries Q into a set of sub-queries Q_k such that $Th(Q, \mathcal{D}, \mathcal{L}) = \bigcup_i Th(Q_i, \mathcal{D}, \mathcal{L})$, and each $Th(Q_i, \mathcal{D}, \mathcal{L})$ can be represented using a single version space. This way we obtain a query plan, in that to obtain the answer to the overall query Q all of the sub-queries Q_i need to be answered. As these Q_i yield version spaces, they can be computed by existing algorithms such as the level-wise version space algorithm of [8]. A key technical contribution is that we also introduce a canonical decomposition in which the number of needed subqueries k is minimal.

This motivates us also to extend the notion of version spaces into generalized version spaces (GVSes) [18] to encapsulate solution sets to such general queries. It is interesting that GVSes form an algebraic space that is closed under the usual set operations: union, intersection and complementation. We prove some theorems that characterize the effect on the dimensions of such operation. Because GVSes are closed under these operations, the concept of GVSes gives us the flexibility to rewrite queries in various forms, find the solutions of subqueries separately, and eventually combine the solutions to obtain the solution of the original query. This opens up many opportunites for query optimization.

This chapter is organized as follows. In Section 4.2, we define the inductive query evaluation problem and illustrate it on the pattern domains of strings and itemsets. We model the solution sets with GVSes, which are introduced in Section 4.3. In Section 4.4, we introduce a decomposition approach to reformulate the original query in simpler sub-queries. Finally, we give our conclusions in Section 4.6.

4.2 Boolean Inductive Queries

We begin with describing more accurately the notions of patterns and pattern languages, as we use them in this chapter. We always assume that datasets consist of a list of data items from a set $\mathcal{U}$, called the *domain* or the *universe* of the dataset.

A *pattern* or *concept* ϕ for $\mathcal{U}$ is some formal expression that defines a subset ϕ_e of $\mathcal{U}$. When $u \in \phi_e$ we say that ϕ *matches* or *covers* u. A *pattern language* $\mathcal{L}$ for $\mathcal{U}$ is a formal language of patterns. The terminology used here is applicable to both concept-learning and pattern mining. In concept-learning, $\mathcal{U}$ would be the space of examples, $2^{\mathcal{U}}$ the set of possible concepts (throughout, we use 2^X to denote the powerset of X), and $\mathscr{L}$ the set of concept-descriptions. However, for simplicity we shall throughout the chapter largely employ the terminology of pattern mining. It is, however, important to keep in mind that it also applies to concept-learning and other machine learning tasks.

Example 4.2.1 Let $\mathscr{I} = \{i_1, \ldots, i_n\}$ be a finite set of possible items, and $\mathscr{U}_{\mathscr{I}} = 2^{\mathscr{I}}$ be the universe of itemsets over $\mathscr{I}$. The traditional pattern language for this domain is $\mathscr{L}_{\mathscr{I}} = \mathscr{U}_{\mathscr{I}}$. A pattern $\phi \in \mathscr{L}_{\mathscr{I}}$ covers the set $\phi_e := \{\mathscr{H} \subseteq \mathscr{I} \mid \phi \subseteq \mathscr{H}\}$.

Instead of using $\mathscr{L}_{\mathscr{I}}$ one might also consider more restrictive languages, e.g., the sublanguage $\mathscr{L}_{\mathscr{I},k} \subseteq \mathscr{L}_{\mathscr{I}}$ that contains the patterns in $\mathscr{L}_{\mathscr{I}}$ of size at most k.

Alternatively, one can also use more expressive languages, the maximally expressive one being the language $2^{\mathscr{U}_{\mathscr{I}}}$ of all subsets of the universe, or as is common in machine learning the language of conjunctive concepts, $\mathscr{L}_{\overline{\mathscr{I}}}$, which consists of all conjunctions of literals over $\mathscr{I}$, that is, items or their negation. This language can be represented using itemsets that may contain items from $\overline{\mathscr{I}} = \mathscr{I} \cup \{\neg i \mid i \in \mathscr{I}\}$. It is easy to see that the basic definitions for itemsets carry over for this language provided that the universe of itemsets is $\mathscr{U}_{\overline{\mathscr{I}}}$.   □

Example 4.2.2 Let Σ be a finite alphabet and $\mathscr{U}_{\Sigma} = \Sigma^*$ the universe of all strings over Σ. We will denote the empty string with ε. The traditional pattern language in this domain is $\mathscr{L}_{\Sigma} = \mathscr{U}_{\Sigma}$. A pattern $\phi \in \mathscr{L}_{\Sigma}$ covers the set $\phi_e = \{\sigma \in \Sigma^* \mid \phi \sqsubseteq \sigma\}$, where $\phi \sqsubseteq \sigma$ denotes that ϕ is a substring of σ.   □

One pattern ϕ for U is *more general* than a pattern ψ for U, written $\phi \succeq \psi$, if and only if $\phi_e \supseteq \psi_e$. For two itemset patterns $\phi, \psi \in \mathscr{L}_{\mathscr{I}}$, for instance, we have $\phi \succeq \psi$ iff $\phi \subseteq \psi$. For two conjunctive concepts $\phi, \psi \in \mathscr{L}_{\mathscr{C}}$, for instance, we have $\phi \succeq \psi$ iff $\phi \models \psi$. For two string patterns $\phi, \psi \in \mathscr{L}_{\Sigma}$ we have $\phi \succeq \psi$ iff $\phi \sqsubseteq \psi$. A pattern language $\mathscr{L}'$ is *more expressive* than a pattern language $\mathscr{L}$, written $\mathscr{L}' \succeq \mathscr{L}$, iff for every $\phi \in \mathscr{L}$ there exists $\phi' \in \mathscr{L}'$ with $\phi_e = \phi'_e$.

A pattern *predicate* defines a primitive property of a pattern, often relative to some data set D (a set of examples). For any given pattern or concept, a pattern predicate evaluates to either *true* or *false*. Pattern predicates are the basic building blocks for building inductive queries. We will be mostly interested in *monotonic* and *anti-monotonic* predicates. A predicate p is monotonic, if $p(\phi)$ and $\psi \succeq \phi$ implies $p(\psi)$, i.e., p is closed under generalizations of concepts. Similarly, anti-monotonic predicates are defined by closure under specializations.

4.2.1 Predicates

We now introduce a number of pattern predicates that will be used for illustrative purposes throughout this chapter. Throughout the section we will introduce predicates that have been inspired by a data mining setting, in particular by the system

MolFea [15], as well as several predicates that are motivated from a machine learning perspective, especially, by Angluin's work on learning concepts from queries [3].

Pattern predicates can be more or less general in that they may be applied to patterns from arbitrary languages $\mathscr{L}$, only a restricted class of languages, or perhaps only are defined for a single language. Our first predicate can be applied to arbitrary languages:

- *minimum_frequency(p,n,D)* evaluates to true iff p is a pattern that occurs in database D with frequency at least $n \in \mathbb{N}$. The frequency $f(\phi, D)$ of a pattern ϕ in a database D is the (absolute) number of data items in D covered by ϕ. Analogously, the predicate *maximum_frequency(p,n,D)* is defined. *minimum_frequency* is a monotonic, *maximum_frequency* an anti-monotonic predicate.

These predicates are often used in data mining, for instance, when mining for frequent itemsets, but they can also be used in the typical concept-learning setting, which corresponds to imposing the constraints *minimum_frequency(p,|P|,P)* $\wedge$ *maximum_frequency(p,0,N)* where P is the set of positive instances and N the set of negative ones, that is, all positive examples should be covered and none of the negatives ones.

A special case of these frequency related predicates is the predicate

- *covers(p,u)* $\equiv$ *minimum_frequency(p,1,{u})*, which expresses that the pattern (or concept) p covers the example u. *covers* is monotonic.

This predicate is often used in a concept-learning setting. Indeed, the result of a membership query (in Angluin's terminology) is a positive or negative example and the resulting constraint corresponds to the predicate *covers* or its negation.

The next predicate is defined in terms of some fixed pattern ψ from a language $\mathscr{L}$. It can be applied to other patterns for $\mathscr{U}$.

- *is_more_general(p,ψ)* is a monotonic predicate that evaluates to true iff p is a pattern for U with $p \succeq \psi$. Dual to the *is_more_general* predicate one defines the anti-monotonic *is_more_specific* predicate.

The *is_more_general(p,ψ)* predicate only becomes specific to the language $\mathscr{L}$ for the fixed universe $\mathscr{U}$ through its parameter ψ. By choice of other parameters, the predicate *is_more_general* becomes applicable to any other pattern language $\mathscr{L}'$. This type of predicate has been used in a data mining context to restrict the patterns of interest [15] to specify that patterns should be sub- or superstrings of a particular pattern. In a concept learning context, these predicates are useful in the context of learning from queries [3]. This is a framework in which the learner may pose queries to an oracle. The answers to these queries then result in constraints on the concept. There are several types of queries that are considered in this framework and that are related to the *is_more_general* and the *is_more_specific* predicates:

- a subset, respectively superset query [3], determines whether a particular concept must cover a subset, respectively a superset, of the positive examples or not. The

answers to these queries directly correspond to constraints using the predicates
is_more_specific and *is_more_general*.

- an equivalence query determines whether a particular concept-description ϕ is
  equivalent to the target concept or not. This can be represented using the predicate
  equivalent(c, ϕ), which can be defined as follows:

$$equivalent(c, \phi) \equiv is\_more\_general(c, \phi) \wedge is\_more\_specific(c, \phi)$$

- a disjointness query determines whether or not a particular concept ϕ overlaps
  with the target concept c, that is, whether there are elements in the universe which
  are covered by both ϕ and c. This can be represented using the anti-monotonic
  predicate *disjoint*(c, ϕ), which evaluates to true iff $c_e \cap \phi_e = \emptyset$. It can be defined
  in terms of generality in case the language of concepts $\mathcal{L}$ is closed under complement:

$$disjoint(c, \phi) \equiv is\_more\_specific(c, \neg\phi)$$

- an exhaustiveness query determines whether a particular concept ϕ together with
  the target concept c covers the whole universe; this can be written using the
  monotonic predicate *exhausts*(c, ϕ), which evaluates to true iff $c_e \cup \phi_e = \mathcal{U}$. It
  can be defined in terms of generality in case the language of concepts $\mathcal{L}$ is
  closed under complement:

$$exhausts(c, \phi) \equiv is\_more\_general(c, \neg\phi)$$

The next pattern predicate is applicable to patterns from many different languages
$\mathcal{L}$. It is required, however, that on $\mathcal{L}$ the length of a pattern is defined.

- *length_at_most*(p,n) evaluates to true for $p \in \mathcal{L}$ iff p has length at most n. Analogously the *length_at_least*(p,n) predicate is defined.

We apply the *length_at_most*-predicate mostly to string patterns, where the length
of a pattern is defined in the obvious way (but note that e.g., for itemset patterns
$\phi \in \mathcal{L}_{\mathcal{I}}$ one can also naturally define the length of ϕ as the cardinality of ϕ, and
then apply the *length_at_most*-predicate).

4.2.2 Illustrations of Inductive Querying

Let us now also look into the use of these predicates for solving a number of machine
learning and data mining problems. First, we look into association rule mining, for
which we introduce a pattern predicate that is applicable only to itemset patterns
$\phi \in \mathcal{L}_{\mathcal{I}}$ for some fixed $\mathcal{I}$. The dependence on $\mathcal{I}$ again comes through the use of
a parameter, here some fixed element $i_j \in \mathcal{I}$.

- *association*(p,i_j,D) evaluates to true for $p \in \mathcal{L}_{\mathcal{I}}$ iff $p \Rightarrow i_j$ is a valid association
  rule in D, i.e., for all data items $d \in D$: if $p \subseteq d$ then $i_j \in d$. *association* is anti-
  monotonic.

The predicate *association*—as defined above—allows only valid association rules, i.e., association rules that have a confidence of 100%. It could also be applied to string patterns. Then the condition would be that $p \Rightarrow i$ is valid iff for all strings $d \in D$: if $d \in \phi_p$ then $d \in \phi_{pi}$, where pi denotes the concatenation of the string p and the character i.

Secondly, let us investigate the use of constraints in clustering, where must-link and cannot-link constraints have been used in machine learning. We can phrase a clustering problem as a concept learning task in our framework by interpreting a clustering of a set of objects $\mathcal{O}$ as a binary relation $cl \subset \mathcal{O} \times \mathcal{O}$, where $cl(o, o')$ means that o and o' are in the same cluster. Thus, with $\mathcal{U} = \mathcal{O} \times \mathcal{O}$, a clustering is just a pattern in our general sense (one may use any suitable pattern language that provides a unique representation for clusterings). According to our general notion of generality of patterns, a clustering c is more general than another clustering c' if $c_e \supseteq c'_e$, i.e., if more pairs of objects belong to the same cluster in c as in c', which, in turn, means that c can be obtained from c' by merging of clusters. Furthermore, the most specific generalizations of c are just the clusterings obtained by merging two clusters of c, whereas the most general specializations are the clusterings obtained by splitting one cluster of c into two.

We can now express as a concept learning task the problem of retrieving all possible clusterings that satisfy certain constraints.

The first two kinds of useful constraints represent generally desirable properties of clusterings:

- *clusters_atmost(cl, k)* evaluates to true if the clustering cl consists of at most k clusters. This predicate is monotonic.
- *within_cluster_distance_atmost(cl, r)* evaluates to true if no two objects with distance $> r$ are in the same cluster. This predicate is anti-monotonic.

Specific constraints as used in constraint based clustering are now:

- *must_link(cl, o$_1$, o$_2$)*, which evaluates to true when the two objects o_i are in the same cluster (monotonic).
- *must_not_link(cl, o$_1$, o$_2$)*, which evaluates to true when the two objects o_i are in different clusters (anti-monotonic)

Using these predicates, we could for a given dataset retrieve with the query

$$clusters\_atmost(cl, 5) \wedge within\_cluster\_distance\_atmost(cl, 0.5) \wedge$$
$$must\_link(cl, o_1, o_2)$$

all possible clusterings of at most 5 clusters, such that no clusters contain points farther apart than 0.5 (in the underlying metric used for the dataset), and such that the two designated objects o_1, o_2 are in the same cluster.

Finally, machine learning has also devoted quite some attention to multi-instance learning. In multi-instance learning examples consist of a set of possible instances and an example is considered covered by a concept, whenever the concept covers at least one of the instances in the set. One way of formalizing multi-instance learning within our framework is to adapt the notion of coverage to have this meaning. Alternatively, a multi-instance learning example could be represented using the predicate *minimum-frequency(c,1,e)* where *c* is the target concept and *e* is the example, represented here as a set of its instances. A negative example then corresponds to the negation of this expression or requiring that *maximum-frequency(c,0,e)* holds.

4.2.3 A General Framework

In all the preceding examples the pattern predicates have the form *pred(p,params)* or *pred(p,D,params)*, where *params* is a tuple of parameter values, *D* is a data set and *p* is a pattern variable.

We also speak a bit loosely of *pred* alone as a pattern predicate, and mean by that the collection of all pattern predicates obtained for different parameter values *params*.

We say that *pred(p,D,params)* is a *monotonic* predicate, if for all pattern languages $\mathscr{L}$ to which *pred(p,D,params)* can be applied, and all $\phi, \psi \in \mathscr{L}$:

$$\phi \succeq \psi \;\Rightarrow\; pred(\psi,D,params) \rightarrow pred(\phi,D,params)$$

We also say that *pred* is monotonic, if *pred(p,D,params)* is monotonic for all possible parameter values *params*, and all datasets *D*. Analogously, we define anti-monotonicity of a predicate by the condition

$$\phi \succeq \psi \;\Rightarrow\; pred(\phi,D,params) \rightarrow pred(\psi,D,params).$$

A pattern predicate *pred(p,D,params)* that can be applied to the patterns from a language $\mathscr{L}$ defines the *solution set*

$$Th(pred(p,D,params),\mathscr{L}) = \{\phi \in \mathscr{L} \mid pred(\phi,D,params) = true\}.$$

Furthermore, for monotonic predicates $m(\ldots)$ these sets will be *monotone*, i.e., for all $\phi \succeq \psi \in \mathscr{L} : \psi \in Th(m(\ldots),\mathscr{L}) \rightarrow \phi \in Th(m(\ldots),\mathscr{L})$. Similarly, anti-monotonic predicates define anti-monotone solution sets.

Figure 4.1 illustrates the definitions given so far. It gives a schematic representation of a universe $\mathscr{U}$ and two pattern languages $\mathscr{L}, \mathscr{L}'$ for $\mathscr{U}$. The $\succeq$ relation between patterns is represented by lines connecting immediate neighbors in the $\succeq$ relation, with the more general patterns being above the more specific ones. For two patterns from $\mathscr{L}$ and one pattern from $\mathscr{L}'$ the subsets of the universe covered by the patterns are indicated. For the pattern $\psi \in \mathscr{L}$ and $\mu \in \mathscr{L}'$ the figure shows the interpretation of the pattern predicates *is_more_general(p, ψ)*, respectively *is_more_general(p, μ)* by filled nodes corresponding to patterns for which these predicates are true.

Example 4.2.3 Consider the string data set $D = \{\texttt{abc}, \texttt{abd}, \texttt{cd}, \texttt{d}, \texttt{cd}\}$. Here we have pattern frequencies $f(\texttt{abc},D) = 1$, $f(\texttt{cd},D) = 2$, $f(\texttt{c},D) = 3$, $f(\texttt{d},D) = 4$, is_more_general(p, ψ) is_more_general(p, μ)

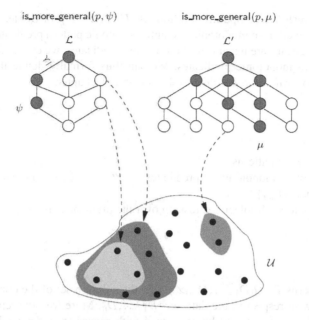

Fig. 4.1 Pattern languages and pattern predicates

$f(\texttt{abcd}, D) = 0$. And trivially, $f(\varepsilon, D) = |D| = 5$. Thus, the following predicates evaluate to true: *minimum_frequency*$(\texttt{c}, 2, D)$, *minimum_frequency*$(\texttt{cd}, 2, D)$, *maximum_frequency*$(\texttt{abc}, 2, D)$.

The pattern predicate $m :=$ *minimum_frequency*$(p, 2, D)$ defines $Th(m, \mathcal{L}_\Sigma) = \{\varepsilon, a, b, c, d, ab, cd\}$, and the predicate $a :=$ *maximum_frequency*$(p, 2, D)$ defines the infinite set $Th(a, \mathcal{L}_\Sigma) = \mathcal{L}_\Sigma \setminus \{\varepsilon, c, d\}$.   □

The definition of $Th(pred(\texttt{p}, D, params), \mathcal{L})$ is extended in the natural way to a definition of the solution set $Th(Q, \mathcal{L})$ for Boolean combinations Q of pattern predicates: $Th(\neg Q, \mathcal{L}) := \mathcal{L} \setminus Th(Q, \mathcal{L})$, $Th(Q_1 \vee Q_2, \mathcal{L}) := Th(Q_1, \mathcal{L}) \cup Th(Q_2, \mathcal{L})$. The predicates that appear in Q may reference one or more data sets $D_1, \ldots, D_n$. To emphasize the different data sets that the solution set of a query depends on, we also write $Th(Q, D_1, \ldots, D_n, \mathcal{L})$ or $Th(Q, \mathcal{D}, \mathcal{L})$ for $Th(Q, \mathcal{L})$.

Example 4.2.4 Let D_1, D_2 be two datasets over the domain of itemsets $\mathcal{U}_\mathcal{I}$. Let $i \in I$, and consider the query

$$Q = association(p, i, D_1) \wedge minimum\_frequency(p, 10, D_1)$$
$$\wedge \neg association(p, i, D_2).$$

The solution $Th(Q, D_1, D_2, \mathcal{L}_\mathcal{I})$ consists of all $p \in \mathcal{L}_\mathcal{I} = 2^\mathcal{I}$ for which $p \Rightarrow i$ is a valid association rule with support at least 10 in D_1, but $p \Rightarrow i$ is not a valid association rule in D_2.   □

We are interested in computing solution sets $Th(Q, \mathscr{D}, \mathscr{L})$ for Boolean queries Q that are constructed from monotonic and anti-monotonic pattern predicates. As anti-monotonic predicates are negations of monotonic predicates, we can, in fact, restrict our attention to monotonic predicates. We can thus formally define the *Boolean inductive query evaluation problem* addressed in this chapter.

Given

- a language $\mathscr{L}$ of patterns,
- a set of monotonic predicates $\mathscr{M} = \{m_1(p, D_1, params_1), \ldots, m_n(p, D_n, params_n)\}$,
- a query Q that is a Boolean expression over the predicates in $\mathscr{M}$,

Find

the set of patterns $Th(Q, D_1, \ldots, D_n, \mathscr{L})$, i.e., the solution set of the query Q in the language $\mathscr{L}$ with respect to the data sets $D_1, \ldots, D_n$. Moreover, the representation of the the solution set should be optimized with regard to understandability and representation size.

4.3 Generalized Version Spaces

We next investigate the structure of solution sets $Th(Q, \mathscr{D}, \mathscr{L})$ based on the classic notion of version spaces [22][12].

Definition 4.3.1 Let $\mathscr{L}$ be a pattern language, and $I \subseteq \mathscr{L}$. If for all $\phi, \phi', \psi \in \mathscr{L}$ it holds that $\phi \preceq \psi \preceq \phi'$ and $\phi, \phi' \in I$ implies $\psi \in I$, then I is called a *version space* (or a *convex set*). The set of all version spaces for $\mathscr{L}$ is denoted $\mathscr{VS}^1(\mathscr{L})$.

A *generalized version space (GVS)* is any finite union of version spaces. The set of all generalized version spaces for $\mathscr{L}$ is denoted by $\mathscr{VS}^{\mathbb{Z}}(\mathscr{L})$.

The *dimension* of a generalized version space I is the minimal k, such that I is the union of k version spaces.

Version spaces are particularly useful when they can be represented by boundary sets, i.e., by the sets $G(Q, \mathscr{D}, \mathscr{L})$ of their maximally general elements, and $S(Q, \mathscr{D}, \mathscr{L})$ of their most specific elements. Generalized version spaces can then be represented simply by pairs of boundary sets for their convex components. Our theoretical results do not require boundary representations for convex sets. However, in most cases our techniques will be more useful for pattern languages in which convexity implies boundary representability. This is guaranteed for finite languages [12].

Definition 4.3.2 The dimension of a query Q is the dimension of the generalized version space $Th(Q, \mathscr{D}, \mathscr{L})$.

Example 4.3.3 Let $\Sigma = \{a, b\}$ and $\mathscr{L}_\Sigma$ as in Example 4.2.2. Let

$$Q = length\_at\_most(p, 1) \vee (is\_more\_specific(p, ab) \wedge is\_more\_general(p, ab)).$$

When evaluated over $\mathscr{L}_\Sigma$, the first disjunct of Q gives the solution $\{\varepsilon, a, b\}$, the second $\{ab\}$, so that $Th(Q, \mathscr{L}_\Sigma) = \{\varepsilon, a, b, ab\}$, which is convex in $\mathscr{L}_\Sigma$. Thus, $\dim(Q) = 1$ (as Q does not reference any datasets, the maximization over $\mathscr{D}$ in the definition of dimension here is vacuous).

$Th(Q, \mathscr{L}_\Sigma)$ can be represented by $S(Q, \mathscr{L}_\Sigma) = \{ab\}$ and $G(Q, \mathscr{L}_\Sigma) = \{\varepsilon\}$. $\qquad \square$

With the following definitions and theorem we provide an alternative characterization of dimension k sets.

Definition 4.3.4 Let $I \subseteq \mathscr{L}$. Call a chain $\phi_1 \preceq \phi_2 \preceq \cdots \preceq \phi_{2k-1} \subseteq \mathscr{L}$ an *alternating chain (of length k) for I* if $\phi_i \in I$ for all odd i, and $\phi_i \notin I$ for all even i.

Definition 4.3.5 Let $I \subseteq \mathscr{L}$. We define two operators on I,

$$I^- = \{\phi \in I \mid \exists \psi \in \mathscr{L} \setminus I, \phi' \in I : \phi \preceq \psi \preceq \phi'\}$$
$$I^+ = I \setminus I^-.$$

Thus, I^- is constructed from I by removing all elements that only appear as the maximal element in alternating chains for I. I^+ is the set of such removed elements. Note that since $I^- \subset I$ by definition, we have $I = I^+ \cup I^-$ and $I^+ \cap I^- = \emptyset$.

Theorem 4.3.6 Let I be a generalized version space. Then $\dim(I)$ is equal to the maximal k for which there exists in $\mathscr{L}$ an alternating chain of length k for I.

Proof: By induction on k: if I only has alternating chains of length 1, then $I \in \mathscr{VS}^1$ and $\dim(I) = 1$ by definition. Assume, then, that $k \geq 2$ is the length of the longest alternating chain for I. As there are chains of length ≥ 2, both I^- and I^+ are nonempty.

It is clear from the definition of I^- that I^- has alternating $\mathscr{L}$-chains of length $k-1$, but not of length k. By induction hypothesis, thus $\dim(I^-) = k-1$. The set I^+, on the other hand, has dimension 1. It follows that $\dim(I = I^+ \cup I^-)$ is at most k. That $\dim(I)$ is at least k directly follows from the existence of an alternating chain $\phi_1 \preceq \phi_2, \preceq \cdots \preceq \phi_{2k-1}$ for I, because $\phi_1, \phi_3, \ldots, \phi_{2k-1}$ must belong to distinct components in every partition of I into convex components. $\qquad \square$

The operator I^+ allows us to define a *canonical decomposition* of a generalized version space. For this, let I be a generalized version space of dimension k. Define

$$I_0 = I^+$$
$$I_i = (I \setminus I_0 \cup \cdots \cup I_{i-1})^+ \quad (1 \leq i \leq k)$$

The version spaces I_i then are convex, disjoint, and $I = \cup_{i=1}^k I_i$.

Our results so far relate to the structure of a fixed generalized version space. Next, we investigate the behavior of GVSs under set-theoretic operations. Hirsh [13] has shown that $\mathcal{VS}^1$ is closed under intersections, but not under unions. Our following results show that $\mathcal{VS}^{\mathbb{Z}}$ is closed under all set-theoretic operations, and that one obtains simple bounds on growth in dimension under such operations.

Theorem 4.3.7 Let $V \in \mathcal{VS}^{\mathbb{Z}}(\mathcal{L})$. Then $\mathcal{L} \setminus V \in \mathcal{VS}^{\mathbb{Z}}(\mathcal{L})$, and $\dim(V) - 1 \leq \dim(\mathcal{L} \setminus V) \leq \dim(V) + 1$.

Proof: Any alternating chain of length k for $\mathcal{L} \setminus V$ defines an alternating chain of length $k - 1$ for $\mathcal{L}$. It follows that $\dim(\mathcal{L} \setminus V) \leq \dim(V) + 1$. By a symmetrical argument $\dim(V) \leq \dim(\mathcal{L} \setminus V) + 1$. □

By definition, generalized version spaces are closed under finite unions, and $\dim(V \cup W) \leq \dim(V) + \dim(W)$. Combining this with the dimension bound for complements, we obtain $\dim(V \cap W) = \dim((V^c \cup W^c)^c) \leq \dim(V) + \dim(W) + 3$. However, a somewhat tighter bound can be given:

Theorem 4.3.8 Let $V, W \in \mathcal{VS}^{\mathbb{Z}}(\mathcal{L})$. Then $\dim(V \cap W) \leq \dim(V) + \dim(W) - 1$.

Proof: Let $\phi_1 \preceq \phi_2, \preceq \cdots \preceq \phi_{2k-1}$ be an alternating chain for $V \cap W$. Let $I_V := \{i \in 1, \ldots k - 1 \mid \phi_{2i} \notin V\}$, and $I_W := \{i \in 1, \ldots k - 1 \mid \phi_{2i} \notin W\}$. Then $I_V \cup I_W = \{1, \ldots, k - 1\}$. Deleting from the original alternating chain all ϕ_{2i}, ϕ_{2i-1} with $i \notin I_V$ gives an alternating chain of length $|I_V| + 1$ for V. Thus, $|I_V| \leq \dim(V) - 1$. Similarly, $|I_W| \leq \dim(W) - 1$. The theorem now follows with $|I_V| + |I_W| \geq k - 1$. □

4.4 Query Decomposition

In the previous section we have studied the structure of the solution sets $Th(Q, \mathcal{D}, \mathcal{L})$. We now turn to the question of how to develop strategies for the computation of solutions so that, first, the computations for complex Boolean queries can be reduced to computations of simple version spaces using standard level-wise algorithms, and second, the solutions obtained have a parsimonious representation in terms of the number of their convex components, and/or the total size of the boundaries needed to describe the convex components.

A first approach to solving a Boolean query Q using level-wise algorithms is to transform Q into disjunctive normal form (DNF). Each disjunct then will be a conjunction of monotonic or anti-monotonic predicates, and thus define a convex solution set. The solution to the query then is simply the union of the solutions of the disjuncts. This approach, however, will often not lead to a parsimonious representation: the number of disjuncts in Q's DNF can far exceed the dimension of Q, so that the solution is not minimal in terms of the number of convex components. The solutions of the different disjunctions also may have a substantial overlap, which can lead to a greatly enlarged size of a boundary representation.

In this section we introduce two alternative techniques for decomposing a Boolean query into one-dimensional sub-queries. The first approach is based on user-defined query plans which can improve the efficiency by a reduction to simple and easy to evaluate convex sub-queries. The second approach, which we call the canonical decomposition, is fully automated and guaranteed to lead to solutions given by convex components that are minimal in number, and non-overlapping.

4.4.1 Query Plans

The solution set $Th(Q, \mathscr{D}, \mathscr{L})$ can be constructed incrementally from basic convex components using algebraic union, intersection and complementation operations. Using Theorems 4.3.7 and 4.3.8 one can bound the number of convex components needed to represent the final solution. For any given query, usually multiple such incremental computations are possible. A query plan in the sense of the following definition represents a particular solution strategy.

Definition 4.4.1 A query plan is a Boolean formula with some of its subqueries marked using the symbol $\underbrace{\qquad}$. Furthermore, all marked subqueries are the conjunction of a monotonic and an anti-monotonic subquery.

Example 4.4.2 Consider the query

$$Q_1 = (a_1 \vee a_2) \wedge (m_1 \vee m_2).$$

Since this is a conjunction of a monotonic and an anti-monotonic part, it can be solved directly, and $\underbrace{(a_1 \vee a_2) \wedge (m_1 \vee m_2)}$ is the corresponding query plan.

A transformation of Q_1 into DNF gives

$$(a_1 \wedge m_1) \vee (a_1 \wedge m_2) \vee (a_2 \wedge m_1) \vee (a_2 \wedge m_2),$$

for which $\underbrace{(a_1 \wedge m_1)} \vee \underbrace{(a_1 \wedge m_2)} \vee \underbrace{(a_2 \wedge m_1)} \vee \underbrace{(a_2 \wedge m_2)}$ is the only feasible query plan, which now requires four calls to the basic inductive query solver. $\square$

For any inductive query Q, we can rewrite it in many different forms. One can thus construct a variety of different query plans by annotating queries that are logically equivalent to Q. The question then arises as to which query plan is optimal, in the sense that the resources (i.e., memory and cpu-time) needed for computing its solution set are as small as possible. A general approach to this problem would involve the use of cost estimates that for each call to a conjunctive solver and operation. One example of a cost function for a call to a conjunctive solver could be *Expected Number of Scans of Data × Size of Data Set*. Another one could be the

Expected Number of Covers Tests. In this chapter, we have studied the query optimization problem under the assumption that each call to a conjunctive solver has
unit cost and that the only set operation allowed is union. Under this assumption,
decomposing a query Q into k subqueries of the form $Q_{a,i} \wedge Q_{m,i}$ (with $Q_{a,i}$ anti-
monotonic and $Q_{m,i}$ monotonic) and $\dim(Q) = k$ is an optimal strategy. We will
leave open the challenging question as to which cost-estimates to use in practice.
However, what should be clear is that given such cost-estimates, one could optimize
inductive queries by constructing all possible query plans and then selecting the best
one. This is effectively an optimization problem, not unlike the query optimization
problem in relational databases.

The optimization problem becomes even more interesting in the light of interactive querying sessions [4], which should be quite common when working with
inductive databases. In such sessions, one typically submits a rough query to get
some insight in the domain, and when the results of this query are available, the
user studies the results and refines the query. This often goes through a few iterations until the desired results are obtained.

4.4.2 Canonical Decomposition

As in the simple DNF decomposition approach, Q will be decomposed into k sub-
queries Q_i such that Q is equivalent to $Q_1 \vee \cdots \vee Q_k$, and each Q_i is convex. Furthermore, the Q_i will be mutually exclusive.

We develop this technique in two stages: in the first stage we do not take the con-
crete pattern language $\mathscr{L}$ into account, and determine Q_i such that $Th(Q, \mathscr{D}, \mathscr{L}) = \cup Th(Q_i, \mathscr{D}, \mathscr{L})$ for all $\mathscr{L}$ to which the predicates in Q can be applied. This step only
uses the monotonicity of the predicates and the Boolean structure of Q. In a second
step we refine the approach in order to utilize structural properties of $\mathscr{L}$ that can
reduce the number of components Q_i needed to represent $Th(Q, \mathscr{D}, \mathscr{L})$.

Applied to the query from Example 4.3.3, for instance, the first step will result
in a decomposition of Q into two components (essentially corresponding to the two
disjuncts of Q), which yields a bound of 2 for the dimension of $Th(Q, \mathscr{D}, \mathscr{L})$ for all
$\mathscr{L}$. The second step then is able to use properties of $\mathscr{L}_\Sigma$ in order to find the tighter
bound 1 for the dimension of $Th(Q, \mathscr{L}_\Sigma)$.

The idea for both stages of the decomposition is to first evaluate Q in a reduced
pattern language $\mathscr{L}'$, so that the desired partition $\vee Q_i$ can be derived from the structure of $Th(Q, \mathscr{L}')$. The solution set $Th(Q, \mathscr{L}')$ does not depend on the datasets $\mathscr{D}$
that Q references, and the complexity of its computation only depends on the size
of Q, but not on the size of any datasets.

In the following we always assume that Q is a query that contains n distinct
predicates $m_1, \ldots, m_n$, and that the m_i are monotonic for all pattern languages $\mathscr{L}$
for which Q can be evaluated (recall that we replace anti-monotonic predicates by
negated monotonic ones).

Definition 4.4.3 Let $\mathcal{M}(Q) = \{m_1, \ldots, m_n\}$, $\mathscr{L}_{\mathcal{M}(Q)} = 2^{\mathcal{M}(Q)}$, and for $\mu \in \mathscr{L}_{\mathcal{M}(Q)}$:

$$\mu_e := \{M \subseteq \mathcal{M}(Q) \mid M \subseteq \mu\}.$$

The predicates m_i are interpreted over $\mathscr{L}_{\mathcal{M}(Q)}$ as

$$Th(m_i, \mathscr{L}_{\mathcal{M}(Q)}) := \{\mu \in \mathscr{L}_{\mathcal{M}(Q)} \mid m_i \in \mu\}.$$

Thus, the definitions of $\mathscr{L}_{\mathcal{M}(Q)}$ and μ_e are similar to the ones for itemsets with $\mathcal{M}(Q)$ the set of possible items (cf. Example 4.2.1). Alternatively, each $\mu \in \mathscr{L}_{\mathcal{M}(Q)}$ can be viewed as an interpretation for the propositional variables $\mathcal{M}(Q)$ (see also Section 4.5). However, the inclusion condition in the definition of μ_e here is the converse of the inclusion condition in Example 4.2.1. In particular, here, $\mu' \succeq \mu$ iff $\mu' \supseteq \mu$. The predicates m_i are interpreted with respect to $\mathscr{L}_{\mathcal{M}(Q)}$ like the predicates *is_more_general*$(p, \{m_i\})$. By the general definition, with $Th(m_i, \mathscr{L}_{\mathcal{M}(Q)})$ $(1 \le i \le k)$ also $Th(Q, \mathscr{L}_{\mathcal{M}(Q)})$ is defined.

Theorem 4.4.4 Let $\mathscr{L}$ be a pattern language for which the predicates m_i in Q are monotone. The dimension of $Th(Q, \mathscr{D}, \mathscr{L})$ is less than or equal to the dimension of $Th(Q, \mathscr{L}_{\mathcal{M}(Q)})$.

Proof: Let $\mathscr{L}$ be given and $\mathscr{D}$ be any dataset. Define a mapping

$$\begin{aligned} h_{\mathscr{D}} : \mathscr{L} &\to \mathscr{L}_{\mathcal{M}(Q)} \\ \phi &\mapsto \{m \in \mathcal{M}(Q) \mid \phi \in Th(m, \mathscr{D}, \mathscr{L})\} \end{aligned} \tag{4.1}$$

First we observe that $h_{\mathscr{D}}$ is order preserving:

$$\phi \succeq \psi \quad \Rightarrow \quad h_{\mathscr{D}}(\phi) \succeq h_{\mathscr{D}}(\psi). \tag{4.2}$$

This follows from the monotonicity of the predicates m, because $\phi \succeq \psi$ and $\psi \in Th(m, \mathscr{D}, \mathscr{L})$ implies $\phi \in Th(m, \mathscr{D}, \mathscr{L})$, so that $h_{\mathscr{D}}(\phi)$ is a superset of $h_{\mathscr{D}}(\psi)$, which, in the pattern language $\mathscr{L}_{\mathcal{M}(Q)}$ just means $h_{\mathscr{D}}(\phi) \succeq h_{\mathscr{D}}(\psi)$.

Secondly, we observe that $h_{\mathscr{D}}$ preserves solution sets:

$$\phi \in Th(Q, \mathscr{D}, \mathscr{L}) \quad \Leftrightarrow \quad h_{\mathscr{D}}(\phi) \in Th(Q, \mathscr{L}_{\mathcal{M}(Q)}). \tag{4.3}$$

To see (4.3) one first verifies that for $i = 1, \ldots, n$:

$$\phi \in Th(m_i, \mathscr{D}, \mathscr{L}) \quad \Leftrightarrow \quad m_i \in h_{\mathscr{D}}(\phi) \quad \Leftrightarrow \quad h_{\mathscr{D}}(\phi) \in Th(m_i, \mathscr{L}_{\mathcal{M}(Q)}).$$

Then (4.3) follows by induction on the structure of queries Q constructed from the m_i.

Now suppose that $\phi_1 \preceq \cdots \preceq \phi_{2k-1} \subseteq \mathscr{L}$ is an alternating chain of length k for $Th(Q, \mathscr{D}, \mathscr{L})$. From (4.2) it follows that $h_{\mathscr{D}}(\phi_1) \preceq \cdots \preceq h_{\mathscr{D}}(\phi_{2k-1}) \subseteq \mathscr{L}_{\mathcal{M}(Q)}$, and from (4.3) it follows that this is an alternating chain of length k for $Th(Q, \mathscr{L}_{\mathcal{M}(Q)})$. From Theorem 4.3.6 it now follows that the dimension of $Th(Q, \mathscr{D}, \mathscr{L})$ is at most the dimension of $Th(Q, \mathscr{L}_{\mathscr{M}(Q)})$. □

Example 4.4.5 Let $\Sigma = \{a, b, \ldots, z\}$. Let

$$m_1 = not\text{-}is\_more\_specific(p, ab)$$
$$m_2 = not\text{-}is\_more\_specific(p, cb)$$
$$m_3 = not\text{-}length\_at\_least(p, 4)$$
$$m_4 = minimum\_frequency(p, 3, D)$$

These predicates are monotonic when interpreted in the natural way over pattern languages for the string domain. The first three predicates are the (monotonic) negations of the (anti-monotonic) standard predicates introduced in Section 4.2 (note that e.g., *not-is_more_specific* is distinct from *is_more_general*). Let

$$Q = \neg m_1 \wedge \neg m_2 \wedge (\neg m_3 \vee m_4). \tag{4.4}$$

Figure 4.2 (a) shows $\mathscr{L}_{\mathscr{M}(Q)}$ for this query. The solution set $Th(Q, \mathscr{L}_{\mathscr{M}(Q)})$ is $\{\emptyset, \{m_4\}, \{m_3, m_4\}\}$, which is of dimension 2, because $\emptyset \preceq \{m_3\} \preceq \{m_3, m_4\}$ is a (maximal) alternating chain of length 2. □

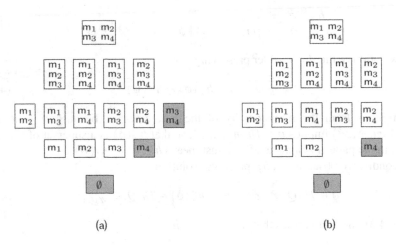

(a)   (b)

Fig. 4.2 Pattern languages $\mathscr{L}_{\mathscr{M}(Q)}$ and $\mathscr{L}_{\mathscr{M}(Q), \mathscr{L}}$

Given Q we can construct $\mathscr{L}_{\mathscr{M}(Q)}$ and $Th(Q, \mathscr{L}_{\mathscr{M}(Q)})$ in time $O(2^n)$. We can then partition $Th(Q, \mathscr{L}_{\mathscr{M}(Q)})$ into a minimal number of convex components $I_1, \ldots, I_k$ by iteratively removing from $Th(Q, \mathscr{L}_{\mathscr{M}(Q)})$ elements that are maximal in alternating chains. More precisely, let:

$$\begin{aligned} R_0 &:= Th(Q, \mathscr{L}_{\mathscr{M}(Q)}) \\ I_{i+1} &:= R_i^+ \\ R_{i+1} &:= R_i^- \end{aligned} \qquad (4.5)$$

for $i = 1, 2, \ldots, k$. The I_h are defined by their sets of maximal and minimal elements, $G(I_h)$ and $S(I_h)$, and define queries Q_h with the desired properties:

Theorem 4.4.6 Let $Th(Q, \mathscr{L}_{\mathscr{M}(Q)}) = I_1 \cup \cdots \cup I_k$ with convex I_h.
Given an element μ from $\mathscr{L}_{\mathscr{M}(Q)}$ define

$$Q_{M,\mu} = \bigwedge_{m \in \mu} m \quad \text{and} \quad Q_{A,\mu} = \bigwedge_{m \notin \mu} \neg m$$

For $h = 1, \ldots, k$ let

$$Q_{h,M} = \bigvee_{\mu \in S(I_h)} Q_{M,\mu} \quad \text{and} \quad Q_{h,A} = \bigvee_{\mu \in G(I_h)} Q_{A,\mu}.$$

Finally, let $Q_h = Q_{h,M} \wedge Q_{h,A}$.
   Then $Th(Q_h, \mathscr{D}, \mathscr{L})$ is convex for all $\mathscr{L}$ and $\mathscr{D}$, and $Th(Q, \mathscr{D}, \mathscr{L}) = Th(\bigvee_{h=1}^{k} Q_h, \mathscr{D}, \mathscr{L}) = \cup_{h=1}^{k} Th(Q_h, \mathscr{D}, \mathscr{L})$.

Proof: The Q_h are constructed so that

$$Th(Q_h, \mathscr{L}_{\mathscr{M}(Q)}) = I_h \quad (h = 1, \ldots, k).$$

Using the embedding $h_{\mathscr{D}}$ from the proof of Theorem 4.4.4 we then obtain for $\phi \in \mathscr{L}$:

$$\begin{aligned} \phi \in Th(Q, \mathscr{D}, \mathscr{L}) &\Leftrightarrow h_{\mathscr{D}}(\phi) \in Th(Q, \mathscr{L}_{\mathscr{M}(Q)}) \\ &\Leftrightarrow h_{\mathscr{D}}(\phi) \in Th(\bigvee_{h=1}^{k} Q_h, \mathscr{L}_{\mathscr{M}(Q)}) \\ &\Leftrightarrow \phi \in Th(\bigvee_{h=1}^{k} Q_h, \mathscr{D}, \mathscr{L}) \\ &\Leftrightarrow \phi \in \cup_{h=1}^{k} Th(Q_h, \mathscr{D}, \mathscr{L}) \end{aligned}$$

$\square$

Example 4.4.7 (continued from Example 4.4.5) Using (4.5) we obtain the partition of $Th(Q, \mathscr{L}_{\mathscr{M}(Q)})$

$$I_1 = \{\{m_4\}, \{m_3, m_4\}\}, \ I_2 = \{\emptyset\},$$

so that

$$\begin{aligned} G(I_1) &= \{\{m_3, m_4\}\} \ S(I_1) = \{\{m_4\}\} \\ G(I_2) &= \{\emptyset\} \qquad\qquad S(I_2) = \{\emptyset\}. \end{aligned}$$

These boundary sets define the queries

$$\begin{aligned} Q_1 &= (\neg m_1 \wedge \neg m_2) \wedge m_4 \\ Q_2 &= (\neg m_1 \wedge \neg m_2 \wedge \neg m_3 \wedge \neg m_4) \end{aligned}$$

When we view Q, Q_1, Q_2 as propositional formulas over propositional variables $m_1, \ldots, m_4$, we see that $Q \leftrightarrow Q_1 \vee Q_2$ is a valid logical equivalence. In fact, one can interpret the whole decomposition procedure we here developed as a method for computing a certain normal form for propositional formulas. We investigate this perspective further in Section 4.5. $\qquad\square$

Example 4.4.8 (continued from Example 4.4.2) Introducing $\tilde{m}_1 = \neg a_1$, $\tilde{m}_2 = \neg a_2$, we can express query Q_1 from Example 4.4.2 using monotone predicates only as

$$\tilde{Q}_1 = (\neg \tilde{m}_1 \vee \neg \tilde{m}_2) \wedge (m_1 \vee m_2)$$

The boundary sets of the single convex component I of $\tilde{Q}_1$ are

$$G(I) = \{\{\tilde{m}_1, m_1, m_2\}, \{\tilde{m}_2, m_1, m_2\}\} \quad S(I) = \{\{m_1\}, \{m_2\}\}.$$

The query construction of Theorem 4.4.6 yields the sub-queries $Q_A = \neg \tilde{m}_1 \vee \neg \tilde{m}_2$ and $Q_M = m_1 \vee m_2$, i.e., the same decomposition as given by the query plan in Example 4.4.2.

Now consider the query

$$Q' = m_1 \vee (m_2 \wedge \neg m_3).$$

This query has dimension two, and can be solved, for example, using the query plan $\underbrace{m_1}\ \vee\ \underbrace{(m_2 \wedge \neg m_3)}$.

The canonical decomposition gives the following boundary sets for two convex components:

$$G(I_1) = \{\{m_1, m_2, m_3\}\}\ S(I_1) = \{\{m_1\}\}$$
$$G(I_2) = \{\{m_2\}\}\qquad S(I_2) = \{\{m_2\}\},$$

which leads to the sub-queries

$$Q'_1 = m_1, \qquad Q'_2 = \neg m_1 \wedge \neg m_3 \wedge m_2.$$

Thus, the we obtain a different solution strategy than from the simple query plan, and the solution will be expressed as two disjoint convex components, whereas the query plan would return overlapping convex components. $\qquad\square$

When we evaluate the query Q from Example 4.4.5 for $\mathscr{L}_\Sigma$, we find that $Th(Q, \mathscr{D}, \mathscr{L}_\Sigma)$ actually has dimension 1. The basic reason for this is that in $\mathscr{L}_\Sigma$ there does not exist any pattern that satisfies $\neg m_1 \wedge \neg m_2 \wedge m_3 \wedge \neg m_4$, i.e., that corresponds to the pattern $\{m_3\} \in \mathscr{L}_{\mathscr{M}(Q)}$ that is the only "witness" for the non-convexity of $Th(Q, \mathscr{L}_{\mathscr{M}(Q)})$. To distinguish patterns in $\mathscr{L}_{\mathscr{M}(Q)}$ that we need not take into account when working with a pattern language $\mathscr{L}$, we introduce the concept of $\mathscr{L}$-admissibility:

Definition 4.4.9 Let $\mathscr{L}$ be a pattern language. A pattern $\mu \in \mathscr{L}_{\mathscr{M}(Q)}$ is called $\mathscr{L}$-admissible if there exists $\phi \in \mathscr{L}$ and datasets $\mathscr{D}$ such that $\mu = h_{\mathscr{D}}(\phi)$, where $h_{\mathscr{D}}$

is as defined by (4.1). Let $\mathscr{L}_{\mathcal{M}(Q),\mathscr{L}} \subseteq \mathscr{L}_{\mathcal{M}(Q)}$ be the language of $\mathscr{L}$-admissible patterns from $\mathscr{L}_{\mathcal{M}(Q)}$. As before, we define

$$Th(m_i, \mathscr{L}_{\mathcal{M}(Q),\mathscr{L}}) = \{\mu \in \mathscr{L}_{\mathcal{M}(Q),\mathscr{L}} \mid m_i \in \mu\}.$$

An alternative characterization of admissibility is that μ is $\mathscr{L}$-admissible if there exists $\mathscr{D}$ such that

$$Th(\bigwedge_{m_i \in \mu} m_i \wedge \bigwedge_{m_j \notin \mu} \neg m_j, \mathscr{D}, \mathscr{L}) \neq \emptyset.$$

Theorem 4.4.10 Let $\mathscr{L}$ and Q be as in Theorem 4.4.4. The dimension of $Th(Q, \mathscr{D}, \mathscr{L})$ is less than or equal to the dimension of $Th(Q, \mathscr{L}_{\mathcal{M}(Q),\mathscr{L}})$.

Proof: The proof is as for Theorem 4.4.4, by replacing $\mathscr{L}_{\mathcal{M}(Q)}$ with $\mathscr{L}_{\mathcal{M}(Q),\mathscr{L}}$ throughout. We only need to note that according to the definition of admissibility, the mapping $h_{\mathscr{D}}$ defined by (4.1) actually maps $\mathscr{L}$ into $\mathscr{L}_{\mathcal{M}(Q),\mathscr{L}}$, so that $h_{\mathscr{D}}$ still is well-defined. □

Example 4.4.11 (continued from Example 4.4.7) Consider the language $\mathscr{L}_{\Sigma}$. Of the patterns in $\mathscr{L}_{\mathcal{M}(Q)}$ two are not $\mathscr{L}_{\Sigma}$-admissible: as *is_more_specific*$(p,ab) \wedge$ *is_more_specific*(p,cb) implies *length_at_least*$(p,4)$, we have that $Th(\neg m_1 \wedge \neg m_2 \wedge m_3 \wedge m_4, \mathscr{D}, \mathscr{L}_{\Sigma}) = Th(\neg m_1 \wedge \neg m_2 \wedge m_3 \wedge \neg m_4, \mathscr{D}, \mathscr{L}_{\Sigma}) = \emptyset$ for all $\mathscr{D}$, so that the two patterns $\{m_3\}$ and $\{m_3, m_4\}$ from $\mathscr{L}_{\mathcal{M}(Q)}$ are not $\mathscr{L}_{\Sigma}$-admissible.

Figure 4.2 (b) shows $\mathscr{L}_{\mathcal{M}(Q),\mathscr{L}_{\Sigma}}$. Now $Th(Q, \mathscr{L}_{\mathcal{M}(Q),\mathscr{L}_{\Sigma}}) = \{\emptyset, \{m_4\}\}$ has dimension 1, so that by Theorem 4.4.10 $Th(Q, \mathscr{D}, \mathscr{L}_{\Sigma})$ also has dimension 1.

With Theorem 4.4.6 we obtain the query

$$Q_1 = \neg m_1 \wedge \neg m_2 \wedge \neg m_3$$

with $Th(Q, \mathscr{D}, \mathscr{L}_{\Sigma}) = Th(Q_1, \mathscr{D}, \mathscr{L}_{\Sigma})$.

□

Table 4.1 summarizes the decomposition approach to inductive query evaluation as derived from Theorems 4.4.10 and 4.4.6. A simplified procedure based on Theorems 4.4.4 and 4.4.6 can be used simply by omitting the first step.

Assuming that $\mathscr{L}$-admissibility is decidable in time exponential in the size s of the query (for the pattern languages and pattern predicates we have considered so far this will be the case), we obtain that steps 1–4 can be performed naively in time $O(2^s)$. This exponential complexity in s we consider uncritical, as the size of the query will typically be very small in comparison to the size of $\mathscr{D}$, i.e., $s \ll |\mathscr{D}|$, so that the time critical step is step 5, which is the only step that requires inspection of $\mathscr{D}$.

Theorem 4.4.10 is stronger than Theorem 4.4.4 in the sense that for given $\mathscr{L}$ it yields better bounds for the dimension of $Th(Q, \mathscr{D}, \mathscr{L})$. However, Theorem 4.4.4 is stronger than Theorem 4.4.10 in that it provides a uniform bound for all pattern languages for which the m_i are monotone. For this reason, the computation

Table 4.1 The schema of the Query Decomposition Approach

Input:

- Query Q that is a Boolean combination of monotone predicates: $\mathcal{M}(Q) = \{m_1, \ldots, m_n\}$
- datasets $\mathcal{D}$
- pattern language $\mathcal{L}$

Step 1: Construct $\mathcal{L}_{\mathcal{M}(Q),\mathcal{L}}$: for each $\mu \subseteq \mathcal{M}(Q)$ decide whether μ is $\mathcal{L}$-admissible.
Step 2: Construct $Th(Q, \mathcal{L}_{\mathcal{M}(Q),\mathcal{L}})$: for each $\mu \in \mathcal{L}_{\mathcal{M}(Q),\mathcal{L}}$ decide
       whether $\mu \in Th(Q, \mathcal{L}_{\mathcal{M}(Q),\mathcal{L}})$.
Step 3: Determine convex components: compute partition $Th(Q, \mathcal{L}_{\mathcal{M}(Q),\mathcal{L}}) = I_1 \cup \cdots \cup I_k$
       into a minimal number of convex components.
Step 4: Decompose Q: compute queries $Q_1, \ldots, Q_k$.
Step 5: For $i = 1, \ldots, k$ determine $Th(Q_i, \mathcal{D}, \mathcal{L})$ by computing the boundaries $G(Q_i, \mathcal{D}, \mathcal{L})$
       and $S(Q_i, \mathcal{D}, \mathcal{L})$.

---

of $Th(Q, \mathcal{D}, \mathcal{L})$ using the simpler approach given by Theorems 4.4.4 and 4.4.6 can also be of interest in the case where we work in the context of a fixed language $\mathcal{L}$, because the solutions computed under this approach are more robust in the following sense: suppose we have computed $Th(Q, \mathcal{D}, \mathcal{L})$ using the decomposition provided by Theorems 4.4.4 and 4.4.6, i.e., by the algorithm shown in Table 4.1 omitting step 1. This gives us a representation of $Th(Q, \mathcal{D}, \mathcal{L})$ by boundary sets $G(Q_h, \mathcal{D}, \mathcal{L}), S(Q_h, \mathcal{D}, \mathcal{L})$. If we now consider any refinement $\mathcal{L}' \succeq \mathcal{L}$, then our boundary sets still define valid solutions of Q in $\mathcal{L}'$, i.e., for all $\psi \in \mathcal{L}'$, if $\phi \preceq \psi \preceq \phi'$ for some $\phi \in S(Q_h, \mathcal{D}, \mathcal{L}), \phi' \in G(Q_h, \mathcal{D}, \mathcal{L})$, then $\psi \in Th(Q, \mathcal{D}, \mathcal{L}')$ (however, the old boundary may not completely define $Th(Q, \mathcal{D}, \mathcal{L}')$, as the maximal/minimal solutions of Q_h in $\mathcal{L}$ need not be maximal/minimal in $\mathcal{L}'$). A similar preservation property does not hold when we compute $Th(Q, \mathcal{D}, \mathcal{L})$ according to Theorem 4.4.10.

4.5 Normal Forms

In this section we analyze some aspects of our query decomposition approach from a propositional logic perspective. Central to this investigation is the following concept of certain syntactic normal forms of propositional formulas.

Definition 4.5.1 Let Q be a propositional formula in propositional variables $m_1, \ldots, m_n$. We say that Q belongs to the class Θ_1 if Q is logically equivalent to a formula of the form

$$\left(\bigvee_{i=1}^{h} M_i \right) \wedge \left(\bigvee_{j=1}^{k} A_j \right), \tag{4.6}$$

where M_i's are conjunctions of positive atoms and A_j's are conjunctions of negative atoms. We say that Q belongs to the class Θ_k if Q is equivalent to the disjunction of k formulas from Θ_1.

The formulas Q_h defined in Theorem 4.4.6 are in the class Θ_1 when read as propositional formulas in $m_1, \ldots, m_n$, and were constructed so as to define convex sets. The following theorem provides a general statement on the relation between the Θ_k-normal form of a query and its dimension. As such, every formula belongs to some Θ_k, as can easily be seen from the disjunctive normal forms.

Theorem 4.5.2 Let Q be a query containing pattern predicates $m_1, \ldots, m_n$. The following are equivalent:

(i)   When interpreted as a Boolean formula over propositional variables $m_1, \ldots, m_n$, Q belongs to Θ_k.
(ii) The dimension of Q with respect to any pattern language $\mathscr{L}$ for which $m_1, \ldots, m_n$ are monotone is at most k.

Proof: (i)$\Rightarrow$(ii): We may assume that Q is written in Θ_k-normal form, i.e., as a disjunction of k subformulas of the form (4.6). As both unions and intersections of monotone sets are monotone, we obtain that the left conjunct of (4.6) defines a monotone subset of $\mathscr{L}$ (provided the m_i define monotone sets in $\mathscr{L}$). Similarly, the right conjunct defines an anti-monotone set. Their conjunction, then, defines a convex set, and the disjunction of k formulas (4.6) defines a union of k convex sets.

(ii)$\Rightarrow$(i): This follows from the proofs of Theorems 4.4.4 and 4.4.6: let $\mathscr{L} = \mathscr{L}_{\mathscr{M}(Q)}$. We can view $\mathscr{L}_{\mathscr{M}(Q)}$ as the set of all truth assignments to the variables m_i by letting for $\mu \in \mathscr{L}_{\mathscr{M}(Q)}$:

$$\mu: \quad m_i \mapsto \begin{cases} true & m_i \in \mu \\ false & m_i \notin \mu \end{cases}$$

Then for all μ and all Boolean formulas $\tilde{Q}$ in $m_1, \ldots, m_n$:

$$\mu \in Th(\tilde{Q}, \mathscr{L}_{\mathscr{M}(Q)}) \Leftrightarrow \mu: \tilde{Q} \mapsto true.$$

Therefore

$$\mu: Q \mapsto true \Leftrightarrow \mu \in Th(Q, \mathscr{L}_{\mathscr{M}(Q)})$$
$$\Leftrightarrow \mu \in \cup_{h=1}^{k} Th(Q_h, \mathscr{L}_{\mathscr{M}(Q)})$$
$$\Leftrightarrow \mu: \vee_{h=1}^{k} Q_h \mapsto true$$

$\square$

In the light of Theorem 4.5.2 we can interpret the decomposition procedure described by Theorems 4.4.4 and 4.4.6 as a Boolean transformation of Q into Θ_k-normal form. This transformation takes a rather greedy approach by explicitly constructing the exponentially many possible truth assignments to the propositional variables in Q. It might seem possible to find a more efficient transformation based on purely syntactic manipulations of Q. The following result shows that this is unlikely to succeed.

Theorem 4.5.3 The problem of deciding whether a given propositional formula Q belongs to Θ_1 is co-NP complete.

Proof: The class Θ_1 is in co-NP: let Q be a formula in variables $m_1, \ldots, m_n$. From Theorems 4.4.4 and 4.5.2 we know that $Q \notin \Theta_1$ iff the dimension of $Th(Q, \mathscr{L}_{\mathscr{M}(Q)})$ is at least 2. This, in turn, is equivalent to the existence of an alternating chain of length 2 for $Th(Q, \mathscr{L}_{\mathscr{M}(Q)})$. The existence of such a chain can be determined in nondeterministic polynomial time by guessing three elements $\mu_1, \mu_2, \mu_3 \in \mathscr{L}_{\mathscr{M}(Q)}$, and checking whether $\mu_1 \succeq \mu_2 \succeq \mu_3$ and $\mu_1, \mu_3 \in Th(Q, \mathscr{L}_{\mathscr{M}(Q)})$ and $\mu_2 \notin Th(Q, \mathscr{L}_{\mathscr{M}(Q)})$.

To show co-NP hardness we reduce the satisfiability problem to the complement of Θ_1. For this let F be a propositional formula in propositional variables $m_1, \ldots, m_k$. Define

$$Q := (F \wedge \neg x_1 \wedge \neg y_1) \vee (m_1 \wedge m_2 \wedge \cdots \wedge m_k \wedge x_1 \wedge y_1),$$

(where x_1, y_1 are new propositional variables). Then $Th(Q, \mathscr{L}_{\mathscr{M}(Q)})$ has dimension ≥ 2 (i.e., $Q \notin \Theta_1$) iff F is satisfiable: If F is not satisfiable, then $(F \wedge \neg x_1 \wedge \neg y_1)$ is not satisfiable. So, Q can only be satisfied when all variables m_i, x_1, y_1 are true. Consequently, $Th(Q, \mathscr{L}_{\mathscr{M}(Q)})$ has only one element, namely $\{m_1, \ldots, m_k, x_1, y_1\}$ and $\dim(Th(Q, \mathscr{L}_{\mathscr{M}(Q)})) = 1$. On the other hand, if F is satisfiable, then $Th(Q, \mathscr{L}_{\mathscr{M}(Q)})$ contains a set $\phi \subseteq \{m_1, \ldots, m_k\}$, and then $\phi \subseteq \phi \cup \{x_1\} \subseteq \{m_1, \ldots, m_k, x_1, y_1\}$ is an alternating chain of length 2, because $\phi \cup \{x_1\} \notin Th(Q, \mathscr{L}_{\mathscr{M}(Q)})$. $\square$

The sub-queries to which the original query Q is reduced not only are known to have convex solution sets $Th(Q_h, \mathscr{D}, \mathscr{L})$, they also are of a special syntactic form $Q_h = Q_{h,M} \wedge Q_{h,A}$, where $Q_{h,M}$ defines a monotone set $Th(Q_{h,M}, \mathscr{D}, \mathscr{L})$, and $Q_{h,A}$ defines an anti-monotone set $Th(Q_{h,A}, \mathscr{D}, \mathscr{L})$. This factorization of Q_h facilitates the computation of the border sets $G(Q_h, \mathscr{D}, \mathscr{L})$ and $S(Q_h, \mathscr{D}, \mathscr{L})$, for which the level wise version space algorithm [8, 15] can be used.

4.6 Conclusions

We have described an approach to inductive querying, which generalizes both the pattern discovery problem in data mining and the concept-learning problem in machine learning. The method is based on the decomposition of the answer set to a collection of components defined by monotonic and anti-monotonic predicates. Each of the components is a convex set or version space, the borders of which can be computed using, for instance, the level wise version space algorithm or—for the pattern domain of strings—using the VSTmine algorithm [17], which employs a data structure called the version space tree.

The work presented is related to several research streams within data mining and machine learning. In machine learning, there has been an interest in version spaces ever since Tom Mitchell's seminal Ph.D. thesis [21]. The key complaint about standard version spaces was for a long time that it only allowed one to cope with essentially conjuctive concept-learning, that is, the induced concepts or patterns need to be conjuctive (which holds also for item sets). There has been quite some work in the machine learning literature on accomodating also disjunctive concepts (as is required in a general rule- or concept-learning setting), for instance, [26, 27]. While such disjunctive version space techniques sometimes also work with multiple borders set and version spaces, the way that this is realized differs from our approach. Indeed, in a disjunctive version space, a single solution consists of a disjunction of patterns, of which each pattern must belong to a single version space in the traditional sense. This differs from our approach in which each member of a single version space is a solution in itself. Algebraic properties of version spaces have also been investigated in the machine learning literature by, for instance, Haym Hirsh [11] who has investigated the properties of set theoretic operations on version spaces, and [16] who have developed a special version space algebra to represent functions, and used it for programming by demonstration.

In data mining, the structure on the search space was already exploited by early algorithms for finding frequent itemsets and association rules [1] leading soon to the concept of border sets [20]. Whereas initially the focus was on the use of the most specific borders, this was soon extended towards using also the most general borders to cope with multiple data sets, with conjunctive inductive queries or emerging patterns [8, 19]. The resulting version space structure was further analyzed by, for instance, Bucila et al. [6] for the case of itemsets. Each of these developments has resulted in new algorithms and techniques for finding solutions to increasingly complex inductive queries. The contribution of our work is that it has generalized this line of ideas in accomodating also non-convex solution sets, that is, generalized version spaces. It also allows one to cope with arbitrary Boolean inductive queries. Finally, as already mentioned in the introduction, the present paper also attempts to bridge the gap between the machine learning and data mining perspective both at the task level – through the introduction of Boolean inductive querying – and at the representation level – through the introduction of generalized version spaces to represent solution sets.

The results we have presented in this chapter are by no means complete, a lot of open problems and questions remain. A first and perhaps most important question is as to an experimental evaluation of the approach. Although some initial results in this direction have been presented in [17, 25, 18], the results are not yet conclusive and a deeper and more thorough evaluation is needed. A second question is concerned with further extending the framework to cope with other primitives, which are neither monotonic nor anti-monotonic. A typical example of such primitives are the questions that ask for the top-k patterns w.r.t. a particular optimization function such as χ^2. This is known as correlated pattern mining [23]. A third question is how to perform a more quantitative query optimization, which would estimate the resources needed to execute particular query plan.

Although there are many remaining questions, the authors hope that the introduced framework provides a sound theory for studying these open questions.

Acknowledgements This work was partly supported by the European IST FET project cInQ.

References

1. R. Agrawal, H. Mannila, R. Srikant, H. Toivonen, and A.I. Verkamo. Fast discovery of association rules. In U. Fayyad, G. Piatetsky-Shapiro, P. Smyth, and R. Uthurusamy, editors, *Advances in Knowledge Discovery and Data Mining*, pages 307–328. MIT Press, 1996.
2. R. Agrawal, T. Imielinski, and A. N. Swami. Mining association rules between sets of items in large databases. In Peter Buneman and Sushil Jajodia, editors, *Proceedings of the 1993 ACM SIGMOD International Conference on Management of Data*, pages 207–216, Washington, D.C., U.S.A., 25–28 May 1993.
3. D. Angluin. Queries and concept-learning. *Machine Learning*, 2:319–342, 1987.
4. E. Baralis and G. Psaila. Incremental refinement of mining queries. In M. K. Mohania and A. Min Tjoa, editors, *Proceedings of the First International Conference on Data Warehousing and Knowledge Discovery (DaWaK'99)*, volume 1676 of *Lecture Notes in Computer Science*, pages 173–182, Florence, Italy, August 30–September 1 1999. Springer.
5. R. Bayardo. Efficiently mining long patterns from databases. In *Proceedings of ACM SIGMOD Conference on Management of Data*, 1998.
6. C. Bucilă, J. Gehrke, D. Kifer, and W. White. Dualminer: A dual-pruning algorithm for itemsets with constraints. *Data Min. Knowl. Discov.*, 7(3):241–272, 2003.
7. L. De Raedt. A perspective on inductive databases. *SIGKDD Explorations: Newsletter of the Special Interest Group on Knowledge Discovery and Data Mining, ACM*, 4(2):69–77, January 2003.
8. L. De Raedt and S. Kramer. The levelwise version space algorithm and its application to molecular fragment finding. In *IJCAI01: Seventeenth International Joint Conference on Artificial Intelligence*, August 4–10 2001.
9. B. Goethals and J. Van den Bussche. On supporting interactive association rule mining. In *Proceedings of the Second International Conference on Data Warehousing and Knowledge Discovery*, volume 1874 of *Lecture Notes in Computer Science*, pages 307–316. Springer, 2000.
10. J. Han, L. V. S. Lakshmanan, and R. T. Ng. Constraint-based multidimensional data mining. *IEEE Computer*, 32(8):46–50, 1999.
11. H. Hirsh. *Incremental Version-Space Merging: A General Framework for Concept Learning*. Kluwer Academic Publishers, 1990.
12. H. Hirsh. Theoretical underpinnings of version spaces. In *Proceedings of the Twelfth International Joint Conference on Artificial Intelligence (IJCAI91)*, pages 665–670. Morgan Kaufmann Publishers, 1991.
13. H. Hirsh. Generalizing version spaces. *Machine Learning*, 17(1):5–46, 1994.
14. M. Kearns and U. Vazirani. *An Introduction to Computational Learning Theory*. MIT, 1994.
15. S. Kramer, L. De Raedt, and C. Helma. Molecular feature mining in HIV data. In *KDD-2001: The Seventh ACM SIGKDD International Conference on Knowledge Discovery and Data Mining*. Association for Computing Machinery, August 26–29 2001. ISBN: 158113391X.
16. T. Lau, S. A. Wolfman, P. Domingos, and D. S. Weld. Programming by demonstration using version space algebra. *Mach. Learn.*, 53(1-2):111–156, 2003.
17. S. D. Lee. *Constrained Mining of Patterns in Large Databases*. PhD thesis, Albert-Ludwigs-University, 2006.

18. S. D. Lee and L. De Raedt. An algebra for inductive query evaluation. In X. Wu, A. Tuzhilin, and J. Shavlik, editors, *Proceedings of The Third IEEE International Conference on Data Mining (ICDM'03)*, pages 147–154, Melbourne, Florida, USA, November 19–22 2003. Sponsored by the IEEE Computer Society.

19. J. Li, K. Ramamohanarao, and G. Dong. The space of jumping emerging patterns and its incremental maintenance algorithms. In *ICML '00: Proceedings of the Seventeenth International Conference on Machine Learning*, pages 551–558, San Francisco, CA, USA, 2000. Morgan Kaufmann Publishers Inc.

20. H. Mannila and H. Toivonen. Levelwise search and borders of theories in knowledge discovery. *Data Mining and Knowledge Discovery*, 1(3):241–258, 1997.

21. T.M. Mitchell. *Version Spaces: An Approach to Concept Learning*. PhD thesis, Stanford University, 1978.

22. T.M. Mitchell. Generalization as search. *Artificial Intelligence*, 18(2):203–226, 1980.

23. S. Morishita and J. Sese. Traversing itemset lattice with statistical metric pruning. In *Proceedings of the 19th ACM SIGACT-SIGMOD-SIGART Symposium on Principles of Database Systems*, pages 226–236. ACM Press, 2000.

24. R. T. Ng, L. V. S. Lakshmanan, J. Han, and A. Pang. Exploratory mining and pruning optimizations of constrained associations rules. In *Proceedings ACM-SIGMOD Conference on Management of Data*, pages 13–24, 1998.

25. L. De Raedt, M. Jaeger, S. D. Lee, and H. Mannila. A theory of inductive query answering. In *Proceedings of the 2002 IEEE International Conference on Data Mining*, pages 123–130, 2002.

26. G. Sablon, L. De Raedt, and M. Bruynooghe. Iterative versionspaces. *Artificial Intelligence*, 69:393–409, 1994.

27. M. Sebag. Delaying the choice of bias: A disjunctive version space approach. In *Proceedings of the 13th International Conference on Machine Learning*, pages 444–452, 1996.

Part II
Constraint-based Mining:
Selected Techniques

Chapter 5
Generalizing Itemset Mining in a Constraint Programming Setting

Jérémy Besson, Jean-François Boulicaut, Tias Guns, and Siegfried Nijssen

Abstract In recent years, a large number of algorithms have been proposed for finding set patterns in boolean data. This includes popular mining tasks based on, for instance, frequent (closed) itemsets. In this chapter, we develop a common framework in which these algorithms can be studied thanks to the principles of constraint programming. We show how such principles can be applied both in specialized and general solvers.

5.1 Introduction

Detecting local patterns has been studied extensively during the last decade (see, e.g., [18] and [22] for dedicated volumes). Among others, many researchers have considered the discovery of relevant set patterns (e.g., frequent itemsets and association rules, maximal itemsets, closed sets) from transactional data (i.e., collections of transactions that are collections of items). Such data sets are quite common in many different application domains like, e.g., basket data analysis, WWW usage mining, biomedical data analysis. In fact, they correspond to binary relations that

Jérémy Besson
Vilnius University, Faculty of Mathematics and Informatics
Naugarduko St. 24, LT-03225 Vilnius, Lithuania
e-mail: contact.jeremy.besson@gmail.com

Jean-François Boulicaut
Université de Lyon, CNRS, INRIA
INSA-Lyon, LIRIS Combining, UMR5205, F-69621, France
e-mail: jean-francois.boulicaut@insa-lyon.fr

Tias Guns · Siegfried Nijssen
Department of Computer Science
Katholieke Universiteit Leuven, Celestijnenlaan 200A, B-3001 Leuven, Belgium
e-mail: firstname.lastname@cs.kuleuven.be

encode whether a given set of objects satisfies a given set of Boolean properties or not.

In the last few years, it appears that such 0/1 data mining techniques have reached a kind of ripeness from both an algorithmic and an applicative perspective. It is now possible to process large amounts of data to reveal, for instance, unexpected associations between subsets of objects and subsets of properties which they tend to satisfy. An important breakthrough for the frequent set mining technology and its multiple uses has been the understanding of efficient mechanisms for computing the so-called condensed representations on the one hand, and the huge research effort on safe pruning strategies when considering user-defined constraints on the other hand.

Inspired by the pioneering contribution [23], frequent closed set mining has been studied extensively by the data mining community (see, e.g., the introduction to the FIMI Workshop [16]). A state-of-the-art algorithm like LCM [27] appears to be extremely efficient and it is now possible, for relevant frequency thresholds, to extract every frequent closed set from either sparse or dense data. The analogy between Formal Concept Analysis (see, e.g., [15]) and frequent closed set mining is well understood and this has motivated the design of new algorithms for computing closed sets and concept lattices. Closed set mining has been also studied as a very nice example of a condensed representation for frequency queries and this topic has motivated quite a large number of contributions the last 5 years (see [10] for a survey). In the same time, the active use of user-defined constraints has been studied a lot (see, e.g., [8, 2, 5]). Most of the recent set pattern mining algorithms can exploit constraints that are not limited to the simple cases of monotonic and/or anti-monotonic ones as described in, for instance, [9]. New concepts have emerged like "flexible constraints" [25], "witnesses" [19] or "soft constraints" [4]. Also, the specific problems of noisy data sets has inspired a constraint-based mining view on fault-tolerance (see, e.g., [24, 28, 3]).

While a few tens of important algorithms have been proposed, we lack a clear abstraction of their principles and implementation mechanisms. We think that a timely challenge is to address this problem. Our objective is twofold. First, we want to elucidate the essence of the already published pattern discovery algorithms which process binary relations. Next, we propose a high-level abstraction of them. To this end, we adopt a constraint programming approach which both suits well with the type of problems we are interested in and can help to identify and to describe all basic steps of the constraint-based mining algorithms. Algorithms are presented without any concern about data structures and optimization issues. We would like to stay along the same lines of [21] which introduced in 1997 the level-wise algorithm. This was an abstraction of severals algorithms already published for frequent itemset mining (typically APRIORI) as well as several works around inclusion dependencies and functional dependencies. The generality of the levelwise algorithm inspired a lot of work concerning the use of the border principle and its relations with classical version spaces, or the identification of fundamental mechanisms of enumeration and pruning strategies. Even though this appears challenging, we consider this paper as a major step towards the definition of a flexible, generic though efficient local pattern mining algorithm.

To illustrate the general approach, we will describe two instances of it: one is a specialized, but sufficiently general pattern mining solver in which pattern mining constraints are the main primitives; the other involves the use of existing constraint programming systems, similar to [13]. Both approaches will be illustrated on a problem of fault tolerant mining to make clear the possible advantages and disadvantages.

The rest of this article is organized as follows. The next section introduces some notations and the needed concepts. Section 3 discusses the principles of several specialized algorithms that have been proposed to support set pattern discovery from 0/1 data. Then, we propose in Section 4 an abstraction of such algorithms. Given such a generalization, both the dedicated solver (Section 5) and an implementation scheme within constraint programming systems (Section 6) are given. Section 7 briefly concludes.

5.2 General Concepts

Let $\mathscr{T} = \{t_1, \ldots, t_m\}$ and $\mathscr{I} = \{i_1, \ldots, i_n\}$ be two sets of respectively transactions and items. Let $\mathbf{r}$ be a boolean matrix in which $\mathbf{r}_{ti} \in \{0, 1\}$, for $t \in \mathscr{T}$ and $i \in \mathscr{I}$. An illustration is given in Figure 5.1 for the sets $\mathscr{T} = \{t_1, t_2, t_3, t_4, t_5\}$ and $\mathscr{I} = \{i_1, i_2, i_3, i_4\}$.

Fig. 5.1 Boolean matrix where $\mathscr{T} = \{t_1, t_2, t_3, t_4, t_5\}$ and $\mathscr{I} = \{i_1, i_2, i_3, i_4\}$

	i_1	i_2	i_3	i_4
t_1	1	1	1	1
t_2	0	1	0	1
t_3	0	1	1	0
t_4	0	0	1	0
t_5	0	0	0	0

In such data sets, we are interested in finding local patterns. A local pattern P is a pair of an itemset and a transaction set, (X, Y), which satisfies user-defined local constraints $\mathscr{C}(P)$. We can consider two types of constraints. First, we have constraints on the *pattern types*, which refers to how the pattern is defined with respect to the input data set. Second, we have constraints on the *pattern form*, which do not take into account the data. Hence the constraint $\mathscr{C}$ can be expressed in the form of a conjunction of two constraints $\mathscr{C}(P) = \mathscr{C}_{type}(P) \wedge \mathscr{C}_{form}(P)$; $\mathscr{C}_{type}$ and $\mathscr{C}_{form}$ can themselves be conjunctions of constraints, i.e., $\mathscr{C}_{type} \equiv \mathscr{C}_{1-type} \wedge \cdots \wedge \mathscr{C}_{k-type}$ and $\mathscr{C}_{form} \equiv \mathscr{C}_{1-form} \wedge \cdots \wedge \mathscr{C}_{l-form}$.

The typical pattern type constraints are given below.

Definition 5.1 (Main Pattern Types). Let $P = (X, Y) \in 2^{\mathscr{T}} \times 2^{\mathscr{I}}$ be a pattern.

(a) P is an itemset with its support set, satisfying $\mathcal{C}_{itemset}(P)$, iff the constraint $(X = \{x \in \mathcal{T} \mid \forall y \in Y, \mathbf{r}_{xy} = 1\})$ is satisfied.

(b) P is a maximal itemset, satisfying $\mathcal{C}_{max-itemset}(P)$, iff $\mathcal{C}_{itemset}(P)$ is satisfied and there does not exist any itemset $P' = (X', Y')$ satisfying $\mathcal{C}_{itemset}(P')$ such that $X' \subseteq X$ and $Y \subset Y'$. Note that $\mathcal{C}_{itemset}$ is usually substituted with a conjunction of more complex constraints[1] with respect to additional constraints on the patterns P'.

(c) P is a formal concept, satisfying $\mathcal{C}_{fc}(P)$, iff $\mathcal{C}_{itemset}(P)$ is satisfied and there does not exist any itemset $P' = (X', Y')$ such that $X = X'$ and $Y \subset Y'$.

These constraints are related to each other:

$$\mathcal{C}_{max-itemset}(P) \implies \mathcal{C}_{fc}(P) \implies \mathcal{C}_{itemset}(P)$$

Example 5.1. Referring to Figure 5.1, $(t_1 t_2, i_4)$ and $(t_1, i_2 i_3 i_4)$ are examples of itemsets with their support sets in $\mathbf{r}_1$. $(t_1 t_2, i_2 i_4)$ and $(t_1 t_3, i_2 i_3)$ are two examples of formal concepts in $\mathbf{r}_1$.

The most well-known form constraints are given in Figure 5.2. The first one is usually called the minimum frequency constraint on itemsets. The second one imposes that both sets of the pattern have a minimal size. The third one is called the "minimal area constraint" and ensures that extracted patterns cover a minimal number of "1" values of the boolean matrix. The next constraint requires that the mean of positive real values associated to each item of the itemset is greater than a given threshold. Constraint $\mathcal{C}_{membership}$ imposes that patterns contain certain elements, for instance $a \in \mathcal{T}$ and $b \in \mathcal{I}$. Emerging patterns satisfying $\mathcal{C}_{emerging}$ must be frequent with respect to a transaction set and infrequent with respect to another one. $\mathcal{C}_{division}$ is another example of a form constraint.

There is a wide variety of combinations of constraints that one could wish to express. For instance, we may want to find itemsets with their support sets (pattern type), for which the support set is greater than 10% (pattern form); or we may wish to find the formal concepts (pattern type) containing at least 3 items and 4 transactions (pattern form) or a fault-tolerant pattern (pattern type) having an area of size at least 20 (pattern form). Fault-tolerant extensions of formal concepts were previously studied in [30, 12, 20, 3], and will be discussed later in more detail.

Fig. 5.2 Examples of interesting pattern form constraints on patterns.

$\mathcal{C}_{form}(X,Y)$		
$\mathcal{C}_{size}$	$\equiv$	$\|X\| > \alpha$
$\mathcal{C}_{min\_rect}$	$\equiv$	$\|X\| > \alpha \wedge \|Y\| > \beta$
$\mathcal{C}_{area}$	$\equiv$	$\|X\| \times \|Y\| > \alpha$
$\mathcal{C}_{mean}$	$\equiv$	$\sum_{t \in X} Val^+(t)/\|X\| > \alpha$
$\mathcal{C}_{membership}$	$\equiv$	$a \in X \wedge b \in Y$
$\mathcal{C}_{emerging}$	$\equiv$	$\|X \cap E_1\| > \alpha \wedge \|X \cap E_2\| < \beta$
$\mathcal{C}_{division}$	$\equiv$	$\|X\|/\|Y\| > \alpha$

---

[1] Indeed, if we would only consider the $\mathcal{C}_{itemset}$ constraint, the only maximal itemset is the itemset containing all items.

As a further example, if the items represent the books of "Antoine De Saint-Exupery" and the transactions people who have read some of these books (a '1' value in the boolean matrix), we may want the groups of at least three people who have read at least three books in common including "The little prince". This extraction task can be declaratively defined by the means of the constraint $\mathscr{C}_{EP} = \mathscr{C}_{type} \wedge \mathscr{C}_{form}$ where $\mathscr{C}_{type} = \mathscr{C}_{fc}$ and $\mathscr{C}_{form} = \mathscr{C}_{min\_rect} \wedge \mathscr{C}_{membership}$ ($|X| > 3$, $|Y| > 3$ and "The little prince" $\in Y$).

For the development of algorithms it is important to study the properties of the constraints. Especially monotonic, anti-monotonic and convertible constraints play a key role in any combinatorial pattern mining algorithm to achieve extraction tractability.

Definition 5.2 ((Anti)-monotonic Constraints). A constraint $\mathscr{C}(X,Y)$ is said to be anti-monotonic with respect to an argument X iff $\forall X, X', Y$ such that $X \subseteq X'$: $\neg\mathscr{C}(X,Y) \implies \neg\mathscr{C}(X',Y)$. A constraint is monotonic with respect to an argument X iff $\forall X, X', Y$ such that $X \supseteq X'$: $\neg\mathscr{C}(X,Y) \implies \neg\mathscr{C}(X',Y)$ We will use the term "(anti)-monotonic" to refer to a constraint which is either monotonic or anti-monotonic.

Example 5.2. The constraint $\mathscr{C}_{size}$ is monotonic. Indeed, if a set X does not satisfy $\mathscr{C}_{size}$ then none of its subsets also satisfies it.

Some constraints are neither monotonic nor anti-monotonic, but still have good properties that can be exploited in mining algorithms. One such class is the class of "convertible constraints" that can be used to safely prune a search-space while preserving completeness.

Definition 5.3 (Convertible Constraints). A constraint is said to be convertible with respect to X iff it is not (anti)-monotonic and if there exists a total order on the domain of X such that if a pattern satisfies the constraint, then every prefix (when sorting the items along the chosen order) also satisfies it. This definition implies that whenever a pattern does not satisfy the constraint, then every other pattern with this pattern as a prefix does not satisfy the constraint either.

Example 5.3. Constraint $\mathscr{C}_{mean}$ of Figure 5.2 is a convertible constraint where $Val^+ : \mathscr{T} \to \mathscr{R}$ associates a positive real value to every transaction. Let the relation order $\leq_{conv}$ such that $\forall t_1, t_2 \in \mathscr{T}$, we have $t_1 \leq_{conv} t_2 \Leftrightarrow Val^+(t_1) \leq Val^+(t_2)$. Thus when the transaction set $\{t_i, t_j, ..., t_k\}$ ordered by $\leq_{conv}$ does not satisfy $\mathscr{C}_{mean}$ then all the ordered transaction sets $\{t_i, t_j, ..., t_l\}$ such that $t_k \leq_{conv} t_l$ do not satisfy $\mathscr{C}_{mean}$ either.

5.3 Specialized Approaches

Over the years, several algorithms have been proposed to find itemsets under constraints. In this section, we review some important ones. In the next section we will generalize these methods.

In all well-known itemset mining algorithms, it is assumed that the constraint $\mathscr{C}_{itemset}$ must be satisfied. Hence, for any given set of items Y, it is assumed we can unambiguously compute its support set X. Most were developed with unbalanced market-basket data sets in mind, in which transactions contain few items, but items can occur in many transactions. Hence, they search over the space of itemsets Y and propose methods for deriving $support(Y)$ for these itemsets.

The most famous algorithm for mining itemsets with high support is the Apriori algorithm [1]. The Apriori algorithm lists patterns increasing in itemset size. It operates by iteratively generating candidate itemsets and determining their support sets in the data. For an itemset Y candidates of size $|Y|+1$ are generated by creating sets $Y \cup \{e\}$ with $e \notin Y$. For example, from the itemset $Y = \{i_2, i_3\}$ the itemset candidate $Y' = \{i_2, i_3, i_4\}$ is generated. All candidates of a certain size are collected; the support sets of these candidates are computed by traversing the boolean matrix in one pass. Exploiting the anti-monotonicity of the size constraint, only patterns whose support set exceeds the required minimum size, are extended again, and so on.

While Apriori searches breadth-first, alternative methods, such as Eclat [29] and FPGrowth [17], traverse the search space depth-first, turning the search into an enumeration procedure which has an enumeration search tree. Each node in the enumeration tree of a depth-first algorithm corresponds to a pattern $(support(Y), Y)$. For each node in the enumeration tree we compute a triple $\langle Y, support(Y), CHILD \cup Y \rangle$, where $CHILD$ is the set of items i such that $support(Y \cup \{i\}) \geq size$. Hence, $CHILD$ contains all items i that can be added to Y such that the resulting itemset is still frequent. A child $\langle support(Y'), Y', IN' \cup Y' \rangle$ in the enumeration tree is obtained by (i) adding an item $i \in CHILD$ to Y; (ii) computing the set $X' = support(Y')$; and (iii) computing the set IN' of items i' from the items $i \in IN$ for which $support(Y' \cup \{i\}) \geq size$. The main efficiency of the depth-first algorithms derives from the fact that the sets $support(Y')$ and $CHILD$ can be computed incrementally. In our example, the support set of itemset $Y' = \{i_2, i_3, i_4\}$ can be computed from the support set of $Y = \{i_2, i_3\}$ by $support(Y') = \{t_1, t_3\} \cap support(i_4) = \{t_1, t_3\} \cap \{t_1, t_2\} = \{t_1\}$, hence only scanning $support(i_4)$ (instead of $support(i_2)$, $support(i_3)$ and $support(i_4)$). The $CHILD'$ set is incrementally computed from the $CHILD$ set, as the monotonicity of the size constraint entails that elements in $CHILD'$ must also be in $CHILD$. Compared to the level-wise Apriori algorithm, depth-first algorithms hence require less memory to store candidates.

The most well-known algorithm for finding formal concepts is Ganter's algorithm [14], which presents the first formal concept mining algorithm based on a depth-first enumeration as well as an efficient way to handle the closure constraint $\mathscr{C}_{fc}$ by enabling "jumps" between formal concepts. Each itemset is represented in the form of a boolean vector. For instance if $|\mathscr{I}| = 4$ then the pattern $(1, 0, 1, 0)$ stands for the pattern $\{i_1, i_3\}$ ("1" for presence and "0" for absence). Formal concepts are enumerated in the lexicographic order of the boolean vectors. For example with three items $\mathscr{I} = \{i_1, i_2, i_3\}$, itemsets are ordered in the following way: $\emptyset$, $\{i_3\}$,$\{i_2\}$, $\{i_2, i_3\}$, $\{i_1\}$, $\{i_1, i_3\}$, $\{i_1, i_2\}$ and $\{i_1, i_2, i_3\}$. Assume that we have given two boolean vectors $A = (a_1, \cdots, a_m)$, $B = (b_1, \cdots, b_m)$, both representing itemsets, then

Ganter defines (i) an operator $A_i^+ = (a_1, \cdots, a_{i-1}, 1, 0, \cdots, 0)$; (ii) a relation $A <_i B$ which holds iff $a_i < b_i$ and $\forall j < i : a_j = a_i$; and (iii) an operator $A \oplus i = (A_i^+)''$, where $''$ is the closure operator, which adds items included in all transactions covered by (A_i^+). These operators allow us to enumerate the formal concepts in a given order: the smallest formal concept after A is $A \oplus i$ where i is the largest integer satisfying $A \leq_i A \oplus i$. This principle is called prefix preserving closure (PPC) extension [26].

In the example of Figure 5.2, if we consider the pattern $\{i_2, i_3\}$ corresponding to the boolean vector $A = (0, 1, 1, 0)$, $i = 1$ is the largest integer satisfying $A \leq_i A \oplus i$. Thus from the itemset $\{i_2, i_3\}$ we can "jump" to the formal concept $\{i_1, i_2, i_3, i_4\} = A \oplus 1$.

A disadvantage with this method is that an itemset A and its successor B in the lexicographic order may be completely different, i.e., $A \cap B = \emptyset$. This means that there is no way to use information of a pattern (A in the example) to compute its successor (B in the example). This may be a problem when we are computing formal concepts in large boolean matrices.

This efficiency problem was addressed in algorithms that combine the frequent itemset mining problem with formal concept analysis (usually called frequent closed itemset miners). Most algorithms for mining closed itemsets borrow ideas both from Ganter's algorithm and depth-first itemset miners like LCM [26]. Similar to depth-first itemset miners, they traverse the search-space depth first, and incrementally compute both support and itemsets; to restrict the search space to formal concepts, they apply Ganter's enumeration order; the main idea is to recursively consider only those candidate itemsets which are PPC extensions.

To address the problem of extracting maximal itemset mining under constraints, the DualMiner algorithm was proposed in [9]. It extends the depth-first frequent itemset mining approach to deal with maximality and monotonic constraints on the itemsets. When we need to satisfy monotonic itemset constraints, no longer every node in the enumeration tree corresponds to a pattern; for instance, if we have the constraint $|Y| \geq 3$, all itemsets of size < 3 are no longer patterns, although we might need to traverse these nodes to reach patterns that do satisfy the constraints. The first modification proposed in DualMiner is to test for each node in the enumeration tree if $Y \cup CHILD$ satisfies the monotonic constraint on the itemset: in our example, assume that $|Y \cup CHILD| < 3$ for a certain node in the enumeration tree, then we no longer need to search further below this node, as none of the itemsets we will be creating can satisfy the monotonic constraint, as they are subsets of $Y \cup CHILD$. To speed-up the search for maximal itemsets a similar observation is used: assume that we find that $Y \cup CHILD$ satisfies the anti-monotonic constraint on the itemsets, then we can skip the enumeration of all itemsets Y' for which $Y \subseteq Y' \subset Y \cup CHILD$, as they cannot be maximal.

One way to think of this algorithm is that every node in its enumeration tree corresponds to a tuple $\langle support(\top_{\mathscr{I}}), \bot_{\mathscr{I}}, support(\bot_{\mathscr{I}}), \top_{\mathscr{I}} \rangle$ such that all patterns (X, Y) found below that node will have $\bot_{\mathscr{I}} \subseteq Y \subseteq \top_{\mathscr{I}}$ and $support(\top_{\mathscr{I}}) \subseteq X \subseteq support(\bot_{\mathscr{I}})$. The pair $\langle \bot_{\mathscr{I}}, \top_{\mathscr{I}} \rangle$ represents a search space (also called a subalgebra, or a sublattice). The set $\top_{\mathscr{I}}$ is defined to be the set $\{i \mid \mathscr{C}(support(\bot_{\mathscr{I}} \cup$

$\{i\}), \perp_{\mathscr{I}} \cup \{i\})\}$, for constraints that are monotonic in X or anti-monotonic in Y. The sets $\top_{\mathscr{I}}$ and $\perp_{\mathscr{I}}$ are used as *witnesses*: we use them to either prune an entire search tree, or to jump to a maximal solution.

This idea was generalized in [19, 25, 6]. In [19] it was shown how to compute witnesses for the more difficult "variance" constraint, a problem that remained opened for several years in the data mining community. Soulet et al. [25] extend the idea to deal with more difficult constraints such as the area constraint, which take into account both support sets and itemsets. For example, if we want to compute all the patterns (X, Y) satisfying $\mathscr{C}_{itemset}$ with an area greater than 3 ($\mathscr{C}_{area}$ where $\alpha = 3$), knowing that $\perp_{\mathscr{I}} \subseteq Y \subseteq \top_{\mathscr{I}}$ and hence $support(\top_{\mathscr{I}}) \subseteq X \subseteq support(\perp_{\mathscr{I}})$, then we can bound the area of (X, Y) by $|support(\top_{\mathscr{I}})| \times |\perp_{\mathscr{I}}| \leq |X| \times |Y| \leq |support(\perp_{\mathscr{I}})| \times |\top_{\mathscr{I}}|$. If $|support(\perp_{\mathscr{I}})| \times |\top_{\mathscr{I}}| < 4$, any pattern of the current search space has an area lower than 4, and one can safely stop considering itemsets below (X, Y). A more sophisticated extension was proposed by Bonchi et al. based on the ExAnte property [6, 7]. The ExAnte property states that if a transaction does not satisfy an easily computable monotonic itemset constraint (such as that the itemset has at least size 3), then this transaction can never be in $support(X)$ (as the transaction can only be in the support set of itemsets with less than 3 items). Hence, we can remove this transaction from consideration. The consequence of this removal is that we are reducing the support counts of all items included in the transaction, which may turn some of them infrequent. Removing infrequent items from consideration, some transactions may no longer satisfy the constraint on the itemsets, and can be removed again; this procedure can be repeated till we reach a fixed point in which both support sets and itemsets do not change any more.

Originally, this was proposed as a pre-processing procedure that can be applied on any dataset to obtain a smaller dataset [6]. However, later it was observed that we can perform such pruning in every node of the enumeration tree of a depth-first itemset miner [7]. Essentially, in every node of the enumeration tree, we have a tuple

$$\langle support(\top_{\mathscr{I}}), \perp_{\mathscr{I}}, \top_{\mathscr{T}}, \top_{\mathscr{I}}, \rangle$$

where $\top_{\mathscr{T}}$ denotes the set of transactions that can still be covered by a pattern. Compared to the DualMiner and other itemset miners, the key observation is that we also evaluate constraints on the transactions, and allow these evaluations to change the set $\top_{\mathscr{I}}$; no longer do we implicitly assume the set $\top_{\mathscr{T}}$ to equal $support(\perp_{\mathscr{I}})$.

5.4 A Generalized Algorithm

We propose to generalize the various approaches that have been sketched in the previous section. It is based on a depth-first search in which every node of the enumeration tree has a tuple

$$SP = \langle \perp, \top \rangle = \langle \perp_{\mathscr{T}} \cup \perp_{\mathscr{I}}, \top_{\mathscr{T}} \cup \top_{\mathscr{I}} \rangle,$$

Table 5.1 Skeleton of a binary backtracking algorithm

SEARCH(A search-space $\langle \bot, \top \rangle$, a data set **r** and a constraint $\mathscr{C}$ on $2^{\mathscr{T}} \times 2^{\mathscr{I}}$)

---

repeat for variables $e \in \mathscr{I} \cup \mathscr{T}$ with $e \notin \bot$ and $e \in \top$ till fixpoint: (*Propagation*)
   if $\neg Upper_{\mathscr{C}}(\langle \bot \cup \{e\}, \top \rangle)$ **then** $\langle \bot, \top \rangle \leftarrow \langle \bot, \top \setminus \{e\} \rangle$
   if $\neg Upper_{\mathscr{C}}(\langle \bot, \top \setminus \{e\} \rangle)$ **then** $\langle \bot, \top \rangle \leftarrow \langle \bot \cup \{e\}, \top \rangle$
if $Upper_{\mathscr{C}}(\langle \bot, \top \rangle)$ **then**   (*Consistency check*)
   if $\bot = \top$ **then**
      PRINT$\langle \bot, \top \rangle$ (*Solution found*)
   else
      Let $e \in \mathscr{I} \cup \mathscr{T}$ with $e \notin \bot$ and $e \in \top$
      SEARCH($\langle \bot, \top \setminus \{e\} \rangle$,**r**,$\mathscr{C}$)
      SEARCH($\langle \bot \cup \{e\}, \top \rangle$,**r**,$\mathscr{C}$)

---

such that below this node we will only find patterns $P = (X,Y)$ in which $\bot_{\mathscr{T}} \subseteq X \subseteq \top_{\mathscr{T}}$ and $\bot_{\mathscr{I}} \subseteq Y \subseteq \top_{\mathscr{I}}$. We will abbreviate this to $P \in SP$; one can say that sets $\bot_{\mathscr{T}}$ and $\top_{\mathscr{T}}$ define the domain of X, and sets $\bot_{\mathscr{I}}$ and $\top_{\mathscr{I}}$ the domain of Y.

Constraints express properties that should hold between X and Y. During the search, this means that if we change the domain for X, this can have an effect on the possible domain for Y, and the other way around. This idea of *propagating* changes in the domain of one set to the domain of another set, is essential in *constraint programming*. The general outline of the search that we wish to perform for itemset mining is given in Table 5.1.

Table 5.1 presents a skeleton of a binary depth-first search algorithm. According to the data set and the constraint $\mathscr{C}$, the domains are first of all reduced through propagation. If the domain is still consistent then the enumeration process keeps going. If a solution is found, it is printed. Otherwise the algorithm selects an element to be enumerated and generates two new nodes (with the function **Search**). Finally, the algorithm is recursively called on the two newly generated nodes. Let us now study in more details what we mean by propagation and consistency.

Definition 5.4 (Constraint Upper bounds). Let SP be the search-space and $\mathscr{C}$ a constraint over $2^{\mathscr{T}} \times 2^{\mathscr{I}}$. Then an upper-bound of $\mathscr{C}$ on SP is a predicate $Upper_{\mathscr{C}}(SP)$ such that if there exists $P \in SP$ for which $\mathscr{C}(P)$ is true, then predicate $Upper_{\mathscr{C}}(SP)$ is true; furthermore, if $\bot = \top$, it should hold that $Upper_{\mathscr{C}}(SP) = \mathscr{C}(\bot)$. Informally, if SP violates an upper-bound of $\mathscr{C}$ then SP is not consistent, i.e., no valid pattern can be generated from this search-space.

From constraint upper-bounds $Upper_{\mathscr{C}}$ and search-space lower- and upper-bounds $SP = \langle \bot, \top \rangle$, propagation can be applied thanks to the following observation:

- if an element $e \in (\top \setminus \bot)$, once added to the lower-bound of its set, violates the predicate $Upper_{\mathscr{C}}(\langle \bot \cup \{e\}, \top \rangle)$, then $\mathscr{C}$ can never be satisfied if e is included in the solution and e should be remove from the upper-bound of the set.

- if an element $e \in (\top \setminus \bot)$, once removed from the upper-bound of its set, violates the predicate $Upper_{\mathscr{C}}(\langle \bot, \top \setminus \{e\} \rangle)$, then $\mathscr{C}$ can never be satisfied if e is not included in the solution, and e should be added to the lower-bound of the set.

Our algorithm generalizes methods such as Eclat, DualMiner and ExAnte. It maintains all bounds that are also maintained in such algorithms; if a constraint $\mathscr{C}_{itemset}$ is enforced, its propagation ensures that the bounds for the transaction set correspond to the appropriate support set, as in these algorithms. The iterative application of propagation is borrowed from the ExAnte algorithm if monotonic constraints on the itemsets are used.

Even though the algorithm in Table 5.1 shows how we would like to perform the search, there are multiple ways of formalizing itemset mining problems and implementing propagation, pruning and search. We present two ways to deal with such issues in the next two sections.

5.5 A Dedicated Solver

5.5.1 Principles

Our first option is to build a dedicated, but still generic enough algorithm for itemset mining. In such a system, the key idea is that the system provides for the search, as indicated before, but the user has the ability to plug in algorithms for evaluating the constraints. These algorithms can be implemented in arbitrary programming languages, and our main problem here is to decide how the search procedure may exploit such plug in algorithms.

Let us first consider the simple case, in which we assume that we have an algorithm for evaluating a constraint $\mathscr{C}(X,Y)$ which is (anti-)monotonic in each of its parameters.

Definition 5.5 (Upperbound for (Anti-)Monotonic Constraints). Let $\mathscr{C}$ be an (anti-)monotonic constraint both on itemsets and transaction sets, i.e., the so-called bisets. The following is a valid upper bound:

$$Upper_{\mathscr{C}}(SP) = \mathscr{C}(M_1(SP), M_2(SP)),$$

where $M_1(SP)$ equals $\top_{\mathscr{T}}$ (resp. $\bot_{\mathscr{T}}$) if $\mathscr{C}$ is monotonic (resp. anti-monotonic) on the transaction set. $M_2(SP)$ is defined similarly for itemsets.

Example 5.4. The constraint $\mathscr{C}_{division}(X,Y) \equiv |X|/|Y| > \alpha$ is monotonic on X and anti-monotonic on Y with respect to the inclusion order. Indeed let $X_1 \subseteq X_2 \in \mathscr{T}$ and $Y_1 \subseteq Y_2 \in \mathscr{I}$, we have $\mathscr{C}_{division}(X_1,Y) \Rightarrow \mathscr{C}_{division}(X_2,Y)$ and $\mathscr{C}_{division}(X,Y_2) \Rightarrow \mathscr{C}_{division}(X,Y_1)$. Finally, the upperbound of $\mathscr{C}_{division}$ is $Upper_{\mathscr{C}_{division}}(SP) = |\top_{\mathscr{T}}|/|\bot_{\mathscr{I}}| > \alpha$.

It is possible to exploit such an observation to call the constraint evaluation algorithm when needed. We can generalize this to constraints which are not monotonic or anti-monotonic in each of its parameters. Let us start with the definition of a function $\mathscr{P}_{\mathscr{C}}$ which is a simple rewriting of $\mathscr{C}$.

Definition 5.6 ($\mathscr{P}_{\mathscr{C}}$). Let $\mathscr{C}(X,Y)$ be a constraint on $2^{\mathscr{I}} \times 2^{\mathscr{I}}$. We denote by $\mathscr{P}_{\mathscr{C}}$ the constraint obtained from $\mathscr{C}$ by substituting each instance of X (resp. Y) in $\mathscr{C}$ with an other parameter X_i (resp. Y_i) in the constraint $\mathscr{P}_{\mathscr{C}}$.

For example, if we want to compute bisets satisfying $\mathscr{C}_{mean}$, i.e., a mean above a threshold α on a criterion $Val^+ : \mathscr{I} \to \mathbb{R}^+$:

$$\mathscr{C}_{mean}(X) \equiv \frac{\sum_{x \in X} Val^+(x)}{|X|} > \alpha.$$

The argument X appears twice in the expression of $\mathscr{C}_{mean}$. To introduce the notion of piecewise monotonic constraint, we have to rewrite such constraints using a different argument for each occurrence of the same argument. For example, the previous constraint is rewritten as:

$$\mathscr{P}_{\mathscr{C}_{mean}}(X_1, X_2) \equiv \frac{\sum_{x \in X_1} Val^+(x)}{|X_2|} > \alpha.$$

Another example is the constraint specifying that bisets must contain a proportion of a given biset (E, F) larger than a threshold α:

$$\mathscr{C}_{intersection}(X, Y) \equiv \frac{|X \cap E| \times |Y \cap F|}{|X| \times |Y|} > \alpha.$$

This constraint is rewritten as

$$\mathscr{P}_{\mathscr{C}_{intersection}}(X_1, X_2, Y_1, Y_2) \equiv \frac{|X_1 \cap E| \times |Y_1 \cap F|}{|X_2| \times |Y_2|} > \alpha$$

We can now define the class of piecewise (anti-)monotonic constraints for which we can define an $Upper_{\mathscr{C}}$ predicate, which allows us to push the constraint in the generic algorithm:

Definition 5.7 (Piecewise (Anti)-Monotonic Constraint). A constraint $\mathscr{C}$ is piecewise (anti-)monotonic if its associated constraint $\mathscr{P}_{\mathscr{C}}$ is either monotonic or anti-monotonic on each of its arguments. We denote by $\mathscr{X}_m$ (respectively $\mathscr{Y}_m$) the set of arguments X_i (resp. Y_i) of $\mathscr{P}_{\mathscr{C}}$ for which $\mathscr{P}_{\mathscr{C}}$ is monotonic. In the same way $\mathscr{X}_{am}$ (respectively $\mathscr{Y}_{am}$) denotes the set of arguments X_i (resp. Y_i) of $\mathscr{P}_{\mathscr{C}}$ for which $\mathscr{P}_{\mathscr{C}}$ is anti-monotonic.

Example 5.5. The constraint $\mathscr{C}_{mean}(X) \equiv \sum_{i \in X} Val^+(i)/|X| > \alpha$, which is not (anti-)monotonic, is piecewise (anti-)monotonic. We can check that $\mathscr{P}_{\mathscr{C}_{mean}} \equiv \sum_{i \in X_1} Val^+(i)/|X_2| > \alpha$ is (anti-)monotonic for each of its arguments, i.e., X_1 and X_2.

We can now define upper-bounds of piece-wise (anti)-monotonic constraints. An upper-bound of a piecewise (anti)-monotonic constraint $\mathscr{C}$ is:

$$Upper_{\mathscr{C}}(SP) = \mathscr{P}_{\mathscr{C}}(P_1, \cdots, P_m),$$

where

$$P_i = \top_{\mathscr{T}} \quad \text{if } P_i \in \mathscr{X}_m$$
$$P_i = \bot_{\mathscr{T}} \quad \text{if } P_i \in \mathscr{X}_{am}$$
$$P_i = \top_{\mathscr{I}} \quad \text{if } P_i \in \mathscr{Y}_m$$
$$P_i = \bot_{\mathscr{I}} \quad \text{if } P_i \in \mathscr{Y}_{am}$$

Example 5.6. We have $\mathscr{P}_{\mathscr{C}_1}(X_1, X_2) \equiv \sum_{i \in X_1} Val^+(i)/|X_2| > \alpha$ where $\mathscr{X}_m = \{X_1\}$ and $\mathscr{X}_{am} = \{X_2\}$. Thus we obtain $Upper(\mathscr{C}_1) \equiv$

$$\sum_{i \in \top_{\mathscr{T}}} \frac{Val^+(i)}{|\bot_{\mathscr{T}}|} > \alpha.$$

For $\mathscr{P}_{\mathscr{C}_2}(X_1, X_2, Y_1) \equiv |X_1 \cup E| * |Y_1|/|X_2| > \alpha$, we have $\mathscr{X}_m = \{X_1\}$, $\mathscr{Y}_m = \{Y_1\}$ and $\mathscr{X}_{am} = \{X_2\}$. Thus an upper-bound of $\mathscr{C}_2$ is $Upper(\mathscr{C}_2) \equiv$

$$\frac{|\top_{\mathscr{T}} \cup E| * |\top_{\mathscr{I}}|}{|\bot_{\mathscr{T}}|} > \alpha.$$

In [25], the authors present a method to compute the same upper-bounds of constraints, but built from a fixed set of primitives. Notice also that [11] provides an in-depth study of piecewise (anti-)monotonicity impact when considering the more general setting of arbitrary $n-$ary relation mining.

Overall, in this system, a user would implement a predicate $\mathscr{P}_{\mathscr{C}}$, and specify for each parameter of this predicate if it is monotone or anti-monotone.

5.5.2 Case study on formal concepts and fault-tolerant patterns

Let us now illustrate the specialized approach on concrete tasks like formal concept analysis and fault-tolerant pattern mining. In the next section, the same problems will be addressed using an alternative approach.

We first show that $\mathscr{C}_{fc}$, the constraint which defines formal concepts, is a piecewise (anti)-monotonic constraint. Then, we introduce a new fault-tolerant pattern type that can be efficiently mined thanks to the proposed framework.

We can rewrite $\mathscr{C}_{fc}$ (see Section 5.2) to get $\mathscr{P}_{\mathscr{C}_{fc}}$:

$$\mathscr{P}_{\mathscr{C}_{fc}}(X_1, X_2, X_3, Y_1, Y_2, Y_3) =$$

$$\bigwedge_{t \in X_1} \bigwedge_{i \in Y_1} \mathbf{r}_{ti}$$
$$\bigwedge_{t \in \mathcal{T} \setminus X_2} \bigvee_{i \in Y_2} \neg \mathbf{r}_{ti}$$
$$\bigwedge_{i \in \mathcal{I} \setminus Y_3} \bigvee_{t \in X_3} \neg \mathbf{r}_{ti}$$

Analysing the monotonicity of $\mathscr{P}_{\mathscr{C}_{CF}}$ we can check that $\mathscr{C}_{fc}$ is a piecewise (anti)-monotonic constraint where $\mathscr{X}_m = \{X_2\}$, $\mathscr{X}_{am} = \{X_1, X_3\}$, $\mathscr{Y}_m = \{Y_3\}$ and $\mathscr{Y}_{am} = \{Y_1, Y_2\}$. Finally we can compute an upper-bound of $\mathscr{C}_{fc}$:

$Upper_{\mathscr{C}_{fc}}(SP) = \mathscr{P}_{\mathscr{C}_{CF}}(\bot_{\mathscr{T}}, \top_{\mathscr{T}}, \bot_{\mathscr{T}}, \bot_{\mathscr{I}}, \bot_{\mathscr{I}}, \top_{\mathscr{I}}) =$

$$\bigwedge_{t \in \bot_{\mathscr{T}}} \bigwedge_{i \in \bot_{\mathscr{I}}} \mathbf{r}_{ti}$$
$$\bigwedge_{t \in \mathscr{T} \setminus \top_{\mathscr{T}}} \bigvee_{i \in \bot_{\mathscr{I}}} \neg \mathbf{r}_{ti}$$
$$\bigwedge_{i \in \mathscr{I} \setminus \top_{\mathscr{I}}} \bigvee_{t \subset \bot_{\mathscr{T}}} \neg \mathbf{r}_{ti}$$

Besides the well-known and well-studied formal concepts, there is an important challenge which concerns the extraction of combinatorial fault-tolerant patterns (see, e.g., [24, 3]). The idea is to extend previous patterns to enable some false values (seen as exceptions) in patterns. We present here the pattern type introduced in [3].

Definition 5.8 (Fault-tolerant pattern). A biset (X, Y) is a fault-tolerant pattern iff it satisfies the following constraint $\mathscr{C}_{DRBS}$:

$\mathscr{L}_{\mathscr{T}}(t, Y) = |\{i \in Y \mid \neg \mathbf{r}_{ti}\}|$
$\mathscr{L}_{\mathscr{I}}(i, X) = |\{t \in X \mid \neg \mathbf{r}_{ti}\}|$

$$\mathscr{C}_{DRBS}(X, Y) \equiv \begin{cases} \bigwedge_{t \in X} \mathscr{L}_{\mathscr{T}}(t, Y) \le \alpha \\ \bigwedge_{i \in Y} \mathscr{L}_{\mathscr{I}}(i, X) \le \alpha \\ \bigwedge_{t \in \mathscr{T} \setminus X} \bigwedge_{t' \in X} \mathscr{L}_{\mathscr{T}}(t', Y) \le \mathscr{L}_{\mathscr{T}}(t, Y) \\ \bigwedge_{i \in \mathscr{I} \setminus Y} \bigwedge_{i' \in Y} \mathscr{L}_{\mathscr{I}}(i', X) \le \mathscr{L}_{\mathscr{I}}(i, X) \end{cases}$$

α stands for the maximal number of tolerated false values per row and per column in the pattern. The two last constraints ensure that elements not included in the patterns contain more false values than those included.

Example 5.7. $(t_1 t_2 t_3, i_2 i_3 i_4)$ and $(t_1 t_2 t_3 t_4, i_2 i_3)$ are examples of fault-tolerant patterns in $\mathbf{r}_1$ with $\alpha = 1$.

We now need to check whether $\mathscr{C}_{DRBS}$ can be exploited within our generic algorithm, i.e., whether it is a piece-wise anti-monotonic constraint. Applying the same principles as described before, we can compute the predicate $\mathscr{P}_{\mathscr{C}_{DRBS}}$ as following:

$\mathscr{P}_{\mathscr{C}_{DRBS}}(X_1, \cdots, X_6, Y_1, \cdots, Y_6) =$

$$\bigwedge_{t \in X_1} \mathscr{L}_{\mathscr{T}}(t, Y_1) \le \alpha$$
$$\bigwedge_{i \in Y_2} \mathscr{L}_{\mathscr{I}}(i, X_2) \le \alpha$$
$$\bigwedge_{t \in \mathscr{T} \setminus X_3} \bigwedge_{t' \in X_4} \mathscr{L}_{\mathscr{T}}(t', Y_3) \le \mathscr{L}_{\mathscr{T}}(t, Y_4)$$
$$\bigwedge_{i \in \mathscr{I} \setminus Y_5} \bigwedge_{i' \in Y_6} \mathscr{L}_{\mathscr{I}}(i', X_5) \le \mathscr{L}_{\mathscr{I}}(i, X_6)$$

According to Definition 5.7, $\mathscr{C}_{DRBS}$ is a piecewise (anti)-monotonic constraint where $\mathscr{X}_m = \{X_3, X_6\}$, $\mathscr{X}_{am} = \{X_1, X_2, X_4, X_5\}$, $\mathscr{Y}_m = \{Y_4, Y_5\}$ and $\mathscr{Y}_{am} = \{Y_1, Y_2, Y_3, Y_6\}$.

Finally we can compute an upper-bound of $\mathscr{C}_{DRBS}$:

$Upper_{\mathscr{C}_{fc}}(SP) =$

$$\bigwedge_{i \in \perp_{\mathscr{T}}} \mathscr{Z}_{\mathscr{T}}(i, \perp_{\mathscr{I}}) \leq \alpha$$
$$\bigwedge_{i \in \perp_{\mathscr{I}}} \mathscr{Z}_{\mathscr{I}}(i, \perp_{\mathscr{T}}) \leq \alpha$$
$$\bigwedge_{t \in \mathscr{T} \setminus \top_{\mathscr{T}}} \bigwedge_{t' \in \perp_{\mathscr{T}}} \mathscr{Z}_{\mathscr{T}}(t', \perp_{\mathscr{I}}) \leq \mathscr{Z}_{\mathscr{T}}(t, \top_{\mathscr{I}})$$
$$\bigwedge_{i \in \mathscr{I} \setminus \top_{\mathscr{I}}} \bigwedge_{i' \in \perp_{\mathscr{I}}} \mathscr{Z}_{\mathscr{T}}(i', \perp_{\mathscr{T}}) \leq \mathscr{Z}_{\mathscr{I}}(i, \top_{\mathscr{T}})$$

5.6 Using Constraint Programming Systems

5.6.1 Principles

An alternative approach is to require that the user specifies constraints in a *constraint programming* language. The constraint programming system is responsible for deriving bounds for the specified constraints. This approach was taken in [13], where it was shown that many itemset mining tasks can be specified using primitives that are available in off-the-shelf constraint programming systems. The essential constraints that were used are the so-called reified summation constraints:

$$(V' = 1) \Leftarrow \sum_k \alpha_k V_k \geq \theta \tag{5.1}$$

and

$$(V' = 1) \Rightarrow \sum_k \alpha_k V_k \geq \theta, \tag{5.2}$$

where V', $V_1 \ldots V_n$ are variables with domains $\{0, 1\}$ and α_k is a constant for each variable V_k within this constraint.

The essential observation in this approach is that an itemset Y can be represented by a set of boolean variables I_i where $I_i = 1$ iff item $i \in Y$. Similarly, we can represent a transaction set X using a set of boolean variables T_t where $T_t = 1$ iff $t \in X$.

For instance, the $\mathscr{C}_{itemset}$ constraint can be specified using the conjunction of the following constraints:

$$T_t = 1 \Leftrightarrow \sum_{i \in \mathscr{I}} I_i (1 - \mathbf{r}_{ti}) = 0, \qquad \text{for all} \qquad t \in \mathscr{T}$$

This constraint states that a transaction t is in the support set of an itemset if and only if all items in the itemset ($I_i = 1$) are not missing in the transaction (($1 - \mathbf{r}_{ti}) = 0$).

As it turns out, many constraint programming systems by default provide propagators for these reified summation constraints, by maintaining domains for boolean variables instead of domains for itemsets and transactions. Let $\perp_V$ denote the lowest element in the domain of a variable, and $\top_V$ denote the highest element. Then for the reified summation constraint of equation (5.2) a propagor computes whether the following condition is true:

$$\sum_{\alpha_k < 0} \alpha_k \perp_{V_k} + \sum_{\alpha_k > 0} \alpha_k \top_{V_k} < \theta;$$

if this condition holds, the sum cannot reach the desired value even in the most optimistic case, and hence the precondition $V' = 1$ cannot be true. Consequently value 1 is removed from the domain of variable V'.

In our running example, if we have transaction 3 with items $\{i_2, i_3\}$, this transaction is represented by the constraint

$$T_3 = 1 \Leftrightarrow I_1 + I_4 = 0.$$

If we set the domain of item I_1 to 1 (or, equivalently, include item i_1 in $\perp_{\mathscr{I}}$), this constraint will be false for $T_3 = 1$. Hence, the evaluation of $Upper_{\mathscr{C}_{itemset}}$ when $t_3 \in \perp_{\mathscr{I}}$ is false, and transaction t_3 will be removed from the domain of $\top_{\mathscr{I}}$.

Consequently, by formalizing the $\mathscr{C}_{itemset}$ constraint using reified implications, we achieve the propagation that we desired in our generalized approach. The search, the propagators and the evaluation of constraints are provided by the constraint programming system; however, the constraints should be specified in the constraint programming language of the system, such as the reified summation constraint.

5.6.2 Case study on formal concepts and fault-tolerant patterns

Let us reconsider the constraints proposed in Section 5.5.2, starting with the constraints that define the formal concepts. Below we show that each of these constraints can be rewritten to an equivalent reified summation constraint:

$$\bigwedge_{i \in \mathscr{I} \setminus I} \bigvee_{t \in \mathscr{T}} \neg \mathbf{r}_{ti} \Leftrightarrow (\forall i \in \mathscr{I} : (\neg I_i) \Rightarrow (\exists t : T_t \wedge \neg \mathbf{r}_{ti}))$$

$$\Leftrightarrow \left(\forall i \in \mathscr{I} : (I_i = 0) \Rightarrow \left(\sum_t T_t (1 - \mathbf{r}_{ti}) > 0 \right) \right)$$

$$\Leftrightarrow \left(\forall i \in \mathscr{I} : (I_i = 1) \Leftarrow \left(\sum_t T_t (1 - \mathbf{r}_{ti}) = 0 \right) \right)$$

$$\bigwedge_{t \in \mathscr{T} \setminus T} \bigvee_{i \in \mathscr{I}} \neg \mathbf{r}_{ti} \Leftrightarrow \left(\forall t \in \mathscr{T} : (T_t = 1) \Leftarrow \left(\sum_i I_i (1 - \mathbf{r}_{ti}) = 0 \right) \right)$$

$$\bigwedge_{t \in \mathcal{T} \backslash T} \bigwedge_{i \in \mathscr{I}} \mathbf{r}_{ti} \Leftrightarrow (\forall t \in \mathcal{T}, i \in \mathscr{I} : \neg T_t \vee \neg I_i \vee \mathbf{r}_{ti})$$

$$\Leftrightarrow (\forall t \in \mathcal{T} : \neg T_t \vee (\forall i \in \mathscr{I} : \neg I_i \vee \mathbf{r}_{ti}))$$

$$\Leftrightarrow (\forall t \in \mathcal{T} : \mathcal{T}_t \Rightarrow (\forall i \in \mathscr{I} : \neg (I_i \wedge \neg \mathbf{r}_{ti})))$$

$$\Leftrightarrow \left(\forall t \in \mathcal{T} : (T_t = 1) \Rightarrow \left(\sum_i I_i(1 - \mathbf{r}_{ti}) = 0 \right) \right)$$

This rewrite makes clear that we can also formulate the formal concept analysis problem in constraint programming systems. The bounds computed by the CP system correspond to those computed by the specialized approach, and the propagation is hence equivalent.

The second problem that we consider is that of mining fault-tolerant formal concepts. We can observe that $\mathscr{L}_{\mathcal{T}}(t, Y) = \sum_i I_i(1 - \mathbf{r}_{ti})$ and $\mathscr{L}_{\mathscr{I}}(i, X) = \sum_t T_t(1 - \mathbf{r}_{ti})$. Hence,

$$\bigwedge_{t \in X} \mathscr{L}_{\mathcal{T}}(t, Y) \leq \alpha \Leftrightarrow \left(\forall t \in \mathcal{T} : T_t = 1 \Rightarrow \sum_i I_i(1 - \mathbf{r}_{ti}) \leq \alpha \right)$$

and

$$\bigwedge_{i \in Y} \mathscr{L}_{\mathscr{I}}(i, X) \leq \alpha \Leftrightarrow \left(\forall i \in \mathscr{I} : I_i = 1 \Rightarrow \sum_t T_t(1 - \mathbf{r}_{ti}) \leq \alpha \right).$$

Note that these formulas are generalizations of the formulas that we developed for the traditional formal concept analysis, the traditional case being $\alpha = 0$.

We can also reformulate the other formulas of fault-tolerant itemset mining.

$$\bigwedge_{t \in \mathcal{T} \backslash X} \bigwedge_{t' \in X} \mathscr{L}_{\mathcal{T}}(t', Y) \leq \mathscr{L}_{\mathcal{T}}(t, Y) \Leftrightarrow$$

$$\left(\forall t, t' \in \mathcal{T} : (T_t = 0 \wedge T_{t'} = 1) \Leftrightarrow \sum_i I_i(1 - \mathbf{r}_{t'i}) \leq \sum_i I_i(1 - \mathbf{r}_{ti}) \right)$$

However, this formulation yields a number of constraints that is quadratic in the number of transactions. Additionally, it is not in the desired form with one variable on the left-hand side.

A formulation with a linear number of constraints can be obtained by further rewriting, defining an additional constraint over an additional variable $\beta_{\mathcal{T}}$:

$$\beta_{\mathcal{T}} = \max_t \sum_i I_i(1 - \mathbf{r}_{ti}) T_t.$$

This corresponds to the maximum number of 1s missing within one row of the formal concept. Then the following linear set of constraints is equivalent to the previous quadratic set:

$$\forall t \in \mathscr{T} : T_t = 1 \Leftarrow \sum_i I_i(1 - \mathbf{r}_{ti}) \leq \beta_{\mathscr{T}}.$$

As we can see, this constraint is very similar to the constraint for usual formal concepts, the main difference being that $\beta_{\mathscr{T}}$ is not a constant, but a variable whose domain needs to be computed. Most constraint programming systems provide the primitives that are required to compute the domain of $\beta_{\mathscr{T}}$.

Observe that adding the reverse implication would be redundant, given how $\beta_{\mathscr{T}}$ is defined. To enforce sufficient propagation, it may be useful to pose additional, redundant constraints, i.e., the conjunction of the following:

$$\beta'_{\mathscr{T}} = \min_t \sum_i I_i(1 - \mathbf{r}_{ti}(1 - T_t))$$

$$\forall t \in \mathscr{T} : T_t = 1 \Rightarrow \sum_i I_i(1 - \mathbf{r}_{ti}) \leq \beta'_{\mathscr{T}}.$$

This constraint considers the number of 1s missing in rows which are (certainly) not part of the formal concept. Its propagators ensure that we can also determine that certain transactions should not be covered in order to satisfy the constraints. Similarly, we can express the constraints over items. Our overall set of constraints becomes:

$$\forall t \in \mathscr{T} : T_t = 1 \Leftarrow \sum_i I_I(1 - \mathbf{r}_{ti}) \leq \beta_{\mathscr{T}}$$

$$\forall t \in \mathscr{T} : T_t = 1 \Rightarrow \sum_i I_i(1 - \mathbf{r}_{ti}) \leq \min(\alpha, \beta'_{\mathscr{T}})$$

$$\forall i \in \mathscr{I} : I_i = 1 \Leftarrow \sum_t T_t(1 - \mathbf{r}_{ti}) \leq \beta_{\mathscr{I}}$$

$$\forall i \in \mathscr{I} : I_i = 1 \Rightarrow \sum_t T_t(1 - \mathbf{r}_{ti}) \leq \min(\alpha, \beta'_{\mathscr{I}})$$

$$\beta_{\mathscr{T}} = \max_t \sum_i I_i(1 - \mathbf{r}_{ti})T_t$$

$$\beta'_{\mathscr{T}} = \min_t \sum_i I_i(1 - \mathbf{r}_{ti}(1 - T_t))$$

$$\beta_{\mathscr{I}} = \max_i \sum_t T_t(1 - \mathbf{r}_{ti})I_i$$

$$\beta'_{\mathscr{I}} = \min_i \sum_t T_t(1 - \mathbf{r}_{ti}(1 - I_i))$$

Clearly, implementing these fault tolerance constraints using CP systems requires the use of a large number of lower-level constraints, such as reified sum constraints, summation constraints, minimization constraints and maximization constraints, which need to be provided by the CP system. To add these constraints in a specialized system, they need to be implemented in the lower-level language in which the system is implemented itself.

5.7 Conclusions

Over the years many specialized constraint-based mining algorithms have been proposed, but a more general perspective is overall missing. In this work we studied the formalization of fundamental mechanisms that have been used in various itemset mining algorithms and aimed to describe high-level algorithms without any details about data structures and optimization issues.

Our guiding principles in this study were derived from the area constraint programming. Key ideas in constraint programming are declarative problem specification and constraint propagation. To allow for declarative problem specification, constraint programming systems provide users a modeling language with basic primitives such as inequalities, sums, and logic operators. For each of these primitives, the system implements propagators; propagators can be thought of as algorithms for computing how the variables in a constraint interact with each other.

Within this general framework, there are still many choices that can be made. We investigated two options. In the first option, we developed a methodology in which data mining constraints are added as basic primitives to a specialized CP system. Advantages of this approach are that users do not need to study lower level modeling primitives (such as summation constraints), that users are provided with a clear path for adding primitives to the system, and that it is possible to optimize the propagation better. To simplify the propagation for new constraints, we introduced the class of piecewise (anti-)monotonic constraints. For constraints within this class it is not needed that a new propagator is introduced in the system; it is sufficient to implement an algorithm for evaluating the constraint. This simplifies the extension of the specialized system with new constraints significantly and makes it possible to add constraints within this class in an almost declarative fashion. However, extending the system with other types of constraints is a harder task that requiring study.

The second option is to implement data mining constraints using lower level modeling primitives provided by existing CP systems. The advantage of this approach is that it is often not necessary to add new primitives to the CP system itself; it is sufficient to formalize a problem using a set of lower level modeling primitives already present in the system. It is also clear how different types of constraints can be combined and how certain non piecewise monotonic constraints can be added. The disadvantage is that it is less clear how to optimize the constraint evaluation, if needed, or how to add constraints that cannot be modeled using the existing primitives in the CP system. This means that the user may still have to implement certain constraints in a lower level programming language. Furthermore, the user needs to have a good understanding of lower level primitives available in CP systems and needs to have a good understanding of the principles of propagation, as principles such as piecewise monotonicity are not used.

Comparing these two approaches, we can conclude that both have advantages and disadvantages; it is likely to depend on the requirements of the user which one is to be preferred.

As indicated, there are several possibilities for extending this work. On the systems side, one could be interested in bridging the gap between these two approaches and build a hybrid approach that incorporates the specialized approach in a general system. This may allow to optimize the search procedure better, where needed.

On the problem specification side, we discussed here only how to apply both approaches to pattern mining. It may be of interest to study alternative problems, ranging from pattern mining problems such as graph mining, to more general problems such as clustering. In these problem settings, it is likely that a hybrid approach will be needed that combines the general approach of constraint programming with more efficient algorithms developed in recent years for specialized tasks. We hope that this work provides inspiration for the development of such future approaches.

Acknowledgements This work has been partly funded by EU contract IST-FET IQ FP6-516169, and ANR BINGO2 (MDCO 2007-2010). Most of this research has been done while J. Besson was affiliated to INSA Lyon.

References

1. R. Agrawal, H. Mannila, R. Srikant, H. Toivonen, and A. Verkamo. Fast discovery of association rules. In *Advances in Knowledge Discovery and Data Mining*, pages 307–328. AAAI Press, 1996.
2. S. Basu, I. Davidson, and K. Wagstaff. *Constrained Clustering: Advances in Algorithms, Theory and Applications*. Chapman & Hall/CRC Press, Data Mining and Knowledge Discovery Series, 2008.
3. J Besson, C. Robardet, and J-F. Boulicaut. Mining a new fault-tolerant pattern type as an alternative to formal concept discovery. In *ICCS'06: Proc. Int. Conf. on Conceptual Structures*, volume 4068 of *LNCS*. Springer, 2006.
4. S. Bistarelli and F. Bonchi. Interestingness is not a dichotomy: Introducing softness in constrained pattern mining. In *PKDD'05: Proc. 9th European Conf. on Principles and Practice of Knowledge Discovery in Databases*, volume 3721 of *LNCS*, pages 22–33. Springer, 2005.
5. Francesco Bonchi, Fosca Giannotti, Claudio Lucchese, Salvatore Orlando, Raffaele Perego, and Roberto Trasarti. A constraint-based querying system for exploratory pattern discovery. *Information Systems*, 34(1):3–27, 2009.
6. Francesco Bonchi, Fosca Giannotti, Alessio Mazzanti, and Dino Pedreschi. Adaptive constraint pushing in frequent pattern mining. In *PKDD'03: Proc. 7th European Conf. on Principles and Practice of Knowledge Discovery in Databases*, volume 2838 of *LNCS*, pages 47–58. Springer, 2003.
7. Francesco Bonchi, Fosca Giannotti, Alessio Mazzanti, and Dino Pedreschi. Examiner: Optimized level-wise frequent pattern mining with monotone constraint. In *ICDM 2003: Proc. 3rd International Conf. on Data Mining*, pages 11–18. IEEE Computer Society, 2003.
8. Jean-François Boulicaut, Luc De Raedt, and Heikki Mannila, editors. *Constraint-Based Mining and Inductive Databases*, volume 3848 of *LNCS*. Springer, 2005.
9. C. Bucila, J. E. Gehrke, D. Kifer, and W. White. Dualminer: A dual-pruning algorithm for itemsets with constraints. *Data Mining and Knowledge Discovery Journal*, 7(4):241–272, Oct. 2003.
10. Toon Calders, Christophe Rigotti, and Jean-François Boulicaut. A survey on condensed representations for frequent sets. In *Constraint-based Mining and Inductive Databases*, volume 3848 of *LNCS*, pages 64–80. Springer, 2005.

11. L. Cerf, J. Besson, C. Robardet, and J.-F. Boulicaut. Closed patterns meet *n*-ary relations. *ACM Trans. on Knowledge Discovery from Data*, 3(1), March 2009.
12. Hong Cheng, Philip S. Yu, and Jiawei Han. Ac-close: Efficiently mining approximate closed itemsets by core pattern recovery. In *ICDM*, pages 839–844, 2006.
13. Luc De Raedt, Tias Guns, and Siegfried Nijssen. Constraint programming for itemset mining. In *KDD'08: Proc. 14th ACM SIGKDD Int. Conf. on Knowledge Discovery and Data Mining*, pages 204–212, 2008.
14. B. Ganter. Two basic algorithms in concept analysis. Technical report, Germany Darmstadt : Technisch Hochschule Darmstadt, Preprint 831, 1984.
15. Bernhard Ganter, Gerd Stumme, and Rudolph Wille. *Formal Concept Analysis, Foundations and Applications*, volume 3626 of *LNCS*. Springer, 2005.
16. B. Goethals and M. J. Zaki, editors. *Frequent Itemset Mining Implementations*, volume 90. CEUR-WS.org, Melbourne, Florida, USA, December 2003.
17. Jiawei Han, Jian Pei, Yiwen Yin, and Runying Mao. Mining frequent patterns without candidate generation: A frequent-pattern tree approach. *Data Mining and Knowledge Discovery*, 8(1):53–87, 2004.
18. David J. Hand, Niall M. Adams, and Richard J. Bolton, editors. *Pattern Detection and Discovery, ESF Exploratory Workshop Proceedings*, volume 2447 of *LNCS*. Springer, 2002.
19. Daniel Kifer, Johannes E. Gehrke, Cristian Bucila, and Walker M. White. How to quickly find a witness. In *Constraint-Based Mining and Inductive Databases*, pages 216–242, 2004.
20. Jinze Liu, Susan Paulsen, Xing Sun, Wei Wang, Andrew B. Nobel, and Jan Prins. Mining approximate frequent itemsets in the presence of noise: Algorithm and analysis. In *SDM*, 2006.
21. H. Mannila and H. Toivonen. Levelwise search and borders of theories in knowledge discovery. In *Data Mining and Knowledge Discovery journal*, volume 1(3), pages 241–258. Kluwer Academic Publishers, 1997.
22. Katharina Morik, Jean-François Boulicaut, and Arno Siebes, editors. *Local Pattern Detection, International Dagstuhl Seminar Revised Selected Papers*, volume 3539 of *LNCS*. Springer, 2005.
23. Nicolas Pasquier, Yves Bastide, Rafik Taouil, and Lotfi Lakhal. Efficient mining of association rules using closed itemset lattices. *Information Systems*, 24(1):25–46, 1999.
24. Jian Pei, Anthony K. H. Tung, and Jiawei Han. Fault-tolerant frequent pattern mining: Problems and challenges. In *DMKD*. Workshop, 2001.
25. A. Soulet and B. Crémilleux. An efficient framework for mining flexible constraints. In *PaKDD'05: Pacific-Asia Conf. on Knowledge Discovery and Data Mining*, volume 3518 of *LNCS*, pages 661–671. Springer, 2005.
26. T. Uno, M. Kiyomi, and H. Arimura. Lcm ver. 2: Efficient mining algorithms for frequent/closed/maximal itemsets. In *FIMI'04, Proceedings of the IEEE ICDM Workshop on Frequent Itemset Mining Implementations*, volume 126 of *CEUR Workshop Proceedings*. CEUR-WS.org, 2004.
27. T. Uno, M. Kiyomi, and H. Arimura. LCM ver.3: collaboration of array, bitmap and prefix tree for frequent itemset mining. In *OSDM'05: Proc. 1st Int. Workshop on Open Source Data Mining*, pages 77–86. ACM Press, 2005.
28. C. Yang, U. Fayyad, and P. S. Bradley. Efficient discovery of error-tolerant frequent itemsets in high dimensions. In *SIGKDD*, pages 194–203, San Francisco, California, USA, August 2001. ACM Press.
29. Mohammed J. Zaki. Scalable algorithms for association mining. *IEEE Trans. Knowl. Data Eng.*, 12(3):372–390, 2000.
30. Mengsheng Zhang, Wei Wang, and Jinze Liu. Mining approximate order preserving clusters in the presence of noise. In *ICDE*, pages 160–168, 2008.

Chapter 6
From Local Patterns to Classification Models

Björn Bringmann, Siegfried Nijssen, and Albrecht Zimmermann

Abstract Using pattern mining techniques for building a predictive model is currently a popular topic of research. The aim of these techniques is to obtain classifiers of better predictive performance as compared to greedily constructed models, as well as to allow the construction of predictive models for data not represented in attribute-value vectors. In this chapter we provide an overview of recent techniques we developed for integrating pattern mining and classification tasks. The range of techniques spans the entire range from approaches that select relevant patterns from a previously mined set for propositionalization of the data, over inducing pattern-based rule sets, to algorithms that integrate pattern mining and model construction. We provide an overview of the algorithms which are most closely related to our approaches in order to put our techniques in a context.

6.1 Introduction

In many applications rule-based classification models are beneficial, as they are not only accurate but also interpretable. Depending on the application, these rules are *propositional*, i.e. of the kind

if	income of a customer **is** high *and* loans **is** low
then predict	the customer is good,

or *relational* or *graph based*, i.e. of the kind

if	carbon **is connected to** a nitrogen in a molecule
then predict	the molecule is active.

Björn Bringmann · Siegfried Nijssen · Albrecht Zimmermann
Department of Computer Science
Katholieke Universiteit Leuven, Celestijnenlaan 200A, 3001 Leuven, Belgium
e-mail: firstname.lastname@cs.kuleuven.be

Finding such rules is a challenge for which many algorithms have been proposed. Initially, most of these approaches were greedy [13, 28]; due to the use of heuristics, however, these algorithms may end up in a local optimum instead of a global one. In particular on structured domains, such as the example in chemoinformatics listed above, certain rules are hard to find by growing rules in small steps. For instance, in molecules, a benzene ring (a ring of 6 carbons) is an important structure with special properties; however, only after adding 6 bonds in a graph does this structure emerge. A greedy algorithm is not likely to find this structure automatically. Algorithms that are not greedy, but instead investigate a larger search space or even provide optimal solutions according to well-chosen criteria, may hence be preferable in these applications.

However, given the large search space of rules in most applications, making an exhaustive search feasible is a major challenge. To address this challenge, *pattern mining algorithms* may be useful. Pattern mining is one of the most studied topics in data mining and focuses mostly on the exhaustive enumeration of structures in databases. A pattern can be thought of as the antecedent of a rule. Even though patterns were originally studied for descriptive tasks [2], an obvious question is how to exploit them also in predictive tasks, where the aim is to exhaustively search through a space of rule antecedents.

Even once we have found a set of patterns (or rules), as mined by pattern mining techniques, these do not immediately correspond to a good classifier. The next question is how we can select and combine patterns for accurate classification models. This is the problem that we study in this chapter.

The probably most simple approach for using patterns in classification models is as follows:

- a pattern mining technique is used to find a large set of patterns;
- a new feature table is created, in which every column corresponds to a pattern, and each row corresponds to an element of the data set;
- a model is learned on this new table, where any learning algorithm can be used that can deal with binary data.

The process is illustrated in Figure 6.1. It is conceptually simple to use as the learning algorithm is treated as a black box. A deeper understanding of classification algorithms is not needed, thus making this approach very useful for non-computer scientists, such as biologists or chemists. Particularly in applications with structured data, such as (bio-)chemistry, this strategy turned out to be attractive and was among the first pattern-based classification approaches [25]. It was used successfully to classify sequences [25, 24], graph structures [24, 15, 20, 8] and also attribute-value data [14]. Its popularity today is attested by the continued use in recent publications [11, 7] as well as a workshop devoted to this topic [21].

Many extensions to the basic procedure are possible. The main complication that was already faced in the early days of pattern-based classification was the large number of patterns produced by pattern mining algorithms [27]. At the time few learning algorithms were commonly known that could deal with large feature spaces (for instance, SVMs for regression were only introduced by Vapnik in 1996, in the same year that the term 'frequent itemset' was introduced). It was necessary to reduce the number of patterns as much as possible. Two approaches can be taken to achieve this.

The direct approach, in which the classifier construction and the feature selection are combined: the selection procedure is made such that the selected patterns are directly inserted in a rule-based classifier [27].

The indirect approach, in which the feature selection is separated from the classifier construction [25]. This step is also indicated in Figure 6.1.

An advantage of the direct approach is that it no longer treats the classifier as a black box. One of the potential advantages of using patterns –interpretable classifiers– is hence not negated in this approach.

This chapter provides an overview of several recent approaches towards classification based on patterns, with a focus on methods that we proposed recently. The chapter is subdivided in sections that roughly correspond to the steps illustrated in Figure 6.1.

The first step in the process is the pattern mining step. The constraint which has traditionally been used to determine if a pattern is relevant and should be passed on

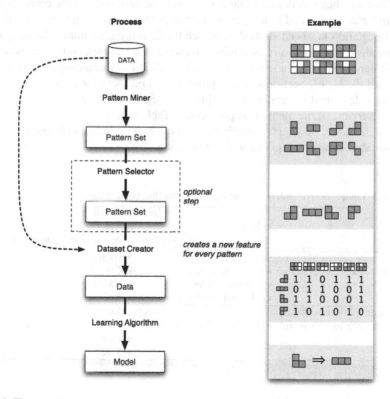

Fig. 6.1 The stepwise approach to use patterns in classification models.

to the next phase, is the minimum support constraint. If the patterns are used for classification purposes, frequent features may however not always be relevant; for instance, consider a pattern which covers all elements of a data set, and hence will never distinguish classes from each other. An alternative is to search for only those patterns which show a correlation with the target attribute. Section 6.3 discusses correlation constraints that can be applied to individual patterns to determine their relevance, and how to enforce them during mining, based on the solution proposed in [29].

Once a pattern mining technique has generated patterns, it is often the case that these patterns are correlated among each other. Section 6.4 discusses *indirect* methods for selecting a subset of useful patterns from a given set of patterns, and places two techniques that we recently proposed [35, 7] in a wider context [22, 23, 4]. These methods are independent of the classifier that is applied subsequently.

Given that patterns can be seen as rules, several algorithms have been proposed that take a set of patterns as input and use these patterns as rules to build a rule-based classification model

Subsequently, Section 6.5 discusses *direct* methods for selecting patterns from a pre-computed set of patterns. Next to traditional methods [27, 26] we summarize a method we proposed [39].

All approaches mentioned so far assume that the patterns, and the corresponding feature table, are created in a separate, first step. However, in more recent papers a tight integration has been studied, in which the learning algorithm calls the pattern mining algorithm *iteratively* as required to create new features, and the construction of a feature table is not a separate phase. In addition to discussing approaches we proposed [9, 31, 6], Section 6.6 also presents an argument for upgrading the third technique described in Section 6.4 [22] to the integrated setting.

An overview of the methods is provided in Table 6.1.

First, however, Section 6.2 briefly reviews the basic principles of pattern mining, and introduces the notation that we will be using.

Table 6.1 Comparison of pattern set mining techniques discussed in this chapter.

	Classifier Construction	
	Indirect	Direct
	(Separate from pat. selection)	*(Integrated with pat. selection)*
Separate Pattern Mining	Section 6.4	Section 6.5
(Post-processing patterns)	• Exhaustive search	• CBA
	• The chosen few	• CMAR
	• Maximally informative sets	• CtC
Integrated Pattern Mining		Section 6.6
(Iterative pattern mining)	• Maximally informative sets	• FitCare
		• DL8
		• Tree^2

6.2 Preliminaries

We assume given a data set $\mathscr{D}$ consisting of tuples $(\mathbf{x}, y)$, where $y \in \mathscr{C}$ is a class label and $\mathbf{x} \in \mathscr{A}$ is a description of an element of the data set. In the simplest case, this description is a tuple of binary attributes, that is, $\mathscr{A} = \{0, 1\}^n$; however, $\mathbf{x}$ may also be more complex, for instance, a labeled graph (V, E, Σ, λ), where V is a set of nodes, $E \subseteq V \times V$ is a set of edges, and λ is a function from V to labels in Σ.

We can partition a dataset according to the class labels; that is, for a given $c \in \mathscr{C}$, we define that $\mathscr{D}^c = \{(\mathbf{x}, y) | (\mathbf{x}, y) \in \mathscr{D}, y = c\}$.

A *pattern* π is a function from $\mathscr{A}$ to $\{0, 1\}$. There are many types of patterns. The simplest type of pattern is an *itemset*. An itemset is represented by a set $I \subseteq \{1, \dots n\}$. When an itemset is applied to a binary vector $\mathbf{x}$, it predicts 1 iff for all $i \in I$: $x_i = 1$. We usually say that pattern I is included in element $\mathbf{x}$. Also patterns can be more complex, for instance, graphs G. Usually, a graph G predicts 1 for a graph that is an element of the data set iff G is subgraph isomorphic with this graph.

Given a dataset $\mathscr{D}$ and a pattern π, by $\pi(\mathscr{D})$ we denote the set of transactions containing the pattern, i.e. $\pi(\mathscr{D}) = \{(\mathbf{x}, y) | \pi(\mathbf{x}) = 1, (\mathbf{x}, y) \in \mathscr{D}\}$.

A pattern mining algorithm is an algorithm that enumerates *all* patterns satisfying certain *constraints* within a certain *pattern language* $\mathscr{L}$. The most popular constraint is the minimum support constraint. The support of a pattern is the number of elements in a data set that includes the pattern, i.e., if π is a pattern, its support is $|\pi(\mathscr{D})|$. A pattern is *frequent* for minimum support threshold θ iff $|\pi(\mathscr{D})| \geq \theta$. Examples of pattern languages are itemsets ($\mathscr{L} = 2^{\mathscr{I}}$) and graphs ($\mathscr{L} = \{G | G \text{ is a graph}\}$).

If we associate to a pattern a class label $c \in \mathscr{C}$ we obtain a class association rule $\pi \Rightarrow c$. The support of a class association rule is defined as

$$|\{(\mathbf{x}, y) \in \mathscr{D}^c | \pi(\mathbf{x}) = 1\}|.$$

To find all patterns satisfying a constraint φ in a space of patterns $\mathscr{P}$, algorithms have been developed which traverse the search space in an efficient way. Despite the large diversity in methods that exist for many pattern domains, they have certain properties in common.

First, they all assume that there is a generality order $\sqsupseteq$ between the patterns. This order satisfies the property that

$$\forall \mathbf{x} : \pi_1(\mathbf{x}) = 1 \wedge \pi_2 \sqsubseteq \pi_1 \rightarrow \pi_2(\mathbf{x}) = 1.$$

In other words, if a pattern is included in an element of the data set, all its generalizations are also included in it.

Consequently, for the minimum support constraint we have that

$$|\pi_1(\mathscr{D})| \geq \theta \wedge \pi_2 \sqsubseteq \pi_1 \rightarrow |\pi_2(\mathscr{D})| \geq \theta$$

for any possible dataset. Any constraint which has a similar property is called *anti-monotonic*. While anti-monotonic constraints were the first ones used and exploited for efficient mining, current pattern mining algorithms exist also for non-monotonic constraints such as *average* value of items in a set or *minimum* χ^2-score.

All descendants of a pattern in the pattern generality order are called its *refinements* or *specializations*. Pattern mining algorithms search for patterns by repeatedly and exhaustively refining them, starting from the most general one(s). Refinements are not generated for patterns that do not satisfy the anti-monotonic constraint. By doing so some patterns are never generated; these patterns are *pruned*.

In the case of itemsets, the subset relation is a suitable generality order. Assuming itemsets are sets $I \subseteq \{1, \ldots, n\}$, the children of an itemset I can be generated by adding an element $1 \leq i \leq n$ to I, creating sets $I \cup \{i\}$. For the subset partial order it holds that

$$I_1 \subseteq I_2 \Rightarrow |I_1(\mathscr{D})| \geq |I_2(\mathscr{D})|.$$

Frequent itemset mining algorithms traverse the generality order either breadth-first (like APRIORI [2]) or depth-first (like ECLAT [38] and FP-GROWTH [19]). When it is found for an itemset I that $|I(\mathscr{D})| < \theta$, the children of I are not generated as all supersets can only be infrequent too.

Similar observations can be exploited for graphs, where usually the subset relation is replaced with the subgraph isomorphism relation.

Frequent pattern mining algorithms for many types of datastructures exist, among which sequences, trees, and relational queries; essential in all these algorithms is that they traverse the pattern space in such a way that the generality order is respected.

We will see in the next section how we can search for patterns under other types of constraints.

In this paper, we will often be using a set of patterns $P = \{\pi_1, \ldots, \pi_n\}$ to create a new binary dataset from a dataset $\mathscr{D} = \{(\mathbf{x}, y)\}$:

$$\{(\mathbf{z}_{\mathbf{x}}^P, y) = (\pi_1(\mathbf{x}), \ldots \pi_n(\mathbf{x}), y) \mid (\mathbf{x}, y) \in \mathscr{D}\}.$$

In this new dataset, every attribute corresponds to a pattern; an attribute has value 1 if the element of the data set includes the pattern.

6.3 Correlated Patterns

In traditional frequent pattern mining the class labels, if present, are ignored. We will refer to a pattern mining technique which takes the class labels into account as a *correlated pattern mining* technique, but many alternative names have also been proposed, among others *emerging pattern mining* [16], *contrast pattern mining* [3] and *discriminative pattern mining* [10, 11] and as subgroup discovery in the first paper in [37]. We will use the notation we proposed in [33].

In statistics, correlation describes a linear dependency between two variables; in our case the class value and the occurrence of a pattern. While in theory more than two possible class values could be handeled, we will restrict ourselves to the binary setting. The aim of correlated pattern mining is to extract patterns π from a data set whose occurrences are *significantly* correlated with the class value y. The main motivation for this approach is that patterns which do not have significant correlation with the target attribute, are not likely to be useful as features for classifiers.

Correlated pattern mining techniques use correlation scores that are computed from contingency tables. Every transaction is either covered by a pattern or not; furthermore, the transaction is either positive or negative. This gives us four disjunct possibilities for each transaction, which can be denoted in a 2×2 contingency table such as Table 6.2. Statistical measures are employed to calculate a correlation score between the class value and the pattern at hand from the number of transactions for each of the four possibilities. Correlated pattern mining aims for the extraction of

Table 6.2 Example of a contingency table.

	Covered by pattern	Not covered by pattern	
Positive example	$p = \lvert \pi(\mathscr{D}^+) \rvert$	$P - p = \lvert \mathscr{D} \setminus \pi(\mathscr{D}^+) \rvert$	P
Negative example	$n = \lvert \pi(\mathscr{D}^-) \rvert$	$N - n = \lvert \mathscr{D} \setminus \pi(\mathscr{D}^-) \rvert$	N
	$p + n$	$P + N - p - n$	$N + P$

patterns that have a high correlation with the target attribute. There are two settings that have been considered:

- find all patterns that reach at least a user-defined score;
- find the k patterns scoring best on the given data set.

We will first provide on how to compute patterns in these settings, before discussing the advantages and disadvantages.

6.3.1 Upper Bound

Obviously it is not efficient to enumerate all possible patterns to find the desired subset. As noted before, the property allowing for an efficient enumeration of all frequent patterns is the anti-monotonicity of the frequency constraint. In correlated pattern mining, we would like to exploit a similar anti-monotonic constraint. Unfortunately, interesting correlation scores are usually not anti-monotonic. A correlation score should at least have the following properties:

(A) a pattern which covers negative and positive elements of the data set in equal proportion as occurs in the full data should not be considered a correlated pattern;
(B) a pattern which covers many elements of one class, and none of the other, should be considered correlated.

As a pattern of type (B) can be a refinement of a pattern of type (A) useful correlation measures cannot be anti-monotonic. Hence,

$$\pi \sqsubseteq \pi' \not\Rightarrow score(\pi) \geq score(\pi')$$

The main approaches proposed to solve this problem involve introducing an *upper bound*. An upper bound is an anti-monotonic measure which bounds the correlation values that refinements of a pattern can reach. Hence, an upper bound allows for an efficient search;

$$\pi \sqsubseteq \pi' \Rightarrow bound(\pi) \geq bound(\pi') \; with \; bound(\pi) \geq score(\pi')$$

The bound allows us to find all correlated patterns exceeding a minimum correlation score in a similar way as we can find frequent patterns; i.e. branches of the search tree are pruned using the anti-monotonic bound constraint, instead of the minimum support constraint. In contrast to frequent pattern mining, not all patterns that exceed the score are output. Only patterns that exceed the threshold on the original score function are part of the result.

6.3.2 Top-k Correlated Pattern Mining

In order to search for the top-k patterns, a naïve approach would be to simply select these patterns in a post-processing step for a given (low) threshold. However, an important question is if we can find the top-k patterns more efficiently if we search for top-k patterns during the mining step.

It turns out that the bounds discussed in the previous section can be used relatively easily to find the top-k patterns. The main idea is to update the correlation treshold θ during the search starting from the lowest possible threshold. A temporary list of the top-k patterns is maintained during the search. Every time a enumerated pattern π exceeds the score of the worst scoring pattern in the top-k list, this newly found pattern π is inserted into the list, which is cropped to the current top-k patterns found. As long as the length of the list is still below the desired k, patterns are simply added. Once the list reaches the desired size of k patterns the threshold used for pruning is always updated to the score of the worst pattern in the list. This threshold is used as a constraint on the bound value of patterns. Thus, while the mining process goes on, the pruning-threshold continues to rise and consequently improves the pruning-power. The final list contains the top-k patterns.[1]

Note that due to the changing threshold the search strategy can influence the enumerated candidates in the top-k setting. This has been investigated and exploited in e.g. [8].

Both constraints – the top-k and the minimum score – can be combined by setting the initial threshold to a user defined threshold. As a result, the patterns extracted will be the k best scoring patterns exceeding the initial user-defined threshold.

---

[1] The list can be shorter if there were less than k patterns exceeding the lowest threshold.

6.3.3 Correlation Measures

As said before, in correlated pattern mining we need a non-trivial upper bound to allow for pruning[2]. This upper bound should be as tight as possible to achieve the maximum amount of pruning.

In this chapter, we will use the PN-space based methodology of [33] to introduce the bounds. The use of bounds in a correlated pattern mining setting was first proposed in [3]; however, we present the bounds introduced in [29] as they are more tight.

As stated earlier, instances in the data set are associated with class labels. We consider a binary class setting with positive labeled instances $\mathscr{D}^+$ and negative labeled instances $\mathscr{D}^-$ such that $\mathscr{D}^+ \cup \mathscr{D}^- = \mathscr{D}$ while $\mathscr{D}^+$ and $\mathscr{D}^-$ are disjunct.

Any pattern π covers $p = |\pi(\mathscr{D}^+)|$ of the positive instances and $n = |\pi(\mathscr{D}^-)|$ of the negative instances and can be represented as point in a *pn*-space, as illustrated in Figure 6.2. Accordingly the total frequency of the pattern π is $|\pi(\mathscr{D})| = |\pi(\mathscr{D}^+)| + |\pi(\mathscr{D}^-)|$. The upper bound has to be defined such that for any specialisation $\pi' \sqsupseteq \pi$ the inequality $bound_{\mathscr{D}}(\pi) \geq score_{\mathscr{D}}(\pi')$ holds, and the bound is anti-monotonic. Due to the anti-monotonicity any specialisation $\pi' \sqsupseteq \pi$ can only cover a subset of π such that $\pi'(\mathscr{D}^+) \subseteq \pi(\mathscr{D}^+)$ and $\pi'(\mathscr{D}^-) \subseteq \pi(\mathscr{D}^-)$. Therefore any π' can only reach values located in the grey area defined by π (Figure 6.2). As a result we need to define the upper bound to be greater or equal to the maximum score over all $p' \leq p = |\pi(\mathscr{D}^+)|$ and $n' \leq n = |\pi(\mathscr{D}^-)|$,

$$bound_{\mathscr{D}}(p,n) \geq \max_{p' \leq p, n' \leq n} score_{\mathscr{D}}(p',n')$$

If the correlation measure is *convex*, the calculation of the upper bound is straightforward. Convex functions are known to reach their extreme values at the borders. Thus, only the scores of these borders have to be computed - the maximum of which specifies the upper bound.

Morishita and Sese [29] introduce the general idea and discuss two popular and frequently used convex functions, χ^2 and *information gain (IG)*. While in these cases convexity is exploited, in general, any function that allows for an efficient cal-

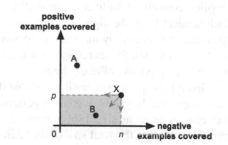

Fig. 6.2 The *pn-space* with patterns *a*, *b*, and *x* covering different parts of the data set.

---

[2] There is always a trivial upper bound that does not allow for any pruning and thus is worthless

culation of an upper bound in a restricted domain can be used to guide the search [5]. One example is the non-convex Hellinger distance. As it still reaches its extrema at the borders, the upper bound can be calculated in the very same way as for χ^2 or Information Gain.

Next to the technique introduced above, other methods to extract correlated patterns based on 'pure' frequent-pattern mining have been proposed. [33] shows that separate mining on each class with a threshold derived from the desired minimum correlation and post-processing can be done efficiently to obtain the desired result. Alternatively, given a threshold on correlation, the minimum of the class-specific support thresholds can be used as a support threshold on the entire dataset. This approach was essentially proposed in [10, 11]; note however that this class-ignorant pruning strategy is suboptimal compared to the approach which takes class labels into account.

6.3.4 Type I Errors

All methods discussed evaluate a correlation score for every pattern. For instance, we used χ^2 to evaluate how much each pattern correlates with a class attribute. In statistics a common recommendation is that if the χ^2 value exceeds 3.84, the correlation is significant. Does this mean that it is always correct to use 3.84 as threshold?

Unfortunately, this is not the case. In statistics it is well-known that if we repeatedly evaluate a significance test, chances are very high that we will incorrectly reject the null hypotheses in some of these tests. This kind of error is called a *type I error*. Given the large number of patterns considered, type I errors are very likely in pattern mining. How to deal with them is an issue, which has not received much attention.

A common way to deal with type I errors is the *direct adjustment approach*, which modifies the minimum correlation score depending on the number of hypotheses evaluated: the more hypotheses considered, the higher the correlation threshold should be.

In bound-based pattern mining, it is however not clear how this method should be applied correctly. Intuitively, one would say that the choice of threshold should be independent from the algorithm used, and hence, independent from the number of patterns enumerated by the search process of the algorithm. However, normalizing by the total size of the search space is not always an option, as some pattern domains (such as graphs) are infinitely large.

Given these problems, until now most direct adjustment approaches take a practical approach. In [3] the score of a pattern is normalized by the number of frequent patterns of the same size; [36] presents an attempt to make this approach more correct by including the total size of the search space; this approach is only applicable on itemset domains however.

An alternative approach is the *holdout* approach. In this case, the dataset is split in two parts. A large set of patterns is first mined in the first part; this set of patterns is evaluated in the second part to determine which are relevant. The advantage is that it is justified in this case to normalize the correlation threshold by the number of patterns returned in the first phase. The approach, however, is not applicable in direct top-*k* pattern mining [36].

6.3.5 Closed and Free Pattern Mining

Pattern sets can often be redundant (see Section 6.4). One such redundancy is caused by considering patterns that are neither closed nor free. The principles of closedness and freeness were originally proposed for unlabeled data sets, in which case a pattern π was called *closed* iff there was no pattern $\pi' \sqsupset \pi$ such that $\pi(\mathscr{D}) = \pi'(\mathscr{D})$; a pattern was called *free* iff there was no pattern $\pi' \sqsubset \pi$ such that $\pi(\mathscr{D}) = \pi'(\mathscr{D})$. Restricting the set of patterns to closed or free ones can often already reduce the number of patterns under consideration significantly and is hence one of the most common approaches to restricting the initial set of patterns.

The extension of this principle to labeled data is possible in multiple ways. The most straightforward approach is to ignore the class labels. However, assume we have two patterns with these properties:

$$|\pi(\mathscr{D}^+)| = |\pi'(\mathscr{D}^+)|, \ |\pi(\mathscr{D}^-)| < |\pi'(\mathscr{D}^-)| \ \text{ and } \ |\pi(\mathscr{D}^-)|/|\mathscr{D}^-| < |\pi(\mathscr{D}^+)|/|\mathscr{D}^+|,$$

where the last condition states that π is more correlated with the positive class than with the negative, then any sensible correlation score would determine that π' is less correlated. This shows that one can also consider a pattern uninteresting if it is closed or free on only one of the two classes. An approach for finding closed patterns, given support thresholds for two classes, was studied in [18]. An alternative approach is to perform top-*k* free pattern mining [8], essentially by not inserting patterns in the queue of the search procedure which are not free compared to the patterns already in the queue.

6.4 Finding Pattern Sets

The preceding section did not only cover top k-correlated pattern mining, but also algorithms to derive all correlated patterns. As we have argued in the introduction, in classifier building an optional second step is often useful in which a set of patterns is filtered. The main goals of this filtering are to remove redundancy, and to bring the potentially large amount of patterns down to a reasonable number. Achieving these goals will benefit both machine learning techniques that use the patterns either directly or indirectly (through data propositionalization) and human users that want to inspect the patterns or the models created from said patterns. For machine learning techniques, a large amount of redundant features means that selecting the relevant features in the process of building a model becomes harder, not to mention the increased computational complexity that goes with it. For humans there is simply the question of human perception – no one can be expected to make sense of hundreds or even thousands of patterns, especially if some of them are largely redundant. In this section we focus on stand-alone techniques that have been proposed for selecting patterns from a set. Techniques which directly use patterns as rules in classifiers are discussed in Section 6.5.

6.4.1 Constrained Pattern Set Mining

There is actually a straightforward way of selecting subsets of (potentially correlating) patterns from an existing large set of mined patterns: search for subsets under constraints as done in pattern mining algorithms. In the same way as *itemsets* can be enumerated in a principled way, *pattern sets* can be assembled. Of these sets, only those are kept that satisfy certain properties, e.g. regarding the redundancy among patterns in the set, the size of the pattern set, or in- or exclusion of particular patterns.

This idea was formalized and the framework and preliminary results were published in [35]. The main insight lies in the fact that the problem of pattern set mining is dual to the problem of pattern mining: while *all* items of an itemset have to occur in an element of the data set to *match*, for pattern sets the most intuitive interpretation is to match an element of the data set if *any* member pattern of the pattern set occurs in the element. This duality leads to an exact reversal of the direction of the (anti-)monotonicity property, which means that the pruning strategies behind the pattern mining algorithms can also be applied in pattern set mining. For instance, the maximum frequency constraint (which is monotonic in itemset mining), is anti-monotonic in pattern set mining, and can be used to prune the search space when growing pattern sets. Useful constraints for pattern set mining were defined, such as *pairwise redundancy* constraints, which allow the user to effectively control the level of redundancy that holds in the data set and together with size constraints allows the selection of compact, informative pattern subsets. These constraints also fulfill the anti-monotonicity property. Finally, the dual nature of the problem allows for the *lower* bounding of interestingness measures such as χ^2 and the effective upper bounding of accuracy, something that is not possible with individual patterns.

The experimental results reported in [35] show the effectiveness of the approach, demonstrated in pattern selection for classifier building, but also the limitations in that pattern set mining faces quite a few of the challenges that local pattern mining encounters. Specifically, depending on the selectiveness of the constraints, many pattern sets satisfy them and enumerating them can quickly exhaust computational resources, given the large amount of patterns forming the basic elements of the language. These observations point towards two promising avenues: 1) it is often not necessary to return *all* pattern sets, a view corresponding to top-k mining for local patterns. Also, 2) the set returned does not have to be the optimal one if optimality can be traded off against efficiency in a reasonable manner. The following two approaches follow these directions.

6.4.2 The Chosen Few

The method we introduced in [7] focusses on optimizing a single pattern set heuristically. The main point of this work is that if two pattern sets partition the data in the same manner, the smaller one of those is considered preferable, carrying the same information as the larger one while being easier to peruse by humans and easier to process by machine learning techniques. Given a set of patterns, instances from which they have been mined can be described in terms of the patterns in the set that are present in the instance and in terms of those are absent. Instances agreeing on all patterns' presence form an equivalence class or block; the set of all blocks makes up the partition on the data. In terms of machine learning techniques, for instance, *all* instances from a particular block appear *equal* to the algorithm since they are encoded in an indistinguishable manner.

Given a set of patterns, and the partition it induces, adding a new pattern to the set can either increase the number of blocks or leave it unchanged. In the latter case, this pattern is clearly redundant, since it can be expressed either by another pattern, by its complement, or by a combination of patterns. In processing the full result set of a local pattern mining operation, it can hence be useful to reject patterns that do not change the partition, whittling the pattern set down to a smaller number of patterns carrying the same amount of information. While it could be argued that there is information which is not recovered by the partition alone, this is not true for all machine learning techniques; in principle, if this is useful, many machine learning techiques could deduce one feature from the others. While the *minimal* number of patterns needed is logarithmic in the number of blocks in the partition, typically more patterns are *actually* needed. Deciding on the minimal set of patterns needed to induce the same partition as the original result set is computationally rather expensive, however, setting the stage for the application of heuristic techniques, for which we developed two alternative approaches.

The first of the two algorithms, BOUNCER, uses a user-defined order and considers each pattern exactly once for potential inclusion. This order is augmented by a measure which evaluates the contribution of a pattern in terms of the granularity of the partition. Using a threshold on the minimal contribution a pattern has to make, the first pattern encountered that exceeds this threshold is added to the set, before patterns that appear later in the order are evaluated further. The combination of the order, the selection of the first pattern for inclusion and the, necessarily, local quality measures allows the efficient mining of locally optimal pattern sets. Notwithstanding their local optimality, the resulting pattern sets were shown to improve on the set of all patterns in terms of utility as features for classification. The size and quality of the pattern sets is strongly influenced by the order and the measure used, however.

Table 6.3 A data set of 10 elements with the coverage of four different patterns.

Elements										
π_1										
π_2										
π_3 Ê										
π_4										

Heuristically optimizing a pattern set will of course have drawbacks that balance the faster execution. Consider the example in Table 6.3: ▪ has an joint entropy of 1, the highest entropy for a single pattern, and would therefore be chosen first. The joint entropy of the set {▪, ▪, ▪} is 2.72193, however, the highest possible with three patterns, and this does not include the highest-scoring individual pattern at all.

While the adoption of a global optimality criterion would prove to be impossible for non-exhaustive search, the heuristic nature of BOUNCER can be improved by changing the used order and the selection of the pattern used for inclusion. This is implemented in the PICKER* algorithm, in which an upper bound on the contribution of individual patterns to the set is calculated and the patterns ordered in descending value of this upper bound. By traversing and evaluating patterns until none of the remaining upper bound values exceeds the contribution of the currently best pattern anymore, a closer approximation of global optimality can be expected. Recalculating the upper bounds and reordering the patterns potentially leads to several evaluations of each individual pattern, as opposed to BOUNCER's approach. The resulting pattern sets do not show the fluctuation in cardinality that different orders caused and in most cases improve on the quality of pattern sets mined by using a user-defined order. Both techniques are able to handle a far larger amount of patterns than the complete method introduced in the preceding section.

6.4.3 Turning Pattern Sets Maximally Informative by Post-Processing

While the approaches in the previous section greedily compute a set of patterns aimed to be diverse, no global optimization criterion is used to measure this diversity. In [22, 23, 4] measures were studied which can be used to measure diversity of a set of patterns. Nevertheless, the BOUNCER algorithm is related to algorithms which optimize under such global measures.

We can distinguish two types of measures. First, there are measures that do not take into account the class labels, such as joint entropy [22]. By picking a set of patterns P we can construct a new representation for the original data, in which for every element $(\mathbf{x}, y)$ we have a new binary feature vector $\mathbf{z}_{\mathbf{x}}^{P}$ (see Section 6.2).

Ideally, in this new representation we can still distinguish two different elements from each other by having different feature vectors. However, in case this is not possible, one may prefer a set of patterns in which any two elements are not likely to have the same feature vector. One way to measure this is by using joint entropy,

$$H(P) = \sum_{\mathbf{b} \in \{0,1\}^{|P|}} -p(\mathbf{b}) \log p(\mathbf{b}),$$

where

$$p(\mathbf{b}) = |\{(\mathbf{x},y)|\mathbf{z}_{\mathbf{x}}^P = \mathbf{b}\}|/|\mathscr{D}|.$$

Hence $p(\mathbf{b})$ denotes the fraction of elements of the data set that have a certain feature vector $\mathbf{b}$ once we have chosen a set of patterns P, and we consider such fractions for all possible feature vectors. Joint entropy has desirable properties for measuring diversity: the larger the number of vectors $\mathbf{b}$ occurring in the data, and the more balanced they occur, the higher the entropy value is. When the entropy is maximized, the patterns are chosen such that elements of the data set are maximally distinguishable from each other.

Another class of measures are the supervised measures. An example of such a measure is [4]:

$$Q(P) = |\{(\mathbf{x},\mathbf{x}')|(\mathbf{x},y) \in \mathscr{D}^+, (\mathbf{x}',y') \in \mathscr{D}^- : \mathbf{z}_{\mathbf{x}}^P = \mathbf{z}_{\mathbf{x}'}^P\}|.$$

This measure calculates the number of pairs of elements in different classes that have the same feature vector when patterns in P are used to build feature vectors.

A set of k patterns that maximizes a global measure is called a maximal informative k-pattern set. To compute such a pattern set, two kinds of approaches have been studied:

- complete approaches [22], which enumerate the space of subsets up to size k, possibly pruning some branches if a bound allows to decide that no solutions can be found. Such an algorithm would be capable of finding the set $\{\blacksquare, \blacksquare, \blacksquare\}$ from Example 6.3.
- a greedy approach, which iteratively adds the pattern that improves the criterion most, and stops when the desired pattern set size is reached [22, 4].

The first approach is very similar to the pattern set mining approach of Section 6.4, while the second approach is similar to the approach of Section 6.4.2. Indeed, one can show that the pattern that is added to a pattern set in each iteration by the PICKER* algorithm is within bounded distance from the pattern chosen by an entropy measure when using difference in entropy as distance measure.

One could wonder how well some of these greedy algorithms are performing: how close to the optimum do they get? In this regard, an interesting property of some optimization criteria is *submodularity*. A criterion $F(P)$ is submodular iff for all $P' \subseteq P$ and patterns π it holds that

$$F(P' \cup \{\pi\}) - F(P') \geq F(P \cup \{\pi\}) - F(P),$$

in other words: there are diminished returns when a pattern is inserted later in a set. For a submodular criterion, it can be shown that the greedy algorithm (operating similar to BOUNCER) is approximates the true optimum: let P_{opt} be the optimal pattern set of size k, then the greedy algorithm will find a set for which $F(P) \geq (1 - \frac{1}{e})F(P_{opt}) \approx 0.63F(P_{opt})$ [30]. Both $Q(P)$ and $H(P)$ are submodular, and hence the greedy algorithm achieves provably good results. Finally, as BOUNCER approximates the choices made by a greedy algorithm that uses entropy as an optimization criterion, also BOUNCER is guaranteed to approximate a global optimum under the joint entropy criterion.

6.5 Direct Predictions from Patterns

Most of the methods discussed in the previous section can be seen as feature selection methods. They ignore the fact that features are actually patterns, and do not construct classifiers. In this section, we study techniques that construct a classifier while taking into account that the used components are in fact patterns. These algorithms are *rule-based*, combining rules of the form *pattern* $\Rightarrow$ *class-label*.

The technique of *associative classification* was first proposed in the work introducing the CBA algorithm [27] in 1998, quickly followed by the CMAR [26] approach in 2001 that extended both the pattern selection step and the actual classification model of CBA (and arguably improved on them). We will therefore discuss these approaches first.

A potential limitation of both of these approaches was however that they were centered on the classical minimum-support, minimum-confidence framework. In recent years, a consensus has developed that these patterns are not necessarily best suited to the task of classification. We developed a new method, called CTC [39], which uses other measures as a starting point. We will discuss it in the third part of this section and elaborate on the differences.

6.5.1 CBA

As mentioned above, CBA was the first algorithm to use the minimum-support, minimum-confidence association rule framework for constructing classifiers, coining the term *associative classification*. The main difference to traditional rule-based machine learning approaches lies in that first a large set of reasonably accurate rules are mined from the data, and in a second step a subset of those is selected that forms the final classifier. The mining step itself is performed using the well-known APRIORI algorithm.

Table 6.4 CBA/CMAR illustrative example

Elements	▦	▦	▦	▦		▦	▦
π_1		▦	▦				
π_2		▦	▦	▦	▦	▦	▦
π_3			▦			▦	▦

In CBA, patterns are used to build class association rules $\pi \Rightarrow c$, and patterns π need to satisfy a minimum relative support constraint $support_{rel}(\pi \Rightarrow c) = \frac{|\pi(\mathcal{D}^c)|}{|\mathcal{D}|} \geq \theta_s$ and a minimum *confidence* constraint $confidence(\pi \Rightarrow c) = \frac{|\pi(\mathcal{D}^c)|}{|\pi(\mathcal{D})|} \geq \theta_c$.

The resulting set of all *class association rules* (CARs) that satisfy the constraints, which we will refer to as $\mathbb{S}$, is then ordered according to a $<_{CBA}$ relation. Given two CARs $car_1 : \pi_1 \Rightarrow c_1$, $car_2 : \pi_2 \Rightarrow c_2$, the relation $<_{CBA}$ between those two rules is

1. Let w.l.o.g. $confidence(car_1) > confidence(car_2)$ then $car_1 <_{CBA} car_2$
2. If $confidence(car_1) = confidence(car_2)$, let w.l.o.g. $support_{rel}(car_1) > support_{rel}(car_2)$ then $car_1 <_{CBA} car_2$
3. If $confidence(car_1 = confidence(car_2)$, and $support_{rel}(car_1) = support_{rel}(car_2)$, let w.l.o.g. $|\pi_1| < |\pi_2|$ then $car_1 <_{CBA} car_2$

The last check holds since both π_1 and π_2 are simply sets of items whose cardinality can be measured. If even the last check fails, the tie is broken arbitrarily. The order used by CBA can thus be summarized as "higher confidence is preferable", "in case of equal confidence, higher support is preferable", and "all things being equal, shorter patterns are preferable".

Using this order, $\mathbb{S}$ is turned into an *ordered* set. Starting from the minimal rule according to this order, rather similar to the kind of pattern selection encountered in Section 6.4.2, $\mathbb{S}$ is traversed and each rule in turn considered for inclusion in the final classifier. For each rule, all elements in the data set it matches are collected, and it is evaluated whether the rule predicts at least one of those elements' class label correctly. If it does, it is included in the final classifier and all covered elements are discarded; if not, the rule is discarded. For classifying an unlabeled element, the minimal rule according to $<_{CBA}$ is used to predict its class label.

To illustrate this, consider the small example in Table 6.4: ▦ $\Rightarrow$ *Dark* predicts the dark class with a confidence of 1.0 and is therefore ranked first. ▦ $\Rightarrow$ *Dark* and ▦ $\Rightarrow$ *Light* have the same confidence (0.66) albeit for different classes, and since ▦'s support is higher, it is ranked before ▦.

CBA will select ▦ as the highest-ranked pattern and remove the elements of the data set it covers (▦ and ▦) from future consideration. The second pattern ▦ still covers and correctly predicts elements of the data set and thus gets selected, removing *all* elements it covers, i.e. all remaining elements. Since this leaves no elements that ▦ could predict correctly, it is discarded. This also illustrates one of the weaknesses of the CBA approach: after removing ▦ and ▦, ▦ in fact shows a confidence of 0.5 on the remaining elements and ▥ should be selected instead of it. Since CBA estimates support and confidence only once – on the training data – however, it makes a sub-optimal decision.

The usual setting for the support threshold θ_s is 0.01, and for the minimum confidence θ_c 0.5. While the second threshold can be justified as only accepting rules that predict their class label more often than not, the first threshold is somewhat arbitrary (and has been shown empirically not to give the best results) [12]. It also has to be pointed out that the selection and classification techniques are rather ad hoc, one-shot techniques. On the other hand, the resulting classifier uses an easily interpretable model – a list of rules ordered according to easily understandable criteria: high confidence characterizes rules that are usually correct in their prediction, high support means that they can be expected not to describe spurious phenomena, and short rules adhere to the principle of *Occam's razor*.

6.5.2 CMAR

CMAR [26] attempted to improve especially on CBA's ad hoc aspects, as well as somewhat on assembling the set of rules who are considered for inclusion in the classifier in the first place. The mining of CARs is performed essentially in the same way. Rules are also ordered according to $<_{CBA}$, but since confidence alone can be a misleading quality measure for rules, CMAR uses the χ^2 statistic to discard rules that do not correlate positively. To give an intuition what this means, consider ▤ from Table 6.4 which covers *all* elements. While this pattern satisfies the minimum confidence constraint, its χ^2-score is 0, denoting that there is actually no correlation with the target class.

The same database coverage approach that was used in CBA is also used in CMAR. There is a notable difference however: instead of removing an element once it is covered by a *single* rule that was included in the final classifier, there has to be more than one such rule. This would be expected to make classification of unseen elements of the data set more reliable since not only one rule would match it. It is suggested in [26] that four rules have to cover an element before it is removed from the data set – there is however no discussion of why this would be a suitable threshold value.

Let us revisit the CBA example (Table 6.4). If the database coverage threshold is set to 2 then selecting ▥▥ does not lead to the exclusion of ▥. More important, however, is that, as mentioned above, ▤ would be discarded before the database coverage pruning even commences.

The second difference lies in the actual classification process. Instead of using the minimal rule according to $<_{CBA}$, the order is discarded, and for each unseen element all the rules are collected that cover it, which should lead to the more reliable classification mentioned above. Additionally, if rules disagree on the class label to be assigned, the impact that each particular rule has on the final decision is based on the rule's quality, measured by trading off its actual χ^2 value against the "maximal" one it could have attained. This weighted voting strategy is however once again chosen ad hoc – based on empirical performance as the authors admit. So for the sake of classification robustness, CMAR replaces CBA's model with a more complex one, albeit still based on confident, frequent rules. The threshold values used for mining are the same as in CBA and also the same for *all* rules mined.

6.5.3 CTC

Considering the order that is imposed on patterns (rules) in both the CBA and CMAR techniques, a notion of importance or desirability emerges: high confidence is valued, leading to rules that are probably useful for classification, but so is large support, leading to rules that can be expected to hold not only on the training data.

Also, as seen in the case of the CMAR approach, a pattern having (relatively) high confidence in connection with a class label does not necessarily correlate positively with the class, if said class label is rather frequent in the first place. Similarly, a frequent rule does not automatically translate into a significant one, especially if the class predicted is the majority class. In CMAR, found rules were subjected to evaluation by the χ^2 statistic and only those accepted that correlated positively.

The χ^2-statistic trades off support against confidence, so to speak, valuing less confident rules highly, if they have only enough support, combining in this way the two criteria of importance expressed in the order used by CBA. An interesting question arising at this point is "Why use minimum-support, minimum-confidence rules at all?", especially if they are not used in a winner-takes-all way, as in CBA, but by weighted voting. Using the principles explained in Section 6.3 it is easily possible to *directly* mine strongly class-correlating predictive patterns, without the detour of mining (and pruning) frequent patterns and assessing their significance *after* enumerating them. The CTC [39] – correlating tree patterns for classification – approach does just that, using the pattern language of labeled rooted trees[3], mining χ^2-quantified patterns instead of ad hoc decided-upon support and confidence thresholds.

There are several important differences between CBA and CMAR on the one hand, and CTC on the other hand.

In CBA and CMAR, first *all* rules are mined that satisfy certain minimum thresholds, and then the database coverage step is used to select patterns. CTC combines these two steps. To achieve this, an order similar to $<_{CBA}$ is used, with a slight change in significance measure:

1. Let w.l.o.g. $\chi^2_{\mathscr{D}}(|\pi_1(\mathscr{D}^+)|, |\pi_1(\mathscr{D}^-)|) > \chi^2_{\mathscr{D}}(|\pi_2(\mathscr{D}^+)|, |\pi_2(\mathscr{D}^-)|)$ then $\pi_1 <_{CtC} \pi_2$
2. If $\chi^2_{\mathscr{D}}(|\pi_1(\mathscr{D}^+)|, |\pi_1(\mathscr{D}^-)|) = \chi^2_{\mathscr{D}}(|\pi_2(\mathscr{D}^+)|, |\pi_2(\mathscr{D}^-)|)$, let w.l.o.g. $|\pi_1| < |\pi_2|$ then $\pi_1 <_{CtC} \pi_2$

---

[3] The extension to other pattern domains is straightforward.

Support has been replaced by χ^2 as a significance measure and *confidence* is not referenced at all anymore. For a two-class problem, obviously if w.l.o.g. $|\pi(\mathscr{D}^+)| \geq |\pi(\mathscr{D}^-)|$ then necessarily $confidence(\pi \Rightarrow +) \geq 0.5$, and the strength of prediction is traded off against coverage of the pattern in the significance measure already.

Folding the two criteria into a single one has an interesting side-effect. The choices for the support threshold θ_s and the confidence threshold θ_c interact to have an effect on the number of rules mined. Increasing the minimum support but lowering the minimum confidence can lead to more rules, for example. It is not clear, however, how many rules will be mined, which gets exacerbated by the use of the data set coverage threshold in CMAR. Using χ^2, on the other hand, allows to make this explicit – CTC takes a single parameter with a clear meaning, k, the number of rules, instead of two or three more opaque ones.

CTC uses the principles described in Section 6.3 to compute the top$-k$ patterns in this order, for example the 1000 highest-scoring patterns. Hence, the selection is not performed *after* but *during* mining. This set is used directly for classification, without further pattern selection.

This has two advantages:

1. Far fewer rules are mined. In fact, most frequent, confident rules do not turn out to be significant.
2. A less complex voting scheme is necessary. CMAR's voting approach is outperformed by the comparably simple *average strength*, i.e. confidences for each class are added up, and a simple majority vote, i.e. each rule predicts its majority class.

In other words, compared to CBA and CMAR, both the heuristic pruning scheme and the ad hoc (or empirically found) classification technique are replaced by more straight-forward and arguably better-founded solutions.

6.6 Integrated Pattern Mining

A common feature of the approaches discussed till now is that they assume that a set of patterns is computed once, either based on a threshold, or on the size of the resulting pattern set, such as in top-k mining. There is no strong interaction between which patterns are mined and how the model is constructed afterwards. The alternative is to perform *integrated pattern mining*, i.e. patterns are mined, potentially refined or re-mined, *while* the classifier or pattern set is constructed, interleaving the mining and the model formation step, without creating an initial pattern set first.

In this section, we will first describe two updates of techniques discussed in the previous section to perform integrated mining; subsequently, we discuss techniques for building one particular type of model, i.e. a decision tree, using integrated mining techniques.

6.6.1 FITCARE

Until now we mostly illustrated methods on binary prediction problems. Good performance on binary problems does not always imply a good performance on multi-class problems however. Turning multi-class problems into binary ones usually strongly increases the computational resources needed. Either a number of *1-vs-all* settings has to be addressed that is equal to the number of classes in the data, or an even greater amount of *1-vs-1* settings. The FITCARE algorithm, on the other hand, was developed specifically for good performance on multi-class problems.

We observed in Section 6.5 that high confidence is not necessarily a good measure for class correlation if the class is a majority class in the first place. The FITCARE algorithm [9] takes this one step further. It extends the observation that CAR-miners usually focus on one-against-all settings, i.e. the confidence of a rule has to be higher w.r.t. the target class than w.r.t. the *union* of all other classes where it applies, towards the problem of badly skewed data sets. In such data sets, high-confidence, high-support rules will be rules correctly classifying the majority class – yet still covering and effectively misclassifying instances from minority classes. Instead of the usual global minimum frequency and confidence thresholds, a new definition of interesting CARs is proposed based on a distinct support threshold for *every* class. Given this vector of thresholds $(\theta_{s_1}, \ldots, \theta_{s_{|\mathscr{C}|}})$, a CAR $\pi \Rightarrow c$ is interesting if

1. $\frac{|\pi(\mathscr{D}^c)|}{|\mathscr{D}^c|} \geq \theta_c$
2. $\forall c' \neq c : \frac{|\pi(\mathscr{D}^{c'})|}{|\mathscr{D}^{c'}|} < \theta_{c'}$
3. $\forall \pi \subset \pi', \exists c' \neq c : \frac{|\pi(\mathscr{D}^{c'})|}{|\mathscr{D}^{c'}|} \geq \theta_{c'}$

The advantage of this technique lies in the fact that the vector of thresholds allows for far better fine-tuning of the differentiating power of a CAR between *any* two classes, finer than *emerging* patterns or *class-correlating* patterns can. The drawback is however, that for any given class y_i, $|\mathscr{C}|$ parameters have to be adjusted – the minimum threshold on the class itself and the $|\mathscr{C}| - 1$ maximum thresholds for the other classes. These $O(|\mathscr{C}|^2)$ parameters make up the threshold matrix Γ whose entries have to be estimated, making the process more expensive than the approaches we have seen so far. This is however traded off against having to break multi-class problems down into several binary problems, as explained above. Using several constraints and a hill-climbing approach, in which pattern mining is repeated, it is possible to estimate these values efficiently. The final matrix is used to extract a set of CARs for each target class in turn. The resulting CARs are

1. highly discriminative between classes, therefore making strongly conflicting predictions unlikely, and
2. highly probable to cover all instances of each target class, thus making default classifications less common.

The resulting rules are once again combined using a rather complex weighted voting scheme that takes into account the reliability of rules when it comes to their contribution to the final prediction. While this is a technique we have seen both in CMAR and in CTC, the focus is now neither on fulfilling a rule's nor on its global confidence but rather on its relative support in the target class. This should, given well-estimated parameters, lead to very small contributions of rules in classes that are not their target class, even if they have globally high support.

6.6.2 Mining Maximally Informative Pattern Sets Directly

In Section 6.4.3 we discussed methods for selecting a subset of k patterns that maximize a global optimization criterion. In these methods it was assumed that we start the search from a set of patterns. However, it is also possible to find such sets without first having to mine an initial set of patterns.

The main trick is to change the greedy step in the greedy algorithm of Section 6.4.3. Instead of iteratively picking from a pre-computed set the pattern which maximizes the optimization criterion, we use branch-and-bound search to determine the pattern that locally optimizes the measure. This branch and bound search employs similar ideas as those used to find correlated patterns (see Section 6.3).

We will illustrate this for the example of entropy. Assume we have a pattern set with entropy $H(P)$, then we are looking for the pattern π which maximizes $H(P \cup \{\pi\})$, or equivalently, $H(P \cup \{\pi\}) - H(P)$. $H(\pi|P) = H(P \cup \{\pi\}) - H(P)$ is known as the conditional entropy of π given P, and can be written as

$$H(\pi|P) = \sum_{\mathbf{b} \in \{0,1\}^{|P|}} p(\mathbf{b}) H(\pi|P = \mathbf{b})$$

where $H(\pi|P = \mathbf{b}) = \sum_{a \in \{0,1\}} -p(\pi = a|P = \mathbf{b}) \log p(\pi = a|P = \mathbf{b})$ and $p(\pi = a|P = \mathbf{b})$ denotes the fraction of elements of the data set characterized by $\mathbf{b}$ also having $\pi(\mathbf{x}) = a$.

The challenge when searching for a pattern that maximizes this score is to determine a bound on the scores of refinements of a pattern; such a bound could allow us to prune parts of the search space that are not promising.

In the case of entropy, we can use the observation that overall entropy is maximized when we maximize the entropy $H(\pi|\mathbf{b})$ in each bin $\mathbf{b}$. Given a pattern π, what is the highest entropy we can achieve in this bin for a pattern π' that is a refinement of π? We can distinguish two cases.

- the pattern covers more than half of the elements in the bin $\mathbf{b}$; then the highest entropy we might obtain for this bin is obtained by covering half of the elements. Hence, the highest entropy is 1.
- the pattern covers less than half of the elements, the best we can hope for is not to lose any of these elements by refining the pattern. Hence, the best we can hope to obtain is $H(\pi|P = \mathbf{b})$.

Combining these observations we achieve the following bound on the quality of any refinement π' of a pattern π:

$$H(\pi'|P) \leq \sum_{\substack{\mathbf{b} \in \{0,1\}^{|P|} \\ p(\pi = 1|P = \mathbf{b}) \geq 0.5}} p(\mathbf{b}) + \sum_{\substack{\mathbf{b} \in \{0,1\}^{|P|}, \\ p(\pi = 1|P = \mathbf{b}) < 0.5}} p(\mathbf{b})H(\pi|P = \mathbf{b}).$$

This bound can be used to prune unpromising branches of a pattern search and makes it possible to find patterns directly without having to post-process a precalculated set of patterns.

Combining this result with that of Section 6.4.3, it follows that we can find a provably good maximally informative pattern set by a combination of branch and bound search and a greedy algorithm.

Similar observations also apply to other measures; for instance, it was also applied in [4] for the supervised $Q(P)$ measure (see Section 6.4.3).

6.6.3 DL8

One of the most popular predictive models is the decision tree. A decision tree is a tree in which each internal node is labeled with a test on an attribute and each leaf is labeled with a prediction [28]. A prediction for a particular element $(\mathbf{x}, y)$ can be obtained by sorting it down the tree starting from the root. The left-hand branch of a node is taken if the specified test on the element of the data set is true; otherwise the right-hand branch is taken. Note that if all attributes are binary, it suffices to label internal nodes with attributes; an element of the data set will be sorted down the left-hand branch of a node labeled with attribute i if its value for $\mathbf{x}_i$ is true; otherwise it is sorted down the right-hand branch.

Many algorithms have been developed for learning decision trees from training data. Most of these algorithms employ the principle of heuristic top-down tree construction [28, 34]: starting from an empty tree, iteratively a leaf of the tree is replaced with a test node. A test is chosen by using a heuristic such as information gain. The advantage of this method is that it is fast and usually obtains sufficiently good results. However, it is not guaranteed to be optimal in many ways: given a bound on tree size, the heuristic method may not find the tree that is either most accurate or most cost-effective, in a setting of cost-based learning. In [31, 32] an algorithm, called DL8, was proposed that addresses the problem of finding optimal decision trees by exploiting a connection between pattern sets and decision trees.

The DL8 algorithm is based on exploiting the relationships between paths in decision trees and itemsets. To make this relationship clear we need to extend traditional itemsets to include *negative items*. Traditionally, an itemset $I \subseteq \{1, \ldots, n\}$ occurs in an element $\mathbf{x}$ of length n iff for all $j \in I$: $\mathbf{x}_j = 1$. Assume now that $I \subseteq \{1, \ldots, n, \neg 1, \ldots, \neg n\}$. Then we can define that an itemset occurs in an element iff for all positive $j \in I$: $\mathbf{x}_j = 1$ and for all negative $\neg j \in I$: $\mathbf{x}_j = 0$.

Fig. 6.3 An example decision
tree corresponding to the
three itemsets $\{B\}$, $\{\neg B, C\}$,
$\{\neg B, \neg C\}$

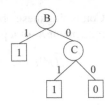

This extension allows us to represent every path in a decision tree as an item-set. For example, consider the decision tree in Figure 6.3. We can determine the leaf to which an element belongs by checking which of the itemsets $\{B\}$, $\{\neg B, C\}$ and $\{\neg B, \neg C\}$ matches. We denote the set of the itemsets corresponding to the leaves of a tree T with $leaves(T)$. Similarly, the itemsets that correspond to paths in the tree are denoted with $paths(T)$. In this case, $paths(T) = \{\emptyset, \{B\}, \{\neg B\}, \{\neg B, C\}, \{\neg B, \neg C\}\}$. A further illustration of the relation between itemsets and decision trees is given in Figure 6.4. In this figure, every node represents an itemset; an edge denotes a subset relation. Highlighted is one possible decision tree.

Given this correspondence, learning a decision tree can be seen as finding a set of class association rules, where the rules should include both positive and negative items and the set of rules should fulfill properties that ensure that it can be represented as a tree.

In the basic setting, the DL8 algorithm can be seen as a post-processing algorithm that can be applied on a lattice of itemsets. For the problem of finding a tree T which minimizes error $error_{\mathscr{T}}(T)$ on a set of examples $\mathscr{T}$, the main property that is exploited by the algorithm is that the error of a decision tree equals the sum of the errors of the left-hand and right-hand subtree of the root of this tree. Hence, we can solve the problem of finding an accurate decision tree by independently and recursively searching the best left-hand and right-hand subtrees of each possible root. By storing the best tree for every itemset, we can avoid that we need to consider every itemset more than once, and the computation is linear in the size of the itemset lattice.

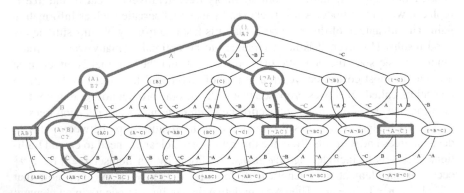

Fig. 6.4 An itemset lattice for items $\{A, \neg A, B, \neg B, C, \neg C\}$.

Fig. 6.5 A decision-tree to
separate light and dark structures based on the shapes.

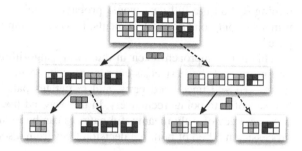

In [31, 32] several extensions are discussed of this general idea:

- the use of condensed representations to limit the number of itemsets that need to be considered;
- how to deal with other constraints and optimisation criteria than minimum support and error, for instance, cost-based constraints;
- how to integrate the decision tree construction with the pattern mining.

This last point is important, as by integrating the pattern mining in the tree construction, a smaller number of patterns needs to be considered and less information of the lattice needs to be stored. Overall, the integrated pattern miner searches for patterns once, but does so as guided by the decision tree construction procedure.

6.6.4 TREE^2

In the preceding discussion on DL8, decision trees were described that iteratively split data on a binary attribute to build an effective classifier. If data is described in terms of attribute values or binary attributes denoting, for instance, an item's presence or absence, this is a straightforward way of building a decision tree. Once more complex data such as trees or graphs needs to be analysed a simple split based on, for instance, the presence or absence of an atom in a molecule will lead to unwieldy classifiers which are unlikely to perform well.

A possible alternative lies in mining a set of patterns, encoding data in terms of their presence or absence and building the decision tree from this re-encoded data. However, we would need to select a constraint under which such patterns need to be mined. Instead of choosing this constraint ad-hoc, we can also integrate the pattern mining step in the decision tree learning algorithm, such that we directly mine for the pattern that the decision tree learning algorithm would select as the best test in post-processing.

To illustrate this, consider the example in Figure 6.5: mining ▦ allows to split the data set into two subsets which each consist 75% of one class, a clear improvement over the original 50–50 split. On these subsets, two more patterns can be mined, ▪ and ⊤, resulting in three pure leaves out of four in total. Especially mining ⬛ on the full data would probably require very lenient constraints since it matches only one of eight elements, leaving the unmatched subset still very "impure".

This is the approach taken in the TREE^2 algorithm [6]. TREE^2 employs top-1 mining to find the best class-correlating subtree in each iteration, splits the data in covered and uncovered parts, and re-iterates pattern mining for each of these two sets of data points recursively. In this way, ad-hoc thresholds are avoided, and compact trees of truly meaningful patterns can be induced, improving on the post-processing method of building decision trees from a set of pre-mined patterns.

6.7 Conclusions

In this chapter we provided an overview of methods we recently proposed for using patterns in classification tasks. We showed that there exists a large variety in methods, ranging from strict step-wise approaches to approaches in which pattern mining and model construction is integrated. We put these methods in context by providing extensive descriptions of related methods.

Despite the amount of work we reported on, this overview is far from complete. Providing a detailed overview of all approaches for pattern-based classification is beyond the scope of this chapter, and is left as future work. Other possibilities for future work include a more detailed invesigation of the merits of algorithms for pattern-based classification. We are not aware that a systematic experimental comparison has been carried out for all pattern-based classification methods. Finally, most methods until now concentrate on rule-based classification. An interesting question is for instance how patterns can be used in graphical models.

References

1. *Proceedings of the 5th IEEE International Conference on Data Mining (ICDM 2005), 27-30 November 2005, Houston, Texas, USA*. IEEE Computer Society, 2005.
2. Rakesh Agrawal, Heikki Mannila, Ramakrishnan Srikant, Hannu Toivonen, and A. Inkeri Verkamo. Fast discovery of association rules. In *Advances in Knowledge Discovery and Data Mining*, pages 307–328. AAAI/MIT Press, 1996.
3. Stephen D. Bay and Michael J. Pazzani. Detecting change in categorical data: Mining contrast sets. In *KDD*, pages 302–306, 1999.
4. Karsten Borgwardt, Xifeng Yan, Marisa Thoma, Hong Cheng, Arthur Gretton, Le Song, Alex Smola, Jiawei Han, Philip Yu, and Hans-Peter Kriegel. Combining near-optimal feature selection with gSpan. In Samuel Kaski, S.V.N. Vishwanathan, and Stefan Wrobel, editors, *MLG*, 2008.
5. Björn Bringmann. *Mining Patterns in Structured Data*. PhD thesis, K.U.Leuven, September 2009. De Raedt, Luc (supervisor).
6. Björn Bringmann and Albrecht Zimmermann. Tree^2 - decision trees for tree structured data. In Alípio Jorge, Luís Torgo, Pavel Brazdil, Rui Camacho, and João Gama, editors, *PKDD*, volume 3721 of *Lecture Notes in Computer Science*, pages 46–58. Springer, 2005.

7. Björn Bringmann and Albrecht Zimmermann. One in a million: picking the right patterns. *Knowl. Inf. Syst.*, 18(1):61–81, 2009.
8. Björn Bringmann, Albrecht Zimmermann, Luc De Raedt, and Siegfried Nijssen. Don't be afraid of simpler patterns. In Fürnkranz et al. [17], pages 55–66.
9. Loïc Cerf, Dominique Gay, Nazha Selmaoui, and Jean-François Boulicaut. A parameter-free associative classification method. In Il-Yeol Song, Johann Eder, and Tho Manh Nguyen, editors, *DaWaK*, volume 5182 of *Lecture Notes in Computer Science*, pages 293–304. Springer, 2008.
10. Hong Cheng, Xifeng Yan, Jiawei Han, and Chih-Wei Hsu. Discriminative frequent pattern analysis for effective classification. In *ICDE*, pages 716–725. IEEE, 2007.
11. Hong Cheng, Xifeng Yan, Jiawei Han, and Philip S. Yu. Direct discriminative pattern mining for effective classification. In *ICDE*, pages 169–178. IEEE, 2008.
12. Frans Coenen and Paul Leng. Obtaining best parameter values for accurate classification. In *ICDM* [1], pages 597–600.
13. William W. Cohen. Fast effective rule induction. In *In Proceedings of the Twelfth International Conference on Machine Learning*, pages 115–123. Morgan Kaufmann, 1995.
14. Mukund Deshpande and George Karypis. Using conjunction of attribute values for classification. In *CIKM*, pages 356–364. ACM, 2002.
15. Mukund Deshpande, Michihiro Kuramochi, Nikil Wale, and George Karypis. Frequent substructure-based approaches for classifying chemical compounds. *IEEE Trans. Knowl. Data Eng.*, 17(8):1036–1050, 2005.
16. Guozhu Dong and Jinyan Li. Efficient mining of emerging patterns: Discovering trends and differences. In *KDD*, pages 43–52, 1999.
17. Johannes Fürnkranz, Tobias Scheffer, and Myra Spiliopoulou, editors. *Knowledge Discovery in Databases: PKDD 2006, 10th European Conference on Principles and Practice of Knowledge Discovery in Databases, Berlin, Germany, September 18-22, 2006, Proceedings*, volume 4213 of *Lecture Notes in Computer Science*. Springer, 2006.
18. Gemma C. Garriga, Petra Kralj, and Nada Lavrac. Closed sets for labeled data. In Fürnkranz et al. [17], pages 163–174.
19. Jiawei Han, Jian Pei, and Yiwen Yin. Mining frequent patterns without candidate generation. In Weidong Chen, Jeffrey F. Naughton, and Philip A. Bernstein, editors, *SIGMOD Conference*, pages 1–12. ACM, 2000.
20. Jeroen Kazius, Siegfried Nijssen, Joost N. Kok, Thomas Bäck, and Adriaan P. IJzerman. Substructure mining using elaborate chemical representation. *Journal of Chemical Information and Modeling*, 46(2):597–605, 2006.
21. Arno Knobbe, Bruno Crémilleux, Johannes Fürnkranz, and Martin Scholz. From local patterns to global models: the LeGo approach to data mining. In Johannes Fürnkranz and Arno Knobbe, editors, *LeGo'08, Proceedings of the ECML PKDD 2008 Workshop 'From Local Patterns to Global Models'*, pages 1–16, 2008.
22. Arno J. Knobbe and Eric K. Y. Ho. Maximally informative k-itemsets and their efficient discovery. In Tina Eliassi-Rad, Lyle H. Ungar, Mark Craven, and Dimitrios Gunopulos, editors, *KDD*, pages 237–244. ACM, 2006.
23. Arno J. Knobbe and Eric K. Y. Ho. Pattern teams. In Fürnkranz et al. [17], pages 577–584.
24. Stefan Kramer and Luc De Raedt. Feature construction with version spaces for biochemical applications. In Carla E. Brodley and Andrea Pohoreckyj Danyluk, editors, *ICML*, pages 258–265. Morgan Kaufmann, 2001.
25. Neal Lesh, Mohammed Javeed Zaki, and Mitsunori Ogihara. Mining features for sequence classification. In *KDD*, pages 342–346, 1999.
26. Wenmin Li, Jiawei Han, and Jian Pei. Cmar: Accurate and efficient classification based on multiple class-association rules. In Nick Cercone, Tsau Young Lin, and Xindong Wu, editors, *ICDM*, pages 369–376. IEEE Computer Society, 2001.
27. Bing Liu, Wynne Hsu, and Yiming Ma. Integrating classification and association rule mining. In *KDD*, pages 80–86, 1998.
28. T.M. Mitchell. *Machine Learning*. McGraw-Hill, New York, 1997.

29. Shinichi Morishita and Jun Sese. Traversing itemset lattice with statistical metric pruning. In *PODS*, pages 226–236. ACM, 2000.
30. G. Nemhauser, L. Wolsey, and M. Fisher. An analysis of the approximations for maximizing submodular set functions. *Mathematical Programming*, 14:265–294, 1978.
31. Siegfried Nijssen and Élisa Fromont. Mining optimal decision trees from itemset lattices. In Pavel Berkhin, Rich Caruana, and Xindong Wu, editors, *KDD*, pages 530–539. ACM, 2007.
32. Siegfried Nijssen and Elisa Fromont. Optimal constraint-based decision tree induction from itemset lattices. *Data Mining and Knowledge Discovery*, 2010. (In press).
33. Siegfried Nijssen and Joost N. Kok. Multi-class correlated pattern mining. In Francesco Bonchi and Jean-François Boulicaut, editors, *KDID*, volume 3933 of *Lecture Notes in Computer Science*, pages 165–187. Springer, 2005.
34. J. Ross Quinlan. *C4.5: Programs for Machine Learning*. Morgan Kaufmann, 1993.
35. Luc De Raedt and Albrecht Zimmermann. Constraint-based pattern set mining. In *SDM*. SIAM, 2007.
36. Geoffrey I. Webb. Layered critical values: a powerful direct-adjustment approach to discovering significant patterns. *Machine Learning*, 71(2-3):307–323, 2008.
37. Stefan Wrobel. An algorithm for multi-relational discovery of subgroups. In Henryk Jan Komorowski and Jan M. Zytkow, editors, *PKDD*, volume 1263 of *Lecture Notes in Computer Science*, pages 78–87. Springer, 1997.
38. Mohammed Javeed Zaki, Srinivasan Parthasarathy, Mitsunori Ogihara, and Wei Li. New algorithms for fast discovery of association rules. In *KDD*, pages 283–286, 1997.
39. Albrecht Zimmermann and Björn Bringmann. CTC - correlating tree patterns for classification. In *ICDM* [1], pages 833–836.

Chapter 7
Constrained Predictive Clustering

Jan Struyf and Sašo Džeroski

Abstract In this chapter, we extend predictive clustering by introducing constraints on the clusters and predictive models. A domain expert is usually not only interested in the most compact clusters or the most accurate model; other factors, such as model size and prediction cost, may also be important. We will see how such factors can be controlled by means of constraints. In predictive clustering trees, constraints can be imposed both from the clustering and the prediction point of view. We present an overview of various constraint types and look into algorithms for enforcing them.

7.1 Introduction

We consider predictive clustering [8], which is an approach to prediction that is based on clustering methods. The inductive step in predictive clustering creates a clustering. To make a prediction for a new data instance, the instance is first assigned to a cluster; the prediction is then computed from that cluster. Section 7.2 reviews clustering and predictive clustering and shows why decision trees, and more generally predictive clustering trees, naturally fit the framework of predictive clustering. It also describes a general algorithm for building clustering trees and lists a number of specific instantiations of predictive clustering.

After reviewing clustering and predictive clustering, we introduce constrained predictive clustering (Section 7.3). This extends predictive clustering by allowing

Jan Struyf
Department of Computer Science, Katholieke Universiteit Leuven
Celestijnenlaan 200A, 3001 Leuven, Belgium
e-mail: Jan.Struyf@struyf-ye.org

Sašo Džeroski
Department of Knowledge Technologies, Jožef Stefan Institute
Jamova cesta 39, 1000 Ljubljana, Slovenia
e-mail: Saso.Dzeroski@ijs.si

user-defined constraints on the clustering. We discuss the constraints that are most relevant to predictive clustering. An important property of such constraints is anti-monotonicity. Anti-monotonicity is discussed in Section 7.4, where we define a search space of clustering trees. This search space, which is structured by the sub-tree relation, is traversed by the algorithms that construct (predictive) clustering trees (PCTs).

Finally, the chapter looks into algorithms for building PCTs that satisfy the given constraints (Section 7.5). We describe two approaches: (1) a two-phase approach that consists of tree induction followed by post pruning, and (2) an approach that is based on beam search. We summarize the results that have been obtained with these algorithms and refer to the relevant specific literature for details.

7.2 Predictive Clustering Trees

7.2.1 Clustering and Intra-cluster Variance

The task in clustering is to partition a given set of data instances into subsets called clusters such that the instances in each cluster are similar. The similarity requirement is usually formulated in terms of a distance measure; it then translates for example into requiring for each cluster a small average pairwise distance between the instances. This is illustrated in Fig. 7.1.a. Here, the data are points in a two dimensional space and the clustering partitions these points into two clusters C_1 and C_2. Points in a given cluster are similar because they are close in terms of the Euclidean distance.

We define clustering formally as follows:

Definition 7.1 (Clustering). Given an instance space Z, a training dataset[1] $T \subseteq Z$, and a loss function $l : (Z \to \mathbb{N}) \times 2^Z \to \mathbb{R}$, with $\mathbb{N}$ the natural numbers, 2^Z the power set of Z, and $\mathbb{R}$ the real numbers, find a cluster assignment function $c : Z \to \mathbb{N}$ such that c minimizes $l(c, T)$.

The cluster assignment function in Def. 7.1 represents the clustering: cluster C_i is the set of instances $\{x \mid x \in T, c(x) = i\}$. Its range is the set of natural numbers. As a result, Def. 7.1 allows clusterings with an arbitrary number of clusters. Later, we will see that clustering size constraints can be used to limit the number of clusters.

The loss function represents the similarity requirement. We define it as a function that takes a clustering c and a dataset T as input and that computes how dissimilar the instances of T are in each cluster induced by c on T. In this chapter, we will mainly consider intra-cluster variance as loss function, which is defined as follows.

---

[1] For convenience, we slightly abuse notation and define the training dataset as a subset of the instance space. Note that in practice, the training set is often not a set but rather a bag in which the same instance may appear multiple times.

Definition 7.2 (Intra-cluster variance). Given an instance space Z, a training set $T \subseteq Z$, a distance measure d on Z, and a cluster assignment function c, the intra-cluster variance according to d is $\mathrm{ICV}^d(c,T) = \sum_i \frac{|C_i|}{|T|} \mathrm{Var}^d(C_i)$, with $C_i = \{x | x \in T, c(x) = i\}$.

We take a general approach in which cluster variance can be measured according to any distance measure and also consider two alternative definitions of variance: the first is computed as the average pairwise distance between cluster members (Def. 7.3), and the second is computed as the average distance between a cluster member and the cluster centroid (Def. 7.5).

Definition 7.3 (Variance based on pairwise distances). Given an instance space Z, a cluster $C \subseteq Z$, and a distance measure d on Z, the variance according to d based on pairwise distances $\mathrm{Var}^d_{\mathrm{pairw}}(C)$ is $\frac{1}{2|C|^2} \sum_{x \in C} \sum_{y \in C} d^2(x,y)$.

Definition 7.4 (Cluster centroid). Given an instance space Z, a cluster $C \subseteq Z$, and a distance measure d on Z, the centroid of C according to d is $\mathrm{centr}^d(C) = \mathrm{argmin}_{y \in Z} \sum_{x \in C} d^2(x,y)$.

Definition 7.5 (Variance based on centroid). Given an instance space Z, a cluster $C \subseteq Z$, and a distance measure d on Z, the variance according to d based on the cluster centroid is $\mathrm{Var}^d_{\mathrm{centr}}(C)$ is $\frac{1}{|C|} \sum_{x \in C} d^2(x, \mathrm{centr}^d(C))$.

For the case that the data instances are vectors in $\mathbb{R}^n$ and the distance measure d is the Euclidean distance, the cluster centroid is the vector mean of the data instances, and Def. 7.5 coincides with the traditional definition of variance used in statistics. In this case Def. 7.3 also yields the same result as Def. 7.5.

For other distance measures, the centroid can often only be approximated by applying, e.g., gradient descent to the expression in Def. 7.4. In such cases, we may want to compute the variance with Def. 7.3 [16]. We may also replace the cluster centroid with the cluster medoid, which is the instance in the cluster with the smallest average distance to the other instances.

7.2.2 Clustering Trees

Clustering trees [6] are decision trees that are used for clustering (Fig. 7.1.b). Each node of a clustering tree represents one cluster: the top node corresponds to a cluster containing all available data, and each test in the tree partitions the local instances into sub-clusters based on the outcome of the test. As such, a clustering tree represents a hierarchy of clusters. A non-hierarchical (flat) clustering can be obtained by only considering the tree leaves.

A clustering tree can serve as a cluster assignment function: a new instance can be assigned to a cluster by sorting it down the tree until it arrives in a leaf. Fig. 7.1.b illustrates this for a non-hierarchical clustering. Here, each leaf corresponds to one cluster and to one value for the cluster assignment function $c(x)$.

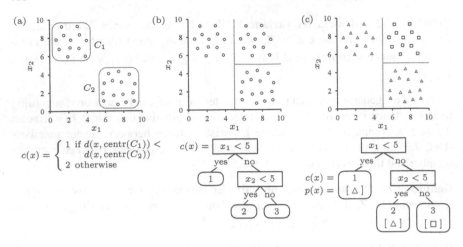

Fig. 7.1 (a) Clustering. (b) Clustering tree. (c) Predictive clustering tree (in this case a traditional classification tree).

Clustering trees belong to the class of conceptual clustering methods [22]. In conceptual clustering, each cluster has a conceptual description in terms of the properties that the instances in the cluster have. For example, cluster C_2 in Fig. 7.1.b is described by $x_0 \geq 5 \wedge x_1 < 5$. Conceptual clustering is important if the user wants to gain insight into the meaning of the clusters. On the other hand, the conceptual descriptions constrain the clustering, which may result in a larger intra-cluster variance compared to non-conceptual methods [16].

7.2.3 Predictive Clustering and Predictive Clustering Trees

Predictive clustering [8] is an approach to prediction that is based on clustering methods. The central idea is to induce a clustering and to use this clustering to make a prediction for a new instance by assigning the instance to a cluster and then making a prediction from that cluster. The underlying assumption is that if the instances in a cluster are similar, their target values will also be similar, and that the target value of a new instance can be accurately predicted from those of the other instances in the cluster.

We define prediction and predictive clustering as follows:

Definition 7.6 (Prediction). Given an instance space $Z = X \times Y$, with X the input subspace of Z and Y the output subspace of Z, a training set $T \subseteq Z$, and a loss function $l : (X \rightarrow Y) \times 2^Z \rightarrow \mathbb{R}$, find a predictive model $f : X \rightarrow Y$, such that $l(f,T)$ is minimal.

Definition 7.7 (Predictive clustering). Predictive clustering is a prediction approach in which $f(x)$ is of the form $p(c(x),x)$, with $c: X \to \mathbb{N}$ the cluster assignment function, and $p: \mathbb{N} \times X \to Y$ the prediction function.

Fig. 7.1.c illustrates these definitions for the case of a traditional classification tree [11]. Here, the input space is $\mathbb{R}^2$, and the output space is the set of class labels $\{\triangle, \square\}$. Given a training set of labeled instances, the goal is to find a cluster assignment function and a prediction function that minimize the loss function (e.g., training set error). Similar to Fig. 7.1.b, the cluster assignment function is represented as a tree in which each leaf corresponds to a cluster. After an instance is assigned to a cluster, the prediction function is used to predict a class for the instance. In classification trees, the prediction function is a constant for each cluster; in model trees it is a linear model [25].

A predictive clustering tree is a predictive clustering model in which the cluster assignment function is represented as a decision tree. The previous example therefore shows that classification trees and model trees are special cases of predictive clustering trees. Viewed in this way, predictive clustering is not a new machine learning method, but rather a framework in which known methods can be explained. Other known learning methods, such as decision rules, can also be cast in the predictive clustering framework [35]. Predictive clustering also leads to new approaches; we will list some in Section 7.2.5.

7.2.4 Learning (Predictive) Clustering Trees

Clustering trees and predictive clustering trees can be constructed with almost the same algorithm. From here on, we will refer to both with (predictive) clustering trees or PCTs. As originally explained by [6], the algorithm is a standard "top-down induction of decision trees" algorithm, similar to that of CART [11], but with a more general heuristic.

Table 7.1 lists the PCT algorithm. It takes as input a training set T. The main loop (Table 7.1, BESTTEST) searches for the best acceptable attribute-value test that can be put in a node. If such a test t^* can be found then the algorithm creates a new internal node labeled t^* and calls itself recursively to construct a subtree for each subset in the partition $\mathscr{P}^*$ induced by t^* on the training instances. To select the best test, the algorithm scores the tests by the reduction in variance they induce on the instances. Maximizing variance reduction locally minimizes intra-cluster variance.

The function ACCEPTABLE is used to test if an attribute-value test is acceptable. It may test different conditions, such as that the clusters in $\mathscr{P}$ are sufficiently large, and that the variance reduction induced by the test is significant (e.g., in terms of a statistical F-test). If ACCEPTABLE fails for all possible attribute-value tests, then the algorithm creates a leaf and labels it with a cluster identifier, and, in the case of predictive clustering trees, with a value for the prediction function. This value is usually the projection of the cluster centroid on the output space, but in general depends on the application domain as we discuss in the next section.

Table 7.1 The top-down induction algorithm for PCTs. T denotes the training instances, t an attribute-value test, $\mathscr{P}$ the partition induced by t on T, and h the heuristic value of t, which is computed based on distance d. $\mathrm{centr}^d(T)$ is defined in Def. 7.4, and $\mathrm{Var}^d(T)$ can be computed with Def. 7.3 or Def. 7.5 (depending on the application domain). $\mathrm{proj}_Y(\cdot)$ projects its argument on the output space Y. The superscript '$*$' indicates the current best test and its corresponding partition and heuristic.

procedure $\mathrm{CLUS}(T)$ **returns** tree	**procedure** $\mathrm{BESTTEST}(T)$
1: $(t^*, \mathcal{P}^*) = \mathrm{BESTTEST}(T)$	1: $(t^*, h^*, \mathcal{P}^*) = (none, 0, \emptyset)$
2: **if** $t^* \neq none$	2: **for each** possible test t
3: **for each** $T_k \in \mathcal{P}^*$	3: $\mathcal{P} = $ partition induced by t on T
4: $tree_k = \mathrm{CLUS}(T_k)$	4: $h = \mathrm{Var}^d(T) - \sum_{T_k \in \mathcal{P}} \frac{|T_k|}{|T|} \mathrm{Var}^d(T_k)$
5: **return** $\mathrm{node}(t^*, \bigcup_k \{tree_k\})$	5: **if** $(h > h^*) \wedge \mathrm{ACCEPTABLE}(t, \mathcal{P})$
6: **else**	6: $(t^*, h^*, \mathcal{P}^*) = (t, h, \mathcal{P})$
7: $i = $ new cluster identifier	7: **return** $(t^*, \mathcal{P}^*)$
8: $p = \mathrm{proj}_Y(\mathrm{centr}^d(T))$	
9: **return** leaf $\left(\begin{bmatrix} i \\ p \end{bmatrix} \right)$	

7.2.5 Instantiations of (Predictive) Clustering Trees

We list a number of instantiations of PCTs, each of which can be obtained by selecting a particular distance measure d.

- Clustering trees [6] are obtained by instantiating d to the Euclidean distance over the instance space $Z = \mathbb{R}^n$. In this case, there is no prediction function. Depending on the application, other distance measures may be used as well. For example, [16] show how PCTs can be used for time series data analysis. They employ a qualitative distance measure [30] and compute variance based on pairwise distances (Def. 7.3).
- Regression trees [11] instantiate d to the Euclidean distance restricted to the target attribute. The heuristic then becomes the traditional variance reduction heuristic used in CART [11], and the prediction function returns for a given leaf the local training instances' mean target value.
- Classification trees [11] instantiate d to the Euclidean distance in a transformed space. For an m class problem, this space is $\mathbb{R}^m$ and a class i instance is mapped to the unit vector in dimension i. It can be shown that the resulting heuristic (using Def. 7.3) is the Gini gain [26]. The prediction function returns for a given leaf the majority class of the local instances.
- Multi-target or multi-task learning [12] is learning a model that predicts multiple output attributes. PCTs can be trivially applied to multi-target learning [6, 7]. To predict m numeric attributes, the only difference with regression trees is that d is now defined on $\mathbb{R}^m$ instead of $\mathbb{R}$. Multi-target classification [20] and multi-target mixed classification and regression can be implemented in a similar way. Fig. 7.2 shows an example of a multi-target PCT.
- Hierarchical multi-label classification is a classification task in which instances may belong to multiple classes and the classes are structured in a class hierarchy.

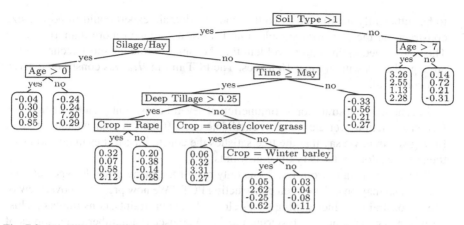

Fig. 7.2 A multi-target PCT predicting the normalized abundances of four organism groups (the mites Cryptostigmata, Prostigmata, Astigmata, and Mesostigmata) in agricultural soil from agricultural events and soil biological parameters [15]. A multi-target PCT provides insight in the relations between the target attributes and makes common factors that are relevant to all targets explicit.

Chapter 15, which is based on [32], explains this approach in detail using gene function prediction as a case study.

The PCT framework is implemented in the CLUS system, which is available as open source software at `http://dtai.cs.kuleuven.be/clus/`. CLUS implements algorithms for constructing PCTs and also for building predictive clustering rule sets [35]. The constrained PCT induction algorithms based on post-pruning and beam search, which are discussed later in this chapter, are also available in CLUS. Of the constraints that we will see, CLUS implements cluster size, clustering size, global loss, depth, syntactic, and instance level constraints (Section 7.3).

7.3 Constrained Predictive Clustering Trees and Constraint Types

So far, clustering and predictive clustering have been defined as an unconstrained optimization problem. That is, the goal is to find the clustering or the predictive clustering model that minimizes a given loss function. From here on, we consider the case where the domain expert is interested in controlling, besides loss, also other properties of the model. To do so, we add constraints to the optimization problem. The goal then becomes to find the model that minimizes the loss function among the models that satisfy the given constraints.

Example 7.1. Consider the task of learning a multi-target PCT. The domain expert may not only be interested in finding the most accurate tree; the tree also needs to be sufficiently interpretable. To this end, the domain expert could impose a size constraint on the tree, for example, that it must not contain more than 10 leaves. The constrained optimization problem then becomes to find the most accurate PCT among all trees with at most 10 leaves. The PCT in Fig. 7.2 was constructed given this constraint.

In general, a constraint or a conjunction of several constraints can be written as a Boolean function over the space of candidate trees. If the function evaluates to true for a given tree, we say that the tree satisfies the constraints; if it evaluates to false, then the tree does not satisfy the constraints.

The size constraint from Ex. 7.1 is only one of the many possible types of constraints that may be useful when constructing PCTs. We now present an overview of useful constraints, which distinguishes cluster level constraints, constraints on clusterings, and constraints on clustering models. We discuss a number of examples of each type.

7.3.1 Cluster Level Constraints

We first consider constraints on individual clusters, which we will call cluster level constraints. A cluster level constraint is a Boolean function that takes a cluster (a set of instances) as input and outputs *true* if the cluster satisfies the constraint. A set of clusters satisfies a cluster level constraint if each cluster in the set satisfies the constraint.

We now present some examples of cluster level constraints.

Cluster size constraint. Cluster level constraints can be constructed by upper or lower bounding a certain numeric property of a cluster. The most simple instantiation of this is a lower bound (or upper bound) on the number of instances in a cluster. For example, enforcing a lower bound on the cluster size is useful in k-means clustering to avoid empty or very small clusters [10], or in PCTs to ensure that each leaf contains enough instances to obtain a good estimate of the centroid. Cluster size constraints can also be used to enforce k-anonymity in privacy preserving data mining [17].

Local loss constraint. Often, the global loss function minimized by the clustering algorithm is additive: the loss of the clustering is the sum of the local losses of the clusters in the clustering. This is for example the case for intra-cluster variance (Def. 7.2). The local loss constraint upper bounds the local loss per cluster instance (e.g., the cluster's variance). It can be used to ensure that the created clusters are sufficiently compact. For classification trees, other local loss functions are used, such as the proportion of misclassified training instances or $1 - \chi^2$ [24].

ε-constraint. The ε-constraint [13] ensures that for each instance in a cluster, there is another instance that is sufficiently close to it, that is, $\forall x_1 \in C, \exists x_2 \in C, x_1 \neq x_2 : d(x_1, x_2) \leq \varepsilon$.

7.3.2 Constraints on Clusterings

We now consider constraints on a set of clusters.

Clustering size constraint. This type of constraints upper or lower bounds the number of clusters in the clustering. Typically, each leaf of a PCT is a cluster. In this case, the tree size constraint from Ex. 7.1 is a clustering size constraint. One can also fix the number of clusters to a particular value. For example, the k-means algorithm enforces the constraint that the number of clusters is precisely equal to its parameter k.

Minimum separation (δ-constraint). Minimum separation [13] ensures that for each pair of instances from different clusters, their distance is at least δ, that is, $\forall C_1, C_2, C_1 \neq C_2, x_1 \in C_1, x_2 \in C_2 : d(x_1, x_2) \geq \delta$. For example, in object recognition tasks, minimum separation can be used to specify a minimum distance between recognized objects.

Instance level constraint. Must-link and cannot-link constraints are constraints about pairs of instances and are therefore called instance level constraints [33]. A must-link constraint $ML(x_1, x_2)$ specifies that instances x_1 and x_2 must belong to the same cluster, and a cannot-link $CL(x_1, x_2)$ specifies that x_1 and x_2 must not be placed in the same cluster.

Instance level constraints provide information about the assignment of instances to clusters and can be used to address, among others, semi-supervised learning tasks. To this end, labeled instance pairs of the same class are must-linked and labeled instance pairs of different classes are cannot-linked. [13] show that ε- and δ-constraints can be converted into instance level constraints.

Balancedness constraint. This constraint ensures that all clusters are of similar size [36]. It is useful, e.g., in marketing applications where one wants to segment the customers in groups of roughly the same size.

Global loss constraint. Normally, the constrained clustering algorithm minimizes the global loss while taking into account the given constraints. Alternatively, one may also minimize a different property of the clustering and constrain the global loss. For example, one could search for the clustering with the smallest number of clusters that has a global loss of at most ε, where ε is chosen by the domain expert. When constructing a classification tree, this translates into finding the smallest tree that has at least a given accuracy. Such constraints are studied by [18] and [28].

Cluster level constraints for hierarchical clusterings. For hierarchical clusterings, such as clusterings created by clustering trees or hierarchical agglomerative clustering, additional constraints on the structure of the hierarchy may be useful.

- **Depth constraint** This constraint upper bounds the depth of the hierarchy. An extreme instantiation of this constraint is to upper bound the depth of the tree to one. In this case, one obtains so-called decision stumps, which are often used in ensemble learning [27].

- **Balancedness constraint** This constraint ensures that each internal node of the hierarchy partitions the local instances into subsets of similar size. In this way, it is a recursive application of the balancedness constraint introduced above. Mathematically, it can be written as a lower bound on the entropy of the partition [24]. Another type of balancedness constraint for hierarchical clusterings is to ensure that for each internal hierarchy node, all subtrees have a similar number of nodes.

7.3.3 Constraints on Clustering Models

We now list constraints that restrict the syntax of PCTs.

Prediction cost constraint. If attributes represent quantities that need to be measured for a new instance at prediction time, then a certain cost may be associated to these measurements (e.g., the cost of a lab test in a medical diagnosis application). Models that predominantly test low cost attributes may in such cases be desired [31, 21]. Such models can be constructed by specifying an upper bound on the prediction cost of the model, which is the sum of the measurement costs of all attributes that are required to make a prediction. For a PCT, these are all the attributes that are tested on a path from the tree root to the leaf where the instance is sorted into. Alternatively, one may also constrain the total cost of all attributes used in the model. The former is a cluster level version of the constraint, while the latter is a global constraint.

Syntactic constraint. Syntactic constraints directly restrict the syntactical part of the model. A syntactic constraint could require that the tree structure is of the form "$\text{node}(X < y_1, \text{node}(c < y_2, \text{Tr}_1, \text{Tr}_2), \text{Tr}_3), X \in \{a, b\}$", that is, a tree in which the root node tests on attribute a or b, and the left subtree of the root tests on attribute c (assuming a binary tree and that y_1 and y_2 are numeric values). Syntactic constraints are useful when the domain expert has a preference for certain model structures. For example, he or she could specify (part of) the structure of a PCT and the system could fill in the PCT's parameters, such as the thresholds used in tests on numeric attributes, or it could further refine the given tree. Syntactic constraints require a language to specify valid syntactic structures, which should ideally be declarative. [2] show how (stochastic) logic programs can be used to this end. Declarative language biases [23] used in inductive logic programming are also examples of syntactic constraints.

7.3.4 Hard Versus Soft Constrained Clustering

The constrained (predictive) clustering tasks defined above are hard constrained: the solution has to satisfy all the given constraints. Nevertheless, sometimes it may be impossible to satisfy all the constraints. In soft constrained clustering, satisfying all the constraints is no longer required.

A first approach to soft constrained clustering is to replace the constraints in the constrained optimization problem by a penalty term in the objective function that counts the number of violated constraints. That is, the objective function is written as $\alpha \cdot Loss + (1 - \alpha) \cdot$ (*Number of constraints violated*), with α a parameter that specifies the relative importance of one unit of loss versus violating one additional constraint. Because this approach only takes the number of violated constraints into account, it is useful in applications with many constraints. For example, [4] have applied such an approach to k-means clustering with instance level constraints.

A second approach takes the degree to which a constraint is violated into account. Consider the constraint that there can be at most five clusters. A clustering with 20 clusters violates this constraint to a larger degree than one with only six clusters. A soft constrained clustering algorithm should trade-off the degree to which constraints are violated versus the loss of the clustering. This can again be accomplished by replacing the constraints by a penalty term in the objective function. Alternatively, it can be accomplished by casting the soft constrained clustering problem into a probabilistic framework [5, 3, 2].

In this chapter, we will present algorithms and corresponding experimental results for both hard constrained PCTs and soft constrained PCTs. In particular, we will treat the clustering size constraint as a hard constraint in Section 7.5.1. Next, we present an algorithm that performs soft instance level constrained clustering (Section 7.5.2).

7.4 A Search Space of (Predictive) Clustering Trees

We now define a search space of PCTs, structured by the subtree relationship. Then we define anti-monotonic constraints. The constrained induction algorithms that we cover in the chapter exploit this property. For ease of notation, we only consider binary trees.

Definition 7.8 (Subtree order). Given two trees Tr_1 and Tr_2,
$$\mathrm{Tr}_1 \leq \mathrm{Tr}_2 \Leftrightarrow \begin{cases} \mathrm{Tr}_1 = \mathrm{leaf}(\cdot), \ or \\ (\mathrm{Tr}_1 = \mathrm{node}(X_1, \mathrm{Tr}_{1l}, \mathrm{Tr}_{1r}) \wedge \mathrm{Tr}_2 = \mathrm{node}(X_2, \mathrm{Tr}_{2l}, \mathrm{Tr}_{2r}) \wedge \\ \quad X_1 = X_2 \wedge \mathrm{Tr}_{1l} \leq \mathrm{Tr}_{2l} \wedge \mathrm{Tr}_{1r} \leq \mathrm{Tr}_{2r}) \end{cases}$$

The subtree order is a partial order: it is possible that neither $\mathrm{Tr}_1 \leq \mathrm{Tr}_2$ nor $\mathrm{Tr}_1 \geq \mathrm{Tr}_2$ holds, e.g., for trees with a different attribute-value test in the top node.

Definition 7.9 (Refinement operator). A refinement operator ρ is a function that maps a tree Tr to a set of trees $\rho(\mathrm{Tr})$ (the refinement set), such that each tree in $\rho(\mathrm{Tr})$ is obtained by replacing precisely one of the leaves of Tr by a new internal node with two new leaves. Each tree in $\rho(\mathrm{Tr})$ is called a refinement of Tr.

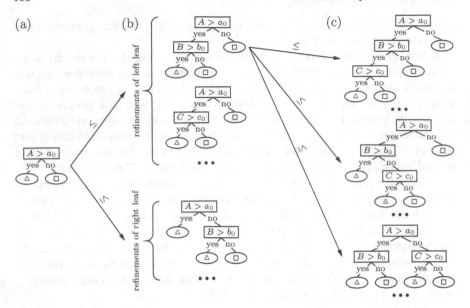

Fig. 7.3 Subtree order on trees. (a) A decision stump. (b) The refinements of tree (a). (c) The refinements of the top-most tree in (b).

A refinement operator returns the immediate successors with regard to the subtree order. That is, for each refinement $\mathrm{Tr}_{\mathrm{ref}}$ in $\rho(\mathrm{Tr})$, it holds that Tr is a subtree of $\mathrm{Tr}_{\mathrm{ref}}$, and that there is no other tree "in between" Tr and $\mathrm{Tr}_{\mathrm{ref}}$ (i.e., $\neg\exists\mathrm{Tr}', \mathrm{Tr}' \neq \mathrm{Tr}, \mathrm{Tr}' \neq \mathrm{Tr}_{\mathrm{ref}}, \mathrm{Tr} \leq \mathrm{Tr}' \leq \mathrm{Tr}_{\mathrm{ref}}$). Fig. 7.3 illustrates the concepts of subtree order and refinement operator.

Definition 7.10 (Anti-monotonic[2] constraint). A constraint $cons : \mathscr{T} \to \{true, false\}$, with $\mathscr{T}$ the set of all possible PCTs, is anti-monotonic with respect to $\leq$ if and only if $\forall \mathrm{Tr}_i, \mathrm{Tr}_j : (\mathrm{Tr}_i \leq \mathrm{Tr}_j \land cons(\mathrm{Tr}_j)) \to cons(\mathrm{Tr}_i)$.

If one considers an increasing sequence of trees according to the refinement order (e.g., going from (a) to (c) in Fig. 7.3) then the value of an anti-monotonic constraint can only decrease along the sequence, that is, change from true to false. This observation can be exploited as follows: If a given tree violates an anti-monotonic constraint, then it is not useful to consider refinements of this tree because any refinement will also violate the constraint (because of its anti-monotonicity).

We list a number of useful anti-monotonic constraints on PCTs.

1. An upper bound on tree size (Ex. 7.1).

2. A lower bound on cluster size.

---

[2] The term anti-monotonic comes from order theory and is sometimes also called order-reversing. It is the opposite of monotonic or order-preserving. (http://en.wikipedia.org/wiki/Monotonic_function.)

3. An upper bound on tree depth.

4. An upper bound on the prediction cost.

5. A hierarchical balancedness constraint.

6. A must-link constraint (if each leaf is a cluster).

7. An ε- and a δ-constraint (if each leaf is a cluster).

8. A syntactic constraint of the form $Tr \geq Tr_{cons}$ (subtree constraint).

9. A conjunction of anti-monotonic constraints.

10. A disjunction of anti-monotonic constraints.

7.5 Algorithms for Enforcing Constraints

Many of the anti-monotonic constraints mentioned in the previous section can be enforced in the PCT induction algorithm (Table 7.1) by simply checking them in the ACCEPTABLE function. As pointed out in the previous section, if an anti-monotonic constraint does not hold, then it will also not hold if the tree is expanded further. Therefore, anti-monotonic constraints can be used as stopping criteria in the function ACCEPTABLE.

For constraints that take the entire tree into account, such as an upper bound on tree size, simple top-down induction (Table 7.1) in combination with a stopping criterion based on the constraint is not a good method. It will result in a tree that satisfies the constraint, but the tree will be imbalanced and may have a far from optimal loss because of the depth-first construction. For example, assume that the algorithm first constructs the left subtree and then the right subtree. If the size bound is the only check in ACCEPTABLE then the resulting tree will always be a "chain" in which only the left child of a node is a test node and the right child is always a leaf (the chain will continue to grow until the maximum size is reached). To avoid this problem, algorithms such as top-down induction in combination with post-pruning (Section 7.5.1), or beam search (Section 7.5.2) are used.

7.5.1 Post Pruning

This approach first top-down induces a large tree (while ignoring the size constraint) and then runs the algorithm from Table 7.2, which computes a minimum loss subtree with at most k nodes, with k the size upper bound. This algorithm was formulated for classification trees by [18] and is based on earlier work by [9] and [1]. [28] extend the algorithm to multi-target PCTs.

The algorithm first calls COMPUTELOSS to find out which nodes are to be included in the solution and then it calls PRUNERECURSIVE to remove the other nodes. COMPUTELOSS employs dynamic programming to compute in $loss[Tr, k]$

Table 7.2 The constrained tree pruning algorithm PRUNETOSIZEK(Tr, k). Tr is the decision tree that is to be pruned. k is the upper bound on the tree's size. Tr_l and Tr_r are the left and right subtrees of Tr. $K_l[\text{Tr}, k]$ stores the maximum size of the left subtree in the minimum loss subtree of at most k nodes rooted at Tr. The corresponding minimum loss is stored in loss[Tr, k].

procedure PRUNETOSIZEK(Tr, k)	**procedure** COMPUTELOSS(Tr, k)
1: COMPUTELOSS(Tr, k)	1: **if** computed[Tr, k]
2: PRUNERECURSIVE(Tr, k)	2: **return** loss[Tr, k]
	3: $K_l[\text{Tr}, k] = -1$
procedure PRUNERECURSIVE(Tr, k)	4: loss[Tr, k] = leaf_loss(Tr)
1: **if** Tr is a leaf	5: **if** $k \geq 3$ **and** Tr is not a leaf
2: **return**	6: **for** $k_l = 1$ **to** $k - 2$
3: **if** $k < 3$ **or** $K_l[\text{Tr}, k] = -1$	7: $k_r = k - k_l - 1$
4: remove children of Tr	8: $e = \text{COMPUTELOSS}(\text{Tr}_l, k_l) + \text{COMPUTELOSS}(\text{Tr}_r, k_r)$
5: **else**	9: **if** $e < $ loss[Tr, k]
6: $k_l = K_l[\text{Tr}, k]$	10: loss[Tr, k] = e
7: $k_r = k - k_l - 1$	11: $K_l[\text{Tr}, k] = k_l$
8: PRUNERECURSIVE(Tr_l, k_l)	12: computed[Tr, k] = true
9: PRUNERECURSIVE(Tr_r, k_r)	13: **return** loss[Tr, k]

the loss of the minimum loss subtree rooted at node Tr containing at most k nodes. This subtree is either the tree in which Tr is pruned to a leaf or a tree in which Tr has two children (we consider binary trees) Tr_l and Tr_r such that Tr_l (Tr_r) is a minimum loss subtree of size at most k_l (k_r) and $k_l + k_r = k - 1$. The algorithm computes the minimum over these possibilities in the for loop starting on line 6. The possibility that Tr is pruned to a leaf is taken into account by initializing the loss in line 4 of the algorithm to the loss that would be incurred if node Tr is replaced by a leaf (e.g., the leaf's variance multiplied by the proportion of instances that are sorted into the leaf). The flag computed[Tr, k] is used to avoid repeated computation of the same information.

After COMPUTELOSS completes, $K_l[\text{Tr}, k]$ stores the maximum size of the left subtree in the minimum loss subtree of at most k nodes rooted at Tr. Note that if $K_l[\text{Tr}, k] = -1$, then this subtree consists of only the leaf Tr. PRUNERECURSIVE is called next to prune nodes that do not belong to the minimum loss subtree.

The algorithm can minimize any loss function that is additive, or that is a monotonically increasing function of an additive loss. Additive means that if a dataset is partitioned into a number of subsets, the loss of the whole set is equal to the sum of the losses of the subsets. Loss functions such as intra-cluster variance (Def. 7.2), mean absolute error, root mean squared error, and classification error all have this property.

PRUNETOSIZEK was used by [15] to construct interpretable PCTs (multi-target regression trees) in the context of an ecological application where the goal is to predict the abundances of organism groups that are present in agricultural soil. Fig. 7.2 shows such a tree where the number of leaves was constrained to 10. Fig. 7.4 illustrates the size/accuracy trade-off of the constructed trees. [28] provide more such experimental results for other application domains. They conclude that multi-target

Fig. 7.4 Tree size versus
cross-validated error trade-off
for a four target PCT on the
same data as was used to learn
the tree in Fig. 7.2.

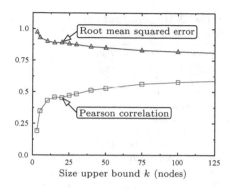

PCTs often attain a better interpretability/accuracy trade-off than traditional single-target regression trees.

7.5.2 Beam Search

[19] propose the PCT induction algorithm CLUS-BS shown in Table 7.3, which implements a beam search. The beam is a set of trees ordered by their heuristic value, and contains at most b trees, with b the beam width. CLUS-BS starts with a beam that contains precisely one tree: a leaf covering all the training data.

Each main loop iteration creates a new beam by refining the trees in the current beam. That is, the algorithm iterates over the trees in the current beam and computes for each tree its set of refinements. A refinement is a copy of the given tree in which one particular leaf is replaced by a new internal node with a particular attribute-value test and two new leaves (Fig. 7.3).

A tree has $|L| \cdot |M|$ refinements, with L its set of leaves, and M the available attribute-value tests. Instead of creating a separate refinement, e.g., for each possible split point of a numeric attribute, CLUS-BS only creates a refinement for the best split (according to its heuristic), that is, $|M|$ is equal to the number of attributes. This limits the number of refinements and increases the diversity of the beam. CLUS-BS only retains refinements that satisfy the given anti-monotonic constraints (tested in the function ACCEPTABLE).

CLUS-BS computes for each generated refinement a heuristic score. The heuristic function used to compute this score differs from the one used in top-down induction (Table 7.1). The heuristic function in the latter is local, i.e., it only depends on the instances local to the node that is being constructed. In CLUS-BS, the heuristic is global and measures the quality of the entire tree. The reason is that beam search needs to compare different trees, whereas top-down induction only needs to rank different tests for the same tree node. CLUS-BS uses the following heuristic:

Table 7.3 The beam search algorithm CLUS-BS. T denotes the training instances, B denotes the beam, and b denotes the beam width.

---

procedure CLUS-BS(T,b)

1: $j = 0$
2: $\text{Tr}_j = \text{leaf}([i,p])$ ▷ i = cluster identifier, p = predicted value
3: $B_j = \{ (h(\text{Tr}_j, T), \text{Tr}_j) \}$ ▷ h = heuristic estimate of the quality of Tr_j
4: **repeat**
5:     $j = j + 1$
6:     $B_j = B_{j-1}$
7:     **for each** $\text{Tr} \in B_{j-1}$
8:         **for each** $\text{Tr}_{\text{cand}} \in \rho(\text{Tr})$ ▷ ρ = refinement operator (Def. 9)
9:             **if** ACCEPTABLE(Tr_{cand})
10:                $h_{\text{cand}} = h(\text{Tr}_{\text{cand}}, T)$
11:                $h_{\text{worst}} = \max_{(h, \text{Tr}) \in B_j} h$
12:                **if** $(h_{\text{cand}} < h_{\text{worst}}) \vee (|B_j| < b)$
13:                    $B_j = B_j \cup \{ (h_{\text{cand}}, \text{Tr}_{\text{cand}}) \}$
14:                **if** $(|B_j| > b)$
15:                    $B_j = B_j \setminus \{ (h_{\text{worst}}, \text{Tr}_{\text{worst}}) \}$ ▷ $(h_{\text{worst}}, \text{Tr}_{\text{worst}}) \in B_j$
16: **until** $B_j = B_{j-1}$
17: **return** B_j

$$h(\text{Tr}, T) = \left(\sum_{\text{leaf}\, l\, \in\, \text{Tr}} \frac{|T_l|}{|T|} \text{Var}^d(T_l) \right) + \alpha \cdot \text{size}(\text{Tr}), \qquad (7.1)$$

with T all training data and T_l the instances sorted into leaf l. It has two components: the first one is the intra-cluster variance of the leaves of the tree, and the second one is a size penalty. The latter biases the search to smaller trees and can be seen as a soft version of a size constraint. Since the heuristic value of a tree is proportional to its loss, CLUS-BS searches for the tree that minimizes the heuristic.

After the heuristic value of a tree is computed, CLUS-BS compares it to the value of the worst tree in the beam. If the new tree is better, or if there are fewer than b trees (b is the beam width), then CLUS-BS adds the new tree to the beam, and if this exceeds the beam width, then it removes the worst tree from the beam. The algorithm ends when the beam no longer changes. This either occurs if none of the refinements of a tree in the beam is better than the current worst tree, or if none of the trees in the beam yields any acceptable refinements.

Eq. 7.1 is similar to the heuristic used in top-down induction. Assume that there are no constraints, $\alpha = 0$ and $b = 1$. In this case, the tree computed by CLUS-BS will be identical to the tree constructed with top-down induction. The only difference is the order in which the leaves are refined: top-down induction refines depth-first, whereas CLUS-BS with $b = 1$ refines best-first. Because CLUS-BS refines best-first, it becomes possible to enforce an upper bound on tree size in the function ACCEPTABLE: the best-first search will not generate a degenerate tree as top-down induction does in this case (start of Section 7.5).

[19] present an experimental evaluation that compares CLUS-BS to top-down induction with PRUNETOSIZEK. This shows that CLUS-BS can generate more accu-

rate trees than top-down induction, mainly because it is less susceptible to myopia. The same paper furthermore investigates how soft similarity constraints can be used to enforce that the trees in the beam are sufficiently diverse; this may allow one to construct an accurate classifier ensemble from the trees in the beam.

7.5.3 Instance Level Constraints

CLUS-BS can be applied for soft constrained clustering in which the constraints are instance level (IL) constraints (must- and cannot-links) by adding a penalty for violating constraints to the heuristic. As proposed by [29], the heuristic then becomes

$$
h(\mathrm{Tr}) = \frac{1-\gamma}{\mathrm{Var}^d(T)} \left(\sum_{\mathrm{leaf}\, l \,\in\, \mathrm{Tr}} \frac{|T_l|}{|T|} \mathrm{Var}^d(T_l) \right) + \gamma \cdot \frac{|\mathrm{violated}(\mathrm{Tr}, IL, T)|}{|IL|}, \qquad (7.2)
$$

with Tr the clustering tree for which the heuristic is to be computed, T the set of instances, IL the set of IL constraints, and T_l the instances in leaf l of Tr. The first term is the normalized intra-cluster variance of the tree and the second term is the proportion of IL constraints that is violated by the tree. The heuristic trades-off both terms by means of a parameter γ.

7.5.3.1 Disjunctive Clustering Trees

A disadvantage of a clustering tree, in which each leaf represents a cluster, is that the description of each cluster is restricted to a conjunction of attribute-value tests. This corresponds to rectangular clusters in the two-dimensional case (Fig. 7.1.b). In many practical problems this assumption is too strong. For example, if clustering with IL constraints is applied to a semi-supervised learning problem, and we know in advance that there are only three classes, then this would restrict the model to a tree with only three leaves. Such a small tree may not be able to accurately model the training data.

Guided by this motivation, [29] propose to adapt clustering trees so that they support disjunctive cluster descriptions. To this end, they define the cluster assignment function so that it can assign the same cluster identifier to multiple leaves. We call a clustering tree with a cluster assignment function that has this property a *disjunctive clustering tree*. For example, the L-shaped cluster C_2 in Fig. 7.5.a is represented by two leaves in Fig. 7.5.b and its disjunctive description is $x_2 \leq 103.5 \vee (x_2 > 103.5 \wedge x_1 > 113.5)$. Note that this is similar to how classification trees represent disjunctive concepts, but here the labels are not given in the data (only IL constraints are given).

When CLUS-BS creates a refined tree, it should decide which cluster identifiers to assign to the two new leaves that are added to the tree. In principle, this can

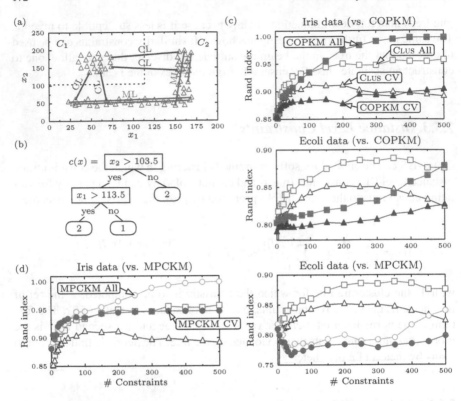

Fig. 7.5 (a) Dataset with instance level constraints. (b) Disjunctive clustering tree for (a). (c) Comparison of CLUS-BS ($b = 1$) and COPKM on datasets with artificially generated constraints (on all data 'All', and cross-validated 'CV'). (d) Similar comparison between CLUS-BS and MPCKM.

be solved by trying all possible pairs of cluster identifiers and picking the one that yields the best heuristic value. It turns out that this can be done more efficiently. In particular, it is possible to compute for a given numeric attribute, the heuristic of each possible split point and the corresponding best cluster identifiers for the new leaves, in one pass over the instances and constraints. After sorting the instances by the value of the numeric attribute, the computational cost of such a pass is $\mathcal{O}(|T_l|(|C| + |A|) + |IL|)$, with C the set of clusters and A the set of attributes. Details of this optimization can be found in [29].

7.5.3.2 Results

Fig. 7.5.c shows a comparison of CLUS-BS to COPKM [34], a well known extension of k-means that performs hard IL constrained clustering. The data are UCI classification tasks in which the class labels have been replaced by a random set of IL constraints that is consistent with the classes. A detailed description of the experimental setup and results can be found in [29].

If COPKM finds a solution, then this solution is always consistent with the constraints (it performs hard constrained clustering). This means that, given sufficient constraints, its training set Rand index is equal to one (Fig. 7.5.c, COPKM, all data). The Rand index measures how well the induced clusters match the original classes in the dataset.

CLUS-BS, which performs soft constrained clustering, does not have this guarantee (CLUS, all data). However, when using constrained clustering for semi-supervised learning, generalization performance to unconstrained instances may be more relevant. The cross-validated results (Fig. 7.5.c, CV) indicate that CLUS-BS has a better generalization performance than COPKM, and more extensive experiments also confirm this (See [29]). This effect can be explained by the high bias of COPKM (it creates spherical clusters). Because it supports disjunctive concepts, CLUS-BS is not limited to rectangular clusters and can approximate complex cluster shapes. Another reason is that CLUS-BS has the option to ignore some of the constraints [14].

Fig. 7.5.d compares CLUS-BS to MPCKM [4], a k-means extension that performs both soft constraint satisfaction and metric learning. MPCKM performs best on Iris, while CLUS-BS performs best on Ecoli. In a larger study, we observed 3 wins for CLUS-BS, 3 wins for MPCKM, and 3 draws. So, in terms of wins/losses both perform equally well, and which system should be preferred depends on the data set at hand and on whether or not one needs conceptual cluster descriptions.

7.6 Conclusion

This chapter reviewed (predictive) clustering trees (PCTs), which are a generally applicable (to both prediction and clustering tasks) and easily interpretable type of models. We described the basic top-down induction algorithm for such trees and showed how it can be instantiated to specific tasks, among others, to multi-task learning. We then addressed the central topic of this chapter, i.e., constrained (predictive) clustering. We presented a review of the constraint types that are most relevant to predictive clustering. Next, we discussed the search space of PCTs, which is traversed by the tree induction algorithms, the subtree order, and anti-monotonic constraints. Finally, we covered two approaches to build trees that satisfy a given set of constraints: top-down induction followed by post-pruning, and beam search. The beam search approach to PCT induction can be applied to instance level constrained clustering. We have illustrated the methods with selected experimental results and provided pointers to the relevant literature for more detailed results.

We have thus shown that PCTs can be used effectively in prediction and clustering applications where constraints are available and interpretable models are required.

The PCT paradigm generalizes prediction and clustering. Consequently both constraints that apply to prediction (e.g., error and size constraints) and constraints that apply to clustering (e.g., instance level constraints) can be taken into account when learning PCTs.

PCTs allow for predicting structured outputs (vectors, hierarchical classes, time series). The present work – constrained induction of PCTs – carries over to those settings. In this sense, this is one of the first works considering constrained structured prediction.

Future work could investigate other types of constraints (e.g., constraints specific to structured prediction) and algorithms for enforcing them.

Acknowledgements The authors are grateful to Celine Vens for providing feedback on an earlier draft of this chapter. Part of the research presented in this chapter was conducted within the project IQ (*Inductive Queries for mining patterns and models*) funded by the European Commission of the EU under contract number FP6-IST 516169. For a complete list of agencies, grants and institutions currently supporting Sašo Džeroski, please consult the Acknowledgements chapter of this volume.

References

1. H. Almuallim. An efficient algorithm for optimal pruning of decision trees. *Artificial Intelligence*, 83(2):347–362, 1996.
2. N. Angelopoulos and J. Cussens. Exploiting informative priors for Bayesian classification and regression trees. In *19th Int'l Joint Conf. on Artificial Intelligence*, pages 641–646, 2005.
3. S. Basu, M. Bilenko, and R.J. Mooney. A probabilistic framework for semi-supervised clustering. In *10th ACM SIGKDD Int'l Conf. on Knowledge Discovery and Data Mining*, pages 59–68, 2004.
4. M. Bilenko, S. Basu, and R.J. Mooney. Integrating constraints and metric learning in semi-supervised clustering. In *21st Int'l Conf. on Machine Learning*, pages 81–88, 2004.
5. S. Bistarelli and F. Bonchi. Extending the soft constraint based mining paradigm. In *5th Int'l Workshop on Knowledge Discovery in Inductive Databases*, pages 24–41, 2007.
6. H. Blockeel, L. De Raedt, and J. Ramon. Top-down induction of clustering trees. In *15th Int'l Conf. on Machine Learning*, pages 55–63, 1998.
7. H. Blockeel, S. Džeroski, and J. Grbović. Simultaneous prediction of multiple chemical parameters of river water quality with Tilde. In *3rd European Conf. on Principles of Data Mining and Knowledge Discovery*, pages 32–40, 1999.
8. Hendrik Blockeel. *Top-down Induction of First Order Logical Decision Trees*. PhD thesis, K.U. Leuven, Dep. of Computer Science, Leuven, Belgium, 1998.
9. M. Bohanec and I. Bratko. Trading accuracy for simplicity in decision trees. *Machine Learning*, 15(3):223–250, 1994.
10. P.S. Bradley, K.P. Bennett, and A. Demiriz. Constrained k-means clustering. Technical Report MSR-TR-2000-65, Microsoft Research, 2000.
11. L. Breiman, J.H. Friedman, R.A. Olshen, and C.J. Stone. *Classification and Regression Trees*. Wadsworth, Belmont, 1984.
12. Rich Caruana. Multitask learning. *Machine Learning*, 28(1):41–75, 1997.
13. I. Davidson and S.S. Ravi. Clustering with constraints: Feasibility issues and the k-means algorithm. In *SIAM Int'l Data Mining Conf.*, 2005.
14. I. Davidson, K. Wagstaff, and S. Basu. Measuring constraint-set utility for partitional clustering algorithms. In *10th European Conf. on Principles and Practice of Knowledge Discovery in Databases*, pages 115–126, 2006.

15. D. Demšar, S. Džeroski, P. Henning Krogh, T. Larsen, and J. Struyf. Using multiobjective classification to model communities of soil microarthropods. *Ecological Modelling*, 191(1):131–143, 2006.

16. S. Džeroski, I. Slavkov, V. Gjorgjioski, and J. Struyf. Analysis of time series data with predictive clustering trees. In *5th Int'l Workshop on Knowledge Discovery in Inductive Databases*, pages 47–58, 2006.

17. A. Friedman, Schuster A., and R. Wolff. k-anonymous decision tree induction. In *10th European Conf. on Principles and Practice of Knowledge Discovery in Databases*, pages 151–162, 2006.

18. M. Garofalakis, D. Hyun, R. Rastogi, and K. Shim. Building decision trees with constraints. *Data Mining and Knowledge Discovery*, 7(2):187–214, 2003.

19. D. Kocev, J. Struyf, and S. Džeroski. Beam search induction and similarity constraints for predictive clustering trees. In *5th Int'l Workshop on Knowledge Discovery in Inductive Databases*, pages 134–151, 2007.

20. D. Kocev, C. Vens, J. Struyf, and S. Džeroski. Ensembles of multi-objective decision trees. In *18th European Conf. on Machine Learning*, pages 624–631, 2007.

21. C. X. Ling, Q. Yang, J. Wang, and S. Zhang. Decision trees with minimal costs. In *21 Int'l Conf on Machine Learning*, pages 544–551, 2004.

22. R.S. Michalski and R.E. Stepp. Learning from observation: Conceptual clustering. In *Machine Learning: An Artificial Intelligence Approach*, volume 1. Tioga Publishing Company, 1983.

23. C. Nédellec, H. Adé, F. Bergadano, and B. Tausend. Declarative bias in ILP. In *Advances in Inductive Logic Programming*, volume 32 of *Frontiers in Artificial Intelligence and Applications*, pages 82–103. IOS Press, 1996.

24. S. Nijssen and E. Fromont. Optimal constraint-based decision tree induction from itemset lattices. *Data Mining and Knowledge Discovery*, 21(1):9–51, 2010.

25. J.R. Quinlan. Learning with continuous classes. In *5th Australian Joint Conf. on Artificial Intelligence*, pages 343–348. World Scientific, 1992.

26. L.E. Raileanu and K. Stoffel. Theoretical comparison between the Gini index and information gain criteria. *Annals of Mathematics and Artificial Intelligence*, 41(1):77–93, 2004.

27. R. E. Schapire and Y. Singer. Improved boosting algorithms using confidence-rated predictions. *Machine Learning*, 37(3):297–336, 1999.

28. J. Struyf and S. Džeroski. Constraint based induction of multi-objective regression trees. In *4th Int'l Workshop on Knowledge Discovery in Inductive Databases*, pages 222–233, 2006.

29. J. Struyf and S. Džeroski. Clustering trees with instance level constraints. In *18th European Conf. on Machine Learning*, pages 359–370, 2007.

30. L. Todorovski, B. Cestnik, M. Kline, N. Lavrač, and S. Džeroski. Qualitative clustering of short time-series: A case study of firms reputation data. In *Integration and Collaboration Aspects of Data Mining, Decision Support and Meta-Learning*, pages 141–149, 2002.

31. P. Turney. Cost-sensitive classification: Empirical evaluation of a hybrid genetic decision tree induction algorithm. *J. of Artificial Intelligence Research*, 2:369–409, 1995.

32. C. Vens, J. Struyf, L. Schietgat, S. Džeroski, and H. Blockeel. Decision trees for hierarchical multi-label classification. *Machine Learning*, 73(2):185–214, 2008.

33. K. Wagstaff and C. Cardie. Clustering with instance-level constraints. In *17th Int'l Conf. on Machine Learning*, pages 1103–1110, 2000.

34. K. Wagstaff, C. Cardie, S. Rogers, and S. Schroedl. Constrained k-means clustering with background knowledge. In *18th Int'l Conf. on Machine Learning*, pages 577–584, 2001.

35. B. Ženko and S. Džeroski. Learning classification rules for multiple target attributes. In *Advances in Knowledge Discovery and Data Mining*, pages 454–465, 2008.

36. S. Zhong and J. Ghosh. Scalable, balanced model-based clustering. In *SIAM Int'l Conf. on Data Mining*, pages 71–82, 2003.

Chapter 8
Finding Segmentations of Sequences

Ella Bingham

Abstract We describe a collection of approaches to inductive querying systems for data that contain segmental structure. The main focus in this chapter is on work done in Helsinki area in 2004-2008. Segmentation is a general data mining technique for summarizing and analyzing sequential data. We first introduce the basic problem setting and notation. We then briefly present an optimal way to accomplish the segmentation, in the case of no added constraints. The challenge, however, lies in adding constraints that relate the segments to each other and make the end result more interpretable for the human eye, and/or make the computational task simpler. We describe various approaches to segmentation, ranging from efficient algorithms to added constraints and modifications to the problem. We also discuss topics beyond the basic task of segmentation, such as whether an output of a segmentation algorithm is meaningful or not, and touch upon some applications.

8.1 Introduction

Segmentation is a general data mining technique for summarizing and analyzing sequential data. It gives a simplified representation of data, giving savings in storage space and helping the human eye to better catch an overall picture of the data. Segmentation problems arise in many data mining applications, including bioinformatics, weather prediction, telecommunications, text processing and stock market analysis, to name a few.

The goal in segmentation is to decompose the sequence, such as a time series or a genomic sequence, into a small number of homogeneous non-overlapping pieces, segments, such that the data in each segment can be described accurately by a

Ella Bingham
Helsinki Institute for Information Technology, University of Helsinki and Aalto University School of Science and Technology
e-mail: ella.bingham@hiit.fi

simple model. In many applications areas, this is a natural representation and reveals the high-level characteristics of the data by summarizing large scale variation. For example, in a measurement time series, each segment s_j could have a different mean parameter μ_j such that the measurement values x in segment s_j are modeled as $x = \mu_j + \text{noise}$.

Segmentation algorithms are widely used for extracting structure from sequences; there exist a variety of applications where this approach has been applied [3, 4, 6, 23, 29, 30, 32, 37, 38]. Sequence segmentation is suitable in the numerous cases where the underlying process producing the sequence has several relatively stable states, and in each state the sequence can be assumed to be described by a simple model. Naturally, dividing a sequence into homogeneous segments does not yield a perfect description of the sequence. Instead, a simplified representation of the data is obtained — and this is often more than welcome.

One should note that in statistics the question of segmentation of a sequence or time series is often called the change-point problem.

If no constraints are made between different segments, finding the optimal segmentation can be done for many model families by using simple dynamic programming [2] in $O(n^2k)$ time, where n is the length of the sequence and k is the number of segments. Thus one challenge lies in adding constraints that relate the segments to each other and make the end result more interpretable for the human eye. Another challenge is to make the computational task simpler. We will discuss both of these challenges in this chapter, and many more.

This chapter is a survey of segmentation work done in the Helsinki area: at Helsinki Institute for Information Technology, which is a joint research institute of University of Helsinki and Aalto University (part of it formerly known as Helsinki University of Technology), roughly between the years 2004 and 2008.

Notation. We assume that our data is a d-dimensional sequence T consisting of n observations, that is, $T = \langle t_1, \ldots, t_n \rangle$ where $t_i \in \mathbb{R}^d$. A k-segmentation S of T is a partition of $\langle 1, 2, \ldots, n \rangle$ into k non-overlapping contiguous subsequences (segments), $S = \langle s_1, \ldots, s_k \rangle$ such that $s_i = \langle t_{b(i)}, \ldots,$
$t_{b(i+1)-1} \rangle$ where $b(i)$ is the beginning of the i:th segment. In its simplest case, segmentation collapses the values within each segment s into a single value μ_s which is e.g. the mean value of the segment. We call this value the *representative* of the segment. Collapsing points into representatives results in a loss of accuracy in the sequence representation. This loss of accuracy is measured by the *reconstruction error*

$$E_p(T,S) = \sum_{s \in S} \sum_{t \in s} ||t - \mu_s||^p.$$

The *segmentation problem* is that of finding the segmentation minimizing this reconstruction error. In practice we consider the cases $p = 1, 2$. For $p = 1$, the optimal representative of each segment is the median of the points in the segment, for $p = 2$ it is the mean of the points.

Depending on the constraints one imposes on the representatives, one can consider several variants of the segmentation problem, and we will discuss many of them later in this chapter. Also, instead of representing the data points in a segment by a single representative, one can consider simple functions of the data points. Extensions of the methods presented in this chapter into functional representatives is often straightforward.

Figure 8.1 shows an example of a signal and its segmentation. In this simple case, the segments are represented by the mean values of the points belonging to a segment.

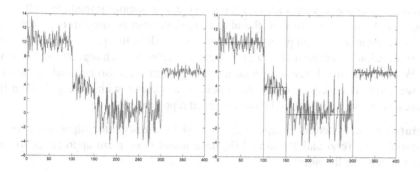

Fig. 8.1 Left: a signal. Right: the result of segmenting into $k = 4$ segments. The vertical bars show the segment boundaries ($b(1) = 0, b(2) = 100, b(3) = 150, b(4) = 300$). The horizontal lines show the representatives of the segments which in this case are the mean values of the points in each segment: $\mu_1 = 10.12, \mu_2 = 3.85, \mu_3 = 0.00, \mu_4 = 5.97$.

Optimal segmentation. Let us start by giving the optimal algorithm for solving the plain segmentation problem. Let $\mathscr{S}_{n,k}$ denote the set of all k-segmentations of sequences of length n. For some sequence T and error measure E_p, we define the optimal segmentation as

$$S_{opt}(T,k) = \arg\min_{S \in \mathscr{S}_{n,k}} E_p(T,S).$$

We sometimes write $E(S)$ instead of $E_p(T,S)$ as the dependence of the data T is obvious, and p is often clear from the context. Finding the optimal segmentation S_{opt} for a given T of length n and for given k and E_p can be done in time $O(n^2 k)$ [2] by a standard dynamic programming (DP) algorithm. The dynamic programming algorithm proceeds in an incremental fashion, using a table A of size $n \times k$, where the entry $A[i, \ell]$ denotes the error of segmenting the sequence $T[1, i]$ using at most ℓ segments. Here $T[1, i]$ denotes the subsequence of T that contains all points between 1 and i: $T[1, i] = \langle t_1, \ldots, t_i \rangle$. Let $E(S_{opt}(T[j,i], 1))$ be the minimum error that can be obtained for the subsequence $T[j, i]$ when representing it as one segment. The computation of the entries of table A is based on the equation ([17, 25] etc.)

$$A[i, \ell] = \min_{1 \leq j \leq i} (A[j-1, \ell-1] + E(S_{opt}(T[j,i], 1)))$$

The table is first initialized with entries $A[i,1] = E(S_{opt}(T[1,i],1))$ for all $i = 1,\ldots,n$. The best k-segmentation is found by storing the choices for j at each step in the recursion in another table B, and by reconstructing the sequence of choices starting from $B[n,k]$. Note that the table B also gives the optimal k-segmentations for all $k' < k$, which can be reconstructed by starting from $B[n,k']$.

The resulting segmentation is optimal in the sense that the representation error (8.1) between the original sequence and a piecewise constant representation with k segments is minimized. A piecewise constant representation is one in which the representatives of the segments are constants (in practice, means or medians of the data points in the segment). In fact, the above algorithm can be used to compute optimal k-segmentations with piecewise polynomial models. In a piecewise polynomial representation, a polynomial of a given degree is fitted to each segment separately.

We note that the dynamic programming algorithm can also be used in the case of weighted sequences in which each point is associated with a weight. Then the representatives are defined to be the weighted representatives.

Related work. There is a large body of work in segmentation algorithms for sequential data. Terzi and Tsaparas [43] have found three main approaches to segmentation in the literature:

1. Heuristics for solving a segmentation problem faster than the optimal dynamic programming algorithm, with promising experimental results but no theoretical guarantees about the quality of the result.
2. Approximation algorithms with provable error bounds, that is, theoretical upper bounds for the error compared to the optimal error.
3. New variations of the basic segmentation problem, imposing some modifications or constraints on the structure of the representatives of the segments.

The majority of the papers published on segmentation fall into Category 1, fast heuristics. The most popular of these algorithms are the top-down and bottom-up greedy algorithms. The *top-down greedy algorithm* is used in e.g. [4, 11, 29, 40] and briefly discussed in [43]: The algorithm starts with an unsegmented sequence and introduces a new boundary at every greedy step. That is, in the i-th step the algorithm introduces the i-th segment boundary by splitting one of the existing i segments into two. The new boundary is selected in such a way that it minimizes the overall error. No changes are made to the existing $i-1$ boundary points. The splitting is repeated until the number of segments reaches k. The running time of the algorithm is $O(nk)$.

In the *bottom-up greedy algorithm*, each point initially forms a segment of its own. At each step, two consecutive segments that cause the smallest increase in the error are merged. The algorithm stops when k segments are formed. The time complexity of the bottom-up algorithm is $O(n\log n)$. The algorithm performs well in terms of error and it has been used widely in time-series segmentation [18, 35, 43].

Yet another fast heuristics is presented by Himberg et al [23]: two slightly different randomized algorithms that start with a random k-segmentation. At each step they pick one segment boundary (randomly or in some order) and search for the best position to put it back. This is repeated until the representation error converges.

Both algorithms run in time $O(In)$ where I is the number of iterations needed until convergence.

For the algorithms in Category 1 there is empirical evidence that their performance is often very good in practice. However, there are no guarantees of their worst-case error ratio. This is in contrast to algorithms in Category 2 for which error bounds can be shown. In Category 2, an interesting contribution is that of Guha et al [16]: a fast segmentation algorithm with provable error bounds. Terzi and Tsaparas [43] have a similar motivation but different point of view, and we will take a closer look at this in Section 8.2. Category 3, variations of the basic segmentation problem, is studied extensively, and several approaches will be described in the following sections.

Online versions of the segmentation problem have also been studied ([26, 36] and others). In this setting, new observations arrive continuously in a streaming manner, making the data a streaming time series.

An interesting restriction on the segmentation problem in the online case is to require more accuracy in the representation of new observations, as opposed to those which arrived further away in the past. This representation is called *amnesic* as the fidelity of approximation decreases with time, and we are willing to answer queries about the recent past with greater precision. Palpanas et al [36] use a piecewise linear segmentation method to this end. The error of the approximation is always kept under some user-specified, time-dependent threshold.

An abstract framework for the study of streaming time series is recently given by Gandhi et al [12]. They present theoretical results for the space-quality approximation bounds. Both data streams, amnesic approximations and out-of-order streams are discussed in their paper. The case of out-of-order time series will also be discussed in Section 8.6 but only in the case of non-streaming, offline segmentation.

A task related to segmentation is time series approximation or summarization. Similarly to the task of segmentation, the goal here is again to simplify the representation of a sequence. Classical signal processing approaches to time series approximation include Discrete Fourier Transform, Discrete Cosine Transform and Discrete Wavelet Transform; common to these tree methods is that a segment-wise presentation is not sought but the characteristics of the sequence are represented using an existing "dictionary" of finer and coarser building blocks. Instead, methods such as Piecewise Aggregate Approximation [46], Adaptive Piecewise Constant Approximation [8], Piecewise Linear Approximation [7, 28] and Piecewise Quadratic Approximation [22] etc. are segmentation methods, and the representatives of the segments are simple functions of the data points in the segment. An interesting comparison on all of these methods is given in Palpanas et al. [36], by measuring their reconstruction accuracy on several real world data sets. A perhaps surprising result was that there was little difference between all the approaches; similar results have been reported elsewhere, too [8, 27, 45]. The take-home message in this respect is that we should not choose the representation method based on approximation fidelity but rather on other features [36]. This is a guiding principle behind the methods described in this chapter, too.

An alternative approach to analyzing sequential data is a Hidden Markov Model (HMM): the observed data is generated by an unknown process that takes several (unobserved) states, and different states output different observations. Churchill [9] was among the first to apply HMMs to sequence segmentation.

Organization. In this chapter, we describe various approaches to segmentation, ranging from efficient algorithms to added constraints and beyond. We start with an efficient approximation algorithm with provable error bounds in Section 8.2. In Sections 8.3 to 8.5 we discuss three different constraints to make the segmentation result more tractable. Sections 8.6 and 8.7 discuss interesting variations in the basic problem setting. Sections 8.8 to 8.10 touch upon other topics related to segmentation, such as determining the goodness of an output of a segmentation algorithm, model selection issues, and bursty event sequences. Finally, Section 17.7 gives a brief conclusion.

8.2 Efficient Algorithms for Segmentation

In the general case, an optimal segmentation for a sequence can be found using dynamic programming [2] in $O(n^2k)$ time, where n is the length of the sequence and k is the number of segments. In practice, sequences are typically very long, and a quadratic algorithm is seldom adequate. Faster heuristics with $O(n \log n)$ or $O(n)$ running time have been presented (see Section 16.1), but there are often no guarantees of the quality of the solutions they produce.

Instead, Terzi and Tsaparas [43] have presented a constant-factor approximation algorithm whose optimal running time is $O(n^{4/3}k^{5/3})$, called the *divide and segment* (DnS) algorithm. The error of the segmentation it produces is provably no more than 3 times that of the optimal segmentation; we thus say that the approximation ratio is 3. The main idea of the algorithm is to divide the problem into smaller subproblems, solve the subproblems optimally and combine their solutions to form the final solution:

- The algorithm starts by partitioning the sequence T into m disjoint subsequences T_i (of equal length, typically).
- Then each T_i is segmented optimally by dynamic programming, yielding a segmentation S_i and a set M_i of k weighted points $M_i = \langle \mu_{i1}, \ldots, \mu_{ik} \rangle$: these are the representatives of the segments (means or medians), weighted by the length of the segment they represent.
- All the mk representatives of the m subsequences are concatenated to form the weighted sequence $T' = \langle \mu_{11}, \ldots, \mu_{1k}, \mu_{21}, \ldots, \mu_{mk} \rangle$, and dynamic programming is then applied on T', outputting the final segmentation.

Assuming that the subsequences are of equal length, the running time of the algorithm depends on m, the number of subsequences. The optimal running time is $2n^{4/3}k^{5/3}$ and it is achieved at $m = (n/k)^{2/3}$ [43].

Terzi and Tsaparas [43] also explore several more efficient variants of the algorithm and quantify the accuracy/efficiency tradeoff. More specifically, they present a recursive application of the DnS algorithm, resulting in a faster algorithm with $O(n \log \log n)$ running time and $O(\log n)$ approximation ratio. All presented algorithms can be made to use a sublinear amount of memory, making them applicable to the case when the data needs to be processed in a streaming fashion (not stored in main memory). Assuming that one has an estimate of n, the size of the sequence, then the algorithm processes the points in batches of size n/m. For each such batch it computes the optimal k-segmentation, and stores the representatives. The space required is $M = n/m + mk$ and this is minimized for $m = \sqrt{n/k}$, resulting in space $M = 2\sqrt{nk}$.

Extensive experiments on both real and synthetic datasets demonstrate that in practice their algorithms perform significantly better than the worst-case theoretical upper bounds, in terms of reconstruction error. Also, the algorithms perform consistently better than fast heuristic algorithms, and the computational costs are comparable [43]. The synthetic datasets are generated by first fixing the dimensionality of the data ($d = 1, 5, 10$) and the segment boudaries ($k = 10$), and then drawing the mean of each segment in each dimension from a Uniform distribution, and adding Gaussian noise whose standard deviation varies from 0.05 to 0.9. The real datasets *balloon, darwin, winding, xrates* and *phone* are from the UCR Time Series Data Mining Archive[1].

8.3 Dimensionality Reduction

Let us then start discussing the various constraints and modifications we add to the problem of segmentation to make the end result more tractable. The first natural constraint that we wish to incorporate in the segmentation arises from dimensionality reduction in multidimensional time series: the multidimensional mean parameters μ_j of the segments should lie within a subspace whose dimensionality is smaller than that of the original space.

Bingham et al. [5] have stated the problem as follows. Given a multidimensional time series, find a small set of latent variables and a segmentation of the series such that the data in each segment can be explained well by some (linear) combination of the latent variables. We call this problem the *basis segmentation problem*.

Our problem formulation allows decomposing the sequences into segments in which the data points are explained by a model unique to the segment, yet the whole sequence can be explained adequately by the vectors of the basis.

Following the notation presented in Section 16.1, our data is a sequence consisting of n observations of d-dimensional vectors. For convenience, we now stack the vectors into a matrix X that contains in its rows the n observations, each of which

---

[1] http://www.cs.ucr.edu/~eamonn/TSDMA/

is d-dimensional, so X is an $n \times d$ matrix. As previously, the n observations will be partitioned into k segments $S = \langle s_1, \ldots, s_k \rangle$.

We will consider basis-vector representations of the data. We denote by $V = \{v_1, \ldots, v_m\}$ the set of m basis vectors $v_\ell \in \mathbb{R}^d$, $\ell = 1, \ldots, m$. The number of basis vectors m is typically significantly smaller than the dimensionality d of the data points. In matrix form, V is an $m \times d$ matrix containing the basis vectors as its rows. Also, for each segment S_j we have a set of coefficients $a_{j\ell} \in \mathbb{R}$ for $\ell = 1, \ldots, m$ that tell us how to represent the data using the basis vectors in V. In matrix notation, $A = (a_{j\ell})$ is a $k \times m$ matrix of coefficients. V and A will be found by Principal Component Analysis (PCA, discussed more in the sequel).

We approximate the sequence with *piecewise constant* linear combinations of the basis vectors, i.e., all observations in segment s_j are represented by a single vector

$$u'_j = \sum_{\ell=1}^{m} a_{j\ell} v_\ell. \tag{8.1}$$

The problem we consider is the following.

Problem 8.1. Denote by $j(i) \in \{1, \ldots, k\}$ the segment to which point i belongs. Given a sequence $T = \langle t_1, \ldots, t_n \rangle$, and integers k and m, find a basis segmentation (S, V, A) that uses k segments and a basis of size m, so that the reconstruction error

$$E(T; S, V, A) = \sum_{i=1}^{n} ||t_i - u'_{j(i)}||^2$$

is minimized. The constant vector $u'_{j(i)}$ for approximating segment S_j is given by Equation (8.1).

To solve the basis segmentation problem, we combine existing methods for sequence segmentation and for dimensionality reduction: (*i*) k-segmentation by dynamic programming, discussed in Section 16.1, and (*ii*) Principal Component Analysis (PCA), one of the most commonly used methods for dimensionality reduction. Given a matrix Z of size $n \times d$ with data points as rows, the goal in PCA is to find a subspace of dimension $r < d$ so that the residual error of the points of Z projected onto the subspace is minimized. The PCA algorithm computes a matrix Y of rank r, and the decomposition $Y = AV$ of Y into the orthogonal basis V of size r, such that

$$Y = \underset{\text{rank}(Y') \le r}{\arg\min} ||Z - Y'||$$

which holds for all matrix norms induced by L_p vector norms. PCA is typically accomplished by Singular Value Decomposition (SVD) on the data matrix Z. The basis vectors $v_1, \ldots, v_m$ are the right singular vectors of the data matrix.

We suggest three different algorithms for solving the basis segmentation problem, all of which combine k-segmentation and PCA in different ways:

- Seg-PCA: First partition into k segments in the full d-dimensional space, to obtain segments $S = \langle s_1, \ldots s_k, \rangle$ and d-dimensional vectors $u_1, \ldots, u_k$ representing the points in each segment. Then consider the set $U_S = \{(u_1, |s_1|), \ldots, (u_k, |s_k|)\}$ where each vector u_j is weighted by $|s_j|$, the length of segment s_j. Perform PCA on the set of weighted points U_S, outputting for each segment vector u_j an approximate representation u'_j as in (8.1). Bingham et al [5] show that the Seg-PCA algorithm yields a solution to the basis segmentation problem such that the reconstruction error is at most 5 times the reconstruction error of the optimal solution. Experiments demonstrate that in practice, the approximation ratios are smaller than 5.

- Seg-PCA-DP: First segment into k segments, then find a basis of size m for the segment means, similarly to above. Then refine the segmentation boundaries by using the discovered basis by a second application of dynamic programming. As the first two steps of the algorithm are identical to the Seg-PCA algorithm, and the last step can only improve the cost of the solution, the same approximation ratio of 5 holds also for Seg-PCA-DP.
- PCA-Seg: First do PCA to dimension m on the whole data set. Then obtain the optimal segmentation of the resulting m-dimensional sequence. This gives computational savings, as the segmentation is not performed on a high-dimensional space.

Experiments on synthetic and real datasets show that all three algorithms discover the underlying structure in the data [5]. Prototype implementations are available to the public at http://www.cs.helsinki.fi/hiit_bru/software/.

A somewhat related problem setting, restricting the complexity of the representatives of the segments, will be considered in the next section.

8.4 Recurrent Models

Whereas in Section 8.3 we represented the segments as different combinations of a small set of global basis vectors, we now wish to use a small set of models to predict the data values in the segments.

Often in a sequence with segmental structure, similar types of segments occur repeatedly: different models are suitable in different segments. For example, high solar radiation implies clear skies, which in the summer means warm temperatures and in the winter cold ones. As another example, the inheritance mechanism of recombinations in chromosomes mean that a genome sequence can be explained by using a small number of ancestral models in a segment-wise fashion. In these examples, the model used to explain the target variable changes relatively seldom, and has a strong effect on the sequence. Moreover, the same models are used repeatedly in different segments: the summer model works in any summer segment, and the same ancestor contributes different segments of the genome.

In an earlier contribution by Gionis and Mannila [13], the idea for searching for recurrent models was used in the context of finding piecewise *constant* approximations in the so called (k, h) segmentation problem. In their paper it was assumed that the sequence can be segmented into k pieces, of which only h are distinct. In other words, there are h hidden sources such that the sequence can be written as a concatenation of $k > h$ pieces, each of which stems from one of the h sources. This problem was shown to be NP-hard, and approximate algorithms were given [13]. The "Segments2Levels" algorithm runs in time $O(n^2(k+h))$ and gives a 3-approximation for $p = 1, 2$ for dimension 1. For higher dimensions, the approximation guarantees are $3 + \varepsilon$ for $p = 1$ and $\alpha + 2$ for $p = 2$ where α is the best approximation factor for the k-means problem. The "ClusterSegments" algorithm yields approximation ratios 5 and $\sqrt{5}$ for $p = 1$ and 2, respectively; the running time is again $O(n^2(k+h))$. The "Iterative" algorithm is inspired by the EM algorithm and provides at least as good approximations as the two previous ones. Its running time is $O(In^2(k+h))$ where I is the number of iterations.

The goal in (k, h) segmentation is similar to, although the technique is different from, using a Hidden Markov Model (HMM) to sequence segmentation, originally proposed by Churchill [9].

In a new contribution by Hyvönen et al [25], this approach was used to arbitrary *predictive models*, which requires considerably different techniques than those in (k, h) segmentation [13]. To find such recurrent predictive models, one must be able to do segmentation based not on the target to be predicted itself, but on *which model* can be used to predict the target variable, given the input measurements. The application areas discussed above, the temperature prediction task and ancestral models in a genome sequence, call for such a recurrent predictive model.

Given a model class $\mathcal{M}$, the task is to search for a small set of h models from $\mathcal{M}$ and a segmentation of the sequence T into k segments such that the behavior of each segment is explained well by a single model. It is assumed that $h < k$, i.e., the same model will be used for multiple segments. More precisely, the data $D = (T, y)$ consist of a multidimensional sequence $T = \langle t_1, \ldots, t_n \rangle$, $t_i \in \mathbb{R}^d$ and corresponding scalar outcome values $y = \langle y_1, \ldots, y_n \rangle$, $y_i \in \mathbb{R}$. We denote a subsequence of the input sequence between the i-th and j-th data point as $D[i, j]$. A model M is a function $M : \mathbb{R}^d \to \mathbb{R}$ that belongs to a model class $\mathcal{M}$. Given a subsequence $D[i, j]$ and a model $M \in \mathcal{M}$ the prediction error of M on $D[i, j]$ is defined as

$$E(D[i, j], M) = \sum_{\ell=i}^{j} ||M(t_\ell) - y_\ell||^2. \tag{8.2}$$

For many commonly used model classes $\mathcal{M}$ one can compute in polynomial time the model $M^* \in \mathcal{M}$ that minimizes the error in (8.2). For example, for the class of linear models, the optimal model can be found using least squares. For probabilistic models one can estimate the maximum likelihood model. For some model classes such as decision trees, finding the optimal model is computationally difficult, but efficient heuristics exist. It is thus assumed that one can always find a good model for a given subsequence.

One should note that the task of predicting a given output value y_i for a multi-dimensional observation t_i using a model $M \in \mathcal{M}$ is now different from the basic segmentation task in which the "output" or the representative of the segment is not given beforehand. In the latter, the task is to *approximate* the sequence rather than *predict*.

Now, let us first define the "easy" problem:

Problem 8.2. Given an input sequence D, a model class $\mathcal{M}$, and a number k, partition D into k segments $D_1, \ldots, D_k$ and find corresponding models $M_1, \ldots, M_k \in \mathcal{M}$ such that the overall prediction error $\sum_{j=1}^{k} E(D_j, M_j)$ is minimized.

The above problem allows for different models in each of the k segments. Our interest, however, is in the *recurrent predictive modeling* problem which is a more demanding task in that it only allows for a small number of h distinct models, $h < k$. Thus, some of the models have to be used in more than one segment. More formally, we define the following problem.

Problem 8.3. Consider a sequence D, a model class $\mathcal{M}$, and numbers k and h, $h < k$. The task is to find a k-segmentation of D into k segments $D_1, \ldots, D_k$, h models $M_1, \ldots, M_h \in \mathcal{M}$, and an assignment of each segment j to a model $M_{m(j)}$, $m(j) \in \{1, \ldots, h\}$ so that the prediction error $\sum_{j=1}^{k} E(D_j, M_{m(j)})$ is minimized.

For any but the simplest model class the problem of finding the best h models is an NP-hard task, so one has to resort to approximate techniques.

Given a sequence D and a class of models $\mathcal{M}$, dynamic programming [2] is first used to find a good segmentation of the sequence into k segments. Thus each segment will have its unique predictive model. The method for finding the model describing a single segment depends, of course, on the model class $\mathcal{M}$. After that, from the k models found in the segmentation step, one selects a smaller number of h models that can be used to model well the whole sequence. In case parameters k and h can be fixed in advance, selecting a smaller number of models is treated as a clustering problem, and solved using the k-median [31] or k-means algorithm. Finally, an *iterative improvement algorithm* that is a variant of the EM algorithm is applied: iteratively fit the current models more accurately in the existing segments, and then find a new segmentation given the improved models. The iteration continues until the error of the solution does not improve any more.

In the more general case, the parameters k and h are not given, but need to be determined from the data. This model selection problem is addressed using the Bayesian Information Criterion (BIC). Selecting a smaller number of models is again a clustering problem, and using the facility location approach [24] one only has to iterate over the number of segments k: For each value of k, the corresponding value of h that minimizes the BIC score is automatically selected by the facility location algorithm.

In [25] the method of recurrent models was applied to two sets of real data, meteorological measurements, and haplotypes in the human genome. The experimental results showed that the method produces intuitive results. For example, in a temperature prediction task, the meteorological time series consisting of 4 consecutive winters and 3 summers was first found to contain $k = 7$ segments — not perhaps surprisingly — and these 7 segments were found to be generated by $h = 2$ recurring models, a winter model and a summer model.

8.5 Unimodal Segmentation

We will discuss another restriction of the basic segmentation problem. In *unimodal* segmentation, the representatives of the segments (for example, means or medians) are required to follow a unimodal curve: the curve that is formed by all representatives of the segments has to change curvature only once. That is, the representatives first increase until a certain point and then decrease during the rest of the sequence, or the other way round. A special case is a *monotonic* curve. Examples of unimodal sequences include (i) the size of a population of a species over time, as the species first appears, then peaks in density and then dies out or (ii) daily volumes of network traffic [18].

In contrast to other segmentation methods discussed in this chapter, the sequence now takes scalar values instead of multidimensional values. Haiminen and Gionis [18] show how this problem can be solved by combining the classic "pool adjacent violators" (PAV) algorithm [1] and the basic dynamic programming algorithm [2] (see Section 16.1). The time complexity of their algorithm is $O(n^2k)$ which is the same as in the unrestricted k-segmentation using dynamic programming.

Haiminen and Gionis [18] also describe a more efficient greedy-merging heuristic that is experimentally shown to give solutions very close to the optimal, and whose time complexity is $O(n \log n)$: the expensive dynamic programming step is replaced with a greedy merging process that starts with m segments and iteratively merges the two consecutive segments that yield the least error, until reaching k segments.

The authors in [18] also give two tests for unimodality of a sequence. The first approach compares the error of an optimal unimodal k-segmentation to the error of an optimal unrestricted k-segmentation. If the sequence exhibits unimodal behaviour, then the error of its unimodal segmentation does not differ very much from the error of its unrestricted segmentation — in other words, requiring for unimodality did not hurt. Instead, if the sequence is not unimodal in nature, then forcing the segments to follow a unimodal curve will increase the representation error. The authors compute the ratio between the error of unrestricted k-segmentation and the error of unimodal segmentation and find a data-dependent threshold value that helps to differentiate between unimodal and non-unimodal sequences.

The second approach for testing for unimodality is to randomly permute the unimodal segments in the data, and to see if the error of unimodal k-segmentation on the permuted sequence is comparable to the error on the original sequence — the random permutation will destroy the unimodal structure of the sequence, if such exists. If the original sequence was indeed unimodal, then the error of the permuted sequences should be larger in a statistically significant way.

After discussing three different restrictions on the representatives of the segments in Sections 8.3, 8.4 and 8.5, we then turn to other modifications of the basic problem of sequence segmentation in the next section.

8.6 Rearranging the Input Data Points

The majority of related work on segmentation primarily focuses on finding a segmentation S of a sequence T taking for granted the order of the points in T. However, more often than not, the order of the data points of a sequence is not clear-cut but some data points actually appear simultaneously or their order is for some other reason observed only approximately correctly. In such a case it might be beneficial to allow for a slight rearrangement of the data points, in order to achieve a better segmentation. This was studied by Gionis and Terzi [15]: in addition to partitioning the sequence they also apply a limited amount of reordering, so that the overall representation error is minimized.

The focus now is to find a rearrangement of the points in T such that the segmentation error of the reordered sequence is minimized. The operations used to rearrange an input sequence consist of bubble-sort swaps and moves (single-element transpositions). The task is to find a sequence of operations O minimizing the reconstruction error on the reordered input sequence T_O:

$$O = \arg\min_{O'} E(S_{opt}(T_{O'}, k))$$

where $S_{opt}(T, k)$ is the optimal segmentation of T into k segments, and there is an upper limit on the number of operations: $|O| \leq C$ for some integer constant C.

The problem of segmentation with rearrangements is shown to be NP-hard to solve or even approximate. However, efficient algorithms are given in [15], combining ideas from linear programming, dynamic programming and outlier-detection algorithms in sequences. The algorithms consist of two steps. In the first step, and optimal segmentation S of the input sequence T into k segments is found. In the second step, a good set of rearrangements is found, such that the total segmentation error or the rearranged sequence is minimized. The latter step, the rearrangement, can be done in several ways, and the authors discuss the task in detail. In one possible formulation, the rearrangement task is a generalization of the well known NP-hard Knapsack problem for which a pseudopolynomial-time algorithm is admittable [44]. For the special case of bubble-sort swaps only, or moves only, a polynomial-time algorithm for the rearrangement is obtained. The authors also present a greedy heuristic with time complexity $O(Ink)$ where I is the number of iterations of the greedy algorithm in [15].

The problem formulation has applications in segmenting data collected from a sensor network where some of the sensors might be slightly out of sync, or in the analysis of newsfeed data where news reports on a few different topics are arriving in an interleaved manner. The authors show experiments on both synthetic data sets and on several real datasets from the UCR time series archive[2].

---

[2] http://www.cs.ucr.edu/~eamonn/TSDMA

8.7 Aggregate Segmentation

Whereas in Section 8.6 we refined the input data, we now turn our attention to the output of one or more segmentation algorithms.

A sequence can often be segmented in several different ways, depending on the choice of the segmentation algorithm, its error function, and in some cases, its initialization. The multitude of segmentation algorithms and error functions naturally raises the question: given a specific dataset, what is the segmentation that best captures the underlying structure of the data?

Thus a natural question is, given a number of possibly contradicting segmentations, how to produce a single *aggregate segmentation* that combines the features of the input segmentations.

Mielikäinen et al [33] adopt a democratic approach that assumes that all segmentations found by different algorithms are correct, each one in its own way. That is, each one of them reveals just one aspect of the underlying true segmentation. Therefore, they aggregate the information hidden in the segmentations by constructing a consensus output that reconciles optimally the differences among the given inputs.

Their approach results in a proof that for a natural formalization of this task, there is an optimal polynomial-time algorithm, and a faster heuristic that has good practical properties. The algorithms were demonstrated in two applications: clustering the behavior of mobile-phone users, and summarizing different segmentations of genomic sequences.

More formally, the input is a set of m different segmentations $S_1, \ldots, S_m$, and the objective is to produce a single segmentation $\hat{S}$ that agrees as much as possible with the input segmentations. The number of segments in $\hat{S}$ is learned during the process. In the discrete case a *disagreement* between two segmentations S and S' is defined as a pair of points (x, y) placed in the same segment by S but in different segments by S', or vice versa. Denoting the total number of disagreements between the sequences S and S' by $D_A(S, S')$, the formal objective is to minimize

$$\sum_{j=1}^{m} D_A(S_j, \hat{S}),$$

the grand total number of disagreements between the input segmentations S_j and the output segmentation $\hat{S}$. In the continuous case, disagreements are defined similarly for intervals instead of discrete points. The polynomial-time exact algorithm is based on the technique of dynamic programming [2], and the approximation algorithm on a greedy heuristic.

Segmentation aggregation can prove useful in many scenarios. We list some of them below, suggested by Mielikäinen et al [33].

In the analysis of genomic sequences of a population one often assumes that the sequences can be segmented into blocks such that in each block, most of the haplotypes fall into a small number of classes. Different segmentation algorithms have successfully been applied to this task, outputting slightly or completely different block structures; aggregating these block structures hopefully sheds light on the underlying truth.

Segmentation aggregation adds to the robustness of segmentation results: most segmentation algorithms are sensitive to erroneous or noisy data, and thus combining their results diminishes the effect of missing or faulty data.

Segmentation aggregation also gives a natural way to cluster segmentations: the representative of a cluster of segmentations is then the aggregate of the cluster. Furthermore, the disagreement distance is now a metric, allowing for various distance-based data mining techniques, together with approximation guarantees for many of them.

Other scenarios where segmentation aggregation can prove useful include segmentation of multidimensional categorical data and segmentation of multidimensional data having both nominal and numerical dimensions; summarization of event sequences; and privacy-preserving segmentations [33].

In this section we assessed and improved the quality of the output of the segmentation by aggregating the output of several segmentation methods. A related point of view is to assess the quality of the segmentations by measuring their statistical significance — this nontrivial task will be considered in Section 8.8.

8.8 Evaluating the Quality of a Segmentation: Randomization

An important question is how to evaluate and compare the quality of segmentations obtained by different techniques and alternative biological features. Haiminen et al [20] apply *randomization* techniques to this end.

Consider a segmentation algorithm that given as input a sequence T outputs a segmentation P. Assume that we a priori know a groundtruth segmentation S^* of T. Then, we can say that segmentation P is good if P is similar to S^*. In more exact terms, P is a good segmentation if the entropy of P given S^*, $H(P \mid S^*)$, and the entropy of S^* given P, $H(S^* \mid P)$, are small. However, a natural question is, how small is small enough?

Before we proceed, let us give some more details on the notation. Consider a segmentation P consisting of k segments $P = \langle p_1, \ldots, p_k \rangle$. If we randomly pick a point t on the sequence, then the probability $t \in p_i$ is $Pr(p_i) = |p_i|/n$ where n is the length of the sequence. The *entropy* of a segmentation P is now

$$H(P) = - \sum_{i=k}^{k} Pr(p_i) \log Pr(p_i).$$

The maximum value that the entropy of a segmentation can have is $\log n$, and this value is achived when all segments are of equal length and thus the probabilities of a random point belonging to any of the segments are equal.

Consider now a pair of segmentations P and Q of sequence S. Assume that P and Q have k_p and k_q segments, respectively: $P = \langle p_1, \ldots, p_{k_p} \rangle$ and $Q = \langle q_1, \ldots, q_{k_q} \rangle$.

The *conditional entropy* [10] of P given Q is defined as

$$H(P \mid Q) = \sum_{j=1}^{k_q} Pr(q_j) H(P \mid q_j)$$

$$= - \sum_{j=1}^{k_q} Pr(q_j) \sum_{i=1}^{k_p} Pr(p_i \mid q_j) \log Pr(p_i \mid q_j)$$

$$= - \sum_{j=1}^{k_q} \sum_{i=1}^{k_p} Pr(p_i, q_j) \log Pr(p_i \mid q_j)$$

That is, the conditional entropy of segmentation P given segmentation Q is the expected amount of information we need to identify the segment of P into which a point belongs, given that we know the segment of this point in Q.

Haiminen et al [20] give an efficient algorithm for computing the conditional entropies between two segmentations: Denote by U the union of two segmentations P and Q, that is, the segmentation defined by the segment boundaries that appear in P or in Q. The conditional entropy of P given Q can be computed as $H(P \mid Q) = H(U) - H(Q)$. The algorithm runs in time $O(k_p + k_q)$.

Let us now return to our original problem setting: Assuming we know a groundtruth segmentation S^* of T, then P is a good segmentation if $H(P \mid S^*)$ and $H(S^* \mid P)$ are small. But how small is small enough? Or, is there a threshold in the values of the conditional entropies below which we can characterize the segmentation P as being correct or interesting? Finally, can we set this threshold universally for all segmentations?

The generic methodology of *randomization techniques* (see [14], [34] among others) that are devised to answer these questions is the following. Given a segmentation P and a ground-truth segmentation S^* of the same sequence, we first compute $H(P \mid S^*)$ and $H(S^* \mid P)$. We compare the values of these conditional entropies with the values of the conditional entropies $H(R \mid S^*)$ and $H(S^* \mid R)$ for a random segmentation R. We conclude that P is similar to S^*, and thus interesting, if the values of both $H(P \mid S^*)$ and $H(S^* \mid P)$ are smaller than $H(R \mid S^*)$ and $H(S^* \mid R)$, respectively, for a large majority of random segmentations R. Typically, 10 000 or 100 000 random segmentations are drawn, and if $H(P \mid S^*) < H(R \mid S^*)$ in all but a couple of cases, then P is deemed interesting. The percentage of cases violating $H(P \mid S^*) < H(R \mid S^*)$ can be interpreted as a p value, and a small value denotes statistical significance.

The example applications in [20] include isochore detection and the discovery of coding-noncoding structure. The authors obtain segmentations of relevant sequences by applying different techniques, and use alternative features to segment on. They show that some of the obtained segmentations are very similar to the underlying true segmentations, and this similarity is statistically significant. For some other segmentations, they show that equally good results are likely to appear by chance.

8.9 Model Selection by BIC and Cross-validation

One of the key questions in segmentation is choosing the number of segments to use. Important features of the sequence may be lost when representing it with too few segments, while using too many segments yields uninformative and overly complex segmentations.

Choosing the number of segments is essentially a model selection task. Haiminen and Mannila [19] present extensive experimental studies on two standard model selection techniques, namely Bayesian Information Criterion (BIC) and cross-validation (CV).

Bayesian Information Criterion (BIC) seeks a balance between model complexity and the accuracy of the model by including a penalty term for the number of parameters. BIC is defined as [39] $BIC = -2\ln L + K\ln N + C$ where L is the maximized likelihood of the model with K free parameters, N is the sample size and C is a small constant that is often omitted. The model with the smallest BIC is optimal in terms of complexity.

Cross-validation is an intuitive iterative method for model selection: A subset of the data is used to train the model. The goodness of fit on the remaining data, also called test data, is then evaluated. This is repeated for a number of times, in each of which the data are randomly split into a training set and test set. In an outer loop, the complexity of the model (here, the number of segments) is varied. When the model complexity is unnecessarily high, the model overfits the training data and fails to represent the test data. Alternatively, if the model complexity is too low, the test data cannot be faithfully represented either. CV is a very general method in that no assumptions regarding the data are made, and any cost function can be used. The use of CV has been discussed by e.g. Stone [42] and Smyth [41].

The results in [19] show that these methods often find the correct number of piecewise constant segments on generated real-valued, binary, and categorical sequences. Also segments having the same means but different variances can be identified. Furthermore, they demonstrate the effect of linear trends and outliers on the results; both phenomena are frequent in real data.

The results indicate that BIC is fairly sensitive to outliers, and that CV in general is more robust. Intuitive segmentation results are given for real DNA sequences with respect to changes in their codon, G+C, and bigram frequencies, as well as copy-number variation from CGH data.

8.10 Bursty Sequences

In the earlier sections, we have assumed some specific constraints on the segments, making the problem more applicable to the human eye. Haiminen et al [21] have also studied constraining the nature of the sequence itself, outside the task of segmenting the sequence. An intuitive subset of sequences is one in which *bursts of activity* occur in time, and a natural question then is, how to formally define and measure this. The problem setting applies to *event sequences*: Given a set of possible event types, an event sequence is a sequence of pairs (r,t), where r is an event type and t is the occurrence location, or time, of the event. Moreover, a *bursty event sequence* is one in which "bursts" of activity occur in time: different types of events often occur close together.

Bursty sequences arise, e.g., when studying potential transcription factor binding sites (events) of certain transcription factors (event types) in a DNA sequence. These events tend to occur in bursts. Tendencies for co-occurrence of binding sites of two or more transcription factors are interesting, as they may imply a co-operative role between the transcription factors in regulatory processes.

Haiminen et al [21] measure the co-occurrence of event types r and r' either by (i) dividing the sequence into non-overlapping windows of a fixed length w and counting the number of windows that contain at least one event of type r and at least one event of type r', or by (ii) counting the number of events of type r that are followed by at least one event of type r' within distance w, or by (iii) counting the number of events of type r that are followed or preceded by at least one event of type r' within distance w.

In order to determine the significance of a co-occurrence score, we need a *null model* to estimate the distribution of the score values and then decide the significance of an individual value. Haiminen et al [21] define three such null models that apply to any co-occurrence score, extending previous work on null models. These models range from very simple ones to more complex models that take the burstiness of sequences into account. The authors evaluate the models and techniques on synthetic event sequences, and on real data consisting of potential transcription factor binding sites.

8.11 Conclusion

In this chapter, we have discussed several variants of the problem of sequence segmentation. An optimal segmentation method, applicable when no specific restrictions are assumed, is segmentation using dynamic programming [2]. However, this is computationally burdensome for very long sequences. Also, it is often the case that by adding some constraints on the output segmentation, or by making small modifications to the problem, the output of the segmentation is more interpretable for the human eye. This chapter is a survey of different approaches for segmentation suggested by researchers at Helsinki Institute for Information Technology during the years 2004 to 2008.

In Section 8.2 we described an efficient segmentation method, with a proven quality of the solution it provides when representing the original data [43]. We then discussed three constraints on the problem setting of segmentation, to make the end result more tractable, in Sections 8.3 to 8.5. Using these constraints we wish to restrict the values that the representatives (that is, means or medians, typically) of the segments can assume. First, the representatives of multidimensional segments can be presented as different combinations of a small set of basis vectors [5]. Secondly, a small set of models can be used to predict the data values in the segments [25]. Thirdly, one can require the representatives of the sequences to follow a unimodal or monotonic curve [18].

Then, instead of constraints on the representatives per se, we discussed small modifications to the basic problem setting in Sections 8.6 and 8.7: By allowing small reorderings of the data points in a sequence, we can decrease the reconstruction error of the segmentation, and in some application areas these reorderings are very natural [15]. In some cases there is a need to choose between several different outputs of segmentation algorithms, and a way to overcome this is to combine the outputs into one aggregate segmentation [33]. A very important and nontrivial task is to characterize the quality of a segmentation in statistical terms, and randomization provides an answer here [20] (Section 8.8). Choosing the number of segments is a question of model selection, and experimental results were discussed in Section 8.9 [19]. Finally in Section 8.10, instead of constraining the representatives of the sequences, we constrained the nature of the sequence itself, when determining when an event sequence is bursty or not [21].

Acknowledgements The author would like to thank Aristides Gionis, Niina Haiminen, Heli Hiisilä, Saara Hyvönen, Taneli Mielikäinen, Evimaria Terzi, Panayiotis Tsaparas and Heikki Mannila for their work in the papers discussed in this survey. The comments given by the anonymous reviewers have greatly helped to improve the manuscript.

References

1. Miriam Ayer, H. D. Brunk, G. M. Ewing, W. T. Reid, and Edward Silverman. An empirical distribution function for sampling with incomplete information. *Annals of Mathematical Statistics*, 26(4):641–647, 1955.
2. Richard Bellman. On the approximation of curves by line segments using dynamic programming. *Communications of the ACM*, 4(6), 1961.
3. K.D. Bennett. Determination of the number of zones in a biostratigraphical sequence. *New Phytologist*, 132(1):155–170, 1996.
4. Pedro Bernaola-Galván, Ramón Román-Roldán, and José L. Oliver. Compositional segmentation and long-range fractal correlations in dna sequences. *Phys. Rev. E*, 53(5):5181–5189, 1996.
5. Ella Bingham, Aristides Gionis, Niina Haiminen, Heli Hiisilä, Heikki Mannila, and Evimaria Terzi. Segmentation and dimensionality reduction. In *2006 SIAM Conference on Data Mining*, pages 372–383, 2006.
6. Harmen J. Bussemaker, Hao Li, and Eric D. Siggia. Regulatory element detection using a probabilistic segmentation model. In *Proceedings of the Eighth International Conference on Intelligent Systems for Molecular Biology*, pages 67–74, 2000.
7. A. Cantoni. Optimal curve fitting with piecewise linear functions. *IEEE Transactions on Computers*, C-20(1):59–67, 1971.
8. K. Chakrabarti, E. Keogh, S. Mehrotra, and M. J. Pazzani. Locally adaptive dimensionality reduction for indexing large time series databases. *ACM Transactions on Database Systems*, 27(2):188–228, 2002.
9. G.A. Churchill. Stochastic models for heterogenous dna sequences. *Bulletin of Mathematical Biology*, 51(1):79–94, 1989.

10. Thomas M. Cover and Joy A. Thomas. *Elements of information theory*. Wiley, 1991.
11. David Douglas and Thomas Peucker. Algorithms for the reduction of the number of points required to represent a digitized line or its caricature. *Canadian Cartographer*, 10(2):112–122, 1973.
12. Sorabh Gandhi, Luca Foschini, and Subhash Suri. Space-efficient online approximation of time series data: Streams, amnesia, and out-of-order. In *Proceedings of the 26th IEEE International Conference on Data Engineering (ICDE)*, 2010.
13. Aristides Gionis and Heikki Mannila. Finding recurrent sources in sequences. In *Proceedings of the Sventh Annual International Conference on Computational Biology (RECOMB 2003)*, 2003.
14. Aristides Gionis, Heikki Mannila, Taneli Mielikäinen, and Panayiotis Tsaparas. Assessing data mining results via swap randomization. *ACM Transactions on Knowledge Discovery from Data (TKDD)*, 1(3), 2007. Article No. 14.
15. Aristides Gionis and Evimaria Terzi. Segmentations with rearrangements. In *SIAM Data Mining Conference (SDM) 2007*, 2007.
16. S. Guha, N. Koudas, and K. Shim. Data-streams and histograms. In *Symposium on the Theory of Computing (STOC)*, pages 471–475, 2001.
17. Niina Haiminen. *Mining sequential data — in search of segmental structure*. PhD Thesis, Department of Computer Science, University of Helsinki, March 2008.
18. Niina Haiminen and Aristides Gionis. Unimodal segmentation of sequences. In *ICDM '04: Proceedings of the Fourth IEEE International Conference on Data Mining*, pages 106–113, 2004.
19. Niina Haiminen and Heikki Mannila. Evaluation of BIC and cross validation for model selection on sequence segmentations. *International Journal of Data Mining and Bioinformatics*. In press.
20. Niina Haiminen, Heikki Mannila, and Evimaria Terzi. Comparing segmentations by applying randomization techniques. *BMC Bioinformatics*, 8(171), 23 May 2007.
21. Niina Haiminen, Heikki Mannila, and Evimaria Terzi. Determining significance of pairwise co-occurrences of events in bursty sequences. *BMC Bioinformatics*, 9:336, 2008.
22. Trevor Hastie, R. Tibshirani, and Jerome Friedman. *The Elements of Statistical Learning*. Springer, 2001.
23. J. Himberg, K. Korpiaho, H. Mannila, J. Tikanmäki, and H. T.T. Toivonen. Time series segmentation for context recognition in mobile devices. In *Proceedings of the 2001 IEEE International Conference on Data Mining*, pages 203–210, 2001.
24. Dorit S. Hochbaum. Heuristics for the fixed cost median problem. *Mathematical Programming*, 22(1):148–162, 1982.
25. Saara Hyvönen, Aristides Gionis, and Heikki Mannila. Recurrent predictive models for sequence segmentation. In *The 7th International Symposium on Intelligent Data Analysis*, Lecture Notes in Computer Science. Springer, 2007.
26. Eamonn Keogh, Selina Chu, David Hart, and Michael Pazzani. An online algorithm for segmenting time series. In *Proceedings of the 2001 IEEE International Conference on Data Mining*, pages 289–296, 2001.
27. Eamonn Keogh and S. Kasetty. On the need for time series data mining benchmarks: A survey and empirical demonstration. In *Proceedings of the ACM SIGKDD '02*, pages 102–111, July 2002.
28. Eamonn Keogh and Michael J. Pazzani. An enhanced representation of time series which allows fast and accurate classification, clustering and relevance feedback. In *Proceedings of the ACM SIGKDD '98*, pages 239–243, August 1998.
29. Victor Lavrenko, Matt Schmill, Dawn Lawrie, Paul Ogilvie, David Jensen, and James Allan. Mining of concurrent text and time series. In *In proceedings of the 6th ACM SIGKDD International Conference on Knowledge Discovery and Data Mining Workshop on Text Mining*, pages 37–44, 2000.
30. W. Li. DNA segmentation as a model selection process. In *Proceedings of the Fifth Annual International Conference on Computational Biology (RECOMB 2001)*, pages 204 – 210, 2001.

31. Jyh-Han Lin and Jeffrey Scott Vitter. ε-approximations with minimum packing constraint violation. In *Proc. ACM Symposium on Theory of Computing (STOC'92)*, pages 771–781, 1992.

32. Jun S. Liu and Charles E. Lawrence. Bayesian inference on biopolymer models. *Bioinformatics*, 15(1):38–52, 1999.

33. Taneli Mielikäinen, Evimaria Terzi, and Panayiotis Tsaparas. Aggregating time partitions. In *The Twelfth ACM SIGKDD International Conference on Knowledge Discovery and Data Mining (KDD 2006)*, pages 347–356, 2006.

34. Markus Ojala, Niko Vuokko, Aleksi Kallio, Niina Haiminen, and Heikki Mannila. Randomization of real-valued matrices for assessing the significance of data mining results. In *Proc. SIAM Data Mining Conference (SDM'08)*, pages 494–505, 2008.

35. T. Palpanas, M. Vlachos, E. Keogh, D. Gunopulos, and W. Truppel. Online amnesic approximation of streaming time series. In *ICDE 2004: Proceedings of the 20th International Conference on Data Engineering*, pages 338–349, 2004.

36. Themis Palpanas, Michail Vlachos, Eamonn Keogh, and Dimitrios Gunopulos. Streaming time series summarization using user-defined amnesic functions. *IEEE Transactions on Knowledge and Data Engineering*, 20(7):992–1006, 2008.

37. V.E. Ramensky, V.J. Makeev, M.A. Roytberg, and V.G. Tumanyan. DNA segmentation through the Bayesian approach. *Journal of Computational Biology*, 7(1-2):215–231, 2000.

38. Marko Salmenkivi, Juha Kere, and Heikki Mannila. Genome segmentation using piecewise constant intensity models and reversible jump MCMC. *Bioinformatics (European Conference on Computational Biology)*, 18(2):211–218, 2002.

39. G. Schwarz. Estimating the dimension of a model. *The Annals of Statistics*, 6(2):461–464, 1978.

40. Hagit Shatkay and Stanley B. Zdonik. Approximate queries and representations for large data sequences. In *ICDE '96: Proceedings of the Twelfth International Conference on Data Engineering*, pages 536–545, 1996.

41. P. Smyth. Model selection for probabilistic clustering using cross-validated likelihood. *Statistics and Computing*, 9:63–72, 2000.

42. M. Stone. Cross-validatory choice and assessment of statistical predictions. *Journal of the Royal Statistical Society: Series B*, 36(2):111–147, 1974.

43. Evimaria Terzi and Panayiotis Tsaparas. Efficient algorithms for sequence segmentation. In *2006 SIAM Conference on Data Mining*, pages 314–325, 2006.

44. V. Vazirani. *Approximation algorithms*. Springer, 2003.

45. Y.-L. Wu, D. Agrawal, and A. El Abbadi. A comparison of DFT and DWT based similarity search in time series databases. In *Proceedings of the Ninth ACM International Conference on Information and Knowledge Management (CIKM'00)*, pages 488–495, November 2000.

46. B. Yi and C. Faloutsos. Fast time sequence indexing for arbitrary LP-norms. In *Proceedings of the 26th International Conference on Very Large Databases (VLDB'00)*, pages 385–394, September 2000.

Chapter 9
Mining Constrained Cross-Graph Cliques in Dynamic Networks

Loïc Cerf, Bao Tran Nhan Nguyen, and Jean-François Boulicaut

Abstract Three algorithms — CUBEMINER, TRIAS, and DATA-PEELER — have been recently proposed to mine closed patterns in ternary relations, i.e., a generalization of the so-called formal concept extraction from binary relations. In this paper, we consider the specific context where a ternary relation denotes the value of a graph adjacency matrix (i. e., a Vertices × Vertices matrix) at different timestamps. We discuss the constraint-based extraction of patterns in such dynamic graphs. We formalize the concept of δ-contiguous closed 3-clique and we discuss the availability of a complete algorithm for mining them. It is based on a specialization of the enumeration strategy implemented in DATA-PEELER. Indeed, the relevant cliques are specified by means of a conjunction of constraints which can be efficiently exploited. The added-value of our strategy for computing constrained clique patterns is assessed on a real dataset about a public bicycle renting system. The raw data encode the relationships between the renting stations during one year. The extracted δ-contiguous closed 3-cliques are shown to be consistent with our knowledge on the considered city.

9.1 Introduction

Mining binary relations (often encoded as Boolean matrices) has been intensively studied. For instance, a popular application domain concerns basket data analysis and mining tasks on *Transactions × Products* relations. In a more general setting, binary relations may denote relationships between objects and a given set of properties giving *Objects × Properties* matrices. Many knowledge discovery processes from potentially large binary relations have been considered. We are interested in

Loïc Cerf · Bao Tran Nhan Nguyen · Jean-François Boulicaut
Université de Lyon, CNRS, INRIA
INSA-Lyon, LIRIS Combining, UMR5205, F-69621, France
e-mail: {lcerf, jboulica}@liris.cnrs.fr, baonhan@pmail.ntu.edu.sg

descriptive approaches that can be based on pattern discovery methods. Pattern types can be frequent itemsets (see, e. g., [1, 22]), closed itemsets or formal concepts (see, e. g., [15, 25, 5]), association rules (see, e. g., [2]) or their generalizations like, e. g., [3]. Interestingly, when looking at the binary relation as the encoding of a bi-partite graph (resp. a graph represented by its adjacency matrix), some of these patterns can be interpreted in terms of graph substructures. A typical example that is discussed in this chapter concerns the analogy between formal concepts and maximal bi-cliques (resp. cliques).

Constraint-based mining is a popular framework for supporting relevant pattern discovery thanks to user-defined constraints (see, e. g., [6]). It provides more interesting patterns when the analyst specifies his/her subjective interestingness by means of a combination of primitive constraints. This is also known as a key issue to achieve efficiency and tractability. Some constraints can be deeply pushed into the extraction process such that it is possible to get complete (every pattern which satisfies the user-defined constraint is computed) though efficient algorithms. As a result, many efficient algorithms are available for computing constrained patterns from binary relations. Among others, this concerns constraint-based mining of closed patterns from binary relations (see, e. g., [23, 28, 4, 25, 26]).

It is clear that many datasets of interest correspond to n-ary relations where $n \geq 3$. For instance, a common situation is that space and time information are available such that we get the generic setting of *Objects* × *Properties* × *Dates* × *Places* 4-ary relations. In this chapter, we consider the encoding of *dynamic graphs* in terms of collections of adjacency matrices, hence a ternary relation *Vertices* × *Vertices* × *Date*. The discovery of closed patterns from ternary relations has been recently studied. From a semantics perspective, such patterns are a straightforward extension of formal concepts. Computing them is however much harder. To the best of our knowledge, the extension towards higher arity relations has given rise to three proposals, namely CUBEMINER [17] or TRIAS [16] for ternary relations, and DATA-PEELER for arbitrary n-ary relations [10, 11]. A major challenge is then to exploit user-defined constraints during the search of application relevant closed patterns. We assume that the state-of-the-art approach is the DATA-PEELER enumeration strategy which can mine closed patterns under a large class of constraints called *piecewise (anti)-monotone constraints* [11].

In this chapter, we consider that data (i. e., a ternary relation) denote a dynamic graph. We assume that the encoded graphs have a fixed set of vertices and that directed links can appear and/or disappear at the different timestamps. Furthermore, we focus on clique patterns which are preserved along almost-contiguous timestamps. For instance, it will provide interesting hypothesis about sub-networks of stations within a bicycle renting system.

We have three related objectives.

- First, we want to illustrate the genericity of the DATA-PEELER algorithm. We show that relevant pattern types can be specified as closed patterns that further satisfy other user-defined constraints in the ternary relation that denotes the dynamic graph. We study precisely the pattern type of δ-contiguous closed 3-cliques, i. e., maximal sets of vertices that are linked to each other and that run along some "almost" contiguous timestamps. To denote a clique pattern, a closed pattern will have to involve identical sets of vertices (using the so-called symmetry constraint). Notice that being a closed pattern will be also expressed in terms of two primitive constraints (namely the connection and closedness constraints) that are efficiently processed by the DATA-PEELER enumeration. We do not provide all the details about the algorithm (see [11] for an in-depth presentation) but its most important characteristics are summarized and we formalize the constraint properties that it can exploit efficiently. Doing so, we show that the quite generic framework of arbitrary n-ary relation mining can be used to support specific analysis tasks in dynamic graphs.

- Next, our second objective is to discuss the specialization of the algorithm to process more efficiently the conjunction of the connection, closedness, symmetry and contiguity constraints, i.e., what can be done to specialize the generic mechanisms targeted to closed pattern discovery from arbitrary n-ary relations when we are looking for preserved cliques in *Vertices* $\times$ *Vertices* $\times$ *Date* ternary relations. This technical contribution enables to discuss efficiency issues and optimized constraint checking.

- Last but not the least, we show that this algorithmic contribution can be used in concrete applications. Graph mining is indeed a popular topic. Many researchers consider graph pattern discovery from large collections of graphs while others focus on data analysis techniques for one large graph. In the latter case, especially in the context of dynamic graphs, we observe two complementary directions of research. On one hand, global properties of such graphs are studied like power-law distribution of node degree or diameters (see, e. g., [20]). On another hand, it is possible to use pattern discovery techniques to identify local properties in the graphs (see, e. g., [27]). We definitively contribute to this later approach. We compute δ-contiguous closed 3-cliques in a real-life dynamic graph related to bicycle renting in a large European city. We illustrate that these usage patterns can be interpreted thanks to domain knowledge and that they provide a feedback on emerging sub-networks.

The rest of the paper is organized as follows. We formalize the mining task and we discuss the type of constraints our algorithm handles in Sect. 9.2. In Section 9.3, we summarize the fundamental mechanisms used in the DATA-PEELER algorithm. Section 9.4 details how the δ-contiguity constraint is enforced. Section 9.5 describes the strategy for computing closed 3-cliques by pushing various primitive constraints into the enumeration strategy. Section 9.6 provides an experimental validation on a real dataset. Related work is discussed in Sect. 9.7, and Sect. 9.8 briefly concludes.

9.2 Problem Setting

Let $\mathscr{T} \in \mathbb{R}^{|\mathscr{T}|}$ a finite set of timestamps. Let $\mathscr{N}$ a set of nodes. A (possibly directed) graph is uniquely defined by its adjacency matrix $A \in \{0, 1\}^{\mathscr{N} \times \mathscr{N}}$. A dynamic graph involving the nodes of $\mathscr{N}$ along $\mathscr{T}$ is uniquely defined by the $|\mathscr{T}|$-tuple $(A_t)_{t \in \mathscr{T}}$

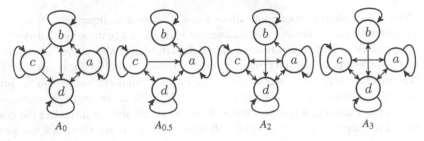

Fig. 9.1 Example of a dynamic directed graph ($\mathscr{N} = \{a,b,c,d\}$, $\mathscr{T} = \{0,0.5,2,3\}$)

gathering the adjacency matrices of the graph at every timestamp $t \in \mathscr{T}$. Visually, such a stack of adjacency matrices can be seen as a $|\mathscr{T}| \times |\mathscr{N}| \times |\mathscr{N}|$ cube of 0/1 values. We write $a_{t,n^1,n^2} = 1$ (resp. $a_{t,n^1,n^2} = 0$) when, at the timestamp t, a link from n^1 to n^2 is present (resp. absent).

Example 9.1. Figure 9.1 depicts a dynamic directed graph involving four nodes a, b, c and d. Four snapshots of this graph are available at timestamps 0, 0.5, 2 and 3. Table 9.1 gives the related 4-tuple $(A_0, A_{0.5}, A_2, A_3)$.

Visually, a closed 3-set $(T, N^1, N^2) \in 2^{\mathscr{T}} \times 2^{\mathscr{N}} \times 2^{\mathscr{N}}$ appears as a combinatorial sub-cube of the data (modulo arbitrary permutations on any dimension) satisfying both the *connection* and the *closedness* primitive constraints. Informally, it means that $T \times N^1 \times N^2$ only contains '1' values (connection), and any "super-cube" of (T, N^1, N^2) violates the connection constraint (closedness). Let us define them more formally.

Definition 9.1 ($\mathscr{C}_{\text{connected}}$). A 3-set (T, N^1, N^2) is said connected, denoted $\mathscr{C}_{\text{connected}}(T, N^1, N^2)$, iff $\forall (t, n^1, n^2) \in T \times N^1 \times N^2, a_{t,n^1,n^2} = 1$.

Definition 9.2 ($\mathscr{C}_{\text{closed}}$). It is said that a 3-set (T, N^1, N^2) is closed, denoted

$$\mathscr{C}_{\text{closed}}(T, N^1, N^2), \text{ iff } \begin{cases} \forall t \in \mathscr{T} \setminus T, \neg\mathscr{C}_{\text{connected}}(\{t\}, N^1, N^2) \\ \forall n^1 \in \mathscr{N} \setminus N^1, \neg\mathscr{C}_{\text{connected}}(T, \{n^1\}, N^2) \\ \forall n^2 \in \mathscr{N} \setminus N^2, \neg\mathscr{C}_{\text{connected}}(T, N^1, \{n^2\}) \end{cases} .$$

A closed 3-set can now be formally defined.

Table 9.1 $(A_0, A_{0.5}, A_2, A_3)$ related to the dynamic graph depicted Fig. 9.1

	a	b	c	d	a	b	c	d	a	b	c	d	a	b	c	d
a	1	1	0	1	1	1	0	1	1	1	1	1	1	0	1	1
b	1	1	1	1	1	1	0	0	0	1	0	1	0	1	0	1
c	0	0	1	1	1	0	1	1	1	0	1	1	1	0	1	1
d	1	1	0	1	1	0	1	1	1	0	1	1	1	1	1	1
	A_0				$A_{0.5}$				A_2				A_3			

Definition 9.3 (Closed 3-set). (T, N^1, N^2) is a closed 3-set iff it satisfies the conjunction $\mathscr{C}_{\text{connected}}(T, N^1, N^2) \wedge \mathscr{C}_{\text{closed}}(T, N^1, N^2)$.

Example 9.2. $(\{0, 2, 3\}, \{a, b, c, d\}, \{d\})$ is a closed 3-set in the toy dataset from Table 9.1: $\forall (t, n^1, n^2) \in \{0, 2, 3\} \times \{a, b, c, d\} \times \{d\}$, we have $a_{t, n^1, n^2} = 1$, and

$$\begin{cases} \forall t \in \{0.5\}, \neg\mathscr{C}_{\text{connected}}(\{t\}, \{a, b, c, d\}, \{d\}) \\ \forall n^1 \in \emptyset, \neg\mathscr{C}_{\text{connected}}(\{0, 2, 3\}, \{n^1\}, \{d\}) \\ \forall n^2 \in \{a, b, c\}, \neg\mathscr{C}_{\text{connected}}(\{0, 2, 3\}, \{a, b, c, d\}, \{n^2\}) \end{cases}.$$

$(\{2, 3\}, \{a, c, d\}, \{a, c, d\})$ and $(\{0, 3\}, \{b, d\}, \{b, d\})$ are two other closed 3-sets. $(\{0.5, 2, 3\}, \{c, d\}, \{c, d\})$ is not a closed 3-set because it violates $\mathscr{C}_{\text{closed}}$. Indeed $\mathscr{C}_{\text{connected}}(\{0.5, 2, 3\}, \{c, d\}, \{a\})$ holds, i.e., the third set of the pattern can be extended with a.

Given $\delta \in \mathbb{R}_+$, a δ-contiguous 3-set is such that it is possible to browse the whole subset of timestamps by jumps from one timestamp to another without exceeding a delay of δ for each of these jumps.

Definition 9.4 (δ-contiguity). A 3-set (T, N^1, N^2) is said δ-contiguous, denoted $\mathscr{C}_{\delta\text{-contiguous}}(T, N^1, N^2)$, iff $\forall t \in [\min(T), \max(T)], \exists t' \in T$ s.t. $|t - t'| \leq \delta$.

Notice that t does not necessarily belong to $\mathscr{T}$ (if $|T| \geq 2$, $[\min(T), \max(T)]$ is infinite). $\mathscr{C}_{\text{connected}} \wedge \mathscr{C}_{\delta\text{-contiguous}}$ being stronger than $\mathscr{C}_{\text{connected}}$ alone, a related and weaker closedness constraint can be defined. Intuitively, a δ-closed 3-set is closed w.r.t. both $\mathscr{N}$ sets and to the timestamps of $\mathscr{T}$ in the vicinity of those inside the 3-set. Hence, a timestamp that is too far away (delay exceeding δ) from any timestamp inside the 3-set, cannot prevent its δ-closedness.

Definition 9.5 (δ-closedness). It is said that a 3-set (T, N^1, N^2) is δ-closed, denoted $\mathscr{C}_{\delta\text{-closed}}(T, N^1, N^2)$, iff

$$\begin{cases} \forall t \in \mathscr{T} \setminus T, (\exists t' \in T \text{ s.t. } |t - t'| \leq \delta \Rightarrow \neg\mathscr{C}_{\text{connected}}(\{t\}, N^1, N^2)) \\ \forall n^1 \in \mathscr{N} \setminus N^1, \neg\mathscr{C}_{\text{connected}}(T, \{n^1\}, N^2) \\ \forall n^2 \in \mathscr{N} \setminus N^2, \neg\mathscr{C}_{\text{connected}}(T, N^1, \{n^2\}) \end{cases}.$$

Definition 9.6 (δ-contiguous closed 3-set). (T, N^1, N^2) is a δ-contiguous closed 3-set iff it satisfies the conjunction $\mathscr{C}_{\text{connected}} \wedge \mathscr{C}_{\delta\text{-contiguous}} \wedge \mathscr{C}_{\delta\text{-closed}}$.

A δ-contiguous closed 3-set is an obvious generalization of a closed 3-set. Indeed, $\forall \delta \geq \max(\mathscr{T}) - \min(\mathscr{T})$, $\begin{cases} \mathscr{C}_{\delta\text{-contiguous}} \equiv true \\ \mathscr{C}_{\delta\text{-closed}} \equiv \mathscr{C}_{\text{closed}} \end{cases}$.

Example 9.3. $(\{2, 3\}, \{a, b, c, d\}, \{d\})$ is a 1.75-contiguous closed 3-set in the toy dataset from Table 9.1. However, it is neither 0.5-contiguous (the timestamps 2 and 3 are not close enough) nor 2-closed (0 can extend the set of timestamps). This illustrates the fact that the number of δ-contiguous closed 3-sets is not monotone in δ.

We want to extract sets of nodes that are entirely interconnected. In this context, a 3-set (T, N^1, N^2) where $N^1 \neq N^2$ is irrelevant and a symmetry constraint must be added.

Definition 9.7 (Symmetry). A 3-set (T, N^1, N^2) is said symmetric, denoted $\mathscr{C}_{\text{symmetric}}(T, N^1, N^2)$, iff $N^1 = N^2$.

Again, let us observe that $\mathscr{C}_{\text{connected}} \wedge \mathscr{C}_{\delta\text{-contiguous}} \wedge \mathscr{C}_{\text{symmetric}}$ being stronger than $\mathscr{C}_{\text{connected}} \wedge \mathscr{C}_{\delta\text{-contiguous}}$, a related and weaker closedness constraint can be defined. Intuitively, if not *both* the row and the column pertaining to a node n can *simultaneously* extend a 3-set without breaking $\mathscr{C}_{\text{connected}}$, the closedness is not violated.

Definition 9.8 (Symmetric δ-closedness). It is said that a 3-set (T, N^1, N^2) is symmetric δ-closed, denoted $\mathscr{C}_{\text{sym-}\delta\text{-closed}}(T, N^1, N^2)$, iff

$$\begin{cases} \forall t \in \mathscr{T} \setminus T, (\exists t' \in T \text{ s. t. } |t - t'| \leq \delta \Rightarrow \neg \mathscr{C}_{\text{connected}}(\{t\}, N^1, N^2)) \\ \forall n \in \mathscr{N} \setminus (N^1 \cap N^2), \neg \mathscr{C}_{\text{connected}}(T, N^1 \cup \{n\}, N^2 \cup \{n\}) \end{cases}$$

Definition 9.9 (δ-contiguous closed 3-clique). It is said that (T, N^1, N^2) is a δ-contiguous closed 3-clique iff it satisfies $\mathscr{C}_{\text{connected}} \wedge \mathscr{C}_{\delta\text{-contiguous}} \wedge \mathscr{C}_{\text{symmetric}} \wedge \mathscr{C}_{\text{sym-}\delta\text{-closed}}$.

Example 9.4. Two out of the three closed 3-sets illustrating Ex. 9.2 are symmetric: $(\{2,3\}, \{a,c,d\}, \{a,c,d\})$ and $(\{0,3\}, \{b,d\}, \{b,d\})$. In Ex. 9.2, it was shown that $(\{0.5, 2, 3\}, \{c,d\}, \{c,d\})$ is not closed w.r.t. $\mathscr{C}_{\text{closed}}$. However it is symmetric 1.75-closed. Indeed, the node a cannot simultaneously extend its second and third sets of elements without violating $\mathscr{C}_{\text{connected}}$.

Problem Setting. Assume $(A_t)_{t \in \mathscr{T}} \in \{0,1\}^{\mathscr{T} \times \mathscr{N} \times \mathscr{N}}$ and $\delta \in \mathbb{R}_+$. This chapter deals with computing the complete collection of the δ-contiguous closed 3-cliques which hold in this data. In other terms, we want to compute every 3-set which satisfies the conjunction of the four primitive constraints defined above, i. e., $\mathscr{C}_{\text{connected}} \wedge \mathscr{C}_{\delta\text{-contiguous}} \wedge \mathscr{C}_{\text{symmetric}} \wedge \mathscr{C}_{\text{sym-}\delta\text{-closed}}$. In practical settings, such a collection is huge. It makes sense to constrain further the extraction tasks (i. e., to also enforce a new user-defined constraint $\mathscr{C}$) to take subjective interestingness into account and to support the focus on more relevant cliques. Thus, the problem becomes the complete extraction of the δ-contiguous closed 3-cliques satisfying $\mathscr{C}$.

Instead of writing, from scratch, an ad-hoc algorithm for computing constrained δ-contiguous closed 3-cliques, let us first specialize the generic closed n-set extractor DATA-PEELER [10, 11]. Its principles and the class of constraints it can exploit are stated in the next section. In Sect. 9.4, we study its adaptation to δ-contiguous closed 3-set mining, and Sect. 9.5 presents how to force the closed 3-sets to be symmetric and symmetric δ-closed.

9.3 DATA-PEELER

9.3.1 Traversing the Search Space

DATA-PEELER [11] aims to extract a complete collection of constrained closed n-sets from an n-ary relation. This section only outlines the basic principles for enumerating the candidates in the particular case $n = 3$. The interested reader would refer to [11] for detailed explanations. To emphasize the generality of DATA-PEELER, the three sets $\mathscr{T}$, $\mathscr{N}$ and $\mathscr{N}$ are, here, replaced by $\mathscr{D}^1$, $\mathscr{D}^2$ and $\mathscr{D}^3$. Indeed, when extracting closed 3-sets, there is no need for $\mathscr{D}^1$ to contain real numbers, and for $\mathscr{D}^2$ and $\mathscr{D}^3$ to be identical. These three sets must only be finite.

Like many complete algorithms for local pattern detection, DATA-PEELER is based on enumerating candidates in a way that can be represented by a binary tree where:

- at every node, an element e is enumerated;
- every pattern extracted from the left child *does contain* e;
- every pattern extracted from the right child does *not* contain e.

This division of the extraction into two sub-problems partitions the search space, i. e., the union of the closed 3-sets found in both enumeration sub-tree are exactly the closed 3-sets to be extracted from the parent node (*correctness*) *and* each of these closed 3-sets is found only once (*uniqueness*). In the case of DATA-PEELER, the enumerated element e can always be freely chosen among all the elements (from all three sets $\mathscr{D}^1$, $\mathscr{D}^2$ and $\mathscr{D}^3$) remaining in the search space.

Three 3-sets $U = (U^1, U^2, U^3)$, $V = (V^1, V^2, V^3)$ and $\mathscr{S} = (\mathscr{S}^1, \mathscr{S}^2, \mathscr{S}^3)$, are attached to every node. The 3-set $U \in 2^{\mathscr{D}^1} \times 2^{\mathscr{D}^2} \times 2^{\mathscr{D}^3}$ contains the elements that are contained in any closed 3-set extracted from the node. The 3-set $V \in 2^{\mathscr{D}^1} \times 2^{\mathscr{D}^2} \times 2^{\mathscr{D}^3}$ contains the elements that may be present in the closed 3-sets extracted from the node, i. e., the search space. The 3-set $\mathscr{S} \in 2^{\mathscr{D}^1} \times 2^{\mathscr{D}^2} \times 2^{\mathscr{D}^3}$ contains the elements that may prevent the 3-sets, extracted from this node, from being closed. To simplify the notations we will often assimilate a 3-set (S^1, S^2, S^3) with $S^1 \cup S^2 \cup S^3$. For example, given two 3-set $A = (A^1, A^2, A^3)$ and $B = (B^1, B^2, B^3)$ and an element e ($e \in \mathscr{D}^1 \cup \mathscr{D}^2 \cup \mathscr{D}^3$), we write:

- $e \in A$ instead of $e \in A^1 \cup A^2 \cup A^3$
- $A \setminus \{e\}$ instead of $\begin{cases} (A^1 \setminus \{e\}, A^2, A^3) \text{ if } e \in \mathscr{D}^1 \\ (A^1, A^2 \setminus \{e\}, A^3) \text{ if } e \in \mathscr{D}^2 \\ (A^1, A^2, A^3 \setminus \{e\}) \text{ if } e \in \mathscr{D}^3 \end{cases}$
- $A \cup B$ instead of $(A^1 \cup B^1, A^2 \cup B^2, A^3 \cup B^3)$

Figure 9.2 depicts the enumeration. The 3-sets attached to a child node are computed from its parent's analogous 3-sets, the enumerated element and the data (for the left children only). In particular, in the left child, DATA-PEELER ensures that U can receive any element from V without breaking $\mathscr{C}_{\text{connected}}$. Hence, at every node,

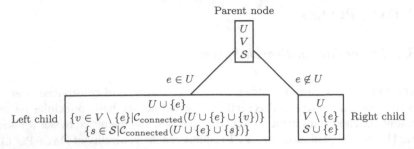

Fig. 9.2 Enumeration of any element $e \in V$

the 3-set U is connected, i. e., $\mathscr{C}_{\text{connected}}(U)$. To ensure that the extracted 3-sets are closed, DATA-PEELER checks, at every node, whether the 3-set $U \cup V$ is closed, i. e., $\mathscr{C}_{\text{closed}}(U \cup V)$. To do so, DATA-PEELER checks whether $\forall s \in \mathscr{S}, \neg \mathscr{C}_{\text{connected}}(U \cup V \cup \{s\})$. If not, every 3-set descendant from this node is not closed. Indeed, $\forall V' \subseteq V, \exists s \in \mathscr{S} | \mathscr{C}_{\text{connected}}(U \cup V \cup \{s\}) \Rightarrow \exists s \in \mathscr{S} | \mathscr{C}_{\text{connected}}(U \cup V' \cup \{s\})$. In this case DATA-PEELER safely prunes the sub-tree rooted by the node.

The enumeration tree is traversed in a depth first way. At the root node, $U = (\emptyset, \emptyset, \emptyset)$, $V = (\mathscr{D}^1, \mathscr{D}^2, \mathscr{D}^3)$ ans $\mathscr{S} = (\emptyset, \emptyset, \emptyset)$. At a given node, if $V = (\emptyset, \emptyset, \emptyset)$ then this node is a leaf and U is a closed 3-set. The algorithm in Table 9.2 sums up DATA-PEELER's principles.

Table 9.2 DATA-PEELER

Input: $U, V, \mathscr{S}$
Output: All closed 3-sets containing the elements in U and, possibly, some elements in V and satisfying $\mathscr{C}$
if $\mathscr{C}$ may be satisfied by a 3-set descending from this node
$\wedge \mathscr{C}_{\text{closed}}(U \cup V)$ **then**
 if $V = (\emptyset, \emptyset, \emptyset)$ **then**
  output(U)
 else
  Choose $e \in V$
  DATA-PEELER$(U \cup \{e\}, \{v \in V \setminus \{e\} | \mathscr{C}_{\text{connected}}(U \cup \{e\} \cup \{v\})\}\}, \{s \in \mathscr{S} | \mathscr{C}_{\text{connected}}(U \cup \{e\} \cup \{s\})\})$
  DATA-PEELER$(U, V \setminus \{e\}, \mathscr{S} \cup \{e\})$
 end if
end if

$\mathscr{C}$ is a user-defined constraint which allows to focus on relevant patterns while decreasing the extraction time by pruning enumeration sub-trees. To be able to efficiently check whether a 3-set descendant from a node satisfies $\mathscr{C}$, $\mathscr{C}$ must be a piecewise (anti)-monotone constraint.

9.3.2 Piecewise (Anti)-Monotone Constraints

DATA-PEELER can efficiently check any piecewise (anti)-monotone constraint $\mathscr{C}$. By "efficiently", we mean it *sometimes* can, from the 3-sets U and V attached to a node (no access to the data), affirm that the enumeration sub-tree rooted by this node is empty of (not necessarily connected or closed) 3-sets satisfying $\mathscr{C}$. When the node is a leaf, it, not only *sometimes*, but *always* can check a piecewise (anti)-monotone constraint, hence ensuring the correctness, i. e., every extracted closed 3-set verifies $\mathscr{C}$. Let us first define the monotonicity and anti-monotonicity per argument.

Definition 9.10 ((Anti)-monotonicity per argument). A constraint $\mathscr{C}$ is said monotone (resp. anti-monotone) w.r.t. the ith argument iff it is monotone (resp. anti-monotone) when all its arguments but the ith are considered constant.

Example 9.5. Consider the following constraint, which forces the patterns to cover at least eight 3-tuples in the relation:

$$\text{A 3-set } (D^1, D^2, D^3) \text{ is 8-large} \Leftrightarrow |D^1 \times D^2 \times D^3| \geq 8 \ .$$

It is monotone on the first argument. Indeed, $\forall (D^1, D^{1'}, D^2, D^3) \in 2^{\mathscr{D}^1} \times 2^{\mathscr{D}^1} \times 2^{\mathscr{D}^2} \times 2^{\mathscr{D}^3}, D^1 \subseteq D^{1'} \Rightarrow (|D^1 \times D^2 \times D^3| \geq 8 \Rightarrow |D^{1'} \times D^2 \times D^3| \geq 8)$. It is monotone on the second and on the third argument too.

When a constraint $\mathscr{C}$ is either monotone or anti-monotone on every argument, DATA-PEELER can efficiently check it. At a given node, it replaces the ith argument by:

- $U^i \cup V^i$ if $\mathscr{C}$ is monotone on this argument;
- U^i if $\mathscr{C}$ is anti-monotone on this argument.

In this way, a 3-set (D^1, D^2, D^3) is obtained $(\forall i \in \{1, 2, 3\}, D^i \in \{U^i, U^i \cup V^i\})$. If $\mathscr{C}(D^1, D^2, D^3)$ then at least this 3-set, descendant from the current node, verifies $\mathscr{C}$. Otherwise the sub-tree rooted by the current node can safely be pruned: it does not contain any 3-set satisfying $\mathscr{C}$.

Example 9.6. Given the two 3-sets $U = (U^1, U^2, U^3)$ and $V = (V^1, V^2, V^3)$ attached to a node, DATA-PEELER checks the 8-large constraint (defined in Ex. 9.5), by testing whether $|U^1 \cup V^1| \times |U^2 \cup V^2| \times |U^3 \cup V^3| \geq 8$.

The class of piecewise (anti)-monotone constraints contains every constraint which is either monotone or anti-monotone on each of its arguments. But it contains many other useful constraints. The definition of piecewise (anti)-monotonicity relies on attributing a separate argument to every occurrence of every variable and, then, proving that the obtained constraint is (anti)-monotone w.r.t. each of its arguments.

Definition 9.11 (Piecewise (anti)-monotonicity). A constraint $\mathscr{C}$ is piecewise (anti)-monotone iff the rewritten constraint $\mathscr{C}'$, attributing a separate argument to every occurrence of every variable in the expression of $\mathscr{C}$, is (anti)-monotone w.r.t. each of its arguments.

To illustrate this class of constraints, the particular context where $\mathscr{D}^1 = \mathscr{T} \in \mathbb{R}_+^{|\mathscr{T}|}$ is chosen:

Example 9.7. Consider the following constraint $\mathscr{C}_{16-\text{small-in-average}}$:

$$\mathscr{C}_{16-\text{small-in-average}}(T, D^2, D^3) \Leftrightarrow T \neq \emptyset \wedge \frac{\sum_{t \in T} t}{|T|} \leq 16 \ .$$

This constraint is both monotone and anti-monotone on the second and the third argument (neither D^2 nor D^3 appearing in the expression of the constraint) but it is neither monotone nor anti-monotone on the first argument. However, giving three different variables T_1, T_2 and T_3 to each of the occurrences of T creates this new constraint which is monotone on the first and third arguments (T_1 and T_3) and anti-monotone on the second one (T_2):

$$\mathscr{C}'_{16-\text{small-in-average}}(T_1, T_2, T_3, D^2, D^3) \equiv T_1 \neq \emptyset \wedge \frac{\sum_{t \in T_2} t}{|T_3|} \leq 16 \ .$$

Therefore $\mathscr{C}_{16-\text{small-in-average}}$ is piecewise (anti)-monotone.

DATA-PEELER can efficiently check any piecewise (anti)-monotone constraint. First, it considers the analogous constraint where every occurrence of the three original attributes is given a different variable. Then, it applies the rules stated previously, i.e., at a given node, it replaces the ith argument by:

- $U^i \cup V^i$ if $\mathscr{C}$ is monotone on this argument;
- U^i if $\mathscr{C}$ is anti-monotone on this argument.

The built assertion is false if, in the enumeration sub-tree that would derive from the node, there is no 3-set satisfying the original constraint. Notice that, in this general setting, the reverse may be false, i.e., the assertion can hold even if no 3-set descendant from the node that verifies the original constraint. Therefore, it can be written that DATA-PEELER relaxes the constraint to efficiently check it.

9.4 Extracting δ-Contiguous Closed 3-Sets

9.4.1 A Piecewise (Anti)-Monotone Constraint...

The constraint $\mathscr{C}_{\delta-\text{contiguous}}$ (see Def. 9.4) is piecewise (anti)-monotone.

Proof. Let $\mathscr{C}'_{\delta-\text{contiguous}}$ the following constraint:

$$\mathscr{C}'_{\delta-\text{contiguous}}(T_1, T_2, T_3, N_1, N_2)$$
$$\equiv \forall t \in [\min(T_1), \max(T_2)], \exists t' \in T_3 \text{ s.t. } |t - t'| \leq \delta \ .$$

The three arguments T_1, T_2 and T_3 substitute the three occurrences of T (in the definition of $\mathscr{C}_{\delta-\text{contiguous}}$). $\mathscr{C}'_{\delta-\text{contiguous}}$ is monotone in on its third argument and anti-monotone on its first and second arguments ($T \subseteq T_1 \Rightarrow \min(T) \geq \min(T_1)$ and $T \subseteq T_2 \Rightarrow \max(T) \leq \max(T_2)$). Moreover, since the two last arguments of $\mathscr{C}'_{\delta\text{-contiguous}}$ do not appear in its expression, this constraint is both monotone and anti-monotone on them. Therefore, by definition, $\mathscr{C}_{\delta\text{-contiguous}}$ is piecewise (anti)-monotone. □

9.4.2 ...Partially Handled in Another Way

Given the 3-sets $U = (U^{\mathscr{T}}, U^{\mathscr{N}^1}, U^{\mathscr{N}^2})$ and $V = (V^{\mathscr{T}}, V^{\mathscr{N}^1}, V^{\mathscr{N}^2})$ attached to the current enumeration node, the proof of Sect. 9.4.1 suggests to check whether it is possible to browse all elements in $[\min(U^{\mathscr{T}}), \max(U^{\mathscr{T}})] \cap (U^{\mathscr{T}} \cup V^{\mathscr{T}})$ by jumps of, at most, δ.

By also taking a look "around" $[\min(U^{\mathscr{T}}, \max(U^{\mathscr{T}})] \cap (U^{\mathscr{T}} \cup V^{\mathscr{T}})$, DATA-PEELER can do better than just telling whether there is no hope in extracting δ-contiguous 3-sets from the current enumeration node. It can prevent the traversal of some of such nodes. More precisely, DATA-PEELER removes from $V^{\mathscr{T}}$ the elements that would, if enumerated, generate left children violating $\mathscr{C}_{\delta\text{-contiguous}}$. To do so, the delay between $t = \min(U^{\mathscr{T}})$ and $\text{before}(t) = \max(\{t' \in V^{\mathscr{T}} | t' < t\})$ is considered. If it is strictly greater than δ then every element in $\{t' \in V^{\mathscr{T}} | t' < t\}$ can be removed from $V^{\mathscr{T}}$. Otherwise, the process goes on with $t = \text{before}(t)$ until a delay greater than δ is found or until $t = \min(V^{\mathscr{T}})$ (in this case no element from $V^{\mathscr{T}}$ lesser than $\min(U^{\mathscr{T}})$ is removed). In a reversed way, the elements in $V^{\mathscr{T}}$ that are too great to be moved to $U^{\mathscr{T}}$ without violating $\mathscr{C}_{\delta\text{-contiguous}}$ are removed as well. Algorithm 9.3 gives a more technical definition of DATA-PEELER's way to purge $V^{\mathscr{T}}$ thanks to $\mathscr{C}_{\delta\text{-contiguous}}$.

In the same way, some elements of $\mathscr{S}^{\mathscr{T}}$ may be too far away from the extrema of $U^{\mathscr{T}} \cup V^{\mathscr{T}}$ to prevent the δ-closedness of any descending 3-set. These elements are those that cannot be added to $U^{\mathscr{T}}$ without making the current enumeration node violate $\mathscr{C}_{\delta\text{-contiguous}}$. Hence, DATA-PEELER removes these elements by applying a procedure PURGE_$\mathscr{S}^{\mathscr{T}}$ to every enumeration node. It is very similar to PURGE_$V^{\mathscr{T}}$ (see Alg. 9.3) except that it is $\mathscr{S}^{\mathscr{T}}$ which is browsed backward from $\text{before}(\min(U^{\mathscr{T}} \cup V^{\mathscr{T}}))$ and forward from $\text{after}(\max(U^{\mathscr{T}} \cup V^{\mathscr{T}}))$.

Example 9.8. Considering the extraction of 1-contiguous 3-sets from the example dataset defined by Table 9.1, if the first enumerated element is 0.5, Fig. 9.3 depicts the root enumeration node and its two children. In the left child, PURGE_$V^{\mathscr{T}}$ removes 2 and 3 from its attached $V^{\mathscr{T}}$ set because $2 - 0.5 > 1$.

These purges of V and $\mathscr{S}$ remind the way DATA-PEELER handles $\mathscr{C}_{\text{connected}}$. $\mathscr{C}_{\text{connected}}$ is anti-monotone on all its arguments, whereas $\mathscr{C}_{\delta\text{-contiguous}}$ is *only* piecewise (anti)-monotone. Hence some enumeration nodes violating $\mathscr{C}_{\delta\text{-contiguous}}$ may be generated despite the calls of PURGE_$V^{\mathscr{T}}$ (whereas a generated enumeration node always complies with $\mathscr{C}_{\text{connected}}$). As a consequence, checking, at every enumeration node, whether $\mathscr{C}_{\delta\text{-contiguous}}$ holds remains necessary. For the same reason, some

Table 9.3 PURGE_$V^{\mathcal{I}}$

Input: $U^{\mathcal{I}}, V^{\mathcal{I}}$
if $U^{\mathcal{I}} \neq \emptyset$ **then**
    $V^{\mathcal{I}} \leftarrow \mathbf{sort}(V^{\mathcal{I}})$
    $t \leftarrow \min(U^{\mathcal{I}})$
    if $t > \min(V^{\mathcal{I}})$ **then**
        before$(t) \leftarrow \max(\{t' \in V^{\mathcal{I}} | t' < t\})$ {Binary search in $V^{\mathcal{I}}$}
        while before$(t) \neq \min(V^{\mathcal{I}}) \wedge t - \text{before}(t) \leq \delta$ **do**
            $t \leftarrow \text{before}(t)$
            before$(t) \leftarrow \mathbf{previous}(V^{\mathcal{I}}, t)$ {$V^{\mathcal{I}}$ is browsed backward}
        end while
        if $t - \text{before}(t) > \delta$ **then**
            $V^{\mathcal{I}} \leftarrow V^{\mathcal{I}} \setminus [\min(V^{\mathcal{I}}), \text{before}(t)]$
        end if
    end if
    $t \leftarrow \max(U^{\mathcal{I}})$
    if $t < \max(V^{\mathcal{I}})$ **then**
        after$(t) \leftarrow \min(\{t' \in V^{\mathcal{I}} | t' > t\})$ {Binary search in $V^{\mathcal{I}}$}
        while after$(t) \neq \max(V^{\mathcal{I}}) \wedge \text{after}(t) - t \leq \delta$ **do**
            $t \leftarrow \text{after}(t)$
            after$(t) \leftarrow \mathbf{next}(V^{\mathcal{I}}, t)$ {$V^{\mathcal{I}}$ is browsed forward}
        end while
        if after$(t) - t > \delta$ **then**
            $V^{\mathcal{I}} \leftarrow V^{\mathcal{I}} \setminus [\text{after}(t), \max(V^{\mathcal{I}})]$
        end if
    end if
end if

---

elements in the 3-sets V and/or $\mathcal{S}$ attached to both left and right children may be purged thanks to $\mathscr{C}_{\delta\text{-contiguous}}$ (whereas $\mathscr{C}_{\text{connected}}$ cannot reduce the search space of a right child).

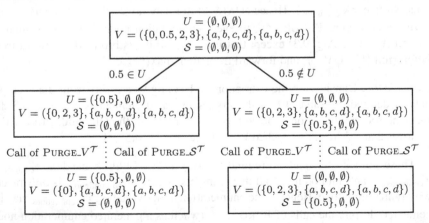

Fig. 9.3 Enumeration of $0.5 \in V$ during the extraction of 1-contiguous 3-sets from the example dataset defined by Table 9.1

9.4.3 Enforcing the δ-Closedness

The constraint $\mathscr{C}_{\delta\text{-closed}}$ (see Def. 9.5) is piecewise (anti-)monotone.

Proof. Let $\mathscr{C}'_{\delta\text{-closed}}$ the following constraint:

$$\mathscr{C}'_{\delta\text{-closed}}(T_1,T_2,T_3,T_4,N_1^1,N_2^1,N_3^1,N_1^2,N_2^2,N_3^2)$$
$$\equiv \begin{cases} \forall t \in \mathscr{T} \setminus T_1, (\exists t' \in T_2 \text{ s.t. } |t-t'| \leq \delta \Rightarrow \neg\mathscr{C}_{\text{connected}}(\{t\},N_1^1,N_1^2)) \\ \forall n^1 \in \mathscr{N} \setminus N_2^1, \neg\mathscr{C}_{\text{connected}}(T_3,\{n^1\},N_2^2) \\ \forall n^2 \in \mathscr{N} \setminus N_3^2, \neg\mathscr{C}_{\text{connected}}(T_4,N_3^1,\{n^2\}) \end{cases}.$$

$\mathscr{C}'_{\delta\text{-closed}}$ is anti-monotone on its second argument and monotone on all its other arguments. Therefore, by definition, $\mathscr{C}_{\delta\text{-closed}}$ is piecewise (anti-)monotone. $\square$

A way to enforce $\mathscr{C}_{\delta\text{-closed}}$ follows from the proof of its piecewise (anti-)monotonicity: an enumeration node, i.e., its attached $U = (U^{\mathscr{T}}, U^{\mathscr{N}^1}, U^{\mathscr{N}^2})$ and $V = (V^{\mathscr{T}}, V^{\mathscr{N}^1}, V^{\mathscr{N}^2})$, may lead to some δ-closed 3-set if $(U^{\mathscr{T}} \cup V^{\mathscr{T}}, U^{\mathscr{N}^1} \cup V^{\mathscr{N}^1}, U^{\mathscr{N}^2} \cup V^{\mathscr{N}^2})$:

- cannot be extended by any element in $\mathscr{T} \setminus (U^{\mathscr{T}} \cup V^{\mathscr{T}})$ distant, by at most δ, from an element in $U^{\mathscr{T}}$;
- cannot be extended by any element in $\mathscr{N} \setminus (U^{\mathscr{N}^1} \cup V^{\mathscr{N}^1})$;
- cannot be extended by any element in $\mathscr{N} \setminus (U^{\mathscr{N}^2} \cup V^{\mathscr{N}^2})$.

As done for $\mathscr{C}_{\text{closed}}$, to avoid useless (and costly) tests, DATA-PEELER maintains the 3-set $\mathscr{S} = (\mathscr{S}^{\mathscr{T}}, \mathscr{S}^{\mathscr{N}^1}, \mathscr{S}^{\mathscr{N}^2})$ containing only the elements that may prevent the closure of the 3-sets descending from the current enumeration node, i.e., the previously enumerated elements and not those that were removed from V thanks to $\mathscr{C}_{\text{connected}} \wedge \mathscr{C}_{\delta\text{-contiguous}}$. Moreover, as explained in Sect. 9.4.2, DATA-PEELER purges $\mathscr{S}$ before checking $\mathscr{C}_{\delta\text{-closed}}$. Since it is used in conjunction with $\mathscr{C}_{\delta\text{-contiguous}}$, $\mathscr{C}_{\delta\text{-closed}}$ can be more strongly enforced: no element in $\mathscr{S}^{\mathscr{T}} \cap [\min(U^{\mathscr{T}}) - \delta, \max(U^{\mathscr{T}}) + \delta]$ is allowed to extend $(U^{\mathscr{T}} \cup V^{\mathscr{T}}, U^{\mathscr{N}^1} \cup V^{\mathscr{N}^1}, U^{\mathscr{N}^2} \cup V^{\mathscr{N}^2})$. Indeed, an element in $\mathscr{S}^{\mathscr{T}} \cap [\min(U^{\mathscr{T}}) - \delta, \max(U^{\mathscr{T}}) + \delta]$ may be distant, by strictly more than δ, from any element in $U^{\mathscr{T}}$ but this will never be the case at the leaves descending from the current enumeration since $U^{\mathscr{T}}$ must then be δ-contiguous. All in all, DATA-PEELER prunes the sub-tree descending from the current enumeration node if $(U^{\mathscr{T}} \cup V^{\mathscr{T}}, U^{\mathscr{N}^1} \cup V^{\mathscr{N}^1}, U^{\mathscr{N}^2} \cup V^{\mathscr{N}^2})$ can be extended by any element in $\mathscr{S}^{\mathscr{T}} \cap [\min(U^{\mathscr{T}}) - \delta, \max(U^{\mathscr{T}}) + \delta]$, $\mathscr{S}^{\mathscr{N}^1}$ or $\mathscr{S}^{\mathscr{N}^2}$.

9.5 Constraining the Enumeration to Extract 3-Cliques

9.5.1 A Piecewise (Anti)-Monotone Constraint...

In a 3-clique, both subsets of $\mathcal{N}$ are identical. An equivalent definition to the symmetry constraint (Def. 9.7) would be as follows: $\mathscr{C}_{\text{symmetric}}(T, N^1, N^2) \equiv N^1 \subseteq N^2 \wedge N^2 \subseteq N^1$. In this form, a piecewise (anti)-monotone constraint is identified.

Proof. Let $\mathscr{C}'_{\text{symmetric}}$ the following constraint:

$$\mathscr{C}'_{\text{symmetric}}(T, N_1^1, N_2^1, N_1^2, N_2^2) \equiv N_1^1 \subseteq N_1^2 \wedge N_2^2 \subseteq N_2^1 .$$

N_1^1 and N_2^1 substitute the two occurrences of N^1 (in the alternative definition of $\mathscr{C}_{\text{symmetric}}$). In the same way, N_1^2 and N_2^2 substitute the two occurrences of N^2. $\mathscr{C}'_{\text{symmetric}}$ is monotone on its third and fourth arguments (N_2^1 and N_1^2) and anti-monotone on its second and fifth arguments (N_1^1 and N_2^2). Moreover, since the first argument (T) does not appear in the expression of $\mathscr{C}'_{\text{symmetric}}$, this constraint is both monotone and anti-monotone on this argument. Therefore, by definition, $\mathscr{C}_{\text{symmetric}}$ is piecewise (anti)-monotone. $\square$

Being piecewise (anti)-monotone, the symmetry constraint can be efficiently exploited by DATA-PEELER. However, the enumeration tree can be further reduced if this constraint is enforced when choosing the element to be enumerated.

9.5.2 ...Better Handled in Another Way

In this section, a distinction between the "first" set of nodes (i.e., the rows of the adjacency matrices) and the "second" one (i.e., the columns of the adjacency matrices) must be made. They are respectively named $\mathcal{N}^1$ and $\mathcal{N}^2$. Intuitively, when an element n^1 from $V^1 \subseteq \mathcal{N}^1$ is chosen to be present (respectively absent) in any 3-clique extracted from the node (see Sect. 9.3.1), the element n^2 from $V^2 \subseteq \mathcal{N}^2$ standing for the same node should be enumerated just after and only to be present (respectively absent) too. Thus, the enumeration tree is not a binary tree anymore (some enumeration nodes only have one child).

When handled as a piecewise (anti)-monotone constraint, the symmetry constraint leads to many more enumeration nodes. When n^2 is chosen to be enumerated, the left (respectively right) child where n^2 is present (respectively absent) is generated even if its counterpart n^1 in the other set was previously set absent (respectively present). Then the symmetry constraint prunes the sub-tree rooted by this node. Since there is no reason for n^2 to be enumerated just after n^1, the intuition tells us that the number of such nodes, whose generation could be avoided by modifying the enumeration (as explained in the previous paragraph), increases exponentially with the average number of enumeration nodes between the enumeration of n^1 and

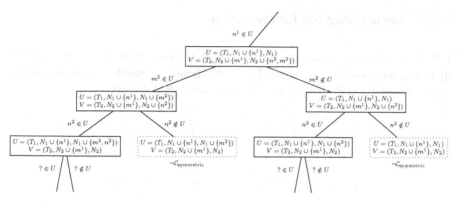

Fig. 9.4 Symmetry handled as an ordinary piecewise (anti)-monotone constraint

that of n^2. This is actually not a theorem because $\mathscr{C}_{\text{sym-}\delta\text{-closed}}$ or $\mathscr{C}$ may prune some descendant sub-trees before n^2 is enumerated. Anyway, in practical settings, handling the symmetry constraint via a modification of the enumeration usually is much more efficient than via the general framework for piecewise (anti)-monotone constraints.

Figures 9.4 and 9.5 informally depict these two approaches (the probable diminutions of the V sets in the left children and the possible pruning due to $\mathscr{C}_{\text{closed}}$ or $\mathscr{C}$ are ignored). T_1 and T_2 are subsets of $\mathscr{T}$. N_1 and N_2 are subsets of $\mathscr{N}$. In both examples, the elements m^2 and n^2 of $\mathscr{N}^2$ are enumerated. The resulting nodes are, of course, the same (the dotted nodes being pruned). However this result is straightforward when the enumeration constraint is handled through a modification of the enumeration (Fig. 9.5), whereas it usually requires more nodes when it is handled as an ordinary piecewise (anti)-monotone constraint (Fig. 9.4). The number of additional nodes in the latter case grows exponentially with the number of elements enumerated between n^1 and n^2 (e. g., m^1 could be enumerated in between).

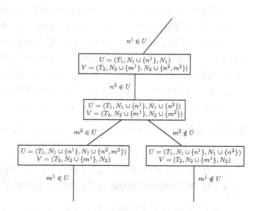

Fig. 9.5 Symmetry handled by a modified enumeration

9.5.3 Constraining the Enumeration

Let $\mathcal{N}^1 = (n_i^1)_{i=1...|\mathcal{N}|}$ and $\mathcal{N}^2 = (n_i^2)_{i=1...|\mathcal{N}|}$ its counterpart, i. e., $\forall i = 1...|\mathcal{N}|$, n_i^1 and n_i^2 stand for the same node. (T, N^1, N^2) being symmetric is a constraint that can be expressed as this list of, so called, *enumeration constraints*:

$$n_1^1 \in N^1 \Rightarrow n_1^2 \in N^2 \qquad n_1^2 \in N^2 \Rightarrow n_1^1 \in N^1$$
$$n_2^1 \in N^1 \Rightarrow n_2^2 \in N^2 \qquad n_2^2 \in N^2 \Rightarrow n_2^1 \in N^1$$
$$\vdots \qquad\qquad \vdots$$
$$n_i^1 \in N^1 \Rightarrow n_i^2 \in N^2 \qquad n_i^2 \in N^2 \Rightarrow n_i^1 \in N^1$$
$$\vdots \qquad\qquad \vdots$$
$$n_{|\mathcal{N}|}^1 \in N^1 \Rightarrow n_{|\mathcal{N}|}^2 \in N^2 \qquad n_{|\mathcal{N}|}^2 \in N^2 \Rightarrow n_{|\mathcal{N}|}^1 \in N^1$$

These constraints belong to a more general class of constraints:

Definition 9.12 (Enumeration constraint). An enumeration constraint $\mathscr{C}_{enum}$ is such that, given a 3-set (T, N^1, N^2), $\mathscr{C}_{enum}(T, N^1, N^2) \equiv \exists k \in \mathbb{N} | a_1 \wedge a_2 \wedge \cdots \wedge a_k \Rightarrow a_{k+1}$, where $\forall i = 1...k+1, a_i$ is of the form $e \in A$ or $e \notin A$, e being an arbitrary element from an arbitrary dimension $A \in \{T, N^1, N^2\}$.

Example 9.9. Here are three examples of enumeration constraints that can be enforced on any 3-set (T, N^1, N^2):

- $t_1 \in T \Rightarrow t_8 \notin T$
- $t_1 \notin T \wedge n_1^1 \in N^1 \Rightarrow t_2 \in T$
- true $\Rightarrow t_1 \notin T$ ($k = 0$ in Def. 9.12)

Notice that the last constraint is not equivalent to removing the element t_1 from the data. Indeed, a closed 3-set in the data set deprived of t_1 may not be closed in the data set containing t_1. Hence it must not be extracted (and it is actually not extracted when the constraint enumeration is used).

Before choosing the element to be enumerated (see algorithm in Table 9.2), DATA-PEELER browses the set of enumeration constraint, and tests whether the left parts of them are true or not. Considered as constraints, these left parts are, again, piecewise (anti-)monotone. Indeed, when there is a term of the form $e \in A$ (respectively $e \notin A$), the left part of the constraint is anti-monotone (respectively monotone) in this occurrence of A. Given the 3-sets U and V attached to the current enumeration node, three cases may arise:

1. The left part will never be fulfilled in the sub-tree rooted by the current enumeration node:
   - if an element in the left part is to be present but it is neither in U not in V.
   - if an element in the right part is to be absent but it is in U.

2. The left part is fulfilled by at least one (but not every) node descending from the current enumeration node.

3. The left part is fulfilled by every node descending from the current enumeration node:

   - if an element in the left part is to be present, it is in U.
   - if an element in the left part is to be absent, it is neither in U nor in V.

DATA-PEELER reacts differently at each of these cases:

1. This enumeration constraint is removed from the set of enumeration constraints when traversing the sub-tree rooted by the current enumeration node. Indeed, it never applies in this sub-tree. Uselessly checking it for every descendant enumeration node would only decrease the performances of DATA-PEELER.
2. This enumeration constraint is kept.
3. The right part of this enumeration constraint is considered.

When the right part of an enumeration constraint is considered, three new cases may arise:

3.1 The right part is already fulfilled:

   - if the element in the right part is to be present, it is already in U.
   - if is to be absent, it is already neither in U nor in V.

3.2 The right part can be fulfilled: the element in the right part is in V.
3.3 The right part cannot be fulfilled:

   - if the element in the right part is to be present, it is neither in U nor in V.
   - if it is to be absent, it is in U.

DATA-PEELER reacts differently at each of these cases:

3.1 This enumeration constraint is removed from the set of enumeration constraints when traversing the sub-tree rooted by the current enumeration node. Indeed, it is satisfied for all 3-sets in this sub-tree. Uselessly checking it for every descendant enumeration node would only decrease the performances of DATA-PEELER.
3.2 The element on the right part of the constraint can be enumerated as specified (one child only).
3.3 The sub-tree rooted by the current enumeration node is pruned. Indeed, none of the 3-sets in this sub-tree verifies the constraint.

In Case 3.2, we write "the element *can* be enumerated" because, at a given enumeration node, several enumeration constraint may be in this case but only one can be applied.

9.5.4 Contraposition of the Enumeration Constraints

If an enumeration constraint holds, its contraposition, logically, holds too. In the general case (conjunction of terms in the left part), the contraposition of an enumeration constraint is not an enumeration constraint (disjunction of terms in the right

Table 9.4 APPEND_CONTRAPOSITION

Input: Set E of enumeration constraints
Output: Set E enlarged with contrapositions
$E' \leftarrow E$
for $a_1 \wedge a_2 \wedge \cdots \wedge a_k \Rightarrow a_{k+1} \in E$ **do**
   if $k = 1$ **then**
      $E' \leftarrow E' \cup \{\neg a_2 \Rightarrow \neg a_1\}$
   end if
end for
return E'

---

part). In the particular case of enumeration constraints of the form $a_1 \Rightarrow a_2$ (see Def. 9.12), e. g., those generated from $\mathscr{C}_{\text{symmetric}}$ (see Sect. 9.5.3), their contrapositions are enumeration constraints too. Thus, DATA-PEELER enforces a larger set of enumeration constraints (the original set of enumeration constraints and the contrapositions of those of the form $a_1 \Rightarrow a_2$) for even faster extractions. The algorithm in Table 9.4 gives a more technical definition of how this larger set is computed.

Example 9.10. Among the enumeration constraints of Ex. 9.9, only the first one ($t_1 \in T \Rightarrow t_8 \notin T$) admits a contraposition ($t_8 \in T \Rightarrow t_1 \notin T$) that is, itself, an enumeration constraint.

9.5.5 Enforcing the Symmetric δ-Closedness

The constraint $\mathscr{C}_{\text{sym-}\delta\text{-closed}}$ (see Def. 9.8) is piecewise (anti)-monotone.

Proof. Let $\mathscr{C}'_{\text{sym-}\delta\text{-closed}}$ the following constraint:

$$\mathscr{C}'_{\text{sym-}\delta\text{-closed}}(T_1, T_2, T_3, N_1^1, N_2^1, N_3^1, N_1^2, N_2^2, N_3^2)$$
$$\equiv \begin{cases} \forall t \in \mathscr{T} \setminus T_1, (\exists t' \in T_2 \text{ s.t. } |t - t'| \leq \delta \Rightarrow \neg \mathscr{C}_{\text{connected}}(\{t\}, N_1^1, N_1^2)) \\ \forall n \in \mathscr{N} \setminus (N_2^1 \cap N_2^2), \neg \mathscr{C}_{\text{connected}}(T, N_3^1 \cup \{n\}, N_3^2 \cup \{n\}) \end{cases}$$

$\mathscr{C}'_{\text{sym-}\delta\text{-closed}}$ is anti-monotone on its second argument (T_2) and monotone on all its other arguments. Therefore, by definition, $\mathscr{C}_{\text{sym-}\delta\text{-closed}}$ is piecewise (anti)-monotone. $\square$

A way to enforce $\mathscr{C}_{\delta\text{-closed}}$ follows from the proof of its piecewise (anti)-monotonicity: an enumeration node, i. e., its attached $U = (U^{\mathscr{T}}, U^{\mathscr{N}^1}, U^{\mathscr{N}^2})$ and $V = (V^{\mathscr{T}}, V^{\mathscr{N}^1}, V^{\mathscr{N}^2})$, may lead to some δ-closed 3-set if $(U^{\mathscr{T}} \cup V^{\mathscr{T}}, U^{\mathscr{N}^1} \cup V^{\mathscr{N}^1}, U^{\mathscr{N}^2} \cup V^{\mathscr{N}^2})$:

- cannot be extended by any element in $\mathscr{T} \setminus (U^{\mathscr{T}} \cup V^{\mathscr{T}})$ distant, by at most δ, from an element in $U^{\mathscr{T}}$;

- cannot be simultaneously extended by any element in $\mathcal{N} \setminus (U^{\mathcal{N}^1} \cup V^{\mathcal{N}^1})$ (row of the adjacency matrices) and its related element in $\mathcal{N} \setminus (U^{\mathcal{N}^2} \cup V^{\mathcal{N}^2})$ (column of the adjacency matrices).

In a similar way to what was done with $\mathcal{C}_{\delta\text{-closed}}$ (see Sect. 9.4.3), DATA-PEELER maintains the 3-set $\mathscr{S} = (\mathscr{S}^{\mathcal{T}}, \mathscr{S}^{\mathcal{N}^1}, \mathscr{S}^{\mathcal{N}^2})$ containing only the elements that may prevent the closure of the 3-sets descending from the current enumeration node and prunes the sub-tree descending from it if $(U^{\mathcal{T}} \cup V^{\mathcal{T}}, U^{\mathcal{N}^1} \cup V^{\mathcal{N}^1}, U^{\mathcal{N}^2} \cup V^{\mathcal{N}^2})$ can be extended by any element in $\mathscr{S}^{\mathcal{T}} \cap [\min(U^{\mathcal{T}}) - \delta, \max(U^{\mathcal{T}}) + \delta]$ or by any element in $\mathscr{S}^{\mathcal{N}^1}$ and its related element in $\mathscr{S}^{\mathcal{N}^2}$. Thus, when $\mathscr{S}^{\mathcal{N}^1}$ (respectively $\mathscr{S}^{\mathcal{N}^2}$) is purged from an element (because it cannot extend $(U^{\mathcal{T}} \cup V^{\mathcal{T}}, U^{\mathcal{N}^1} \cup V^{\mathcal{N}^1}, U^{\mathcal{N}^2} \cup V^{\mathcal{N}^2})$ without violating $\mathcal{C}_{\text{connected}}$), the related element in $\mathscr{S}^{\mathcal{N}^2}$ (respectively $\mathscr{S}^{\mathcal{N}^1}$) is removed as well.

An overall view of the complete extraction of the δ-contiguous closed 3-cliques under constraint can now be presented. The details and justifications of how every identified constraint is handled are present within the two previous sections, hence proving its correctness. The algorithm in Table 9.5 is the main procedure solving the problem presented in Sect. 9.2. It calls the algorithm presented in Table 9.6 which can be regarded as a specialization of the algorithm in Table 9.2.

Table 9.5 MAIN

Input: $(A_t)_{t \in \mathcal{T}} \in \{0,1\}^{\mathcal{T} \times \mathcal{N} \times \mathcal{N}}$, $\delta \in \mathbb{R}_+$ and a user-defined piecewise (anti)-monotone constraint $\mathcal{C}$
Output: All δ-contiguous closed 3-cliques in $(A_t)_{t \in \mathcal{T}}$ satisfying $\mathcal{C}$
$E \leftarrow$ Set of enumeration constraints pertaining to $\mathcal{C}_{\text{symmetric}}$ (see Sect. 9.5.3)
$E' \leftarrow$ APPEND_CONTRAPOSITION(E)
DATA-PEELER$((\emptyset, \emptyset, \emptyset), (\mathcal{T}, \mathcal{N}, \mathcal{N}), (\emptyset, \emptyset, \emptyset))$

9.6 Experimental Results

The experiments were performed on an AMD Sempron[TM] 2600+ computer with 512 MB of RAM and running a GNU/Linux[TM] operating system. DATA-PEELER was compiled with GCC 4.3.2.

9.6.1 Presentation of the Vélo'v Dataset

Vélo'v is a bicycle rental service run by the city of Lyon, France. 338 Vélov stations are spread over this city. At any of these stations, the users can take a bicycle and

Table 9.6 DATA-PEELER specialization

Input: $U, V, \mathscr{S}$
Output: All δ-contiguous closed 3-cliques containing the elements in U and, possibly, some elements in V and satisfying $\mathscr{C}$
PURGE_$V^{\mathscr{I}}$
PURGE_$\mathscr{S}^{\mathscr{I}}$
if $\mathscr{C} \wedge \mathscr{C}_{\delta\text{-contiguous}} \wedge \mathscr{C}_{\text{sym-}\delta\text{-closed}}$ may be satisfied by a 3-set descending from this node **then**
   Process E' as detailed in Sect. 9.5.3
   if Case 3.3 was never encountered **then**
      if $V = (\emptyset, \emptyset, \emptyset)$ **then**
         output(U)
      else
         if Case 3.2 was encountered with an enumeration constraint concluding on a_{k+1} (see Def. 9.12) **then**
            if a_{k+1} is of the form $e \in A$ **then**
               DATA-PEELER($U \cup \{e\}, \{v \in V \setminus \{e\} | \mathscr{C}_{\text{connected}}(U \cup \{e\} \cup \{v\})\}, \{s \in \mathscr{S} | \mathscr{C}_{\text{connected}}(U \cup \{e\} \cup \{s\})\}$)
            else
               a_{k+1} is of the form $e \notin A$
               DATA-PEELER($U, V \setminus \{e\}, \mathscr{S} \cup \{e\}$)
            end if
         else
            Choose $e \in V$
            DATA-PEELER($U \cup \{e\}, \{v \in V \setminus \{e\} | \mathscr{C}_{\text{connected}}(U \cup \{e\} \cup \{v\})\}, \{s \in \mathscr{S} | \mathscr{C}_{\text{connected}}(U \cup \{e\} \cup \{s\})\}$)
            DATA-PEELER($U, V \setminus \{e\}, \mathscr{S} \cup \{e\}$)
         end if
      end if
   end if
end if

---

return it to any other station. Whenever a bicycle is rented or returned, this event is logged. We focus here on the data generated during the year 2006. These data are aggregated to obtain one graph per period of time (we chose a period of 30 minutes). For instance, one of these graphs presents the activity of the network during an average Monday of 2006 between nine o'clock and half past nine. The set of nodes $\mathcal{N}$ of such a graph corresponds to the Vélo'v stations. Its edges are labelled with the total number of rides in 2006 between the two linked stations (whatever their orientation) during the considered period of time. Setting a threshold allows to select the most significant edges. Many statistical tests can be used to fix this threshold (which can be different between the graphs). We opted for the rather simple procedure below.

α-binarization   Given a graph whose edges are labelled by values quantifying them, let m be the maximum of these values. Given a user-defined real number $\alpha \in [0,1]$ (common to all graphs), the threshold is fixed to $(1 - \alpha) \times m$.

Once the thresholds set, all edges linked to some station may be considered insignificant. Such an infrequently used station is removed from the dynamic graph. In our experiments, 204 stations remained after an α-binarization with $\alpha = 0.8$. Unless an experiment requires different datasets (scalability w.r.t. the density), every experiment uses this extraction context.

To filter out the 3-cliques corresponding to frequent rides between two stations only, a monotone constraint pertaining to the number of stations is enforced: the 3-cliques must involve at least 3 nodes to be extracted.

9.6.2 Extracting Cliques Via Enumeration Constraints

To confirm that the use of enumeration constraints actually helps in reducing the extraction time, three different strategies for 3-clique extraction are empirically compared:

1. DATA-PEELER extracts all closed 3-sets. Among them, the 3-cliques are collected by post-processing: all closed 3-sets are browsed and those that are not symmetric are filtered out. Notice that this strategy is correct for this application because the considered dynamic graph is undirected, hence, a 3-set that does not satisfy $\mathscr{C}_{\delta\text{-closed}}$ will not satisfy $\mathscr{C}_{\text{sym-}\delta\text{-closed}}$ either.
2. DATA-PEELER handles the symmetry constraint via "classical" piecewise (anti)-monotone constraints (see Sect. 9.5.1).
3. DATA-PEELER handles the symmetry constraint via enumeration constraints (see Sect. 9.5.2).

Figure 9.6 depicts the extraction times of these three strategies under different minimal size constraints on the number of time periods to be present (abscissa). In this experiment the second strategy is only slightly faster than the extraction of all closed 3-sets (notice however that the required post-treatment is not included in the plotted results), whereas the use of enumeration constraints significantly reduces the extraction time.

This advantage grows with the density of the dataset. To test this, another binarization is used. It directly controls the number of edges kept in the dynamic graph:

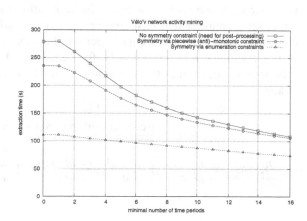

Fig. 9.6 Extraction times for different strategies (variable minimal size constraint)

Fig. 9.7 Extraction times for
different strategies (variable
density)

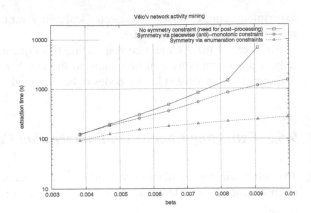

β-binarization   Given a graph whose edges are labelled by values quantifying
them and a user-defined real number $\beta \in [0,1]$ (common to all graphs), the edges
labelled with the $\beta \times |\mathcal{N}^2|$ highest values are kept.

Figure 9.7 shows how handling the symmetry via enumeration constraints more and
more reduces the extraction time when β grows. As explained earlier, the number of
nodes may be changed when we increase the number of edges. In this experiment,
it varies between 201 (when $\beta = 0.0038$) and 240 (when $\beta = 0.01$). In addition
to the minimal size constraint on the number of stations (at least three) involved
in every extracted pattern, each of these patterns is, here, forced to gather at least
two periods of time too. When $\beta = 0.0091$, it takes almost two hours to extract all
1,033,897 closed 3-sets. Among them, the post-process would retain the 18,917 ones
that are symmetric. In contrast, these cliques are directly extracted in less than 20
minutes when the symmetry constraint is enforced as a piecewise (anti)-monotone
constraint. The use of enumeration constraints provides the best performance: the
extraction takes about four minutes.

Adding, to the set of enumeration constraints generated from the symmetry constraint, their contrapositions (see Sect. 9.5.4), is believed to improve the extraction
time. However the cost of checking the application (or the non application) of a
larger set of enumeration constraints brings an overhead. The following experiment
confirms the advantage in using a larger set of enumeration constraints.

For each node n (more precisely, for each $n^1 \in \mathcal{N}^1$ or $n^2 \in \mathcal{N}^2$), one of the
following sets of enumeration constraints is sufficient to enforce the symmetry constraint:

Set 1	Set 2 (contraposition of Set 1)	Set 3 (union of Set 1 and Set 2)
$n^1 \in \mathcal{N}^1 \Rightarrow n^2 \in \mathcal{N}^2$ $n^2 \in \mathcal{N}^2 \Rightarrow n^1 \in \mathcal{N}^1$	$n^2 \in \mathcal{N}^2 \Rightarrow n^1 \in \mathcal{N}^1$ $n^2 \notin \mathcal{N}^2 \Rightarrow n^1 \notin \mathcal{N}^1$	$n^1 \in \mathcal{N}^1 \Rightarrow n^2 \in \mathcal{N}^2$ $n^1 \notin \mathcal{N}^1 \Rightarrow n^2 \notin \mathcal{N}^2$ $n^2 \notin \mathcal{N}^2 \Rightarrow n^1 \notin \mathcal{N}^1$ $n^1 \notin \mathcal{N}^1 \Rightarrow n^2 \notin \mathcal{N}^2$

Fig. 9.8 Running time with different sets of enumeration constraints

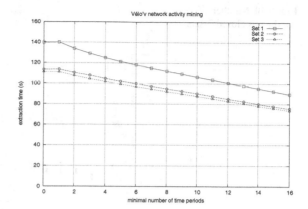

The results are plotted in Figs. 9.8 and 9.9. The experimental context is perfectly identical to that of the experiment depicted in Fig. 9.6. The running times obtained with Set 2 are lower than those obtained with Set 1 because the closed 3-cliques involve small proportions of the nodes in $\mathcal{N}$. That is why what is *not* in the patterns more frequently triggers enumeration constraints. Anyway, the fastest extractions are obtained with the largest set of enumeration constraints. It may look odd that, while being faster, the extractions performed with Set 2 generates many more enumeration nodes than those generated with Set 1. The difference between the costs of generating left enumeration nodes and right enumeration nodes explain it. Indeed, although more enumeration nodes are traversed when using Set 2, these nodes mainly are right nodes (the constraints in Set 2 conclude on such nodes), whereas the constraints in Set 1 impose the creation of left enumeration nodes. The left enumeration nodes do not prune much the search space (hence their numbers) but are very cheap to generate since the cost only is that of moving an element from a vector to another (see Fig 9.2).

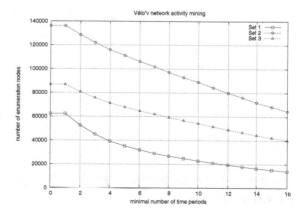

Fig. 9.9 Number of enumeration nodes with different sets of enumeration constraints

Fig. 9.10 Number of δ-contiguous closed 3-sets

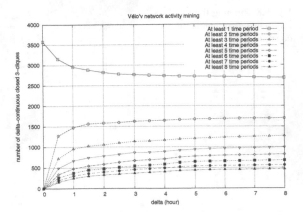

9.6.3 Extraction of δ-Contiguous Closed 3-Cliques

Figure 9.10 depicts the number of δ-contiguous closed 3-cliques when δ varies between 0 and 8 hours. Different minimal size constraints, on the number of time periods to be present in any extracted pattern, are used. When this minimal size is set to 1, the number of δ-contiguous closed 3-cliques decreases while δ increases. This means that this dynamic graph contains many 3-cliques with one time period only. When δ grows, some of these 3-cliques are merged, thus gathering more time periods. That is why, when the patterns are constrained to gather at least two (or more) time periods, the size of the collection of δ-contiguous closed 3-cliques increases with δ. These behaviors are data-dependent. For example, under a size constraint greater or equal to 2, it is possible to find datasets where, when δ increases, the size of the collection would first increase (the involved timestamps were too distant to be extracted with smaller δs) and then decrease (the patterns found with smaller δs merge).

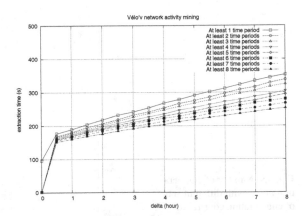

Fig. 9.11 Running time

Figure 9.11 shows that smaller δs mean smaller extraction times. Hence, if a dynamic graph gathers many timestamps, enforcing a δ-contiguity helps a lot in making the knowledge extraction tractable. Furthermore this performance gain, that occurs when δ decreases, is greater when the minimal size constraints (on the number of timestamps) is smaller. Thus the performance gain is even more useful to compensate the difficulty to extract patterns that contain few timestamps. In the figure, the divergence of the curves, when δ increases, illustrates this interesting property.

9.6.4 Qualitative Validation

To assess, by hand, the quality of the extracted δ-contiguous closed 3-cliques, the returned collection must be small. Hence stronger constraints are enforced. The minimal number of Vélov stations that must be involved in a δ-contiguous closed 3-clique is raised to 6 and the minimal number of periods to 4. With $\delta = 0.5$ hours, only three patterns are returned. Two of them take place during the evening (they start at half past 19) and gather stations that are in the center of Lyon (the "2nd and 3rd arrondissement"). They differ by one station (one station is present in the first 0.5-contiguous closed 3-clique and absent from the other and vice versa) and one of them runs during one more time period. An agglomerative post-process, such as [12], would certainly merge these two patterns. The third 0.5-contiguous closed 3-clique is displayed in Fig 9.12. The circles stand for the geographical positions of the Vélov stations. The larger and filled circles are the stations involved in the shown pattern. The disposition of the stations follows one of the main street in Lyon: "Cours Gambetta". Obviously it is much used by the riders during the evening. The outlying Vélov station is, overall, the most frequently used one: "Part-Dieu/Vivier-Merle". At this place, the rider finds the only commercial center in Lyon, the main train station, etc.

Extracting, with the same minimal size constraints, the 1-contiguous closed 3-cliques provides a collection of nine patterns. Among them, the three 0.5-contiguous closed 3-cliques are found unaltered; some slight variations of them are found (one or two stations are changed); one pattern takes place during the morning (to obtain patterns involving night periods the constraints must be weakened a lot: nightly rides do not comply much with a model). The majority of the extracted 1-contiguous closed 3-cliques involves Vélov stations in the "2nd and 3rd arrondissement". Figure 9.13 depicts one of them. The disposition of the stations follows the street connecting the two most active districts in Lyon: "Rue de la Part-Dieu". The outlying Vélov station is, overall, one of the most frequently used: "Opéra". At this place, the rider can find, not only the opera, but also the town hall, the museum of fine arts, a cinema, bars, etc. For the maintenance of the Vélov network, these examples of constrained cliques correspond to relevant sub-networks. More generally, we believe that preserved clique patterns are a priori interesting (i. e., independently from

Fig. 9.12 A 0.5-contiguous
closed 3-clique with $T =$
$\{18.5, 19, 19.5, 20, 20.5\}$

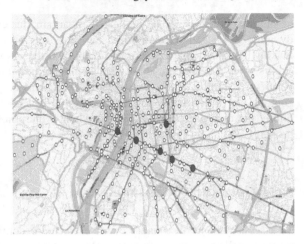

the application context). The possibility to exploit other user-defined constraints
supports the discovery of actionable patterns.

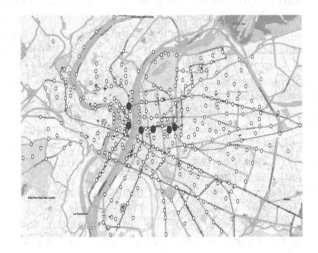

Fig. 9.13 A 1-contiguous
closed 3-clique with $T =$
$\{16, 17, 17.5, 18.5\}$

9.7 Related Work

The harder problem of extracting a complete collection of closed 3-sets directly
from real-valued tensors (e. g., rough kinetic microarray datasets) is not discussed
here. To the best of our knowledge, DATA-PEELER only faces two competitors
able to extract all closed 3-sets from ternary relations: CUBEMINER [17] and
TRIAS [16]. None of them have the generality of DATA-PEELER. In particular, they

cannot deal with n-ary relations and cannot enforce any piecewise (anti)-monotone constraints. This latter drawback makes them harder to specialize in the extraction of δ-contiguous closed 3-cliques. Furthermore, [11] shows that DATA-PEELER outperforms both of them by orders of magnitude. The interested reader will refer to the "Related Work" section of that article for a detailed analysis of what makes DATA-PEELER more efficient than both CUBEMINER and TRIAS.

Extracting every clique in a single graph is a classical problem [7] and algorithms with polynomial delay were designed to extract the maximal (i. e., closed) ones (e. g., [19]). Collections of large graphs were built to help in understanding genetics. These graphs commonly have tens of thousands of nodes and are much noisy. For about four years, extracting knowledge by crossing such graphs has been a hot topic. For example, there is a need to extract patterns that remain valid across several co-expression graphs obtained from microarray data or to cross the data pertaining to physical interactions between molecules (e. g., protein-protein, protein-gene) with more conceptual data (e. g., co-expression of genes, co-occurrence of proteins in the literature). One of the most promising pattern helping in these tasks is the closed 3-clique or, better, the closed quasi-3-clique. CLAN [27] is able to extract closed 3-cliques from collections of large and dense graphs. Crochet+ [18], Cocain* [29] and Quick [21] are the state-of-the-art extractors of closed quasi-3-cliques. They all use the same definition of noise tolerance: every node implied in a pattern must have, in every graph independently from the others, a degree exceeding a user-defined proportion of the maximal degree it would reach if the clique was exact.

As detailed in [9], DATA-PEELER can be generalized towards the tolerance of noise. Combining it with the present work enables the extraction of closed quasi-3-cliques. However the chosen definition for noise tolerance being defined on any n-ary relation, it is different from that of the approaches cited in the previous paragraph. Indeed this tolerance applies to every node across all graph (to be part of a quasi-3-clique, a node must be *globally* much connected to the other nodes of the pattern) and to the graphs themselves. As a consequence our approach does not scale well to graphs connecting thousands of nodes but it can extract closed quasi-3-cliques in large collections of smaller graphs, whereas the previously presented approaches cannot (or they must be used with a very strong minimal size constraint on the number of involved graphs). When the graphs are collected along an ordered dimension (typically the time), the use of the δ-contiguity constraint further increases this difference. Notice that the previous approaches focus on collections of undirected graphs, whereas our approach works on (possibly) directed graphs.

The δ-contiguity stems from an analogous constraint, called *max-gap* constraint, initially applied to sequence mining. It was introduced in the GSP approach [24]. The way the δ-contiguity is enforced in our approach (see Sect. 9.4) is similar to that of this seminal article. The *min-gap* and the *window size* constraints [24] uses could as well be enforced in our approach. Nevertheless, in [24], these constraints modify the enumeration order, whereas, in our approach, they reduce the search space and let the enumeration strategy unaltered. Furthermore, the nature of the mined patterns is much different. In the context of [24], the considered datasets are multiple sequences of itemsets and the extracted patterns are sub-sequences of itemsets whose order (but not position in time) is to be respected in all (1-dimensional) supporting sequences. In our approach, the supporting domain contains (2-dimensional) graphs and their position in time must be aligned.

Notice that the max-gap constraint was used in other contexts too. For example, [8] enforces it to extract episodes (repetition of sub-sequences in one sequence) and [14] somehow aggregates the two tasks by extracting, under a max-gap constraint, frequent sub-sequences whose support is the sum of the number of repetitions in all sequences of the dataset. Finally let us notice that an extended abstract of this chapter was previously published [13].

9.8 Conclusion

This chapter focuses on specializing the DATA-PEELER closed n-set extractor to mine δ-contiguous closed 3-cliques. All the additional constraints imposed to achieve this goal were piecewise (anti)-monotone. Hence, in its original form, DATA-PEELER could handle them all. However, to be able to extract δ-contiguous closed 3-cliques from large dynamic graphs (e. g., hundreds of nodes and of timestamps), ad-hoc strategies must be used. Interestingly, the idea is the same for all of them (and for the connection constraint too): they must be used as soon as possible in the enumeration tree. The symmetry constraint has even been split into many small constraints that are individually exploited as soon as possible. These constraints are particular since they change the structure of the enumeration which does not follow a binary tree anymore. This chapter focuses on the extraction of δ-contiguous closed 3-cliques. However, DATA-PEELER is not restricted to it. It can mine closed n-sets (or cliques) with n an arbitrary integer greater or equal to 2, it can force the contiguity of the patterns on several dimensions at the same time (possibly with different δ values), etc. Furthermore, DATA-PEELER can mine closed n-sets adapted to any specific problem that can be expressed in terms of piecewise (anti)-monotone constraints.

Acknowledgements This work has been partly funded by EU contract IST-FET IQ FP6-516169, and ANR BINGO2 (MDCO 2007-2010). Tran Bao Nhan Nguyen has contributed to this study thanks to a Research Attachment programme between the Nanyang Technological University (Singapore), where he is an undergraduate student, and INSA-Lyon. Finally, we thank Dr. J. Besson for exciting discussions.

References

1. Agrawal, R., Imielinski, T., Swami, A.N.: Mining association rules between sets of items in large databases. In: SIGMOD'93: Proc. SIGMOD Int. Conf. on Management of Data, pp. 207–216. ACM Press (1993)

2. Agrawal, R., Mannila, H., Srikant, R., Toivonen, H., Verkamo, A.I.: Fast discovery of association rules. In: Advances in Knowledge Discovery and Data Mining, pp. 307–328. AAAI/MIT
   Press (1996)
3. Antonie, M.L., Zaïane, O.R.: Mining positive and negative association rules: An approach for
   confined rules. In: PKDD'04: Proc. European Conf. on Principles and Practice of Knowledge
   Discovery in Databases, pp. 27–38. Springer (2004)
4. Besson, J., Robardet, C., Boulicaut, J.F., Rome, S.: Constraint-based formal concept mining
   and its application to microarray data analysis. Intelligent Data Analysis 9(1), 59–82 (2005)
5. Boulicaut, J.F., Besson, J.: Actionability and formal concepts: A data mining perspective. In:
   ICFCA'08: Proc. Int. Conf. on Formal Concept Analysis, pp. 14–31. Springer (2008)
6. Boulicaut, J.F., De Raedt, L., Mannila, H. (eds.): Constraint-Based Mining and Inductive
   Databases, *LNCS*, vol. 3848. Springer (2006)
7. Bron, C., Kerbosch, J.: Finding all cliques of an undirected graph (algorithm 457). Communications of the ACM 16(9), 575–576 (1973)
8. Casas-Garriga, G.: Discovering unbounded episodes in sequential data. In: PKDD'03: Proc.
   European Conf. on Principles and Practice of Knowledge Discovery in Databases, pp. 83–94.
   Springer (2003)
9. Cerf, L., Besson, J., Boulicaut, J.F.: Extraction de motifs fermés dans des relations n-aires
   bruitées. In: EGC'09: Proc. Journées Extraction et Gestion de Connaissances, pp. 163–168.
   Cepadues-Editions (2009)
10. Cerf, L., Besson, J., Robardet, C., Boulicaut, J.F.: DATA-PEELER: Constraint-based closed
    pattern mining in n-ary relations. In: SDM'08: Proc. SIAM Int. Conf. on Data Mining, pp.
    37–48. SIAM (2008)
11. Cerf, L., Besson, J., Robardet, C., Boulicaut, J.F.: Closed patterns meet n-ary relations. ACM
    Trans. on Knowledge Discovery from Data 3(1) (2009)
12. Cerf, L., Mougel, P.N., Boulicaut, J.F.: Agglomerating local patterns hierarchically with AL-
    PHA. In: CIKM'09: Proc. Int. Conf. on Information and Knowledge Management, pp. 1753–
    1756. ACM Press (2009)
13. Cerf, L., Nguyen, T.B.N., Boulicaut, J.F.: Discovering relevant cross-graph cliques in dynamic
    networks. In: ISMIS'09: Proc. Int. Symp. on Methodologies for Intelligent Systems, pp. 513–
    522. Springer (2009)
14. Ding, B., Lo, D., Han, J., Khoo, S.C.: Efficient mining of closed repetitive gapped subse-
    quences from a sequence database. In: ICDE'09: Proc. Int. Conf. on Data Engineering. IEEE
    Computer Society (2009)
15. Ganter, B., Stumme, G., Wille, R.: Formal Concept Analysis, Foundations and Applications.
    Springer (2005)
16. Jaschke, R., Hotho, A., Schmitz, C., Ganter, B., Stumme, G.: TRIAS–an algorithm for min-
    ing iceberg tri-lattices. In: ICDM'06: Proc. Int. Conf. on Data Mining, pp. 907–911. IEEE
    Computer Society (2006)
17. Ji, L., Tan, K.L., Tung, A.K.H.: Mining frequent closed cubes in 3D data sets. In: VLDB'06:
    Proc. Int. Conf. on Very Large Data Bases, pp. 811–822. VLDB Endowment (2006)
18. Jiang, D., Pei, J.: Mining frequent cross-graph quasi-cliques. ACM Trans. on Knowledge
    Discovery from Data 2(4) (2009)
19. Johnson, D.S., Papadimitriou, C.H., Yannakakis, M.: On generating all maximal independent
    sets. Information Processing Letters 27(3), 119–123 (1988)
20. Leskovec, J., Kleinberg, J.M., Faloutsos, C.: Graph evolution: Densification and shrinking
    diameters. ACM Trans. on Knowledge Discovery from Data 1(1) (2007)
21. Liu, G., Wong, L.: Effective pruning techniques for mining quasi-cliques. In: ECML
    PKDD'08: Proc. European Conf. on Machine Learning and Knowledge Discovery in
    Databases - Part II, pp. 33–49. Springer (2008)
22. Mannila, H., Toivonen, H.: Multiple uses of frequent sets and condensed representations. In:
    KDD, pp. 189–194 (1996)
23. Pei, J., Han, J., Mao, R.: CLOSET: An efficient algorithm for mining frequent closed itemsets.
    In: SIGMOD'00: Workshop on Research Issues in Data Mining and Knowledge Discovery,
    pp. 21–30. ACM Press (2000)

24. Srikant, R., Agrawal, R.: Mining sequential patterns: Generalizations and performance improvements. In: EDBT'96: Proc. Int. Conf. on Extending Database Technology, pp. 3–17. Springer (1996)
25. Stumme, G., Taouil, R., Bastide, Y., Pasquier, N., Lakhal, L.: Computing iceberg concept lattices with TITANIC. Data & Knowledge Engineering **42**(2), 189–222 (2002)
26. Uno, T., Kiyomi, M., Arimura, H.: LCM ver.3: Collaboration of array, bitmap and prefix tree for frequent itemset mining. In: OSDM'05: Proc. Int. Workshop on Open Source Data Mining, pp. 77–86. ACM Press (2005)
27. Wang, J., Zeng, Z., Zhou, L.: CLAN: An algorithm for mining closed cliques from large dense graph databases. In: ICDE'06: Proc. Int. Conf. on Data Engineering, pp. 73–82. IEEE Computer Society (2006)
28. Zaki, M.J., Hsiao, C.J.: CHARM: An efficient algorithm for closed itemset mining. In: SDM'02: Proc. SIAM Int. Conf. on Data Mining. SIAM (2002)
29. Zeng, Z., Wang, J., Zhou, L., Karypis, G.: Out-of-core coherent closed quasi-clique mining from large dense graph databases. ACM Trans. on Database Systems **32**(2), 13–42 (2007)

Chapter 10
Probabilistic Inductive Querying Using ProbLog

Luc De Raedt, Angelika Kimmig, Bernd Gutmann, Kristian Kersting, Vítor Santos Costa, and Hannu Toivonen

Abstract We study how probabilistic reasoning and inductive querying can be combined within ProbLog, a recent probabilistic extension of Prolog. ProbLog can be regarded as a database system that supports both probabilistic and inductive reasoning through a variety of querying mechanisms. After a short introduction to ProbLog, we provide a survey of the different types of inductive queries that ProbLog supports, and show how it can be applied to the mining of large biological networks.

10.1 Introduction

In recent years, both probabilistic and inductive databases have received considerable attention in the literature. Probabilistic databases [1] allow one to represent and reason about uncertain data, while inductive databases [2] aim at tight integration of data mining primitives in database query languages. Despite the current interest in these types of databases, there have, to the best of the authors' knowledge, been no attempts to integrate these two trends of research. This chapter wants to contribute to

Luc De Raedt · Angelika Kimmig · Bernd Gutmann
Department of Computer Science
Katholieke Universiteit Leuven, Belgium
e-mail: {firstname.lastname}@cs.kuleuven.be

Kristian Kersting
Fraunhofer IAIS, Sankt Augustin, Germany
e-mail: kristian.kersting@iais.fraunhofer.de

Vítor Santos Costa
Faculdade de Ciências, Universidade do Porto, Portugal
e-mail: vsc@dcc.fc.up.pt

Hannu Toivonen
Department of Computer Science, University of Helsinki, Finland
e-mail: hannu.toivonen@cs.helsinki.fi

a better understanding of the issues involved by providing a survey of the developments around ProbLog [3][1], an extension of Prolog, which supports both inductive and probabilistic querying. ProbLog has been motivated by the need to develop intelligent tools for supporting life scientists analyzing large biological networks. The analysis of such networks typically involves uncertain data, requiring probabilistic representations and inference, as well as the need to find patterns in data, and hence, supporting data mining. ProbLog can be conveniently regarded as a probabilistic database supporting several types of inductive and probabilistic queries. This paper provides an overview of the different types of queries that ProbLog supports.

A ProbLog program defines a probability distribution over logic programs (or databases) by specifying for each fact (or tuple) the probability that it belongs to a randomly sampled program (or database), where probabilities are mutually independent. The semantics of ProbLog is then defined by the success probability of a query, which corresponds to the probability that the query succeeds in a randomly sampled program (or database). ProbLog is closely related to other probabilistic logics and probabilistic databases that have been developed over the past two decades to face the general need of combining deductive abilities with reasoning about uncertainty, see e.g. [4, 5, 6, 7, 8]. The semantics of ProbLog is studied in Section 10.2. In Section 10.10, we discuss related work in statistical relational learning.

We now give a first overview of the types of queries ProbLog supports. Throughout the chapter, we use the graph in Figure 1(a) for illustration, inspired on the application in biological networks discussed in Section 10.9. It contains several nodes (representing entities) as well as edges (representing relationships). Furthermore, the edges are probabilistic, that is, they are present only with the probability indicated.

Probabilistic Inference   *What is the probability that a query succeeds?*
    Given a ProbLog program and a query, the inference task is to compute the success probability of the query, that is, the probability that the query succeeds in a randomly sampled non-probabilistic subprogram of the ProbLog program. As one example query, consider computing the probability that there exists a proof of $path(c,d)$ in Figure 1(a), that is, the probability that there is a path from c to d in the graph, which will have to take into account the probabilities of both possible paths. Computing and approximating the success probability of queries will be discussed in Section 10.3.
Most Likely Explanation   *What is the most likely explanation for a query?*
    There can be many possible explanations (or reasons) why a certain query may succeed. For instance, in the $path(c,d)$ example, there are two explanations, corresponding to the two different paths from c to d. Often, one is interested in the most likely such explanations, as this provides insight into the problem at hand (here, the direct path from c to d). Computing the most likely explanation realizes a form of probabilistic abduction, cf. [9], as it returns the most likely cause for the query to succeed. This task will be discussed in Section 10.3.1.

---

[1] http://dtai.cs.kuleuven.be/problog/

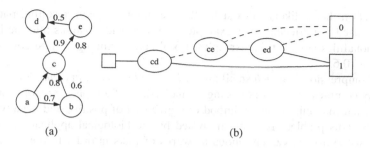

(a)   (b)

Fig. 10.1 (a) Example of a probabilistic graph: edge labels indicate the probability that the edge is part of the graph. (b) Binary Decision Diagram (cf. Sec. 10.4.3) encoding the DNF formula $cd \lor (ce \land ed)$, corresponding to the two proofs of query *path(c,d)* in the graph. An internal node labeled *xy* represents the Boolean variable for the edge between x and y, solid/dashed edges correspond to values true/false.

The above two types of queries are *probabilistic*, that is, they use standard probabilistic inference methods adapted to the context of the ProbLog framework. The types of queries presented next are *inductive*, which means that they start from one or more examples (typically, ground facts such as *path(c,d)*) describing particular relationships, and perform inferences about other examples or about patterns holding in the database.

Analogy and Similarity Based Reasoning via Generalized Explanations
  Which examples are most similar to a given example?
  In explanation based learning the goal is to find a generalized explanation for a particular example in the light of a background theory. Within ProbLog, the traditional approach on explanation based learning is put into a new probabilistic perspective, as in a probabilistic background theory, choosing the most likely explanation provides a fundamental solution to the problem of multiple explanations, and furthermore, the found explanation can be used to retrieve and rank similar examples, that is, to reason by analogy. The most likely explanation thus acts as a kind of local pattern that is specific to the given example(s), thereby allowing the user to get insight into particular relationships. In our example graph, given the definition of *path* in the background theory and an example such as *path(c,d)*, probabilistic explanation based learning finds that a direct connection is the most likely explanation, which can then be used to retrieve and rank other directly connected examples. This type of query is discussed in Section 10.5.
Local Pattern Mining   *Which queries are likely to succeed for a given set of examples?*
  In local pattern mining the goal is to find those patterns that are likely to succeed on a set of examples, that is, instances of a specific relation *key*. This setting is a natural variant of the explanation based learning setting, but without the need for a background theory. The result is a kind of probabilistic relational association rule miner. On our example network, the local pattern miner could start, for instance, from the examples *key(c,d)* and *key(a,c)* and infer that there is a direct

connection that is likely to exist for these examples. Again, resulting patterns can be used to retrieve similar examples and to provide insights into the likely commonalities amongst the examples. Local pattern mining will be covered in Section 10.6.

Theory Compression   *Which* small *theory best explains a set of examples?*

Theory compression aims at finding a small subset of a ProbLog theory (or network) that maximizes the likelihood of a given set of positive and negative examples. This problem is again motivated by the biological application, where scientists try to analyze enormous networks of links in order to obtain an understanding of the relationships amongst a typically small number of nodes. The idea now is to compress these networks as much as possible using a set of positive and negative examples. The examples take the form of relationships that are either interesting or uninteresting to the scientist. The result should ideally be a small network that contains the essential links and assigns high probabilities to the positive and low probabilities to the negative examples. This task is analogous to a form of theory revision [10, 11] where the only operation allowed is the deletion of rules or facts. Within the ProbLog theory compression framework, examples are true and false ground facts, and the task is to find a subset of a given ProbLog program that maximizes the likelihood of the examples. In the example, assume that $path(a,d)$ is of interest and that $path(a,e)$ is not. We can then try to find a small graph (containing k or fewer edges) that best matches these observations. Using a greedy approach, we would first remove the edges connecting e to the rest of the graph, as they strongly contribute to proving the negative example, while positive example still has likely proofs in the resulting graph. Theory compression will be discussed in Section 10.7.

Parameter Estimation   *Which parameters best fit the data?*

The goal is to learn the probabilities of facts from a given set of training examples. Each example consists of a query and target probability. This setting is challenging because the explanations for the queries, namely the proofs, are unknown. Using a modified version of the probabilistic inference algorithm, a standard gradient search can be used to find suitable parameters efficiently. We will discuss this type of query in Section 10.8.

To demonstrate the usefulness of ProbLog for inductive and probabilistic querying, we have evaluated the different types of queries in the context of mining a large biological network containing about 1 million entities and about 7 million edges [12]. We will discuss this in more detail in Section 10.9.

This paper is organized as follows. In Section 10.2, we introduce the semantics of ProbLog and define the probabilistic queries; Section 10.3 discusses computational aspects and presents several algorithms (including approximation and Monte Carlo algorithms) for computing probabilities of queries, while the integration of ProbLog in the well-known implementation of YAP-Prolog is discussed in Section 11.3.1. The following sections in turn consider each of the inductive queries listed above. Finally, Section 10.9 provides a perspective on applying ProbLog on biological network mining, Section 10.10 positions ProbLog in the field of statistical relational learning, and Section 10.11 concludes.

10.2 ProbLog: Probabilistic Prolog

In this section, we present ProbLog and its semantics and then introduce two types of probabilistic queries: *probabilistic inference*, that is, computing the *success probability* of a query, and finding the *most likely explanation*, based on the *explanation probability*.

A ProbLog program consists of a set of labeled facts $p_i :: c_i$ together with a set of definite clauses. Each ground instance (that is, each instance not containing variables) of such a fact c_i is true with probability p_i, where all probabilities are assumed mutually independent. To ensure a natural interpretation of these random variables, no two different facts c_i, c_j are allowed to unify, as otherwise, probabilities of ground facts would be higher than the individual probability given by different non-ground facts. The definite clauses allow the user to add arbitrary *background knowledge* (BK).[2] For ease of exposition, in the following we will assume all probabilistic facts to be ground.

Figure 1(a) shows a small probabilistic graph that we use as running example in the text. It can be encoded in ProbLog as follows:

$$0.8 :: \text{edge}(a, c). \quad 0.7 :: \text{edge}(a, b). \quad 0.8 :: \text{edge}(c, e).$$
$$0.6 :: \text{edge}(b, c). \quad 0.9 :: \text{edge}(c, d). \quad 0.5 :: \text{edge}(e, d).$$

Such a probabilistic graph can be used to sample subgraphs by tossing a coin for each edge. A ProbLog program $T = \{p_1 :: c_1, \cdots, p_n :: c_n\} \cup BK$ defines a probability distribution over subprograms $L \subseteq L_T = \{c_1, \cdots, c_n\}$:

$$P(L|T) = \prod_{c_i \in L} p_i \prod_{c_i \in L_T \setminus L} (1 - p_i).$$

We extend our example with the following background knowledge:

```
path(X, Y) : − edge(X, Y).
path(X, Y) : − edge(X, Z), path(Z, Y).
```

We can then ask for the probability that there exists a path between two nodes, say c and d, in our probabilistic graph, that is, we query for the probability that a randomly sampled subgraph contains the edge from c to d, or the path from c to d via e (or both of these). Formally, the *success probability* $P_s(q|T)$ of a query q in a ProbLog program T is defined as

$$P_s(q|T) = \sum_{L \subseteq L_T, \exists \theta : L \cup BK \models q\theta} P(L|T). \tag{10.1}$$

---

[2] While in early work on ProbLog [3] probabilities were attached to arbitrary definite clauses and all groundings of such a clause were treated as a single random event, we later on switched to a clear separation of logical and probabilistic part and random events corresponding to ground facts. This is often more natural and convenient, but can still be used to model the original type of clauses (by adding a corresponding probabilistic fact to the clause body) if desired.

In other words, the success probability of query q is the probability that the query q is *provable* in a randomly sampled logic program.

In our example, 40 of the 64 possible subprograms allow one to prove $path(c,d)$, namely all those that contain at least edge (c, d) (*cd* for short) or both edge (c, e) and edge (e, d), so the success probability of that query is the sum of the probabilities of these programs: $P_s(path(c,d)|T) = P(\{ab, ac, bc, cd, ce, ed\}|T) + \ldots + P(\{cd\}|T) = 0.94$.

As a consequence, the probability of a *specific* proof, also called *explanation*, corresponds to that of sampling a logic program L that contains all the facts needed in that explanation or proof. The *explanation probability* $P_x(q|T)$ is defined as the probability of the most likely explanation or proof of the query q

$$P_x(q|T) = \max_{e \in E(q)} P(e|T) = \max_{e \in E(q)} \prod_{c_i \in e} p_i, \qquad (10.2)$$

where $E(q)$ is the set of all explanations for query q [13].

In our example, the set of all explanations for $path(c,d)$ contains the edge from c to d (with probability 0.9) as well as the path consisting of the edges from c to e and from e to d (with probability $0.8 \cdot 0.5 = 0.4$). Thus, $P_x(path(c,d)|T) = 0.9$.

The ProbLog semantics is an instance of the distribution semantics [14], where the basic distribution over ground facts is defined by treating each such fact as an independent random variable. Sato has rigorously shown that this class of programs defines a joint probability distribution over the set of possible least Herbrand models of the program, where each possible least Herbrand model corresponds to the least Herbrand model of the background knowledge BK together with a subprogram $L \subseteq L_T$; for further details we refer to [14]. Similar instances of the distribution semantics have been used widely in the literature, e.g. [4, 5, 6, 7, 8]; see also Section 10.10.

10.3 Probabilistic Inference

In this section, we present various algorithms and techniques for performing probabilistic inference in ProbLog, that is computing the success probabilities and most likely explanations of queries. We will discuss the implementation of these methods in Section 11.3.1.

10.3.1 Exact Inference

As computing the *success probability* of a query using Equation (10.1) directly is infeasible for all but the tiniest programs, ProbLog uses a method involving two steps [3]. The first step computes the proofs of the query q in the logical part of the theory T, that is, in $L_T \cup BK$. The result will be a DNF formula. The second step

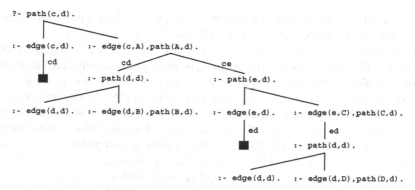

Fig. 10.2 SLD-tree for query path(c,d).

employs Binary Decision Diagrams [15] to compute the probability of this formula. Comparable first steps are performed in pD [6], PRISM [8] and ICL [16], however, as we will see below, these systems differ in the method used to tackle the second step. Let us now explain ProbLog's two steps in more detail.

The first step employs SLD-resolution [17], as in Prolog, to obtain all different proofs. As an example, the SLD-tree for the query *?- path(c,d)*. is depicted in Figure 10.2. Each successful proof in the SLD-tree uses a set of facts $\{p_1 :: d_1, \cdots, p_k :: d_k\} \subseteq T$. These facts are necessary for the proof, and the proof is independent of other probabilistic facts in T.

Let us now introduce a Boolean random variable b_i for each fact $p_i :: c_i \in T$, indicating whether c_i is in logic program, that is, b_i has probability p_i of being true. The probability of a particular proof involving facts $\{p_{i_1} :: d_{i_1}, \cdots, p_{i_k} :: d_{i_k}\} \subseteq T$ is then the probability of the conjunctive formula $b_{i_1} \wedge \cdots \wedge b_{i_k}$. Since a goal can have multiple proofs, the success probability of query q equals the probability that the disjunction of these conjunctions is true. This yields

$$P_s(q|T) = P\left(\bigvee_{e \in E(q)} \bigwedge_{b_i \in cl(e)} b_i \right) \tag{10.3}$$

where $E(q)$ denotes the set of proofs or explanations of the goal q and $cl(e)$ denotes the set of Boolean variables representing ground facts used in the explanation e. Thus, the problem of computing the success probability of a ProbLog query can be reduced to that of computing the probability of a DNF formula. The formula corresponding to our example query *path(c,d)* is $cd \vee (ce \wedge ed)$, where we use xy as Boolean variable representing *edge(x,y)*.

Computing the probability of DNF formulae is an #P-hard problem [18], as the different conjunctions need not be independent. Indeed, even under the assumption of independent variables used in ProbLog, the different conjunctions are not mutually exclusive and may overlap. Various algorithms have been developed to tackle this problem, which is known as the disjoint-sum-problem. The pD-engine

HySpirit [6] uses the inclusion-exclusion principle, which is reported to scale to about ten proofs. PRISM [8] and PHA [7] avoid the disjoint-sum-problem by requiring proofs to be mutually exclusive, while ICL uses a symbolic disjoining technique with limited scalability [16]. As the type of application considered here often requires dealing with hundreds or thousands of proofs, the second step of our implementation employs Binary Decision Diagrams (BDDs) [15], an efficient graphical representation of a Boolean function over a set of variables which scales to tens of thousands of proofs; we will discuss the details in Section 10.4.3. Nevertheless, calculating the probability of a DNF formula remains a hard problem and can thus become fairly expensive, and finally infeasible. For instance, when searching for paths in graphs or networks, even in small networks with a few dozen edges there are easily $O(10^6)$ possible paths between two nodes. ProbLog therefore includes several approximation methods for the success probability. We will come back to these methods from Section 10.3.2 onwards.

Compared to probabilistic inference, computing the most likely explanation is much easier. Indeed, calculating the *explanation probability P_x* corresponds to computing the probability of a conjunctive formula only, so that the disjoint-sum-problem does not arise. While one could imagine to use Viterbi-like dynamic programming techniques on the DNF to calculate the explanation probability, our approach avoids constructing the DNF – which requires examining a potentially high number of low-probability proofs – by using a best-first search, guided by the probability of the current partial proof. In terms of logic programming [17], the algorithm does not completely traverse the entire SLD-tree to find all proofs, but instead uses iterative deepening with a probability threshold α to find the most likely one. Algorithm in Table 10.1 provides the details of this procedure, where *stop* is a minimum threshold to avoid exploring infinite SLD-trees without solution and *resolutionStep* performs the next possible resolution step on the goal and updates the probability p of the current derivation and its explanation *expl* accordingly; backtracking reverts these steps to explore alternative steps while at the same time keeping the current best solution (*max, best*) and the current threshold α.

10.3.2 Bounded Approximation

The first approximation algorithm for obtaining success probabilities, similar to the one proposed in [3], uses DNF formulae to obtain both an upper and a lower bound on the probability of a query. It is related to work by [9] in the context of PHA, but adapted towards ProbLog. The algorithm uses an incomplete SLD-tree, i.e. an SLD-tree where branches are only extended up to a given probability threshold[3], to obtain DNF formulae for the two bounds. The lower bound formula d_1 represents all proofs with a probability above the current threshold. The upper bound formula d_2 additionally includes all derivations that have been stopped due to reaching the threshold,

---

[3] Using a probability threshold instead of the depth bound of [3] has been found to speed up convergence, as upper bounds are tighter on initial levels.

Table 10.1 Calculating the most likely explanation by iterative deepening search in the SLD-tree.

function BESTPROBABILITY(query q)
$\alpha := 0.5$; $max = -1$; $best := false$; $expl := \emptyset$; $p = 1$; $goal = q$;
while $\alpha > stop$ **do**
repeat
$(goal, p, expl) := resolutionStep(goal, p, expl)$
if $p < \alpha$ **then**
backtrack resolution
end if
if $goal = \emptyset$ **then**
$max := p$; $best := expl$; $\alpha := p$; backtrack resolution
end if
until no further backtracking possible
if $max > -1$ **then**
return $(max, best)$
else
$\alpha := 0.5 \cdot \alpha$
end if
end while

as these still *may* succeed. The algorithm proceeds in an iterative-deepening manner, starting with a high probability threshold and successively multiplying this threshold with a fixed shrinking factor until the difference between the current bounds becomes sufficiently small. As $d_1 \models d \models d_2$, where d is the Boolean DNF formula corresponding to the full SLD-tree of the query, the success probability is guaranteed to lie in the interval $[P(d_1), P(d_2)]$.

As an illustration, consider a probability bound of 0.9 for the SLD-tree in Figure 10.2. In this case, d_1 encodes the left success path while d_2 additionally encodes the path up to $path(e, d)$, i.e. $d_1 = cd$ and $d_2 = cd \vee ce$, whereas the formula for the full SLD-tree is $d = cd \vee (ce \wedge ed)$.

10.3.3 K-Best

Using a fixed number of proofs to approximate the success probability allows for better control of the overall complexity, which is crucial if large numbers of queries have to be evaluated e.g. in the context of parameter learning, cf. Section 10.8. [19] therefore introduce the k-probability $P_k(q|T)$, which approximates the success probability by using the k best (that is, most likely) explanations instead of all proofs when building the DNF formula used in Equation (10.3):

$$P_k(q|T) = P\left(\bigvee_{e \in E_k(q)} \bigwedge_{b_i \in cl(e)} b_i \right) \tag{10.4}$$

where $E_k(q) = \{e \in E(q)|P_x(e) \geq P_x(e_k)\}$ with e_k the kth element of $E(q)$ sorted by non-increasing probability. Setting $k = \infty$ and $k = 1$ leads to the success and the explanation probability respectively. Finding the k best proofs can be realized using a simple branch-and-bound approach extending the algorithm presented in Table 10.1; cf. also [7].

To illustrate k-probability, we consider again our example graph, but this time with query $path(a,d)$. This query has four proofs, represented by the conjunctions $ac \wedge cd$, $ab \wedge bc \wedge cd$, $ac \wedge ce \wedge ed$ and $ab \wedge bc \wedge ce \wedge ed$, with probabilities 0.72, 0.378, 0.32 and 0.168 respectively. As P_1 corresponds to the explanation probability P_x, we obtain $P_1(path(a,d)) = 0.72$. For $k = 2$, overlap between the best two proofs has to be taken into account: the second proof only adds information if the first one is absent. As they share edge cd, this means that edge ac has to be missing, leading to $P_2(path(a,d)) = P((ac \wedge cd) \vee (\neg ac \wedge ab \wedge bc \wedge cd)) = 0.72 + (1 - 0.8) \cdot 0.378 = 0.7956$. Similarly, we obtain $P_3(path(a,d)) = 0.8276$ and $P_k(path(a,d)) = 0.83096$ for $k \geq 4$.

10.3.4 Monte Carlo

As an alternative approximation technique without BDDs, [20] propose a Monte Carlo method. The algorithm repeatedly samples a logic program from the ProbLog program and checks for the existence of some proof of the query of interest. The fraction of samples where the query is provable is taken as an estimate of the query probability, and after each m samples the 95% confidence interval is calculated. Although confidence intervals do not directly correspond to the exact bounds used in bounded approximation, the same stopping criterion is employed, that is, the Monte Carlo simulation is run until the width of the confidence interval is at most δ. Such an algorithm (without the use of confidence intervals) was suggested already by Dantsin [4], although he does not report on an implementation. It was also used in the context of networks (not Prolog programs) by [12].

10.4 Implementation

This section discusses the main building blocks used to implement ProbLog on top of the YAP-Prolog system [21] as introduced in [20]. An overview is shown in Figure 10.3, with a typical ProbLog program, including ProbLog facts and background knowledge (BK), at the top.

The implementation requires ProbLog programs to use the `problog` module. Each program consists of a set of labeled facts and of unlabeled *background knowledge*, a generic Prolog program. Labeled facts are preprocessed as described below. Notice that the implementation requires all queries to non-ground probabilistic facts to be ground on calling.

Fig. 10.3 ProbLog Implementation: A ProbLog program (top) requires the ProbLog library which in turn relies on functionality from the tries and array libraries. ProbLog queries (bottom-left) are sent to the YAP engine, and may require calling the BDD library CUDD via SimpleCUDD.

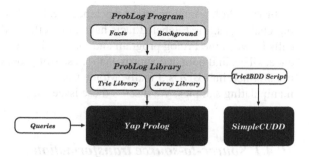

In contrast to standard Prolog queries, where one is interested in answer substitutions, in ProbLog one is primarily interested in a probability. As discussed before, two common ProbLog queries ask for the most likely explanation and its probability, and the probability of whether a query would have an answer substitution. In Section 10.3, we have discussed two very different approaches to the problem:

- In exact inference (Section 10.3.1), *k*-best (Section 10.3.3) and bounded approximation (Section 10.3.2), the engine explicitly reasons about probabilities of proofs. The challenge is how to compute the probability of each individual proof, store a large number of proofs, and compute the probability of sets of proofs.
- In Monte Carlo (Section 10.3.4), the probabilities of facts are used to sample from ProbLog programs. The challenge is how to compute a sample quickly, in a way that inference can be as efficient as possible.

ProbLog programs execute from a top-level query and are driven through a ProbLog query. The inference algorithms discussed in Section 10.3 can be abstracted as follows:

- Initialize the inference algorithm;
- While probabilistic inference did not converge:
  - initialize a new query;
  - execute the query, instrumenting every ProbLog call in the current proof. Instrumentation is required for recording the ProbLog facts required by a proof, but may also be used by the inference algorithm to stop proofs (e.g., if the current probability is lower than a bound);
  - process success or exit substitution;
- Proceed to the next step of the algorithm: this may be trivial or may require calling an external solver, such as a BDD tool, to compute a probability.

Notice that the current ProbLog implementation relies on the Prolog engine to efficiently execute goals. On the other hand, and in contrast to most other probabilistic language implementations, in ProbLog there is no clear separation between logical and probabilistic inference: in a fashion similar to constraint logic programming, probabilistic inference can drive logical inference.

From a Prolog implementation perspective, ProbLog poses a number of interesting challenges. First, labeled facts have to be efficiently compiled to allow mutual calls between the Prolog program and the ProbLog engine. Second, for exact inference, k-best and bounded approximation, sets of proofs have to be manipulated and transformed into BDDs. Finally, Monte Carlo simulation requires representing and manipulating samples. We discuss these issues next.

10.4.1 Source-to-source transformation

We use the `term_expansion` mechanism to allow Prolog calls to labeled facts, and for labeled facts to call the ProbLog engine. As an example, the program:

$$0.715 :: \text{edge}('PubMed\_2196878','MIM\_609065').$$
$$0.659 :: \text{edge}('PubMed\_8764571','HGNC\_5014'). \tag{10.5}$$

would be compiled as:

```
edge(A, B) : − problog_edge(ID, A, B, LogProb),
              grounding_id(edge(A, B), ID, GroundID),
              add_to_proof(GroundID, LogProb).
```
$$\tag{10.6}$$

```
problog_edge(0,'PubMed_2196878','MIM_609065', −0.3348).
problog_edge(1,'PubMed_8764571','HGNC_5014', −0.4166).
```

Thus, the internal representation of each fact contains an identifier, the original arguments, and the logarithm of the probability[4]. The `grounding_id` procedure will create and store a grounding specific identifier for each new grounding of a non-ground probabilistic fact encountered during proving, and retrieve it on repeated use. For ground probabilistic facts, it simply returns the identifier itself. The `add_to_proof` procedure updates the data structure representing the current path through the search space, i.e., a queue of identifiers ordered by first use, together with its probability.

10.4.2 Tries

Manipulating proofs is critical in ProbLog. We represent each proof as a queue containing the identifier of each different ground probabilistic fact used in the proof, ordered by first use. The implementation requires calls to non-ground probabilistic facts to be ground, and during proving maintains a table of groundings used

---

[4] We use the logarithm to avoid numerical problems when calculating the probability of a derivation, which is used to drive inference.

within the current query together with their identifiers. In our implementation, the queue is stored in a backtrackable global variable, which is updated by calling `add_to_proof` with an identifier for the current ProbLog fact. We thus exploit Prolog's backtracking mechanism to avoid recomputation of shared proof prefixes when exploring the space of proofs. Storing a proof is simply a question of adding the value of the variable to a store.

Storing and manipulating proofs is critical in ProbLog. When manipulating proofs, the key operation is often *insertion*: we would like to add a proof to an existing set of proofs. Some algorithms, such as exact inference or Monte Carlo, only manipulate complete proofs. Others, such as bounded approximation, require adding partial derivations too. The nature of the SLD-tree means that proofs tend to share both a prefix and a suffix. Partial proofs tend to share prefixes only. This suggests using *tries* [22] to maintain the set of proofs. We use the YAP implementation of tries for this task, based itself on XSB Prolog's work on tries of terms [23].

10.4.3 Binary Decision Diagrams

To efficiently compute the probability of a DNF formula representing a set of proofs, our implementation represents this formula as a Binary Decision Diagram (BDD) [15]. Given a fixed variable ordering, a Boolean function f can be represented as a full Boolean decision tree, where each node on the ith level is labeled with the ith variable and has two children called low and high. Leaves are labeled by the outcome of f for the variable assignment corresponding to the path to the leaf, where in each node labeled x, the branch to the low (high) child is taken if variable x is assigned 0 (1). Starting from such a tree, one obtains a BDD by merging isomorphic subgraphs and deleting redundant nodes until no further reduction is possible. A node is redundant if the subgraphs rooted at its children are isomorphic.

Figure 10.1b shows the BDD corresponding to $cd \lor (ce \land ed)$, the formula of the example query $path(c,d)$. Given a BDD, it is easy to compute the probability of the corresponding Boolean function by traversing the BDD from the root node to a leaf. At each inner node, probabilities from both children are calculated recursively and combined afterwards as shown in algorithm in Table 10.2. In practice, memorization of intermediate results is used to avoid the recomputation at nodes that are shared between multiple paths, resulting in a time and space complexity linear in the number of nodes in the BDD.

We use SimpleCUDD [24][5] as a wrapper tool for the BDD package CUDD[6] to construct and evaluate BDDs. More precisely, the trie representation of the DNF is translated to a BDD generation script, which is processed by SimpleCUDD to build the BDD using CUDD primitives. It is executed via Prolog's shell utility, and results are reported via shared files.

---

[5]   http://people.cs.kuleuven.be/~theofrastos.mantadelis/tools/simplecudd.html

[6] http://vlsi.colorado.edu/~fabio/CUDD

Table 10.2 Calculating success probability by traversing BDD.

function PROBABILITY(BDD node n)
If n is the 1-terminal return 1
If n is the 0-terminal return 0
let h and l be the high and low children of n
$prob(h) :=$ PROBABILITY(h)
$prob(l) :=$ PROBABILITY(l)
return $p_n \cdot prob(h) + (1 - p_n) \cdot prob(l)$

During the generation of the code, it is crucial to exploit the structure sharing (prefixes and suffixes) already in the trie representation of a DNF formula, otherwise CUDD computation time becomes extremely long or memory overflows quickly. Since CUDD builds BDDs by joining smaller BDDs using logical operations, the trie is traversed bottom-up to successively generate code for all its subtrees. Two types of operations are used to combine nodes. The first creates conjunctions of leaf nodes and their parent if the leaf is a single child, the second creates disjunctions of all child nodes of a node if these child nodes are all leaves. In both cases, a subtree that occurs multiple times in the trie is translated only once, and the resulting BDD is used for all occurrences of that subtree. Because of the optimizations in CUDD, the resulting BDD can have a very different structure than the trie.

10.4.4 Monte Carlo

Monte Carlo execution is quite different from the approaches discussed before, as the two main steps are **(a)** generating a sample program and **(b)** performing standard refutation on the sample. Thus, instead of combining large numbers of proofs, we need to manipulate large numbers of different programs or samples.

One naive approach would be to generate a complete sample, and to check for a proof within the sample. Unfortunately, the approach does not scale to large databases, even if we try to reuse previous proofs: just generating a sample can be fairly expensive, as one would need to visit every ProbLog fact at every sample. In fact, in our experience, just representing and generating the whole sample can be a challenge for large databases. To address this first problem, we rely on YAP's efficient implementation of arrays as the most compact way of representing large numbers of nodes. Moreover, we take advantage of the observation that often proofs are local, i.e. we only need to verify whether facts from a small fragment of the database are in the sample, to generate the sample *lazily*. In other words, we verify if a fact is in the sample only when we need it for a proof. Samples are thus represented as a three-valued array, originally initialized to 0, that means sampling was not asked yet; 1 means that the fact is in the sampled program, and 2 means not in sample. Note that as fact identifiers are used to access the array, the approach cannot directly be used for non-ground facts, whose identifiers are generated on demand.

The current implementation of Monte Carlo therefore uses the internal database to store the result of sampling different groundings of such facts.

The tight integration of ProbLog's probabilistic inference algorithms in the state-of-the-art YAP-Prolog system discussed here includes several improvements over the initial implementation used in [3], thereby enabling the use of ProbLog to effectively query Sevon's Biomine network [12] containing about 1,000,000 nodes and 6,000,000 edges. For experimental results obtained using the various methods in the context of this network as well as for further implementation details, we refer to [25].

10.5 Probabilistic Explanation Based Learning

In this section, we address the question of finding examples that are similar or analogous to a given example. To this end, we combine two types of queries, namely finding the *most likely (generalized) explanation* for an example and *reasoning by analogy*, which is the process of finding (and possibly ranking) examples with a similar explanation. ProbLog's probabilistic explanation based learning technique (PEBL) [13] employs a background theory that allows to compute a most likely explanation for the example and to generalize that explanation. It thus extends the concept of explanation based learning (EBL) to a probabilistic framework. Probabilistic explanation based learning as introduced here is also related to probabilistic abduction, as studied by Poole [7]. The difference with Poole's work however is that we follow the deductive view of EBL to compute *generalized* explanations and also apply them for analogical reasoning.

The central idea of explanation based learning [26, 27] is to compute a generalized explanation from a concrete proof of an example. Explanations use only so-called *operational* predicates, i.e. predicates that capture essential characteristics of the domain of interest and should be easy to prove. Operational predicates are to be declared by the user as such. The problem of probabilistic explanation based learning can be sketched as follows.

> *Given* a positive example e (a ground fact), a ProbLog theory T, and declarations that specify which predicates are operational,
> *Find* a clause c such that $T \models c$ (in the logical sense, so interpreting T as a Prolog program), $body(c)$ contains only operational predicates, there exists a substitution θ such that $head(c)\theta = e$ and $body(c)\theta$ is the most likely explanation for e given T.

Following the work by [28, 29], explanation based learning starts from a definite clause theory T, that is a pure Prolog program, and an example in the form of a ground atom $p(t_1,...,t_n)$. It then constructs a refutation proof of the example using SLD-resolution. Explanation based learning will generalize this proof to obtain a generalized explanation. This is realized performing the same SLD-resolution steps as in the proof for the example, but starting from the variabelized goal, i.e. $p(X_1, ..., X_n)$ where the X_i are different variables. The only difference is that in the general proof atoms $q(s_1, ..., s_r)$ for operational predicates q in a goal $? - g_1, ..., g_i, q(s_1, ..., s_r), g_{i+1}, ..., g_n$ are not resolved away. Also, the proof procedure stops when the goal contains only atoms for operational predicates. The resulting goal provides a *generalized explanation* for the example. In terms of the SLD-resolution proof tree, explanation based learning cuts off branches below operational predicates. It is easy to implement the explanation based proof procedure as a meta-interpreter for Prolog [28, 29].

Reconsider the example of Figure 10.1a, ignoring the probability labels for now. We define `edge/2` to be the only operational predicate, and use `path(c,d)` as training example. EBL proves this goal using one instance of the operational predicate, namely `edge(c,d)`, leading to the explanation `edge(X,Y)` for the generalized example `path(X,Y)`. To be able to identify the examples covered by such an explanation, we represent it as so-called *explanation clause*, where the generalized explanation forms the body and the predicate in the head is renamed to distinguish the clause from those for the original predicate. In our example, we thus get the explanation clause `exp_path(X,Y) ← edge(X,Y)`. Using the second possible proof of `path(c,d)` instead, we would obtain `exp_path(X,Y) ← edge(X,Z), edge(Z,Y)`.

PEBL extends EBL to probabilistic logic representations, computing the generalized explanation from the most likely proof of an example as determined by the explanation probability $P_x(q|T)$ (10.2). It thus returns the first explanation clause in our example.

As we have explained in Section 10.3.1, computing the most likely proof for a given goal in ProbLog is straightforward: instead of traversing the SLD-tree in a left-to-right depth-first manner as in Prolog, nodes are expanded in order of the probability of the derivation leading to that node. This realizes a best-first search with the probability of the current proof as an evaluation function. We use iterative deepening in our implementation to avoid memory problems. The PEBL algorithm thus modifies the algorithm in Table 10.1 to return the generalized explanation based on the most likely proof, which, as in standard EBL, is generated using the same sequence of resolution steps on the variabelized goal. As for the k-probability (Section 10.3.3), a variant of the algorithm can be used to return the k most probable structurally distinct explanations.

The probabilistic view on explanation based learning adopted in ProbLog offers natural solutions to two issues traditionally discussed in the context of explanation based learning [26, 30]. The first one is the *multiple explanation* problem, which is concerned with choosing the explanation to be generalized for examples having multiple proofs. The use of a sound probabilistic framework naturally deals with this issue by selecting the *most likely* proof. The second problem is that of *generalizing from multiple examples*, another issue that received considerable attention in traditional explanation based learning. To realize this in our setting, we modify the best-first search algorithm so that it searches for the most likely generalized explanation shared by the n examples $e_1, ..., e_n$. Including the variabelized atom e, we compute $n + 1$ SLD-resolution derivations in parallel. A resolution step resolving an atom for a non-operational predicate in the generalized proof for e is allowed only when the same resolution step can also be applied to each of the n parallel derivations. Atoms corresponding to operational predicates are – as sketched above – not resolved in the generalized proof, but it is nevertheless required that for each occurrence of these atoms in the n parallel derivations, there exists a resolution derivation.

Consider again our running example, and assume that we now want to construct a common explanation for `path(c,d)` and `path(b,e)`. We thus have to simultaneously prove both examples and the variabelized goal `path(X,Y)`. After resolving all three goals with the first clause for `path/2`, we reach the first instance of the operational predicate `edge/2` and thus have to prove both `edge(c,d)` and `edge(b,e)`. As proving `edge(b,e)` fails, the last resolution step is rejected and the second clause for `path/2` used instead. Continuing this process finally leads to the explanation clause `exp_path(X,Y) ← edge(X,Z),edge(Z,Y)`.

At the beginning of this section, we posed the question of finding examples that are similar or analogous to a given example. The explanation clause constructed by PEBL provides a concrete measure for analogy or similarity based reasoning: examples are considered similar if they can be explained using the general pattern that best explains the given example, that is, if they can be proven using the explanation clause. In our example, using the clause `exp_path(X,Y) ← edge(X,Y)` obtained from `path(c,d)`, five additional instances of `exp_path(X,Y)` can be proven, corresponding to the other edges of the graph. Furthermore, such similar examples can naturally be ranked according to their probability, that is, in our example, `exp_path(a,c)` and `exp_path(c,e)` would be considered most similar to `path(c,d)`, as they have the highest probability.

We refer to [13] for more details as well as experiments in the context of biological networks.

10.6 Local Pattern Mining

In this section, we address the question of finding queries that are likely to succeed on a given set of examples. We show how local pattern mining can be adapted towards probabilistic databases such as ProbLog. Even though local pattern mining is related to probabilistic explanation based learning, there are some important differences. Indeed, probabilistic explanation based learning typically employs a single positive example and a background theory to compute a generalized explanation of the example. Local pattern mining, on the other hand, does not rely on a background theory or declarations of operational predicates, uses a set of examples – possibly including negative ones – rather than a single one, and computes a set of patterns (or clauses) satisfying certain conditions. As in probabilistic explanation based learning, the discovered patterns can be used to retrieve and rank further examples, again realizing a kind of similarity based reasoning or reasoning by analogy.

Our approach to probabilistic local pattern mining [31] builds upon multi-relational query mining techniques [32], extending them towards probabilistic databases. We use ProbLog to represent databases and queries, abbreviating vectors of variables as $\mathbf{X}$. We assume a designated relation *key* containing the set of tuples to be characterized using queries, and restrict the language $\mathscr{L}$ of patterns to the set of conjunctive queries $r(\mathbf{X})$ defined as

$$r(\mathbf{X}) : -key(\mathbf{X}), l_1, ..., l_n \tag{10.7}$$

where the l_i are positive atoms. Additional syntactic or semantic restrictions, called *bias*, can be imposed on the form of queries by explicitly specifying the language $\mathscr{L}$, cf. [33, 34, 32]. *Query Mining* aims at finding all queries satisfying a selection predicate ϕ. It can be formulated as follows, cf. [32, 34]:

Given   a language $\mathscr{L}$ containing queries of the form (10.7), a database $\mathscr{D}$ including the designated relation *key*, and a selection predicate ϕ

Find   all queries $q \in \mathscr{L}$ such that $\phi(q, \mathscr{D}) = true$.

The most prominent selection predicate is minimum frequency, an anti-monotonic predicate, requiring a minimum number of tuples covered. Anti-monotonicity is based on a generality relation between patterns. We employ *OI*-subsumption [35], as the corresponding notion of subgraph isomorphism is favorable within the intended application in network mining.

Correlated Pattern Mining [36] uses both *positive* and *negative* examples, given as two designated relations key^+ and key^- of the same arity, to find the top k patterns, that is, the k patterns scoring best w.r.t. a function ψ. The function ψ employed is convex, e.g. measuring a statistical significance criterion such as χ^2, cf. [36], and measures the degree to which the pattern is statistically significant or unexpected. Thus correlated pattern mining corresponds to the setting

$$\phi(q, \mathscr{D}) = q \in \arg_k \max_{q \in \mathscr{L}} \psi(q, \mathscr{D}) \ . \tag{10.8}$$

Consider the database corresponding to the graph in Figure 1(a) (ignoring probability labels) with $key^+ = \{a, c\}$ and $key^- = \{d, e\}$. A simple correlation function is $\psi(q, \mathscr{D}) = \mathsf{COUNT}(q^+(*)) - \mathsf{COUNT}(q^-(*))$, where $\mathsf{COUNT}(q(*))$ is the number of different provable ground instances of q and q^x denotes query q restricted to key^x. We obtain $\psi(Q1, \mathscr{D}) = 2 - 0 = 2$ and $\psi(Q2, \mathscr{D}) = 1 - 1 = 0$ for queries

$(Q1) \quad q(X) : -key(X), edge(X, Y), edge(Y, Z).$

$(Q2) \quad q(X) : -key(X), edge(X, d).$

Multi-relational query miners such as [32, 34] often follow a level-wise approach for frequent query mining [37], where at each level new candidate queries are generated from the frequent queries found on the previous level. In contrast to Apriori, instead of a "joining" operation, they employ a refinement operator ρ to compute more specific queries, and also manage a set of infrequent queries to take into account the specific language requirements imposed by $\mathscr{L}$. To search for all solutions, it is essential that the refinement operator is optimal w.r.t. $\mathscr{L}$, i.e. ensures that there is exactly one path from the most general query to every query in the search space.

Table 10.3 Counts on key^+ and key^- and ψ-values obtained during the first level of mining in the graph of Figure 1(a). The current minimal score for best queries is 1, i.e. only queries with $\psi \geq 1$ or $c^+ \geq 1$ will be refined on the next level.

	query	c^+	c^-	ψ
1	key(X),edge(X,Y)	2	1	1
2	key(X),edge(X,a)	0	0	0
3	key(X),edge(X,b)	1	0	1
4	key(X),edge(X,c)	1	0	1
5	key(X),edge(X,d)	1	1	0
6	key(X),edge(X,e)	1	0	1
7	key(X),edge(Y,X)	1	2	-1
8	key(X),edge(a,X)	1	0	1
9	key(X),edge(b,X)	1	0	1
10	key(X),edge(c,X)	0	2	-2
11	key(X),edge(d,X)	0	0	0
12	key(X),edge(e,X)	0	1	-1

This can be achieved by restricting the refinement operator to generate queries in a canonical form, cf. [34].

Morishita and Sese [36] adapt Apriori for finding the top k patterns w.r.t. a boundable function ψ, i.e. for the case where there exists a function u (different from a global maximum) such that $\forall g, s \in \mathcal{L} : g \preceq s \rightarrow \psi(s) \leq u(g)$. Again, at each level candidate queries are obtained from those queries generated at the previous level that qualify for refinement, which now means they either belong to the current k best queries, or are still promising as their upper-bound is higher than the value of the current k-th best query. The function $\psi(q, \mathcal{D}) = \mathsf{COUNT}(\mathsf{q}^+(*)) - \mathsf{COUNT}(\mathsf{q}^-(*))$ used in the example above is upper-boundable using $u(q, \mathcal{D}) = \mathsf{COUNT}(\mathsf{q}^+(*))$. For any $g \preceq s$, $\psi(s) \leq \mathsf{COUNT}(\mathsf{s}^+(*)) \leq \mathsf{COUNT}(\mathsf{g}^+(*))$, as $\mathsf{COUNT}(\mathsf{s}^-(*)) \geq 0$ and COUNT is anti-monotonic. To illustrate this, assume we mine for the 3 best correlated queries in our graph database. Table 10.3 shows counts on key^+ and key^- and ψ-values obtained during the first level of mining. The highest score achieved is 1. Queries 1, 3, 4, 6, 8, 9 are the current best queries and will thus be refined on the next level. Queries 5 and 7 have lower scores, but upper bound $c^+ = 1$, implying that their refinements may still belong to the best queries and have to be considered on the next level as well. The remaining queries are pruned, as they all have an upper bound $c^+ = 0 < 1$, i.e. all their refinements are already known to score lower than the current best queries.

The framework for query mining as outlined above can directly be adapted towards *probabilistic* databases. The key changes involved are 1) that the database $\mathcal{D}$ is *probabilistic*, and 2) that the selection predicate ϕ or the correlation measure ψ is based on the probabilities of queries. In other words, we employ a probabilistic membership function. In non-probabilistic frequent query mining, every tuple in the relation *key* either satisfies the query or not. So, for a conjunctive query q and a 0-1 membership function $M(t|q, \mathcal{D})$, we can explicitly write the counting function underlying frequency as a sum:

$$freq(q, \mathcal{D}) = \sum_{t \in key} M(t|q, \mathcal{D})$$

On a more general level, this type of function can be seen as *aggregate* of the membership function $M(t|q, \mathcal{D})$.

To apply the algorithms sketched above with a probabilistic database $\mathcal{D}$, it suffices to replace the deterministic membership function $M(t|q, \mathcal{D})$ with a probabilistic variant. Possible choices for such a probabilistic membership function $P(t|q, \mathcal{D})$ include the success probability $P_s(q(t)|\mathcal{D})$ or the explanation probability $P_x(q(t)|\mathcal{D})$ as introduced for ProbLog in Equations (10.1) and (10.2). Note that using such query probabilities as probabilistic membership function is anti-monotonic, that is, if $q_1 \preceq q_2$ then $P(t|q_1, \mathcal{D}) \geq P(t|q_2, \mathcal{D})$. Again, a natural choice of selection predicate ϕ is the combination of a minimum threshold with an aggregated probabilistic membership function:

$$agg(q, \mathcal{D}) = \mathbf{AGG}_{t \in key} P(t|q, \mathcal{D}). \tag{10.9}$$

Here, **AGG** denotes an aggregate function such as $\sum$, min, max or $\prod$, which is to be taken over all tuples t in the relation *key*. Choosing $\sum$ with a deterministic membership relation corresponds to the traditional frequency function, whereas $\prod$ computes a kind of *likelihood* of the data. Note that whenever the membership function P is anti-monotone, selection predicates of the form $agg(q, \mathcal{D}) > c$ (with $agg \in \{\sum, \min, \max, \prod\}$) are anti-monotonic with regard to *OI*-subsumption, which is crucial to enable pruning.

When working with both positive and negative examples, the main focus lies on finding queries with a high aggregated score on the positives and a low aggregated score on the negatives. Note that using unclassified instances *key* corresponds to the special case where $key^+ = key$ and $key^- = \emptyset$. In the following, we will therefore consider instances of the selection function (10.9) for the case of classified examples key^+ and key^- only. Choosing sum as aggregation function results in a *probabilistic frequency pf* (10.10) also employed by [38] in the context of item-set mining, whereas product defines a kind of *likelihood LL* (10.11). Notice that using the product in combination with a non-zero threshold implies that *all* positive examples must be covered with non-zero probability. We therefore introduce a softened version LL_n (10.12) of the likelihood, where $n < |key^+|$ examples have to be covered with non-zero probability. This is achieved by restricting the set of tuples in the product to the n highest scoring tuples in key^+, thus integrating a deterministic (anti-monotonic) selection predicate into the probabilistic one. More formally, the three functions used are defined as follows:

$$pf(q, \mathcal{D}) = \sum_{t \in key^+} P(t|q, \mathcal{D}) - \sum_{t \in key^-} P(t|q, \mathcal{D}) \tag{10.10}$$

$$LL(q, \mathcal{D}) = \prod_{t \in key^+} P(t|q, \mathcal{D}) \cdot \prod_{t \in key^-} (1 - P(t|q, \mathcal{D})) \tag{10.11}$$

$$LL_n(q, \mathcal{D}) = \prod_{t \in key_n^+} P(t|q, \mathcal{D}) \cdot \prod_{t \in key^-} (1 - P(t|q, \mathcal{D})) \tag{10.12}$$

Here, key_n^+ contains the n highest scoring tuples in key^+. In correlated query mining, we obtain an upper bound on each of these functions by omitting the scores of negative examples, i.e. the aggregation over key^-.

Consider again our graph database, now with probabilities. Using P_x as probabilistic membership function, the query $q(X) : -key(X), edge(X, Y)$ gets probabilistic frequency $pf(q, \mathcal{D}) = P_x(a|q, \mathcal{D}) + P_x(c|q, \mathcal{D}) - (P_x(d|q, \mathcal{D}) + P_x(e|q, \mathcal{D})) = 0.8 + 0.9 - (0 + 0.5) = 1.2$ (with upper bound $0.8 + 0.9 = 1.7$), likelihood $LL(q, \mathcal{D}) = 0.8 \cdot 0.9 \cdot (1 - 0) \cdot (1 - 0.5) = 0.36$ (with upper bound $0.8 \cdot 0.9 = 0.72$), and softened likelihood $LL_1(q, \mathcal{D}) = 0.9 \cdot (1 - 0) \cdot (1 - 0.5) = 0.9$ (with upper bound 0.9).

For further details and experiments in the context of the biological network of Section 10.9, we refer to [31].

10.7 Theory Compression

In this section, we investigate how to obtain a small compressed probabilistic database that contains the essential links w.r.t. a given set of positive and negative examples. This is useful for scientists trying to understand and analyze large networks of uncertain relationships between biological entities as it allows them to identify the most relevant components of the theory.

The technique on which we build is that of theory compression [39], where the goal is to remove as many edges, i.e., probabilistic facts as possible from the theory while still explaining the (positive) examples. The examples, as usual, take the form of relationships that are either interesting or uninteresting to the scientist. The resulting theory should contain the essential facts, assign high probabilities to the positive and low probabilities to the negative examples, and it should be a lot smaller and hence easier to understand and to employ by the scientists than the original theory.

As an illustrative example, consider again the graph in Figure 1(a) together with the definition of the path predicate given earlier. Assume now that we just confirmed that $path(a, d)$ is of interest and that $path(a, e)$ is not. We can then try to find a small graph (containing k or fewer edges) that best matches these observations. Using a greedy approach, we would first remove the edges connecting e to the rest of the graph, as they strongly contribute to proving the negative example, while the positive example still has likely proofs in the resulting graph.

Before introducing the ProbLog theory compression problem, it is helpful to consider the corresponding problem in a purely logical setting, i.e., ProbLog programs where all facts are part of the background knowledge. In this case, the theory compression task coincides with a form of theory revision [10, 11] where the only operation allowed is the deletion of rules or facts: *given* a set of positive and negative examples in the form of true and false facts, *find* a theory that best explains the examples, i.e., one that scores best w.r.t. a function such as accuracy. At the same time, the theory should be *small*, that is it should contain at most k facts. So, *logical* theory compression aims at finding a small theory that best explains the examples. As a result the compressed theory should be a better fit w.r.t. the data but should also be much easier to understand and to interpret. This holds in particular when starting with large networks containing thousands of nodes and edges and then obtaining a small compressed graph that consists of say 20 edges only. In biological databases such as the ones considered in this chapter, scientists can easily analyze the inter-actions in such small networks but have a very hard time with the large networks. The *ProbLog Theory Compression Problem* is now an adaptation of the traditional theory revision (or compression) problem towards probabilistic Prolog programs. Intuitively, we are interested in finding a small number of facts (at most k many) that maximizes the likelihood of the examples. More formally:

Given   a ProbLog theory S, sets P and N of positive and negative examples in the form of independent and identically-distributed (iid) ground facts, and a constant $k \in \mathbb{N}$,

Find   a theory $T \subseteq S$ of size at most k ($|T| \leq k$) that has a maximum likelihood $\mathscr{L}$ w.r.t. the examples $E = P \cup N$, i.e., $T = \arg\max_{T \subseteq S \wedge |T| \leq k} \mathscr{L}(E|T)$, where

$$\mathscr{L}(E|T) = \prod_{e \in P} P(e|T) \cdot \prod_{e \in N} (1 - P(e|T)) \qquad (10.13)$$

In other words, we use a ProbLog theory T to specify the conditional class distribu-tion, i.e., the probability $P(e|T)$ that any given example e is positive[7]. Because the examples are assumed to be iid the total likelihood is obtained as a simple product.

Despite its intuitive appeal, using the likelihood as defined in Eq. (10.13) has some subtle downsides. For an optimal ProbLog theory T, the probability of the positives is as close to 1 as possible, and for the negatives as close to 0 as possible. In general, however, we want to allow for misclassifications (with a high cost in order to avoid overfitting) to effectively handle noisy data and to obtain smaller theories. Furthermore, the likelihood function can become 0, e.g., when a positive example is not covered by the theory at all. To overcome these problems, we slightly redefine $P(e|T)$ in Eq. (10.13) as

$$\hat{P}(e|T) = \max\left(\min[1 - \varepsilon, P(e|T)], \varepsilon\right) \qquad (10.14)$$

for some constant $\varepsilon > 0$ specified by the user.

The compression approach can efficiently be implemented following a two-steps strategy as shown in algorithm in Table 10.4. In a first step, we compute the BDDs for all given examples. Then, we use these BDDs in a second step to greedily remove facts. This compression approach is efficient since the (expensive) construction of the BDDs is performed only once per example.

More precisely, the algorithm starts by calling the approximation algorithm sketched in Section 10.3.2, which computes the DNFs and BDDs for lower and upper bounds (for-loop). In the second step, only the lower bound DNFs and BDDs are employed because they are simpler and, hence, more efficient to use. All facts used in at least one proof occurring in the (lower bound) BDD of some example con-

---

[7] Note that this is slightly different from specifying a distribution over (positive) examples.

Table 10.4 ProbLog theory compression

---

functionCOMPRESS$(S = \{p_1 :: c_1, \ldots, p_n :: c_n\}, E, k, \varepsilon)$

---

for $e \in E$ **do**

    Call APPROXIMATE(e, S, δ) to get $DNF(low, e)$ and $BDD(e)$

               where $DNF(low, e)$ is the lower bound DNF formula for e

               and $BDD(e)$ is the BDD corresponding to $DNF(low, e)$

end for

$R := \{p_i :: c_i \mid b_i \text{ (indicator for fact } i\text{) occurs in a } DNF(low, e)\}$

$BDD(E) := \bigcup_{e \in E}\{BDD(e)\}$

improves := true

while ($|R| > k$ or improves) and $R \neq \emptyset$ **do**

    $ll :=$ LIKELIHOOD$(R, BDD(E), \varepsilon)$

    $i := \arg\max_{i \in R}$ LIKELIHOOD$(R - \{i\}, BDD(E), \varepsilon)$

    improves := $(ll \leq$ LIKELIHOOD$(R - \{i\}, BDD(E), \varepsilon))$

    if improves or $|R| > k$ **then**

        $R := R - \{i\}$

    end if

end while

Return R

---

stitute the set R of possible revision points. All other facts do not occur in any proof contributing to probability computation and hence can immediately be removed.

After the set R of revision points has been determined and the other facts removed the ProbLog theory compression algorithm performs a *greedy* search in the space of subsets of R (while-loop). At each step, the algorithm finds that fact whose deletion results in the best likelihood score, and then deletes it. As explained in more details in [39], this can efficiently be done using the BDDs computed in the preprocessing step: *set the probability of the node corresponding to the fact to 0 and recompute the probability of the BDD*. This process is continued until both $|R| \leq k$ *and* deleting further facts does not improve the likelihood.

Theory compression as introduced here bears some relationships to the PTR approach by [40], where weights or probabilities are used as a kind of bias during the process of revising a logical theory. ProbLog compression is also somewhat related to Zelle and Mooney's work on Chill [41] in that it specializes an overly general theory but differs again in the use of a probabilistic framework. In the context of probabilistic logic languages, PFORTE [42] is a theory revision system using BLPs [43] that follows a hill-climbing approach similar to the one used here, but with a wider choice of revision operators.

For more details including experiments showing that ProbLog compression is not only of theoretical interest but is also applicable to various realistic problems in a biological link discovery domain we refer to [39].

10.8 Parameter Estimation

In this section, we address the question of how to set the parameters of the ProbLog facts in the light of a set of examples. These examples consist of ground queries together with the desired probabilities, which implies that we are dealing with weighted examples such as $0.6 : locatedIn(a,b)$ and $0.7 : interacting(a,c)$ as used by Gupta and Sarawagi [44] and Chen *et al.* [45]. The parameter estimation technique should then determine the best values for the parameters. Our approach as implemented in LeProbLog [19, 46] (Least Square Parameter Estimation for ProbLog) performs a gradient-based search to minimize the error on the given training data. The problem tackled can be formalized as regression task as follows:

Given   a ProbLog database T with unknown parameters and a set of training examples $\{(q_i, \tilde{p}_i)\}_{i=1}^{M}, M > 0$, where each $q_i \in \mathcal{H}$ is a query or proof and $\tilde{p}_i$ is the k-probability of q_i,

Find   the parameters of the database T that minimize the mean squared error:

$$MSE(T) = \frac{1}{M} \sum_{1 \leq i \leq M} \left(P_k(q_i|T) - \tilde{p}_i \right)^2 . \tag{10.15}$$

Gradient descent is a standard way of minimizing a given error function. The tunable parameters are initialized randomly. Then, as long as the error did not converge, the gradient of the error function is calculated, scaled by the learning rate η, and subtracted from the current parameters. To get the gradient of the MSE, we apply the sum and chain rule to Eq. (10.15). This yields the partial derivative

$$\frac{\partial MSE(T)}{\partial p_j} = \frac{2}{M} \sum_{1 \leq i \leq M} \underbrace{\left(P_k(q_i|T) - \tilde{p}_i \right)}_{\text{Part 1}} \cdot \underbrace{\frac{\partial P_k(q_i|T)}{\partial p_j}}_{\text{Part 2}} . \tag{10.16}$$

where part 1 can be calculated by a ProbLog inference call computing (10.4). It does not depend on j and has to be calculated only once in every iteration of a gradient descent algorithm. Part 2 can be calculated as following

$$\frac{\partial P_k(q_i|T)}{\partial p_j} = \sum_{\substack{S \subseteq L_T \\ S \models q_i}} \delta_{jS} \prod_{\substack{c_x \in S \\ x \neq j}} p_x \prod_{\substack{c_x \in L_T \setminus S \\ x \neq j}} (1 - p_x) , \tag{10.17}$$

where $\delta_{jS} := 1$ if $c_j \in S$ and $\delta_{jS} := -1$ if $c_j \in L_T \setminus S$. It is derived by first deriving the gradient $\partial P(S|T)/\partial p_j$ for a fixed subset $S \subseteq L_T$ of facts, which is straightforward, and then summing over all subsets S where q_i can be proven.

To ensure that all p_j stay probabilities during gradient descent, we reparameterize the search space and express each $p_j \in]0, 1[$ in terms of the sigmoid function $p_j = \sigma(a_j) := 1/(1 + \exp(-a_j))$ applied to $a_j \in \mathbb{R}$. This technique has been used for Bayesian networks and in particular for sigmoid belief networks [47]. We derive the partial derivative $\partial P_k(q_i|T)/\partial a_j$ in the same way as (10.17) but we have to apply

Table 10.5 Evaluating the gradient of a query efficiently by traversing the corresponding BDD, calculating partial sums, and adding only relevant ones.

---

function GRADIENT(BDD b, fact to derive for n_j)

$(val, seen) = \text{GRADIENTEVAL}(root(b), n_j)$
If $seen = 1$ return $val \cdot \sigma(a_j) \cdot (1 - \sigma(a_j))$
Else return 0

---

function GRADIENTEVAL(node n, target node n_j)

If n is the 1-terminal return $(1, 0)$
If n is the 0-terminal return $(0, 0)$
Let h and l be the high and low children of n
$(val(h), seen(h)) = \text{GRADIENTEVAL}(h, n_j)$
$(val(l), seen(l)) = \text{GRADIENTEVAL}(l, n_j)$
If $n = n_j$ return $(val(h) - val(l), 1)$
ElseIf $seen(h) = seen(l)$ return $(\sigma(a_n) \cdot val(h) + (1 - \sigma(a_n)) \cdot val(l), seen(h))$
ElseIf $seen(h) = 1$ return $(\sigma(a_n) \cdot val(h), 1)$
ElseIf $seen(l) = 1$ return $((1 - \sigma(a_n)) \cdot val(l), 1)$

---

the chain rule one more time due to the σ function

$$\sigma(a_j) \cdot (1 - \sigma(a_j)) \cdot \sum_{\substack{S \subseteq L_T \\ L \models q_i}} \delta_{jS} \prod_{\substack{c_x \in S \\ x \neq j}} \sigma(a_x) \prod_{\substack{c_x \in L_T \setminus S \\ x \neq j}} (1 - \sigma(a_x)).$$

We also have to replace every p_j by $\sigma(p_j)$ when calculating the success probability. We employ the BDD-based algorithm to compute probabilities as outlined in algorithm in Table 10.2. In the following, we update this towards the gradient and introduce LeProbLog, the gradient descent algorithm for ProbLog.

The following example illustrates the gradient calculation on a simple query.

Example 10.1 (Gradient of a query). Consider a simple coin toss game: One can either win by getting heads or by cheating as described by the following theory:

$$\begin{array}{ll} \text{?? :: heads.} & \text{?? :: cheat\_succesfully.} \\ \text{win} : -\text{cheat\_successfully.} & \\ \text{win} : -\text{heads.} & \end{array}$$

Suppose we want to estimate unknown fact probabilities (indicated by the symbol ??) from the training example $P(\text{win}) = 0.3$.

As a first step the fact probabilities get initialized with some random probabilities:

$$\begin{array}{ll} 0.6 :: \text{heads.} & 0.2 :: \text{cheat\_succesfully.} \\ \text{win} : -\text{cheat\_successfully.} & \\ \text{win} : -\text{heads.} & \end{array}$$

In order to calculate the gradient of the MSE (cf. Equation (10.16)), the algorithm evaluates the partial derivative for every probabilistic fact and every training exam-

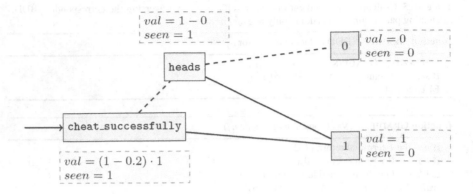

Fig. 10.4 Intermediate results when calculating the gradient $\partial P(\texttt{win})/\partial \texttt{heads}$ using the algorithm in Table 10.5. The result is read off at the root node of the BDD.

ple. Figure 10.4 illustrates the calculation of the partial derivate $\partial P(\texttt{win})/\partial \texttt{heads}$ using the algorithm in Table 10.5.

As described in Section 10.3, BDDs can be used to efficiently calculate the success probability of a query, solving the disjoint-sum problem arising at summing over probabilities in an elegant way. The algorithm in Table 10.2 can be modified straightforwardly such that it calculates the value of the gradient (10.17) of a success probability. The algorithm in Table 10.5 shows the pseudocode. Both algorithms have a time and space complexity of O(number of nodes in the BDD) when intermediate results are cached.

To see why this algorithm calculates the correct output let us first consider a full decision tree instead of a BDD. Each branch in the tree represents a product $n_1 \cdot n_2 \cdot \ldots \cdot n_i$, where the n_i are the probabilities associated to the corresponding variable assignment of nodes on the branch. The gradient of such a branch b with respect to n_j is $g_b = n_1 \cdot n_2 \cdot \ldots n_{j-1} \cdot n_{j+1} \cdot \ldots \cdot n_i$ if n_j is true, and $-g_b$ if n_j is false in b. As all branches in a full decision tree are mutually exclusive, the gradient w.r.t. n_j can be obtained by simply summing the gradients of all branches ending in a leaf labeled 1. In BDDs however, isomorphic sub-parts are merged, and obsolete parts are left out. This implies that some paths from the root to the 1-terminal may not contain n_j, therefore having a gradient of 0. So, when calculating the gradient on the BDD, we have to keep track of whether n_j appeared on a path or not. Given that the variable order is the same on all paths, we can easily propagate this information in our bottom-up algorithm. This is exactly what is described in the algorithm in Table 10.5. Specifically, GRADIENTEVAL(n, n_j) calculates the gradient w.r.t. n_j in the sub-BDD rooted at n. It returns two values: the gradient on the sub-BDD and a Boolean indicating whether or not the target node n_j appears in the sub-BDD. When at some node n the indicator values for the two children differ, we know that n_j does not appear above the current node, and we can drop the partial result from the child with indicator 0. The indicator variable is also used on the top level: GRADIENT returns the value calculated by the bottom-up algorithm if n_j occurred in the BDD and 0 otherwise.

LeProbLog combines the BDD-based gradient calculation with a standard gradient descent search. Starting from parameters $\mathbf{a} = a_1, \ldots, a_n$ initialized randomly, the gradient $\Delta\mathbf{a} = \Delta a_1, \ldots, \Delta a_n$ is calculated, parameters are updated by subtracting the gradient, and updating is repeated until convergence. When using the k-probability with finite k, the set of k best proofs may change due to parameter updates. After each update, we therefore recompute the set of proofs and the corresponding BDD.

One nice side effect of the use of ProbLog is that it naturally combines *learning from entailment* and *learning from proofs*, two learning settings that so far have been considered separately. So far, we have assumed that the examples were ground facts together with their target probability. It turns out that the sketched technique also works when the examples are proofs, which correspond to conjunctions of probabilistic facts, and which can be seen as a conjunction of queries. Therefore, LeProbLog can use examples of both forms, (atomic) queries and proofs, at the same time. For further details and experimental results in the context of the biological network application, we refer to [19, 46].

10.9 Application

As an application of ProbLog, consider link mining in large networks of biological entities, such as genes, proteins, tissues, organisms, biological processes, and molecular functions. Life scientist utilize such data to identify and analyze relationships between entities, for instance between a protein and a disease.

Molecular biological data is available from public sources, such as Ensembl[8], NCBI Entrez[9], and many others. They contain information about various types of objects, such as the ones mentioned above, and many more. Information about known or predicted relationships between entities is also available, e.g., that gene A of organism B codes for protein C, which is expressed in tissue D, or that genes E and F are likely to be related since they co-occur often in scientific articles. Mining such data has been identified as an important and challenging task (cf. [48]).

A collection of interlinked heterogeneous biological data can be conveniently seen as a weighted graph or network of biological concepts, where the weight of an edge corresponds to the probability that the corresponding nodes are related [12]. A ProbLog representation of such a graph can simply consist of probabilistic edge/2 facts, though finer grained representations using relations such as codes/2, expresses/2 are also possible.

We have used the Biomine dataset [12] in our applications. It is an integrated index of a number of public biological databases, consisting of about 1 million ob-

---

[8] http://www.ensembl.org
[9] http://www.ncbi.nlm.nih.gov/Entrez/

jects and about 7 million relations. In this dataset, weights are associated to edges, indicating the probability that the corresponding nodes are related[10].

We next outline different ways of using ProbLog to query the Biomine dataset. We only assume probabilistic edge/3 facts, where the third term indicates the edge type, and a simple background theory that contains the type of individual nodes as node/2 facts and specifies an acyclic, indirected (symmetric) path/2 relation.

Probabilistic inference (Section 10.3) Assume a life scientist has hypothesized that ROBO1 gene is related to Alzheimer disease (AD). The probability that they are related is computed by ProbLog query *?- path('ROBO1', 'AD')*. The results is 0.70, indicating that—under all the assumptions made by ProbLog, Biomine and the source databases—they might be related. Assuming the life scientist has 100 candidate genes for Alzheimer disease, ProbLog can easily be used to rank the genes by their likelihood of being relevant for AD.

Most likely explanation (Section 10.3.1) Obviously, our life scientist would not be happy with the answer 0.70 alone. Knowing the possible *relation* is much more interesting, and could potentially lead to novel insight.

When including node type information in the definition of a path between two nodes, the best (most likely) proof of path('ROBO1','AD') obtained by ProbLog is

```
node('ROBO1', gene),
edge('ROBO1', 'SLIT1',  interacts-with),
node('SLIT1', gene),
edge('SLIT1', 'hsa10q23.3-q24', is-located-in),
node('hsa10q23.3-q24', genomic-context),
edge('hsa10q23.3-q24', 'hsa10q24', contains),
node('hsa10q24', genomic-context),
edge('hsa10q24', 'AD', is-related-to),
node('AD', phenotype).
```

In other words, ROBO1 interacts with SLIT1, which is located in a genomic area related to AD. This proof has probability 0.14.

Most likely generalized explanation (Section 10.5) Explanations obtained by probabilistic explanation based learning within ProbLog are on a more general level, that is, they replace constants occurring in a concrete proof by variables. By defining predicates related to node and edge types as operational, the proof above is generalized to explanation exp_path(A, B) ←

```
node(A, gene), edge(A, C, interacts-with),
node(C, gene), edge(C, D, is-located-in),
node(D, genomic-context), edge(D, E, contains),
node(E, genomic-context),
edge(E, B, is-related-to), node(B, phenotype).
```

---

[10] [12] view this strength or probability as the product of three factors, indicating the *reliability*, the *relevance* as well as the *rarity* (specificity) of the information.

Table 10.6 Additional explanation clauses for `path(A,B)`, connecting gene A to phenotype B, obtained from different examples.

```
e_path(A,B) ← node(A,gene), edge(A,C,belongs_to),
    node(C,homologgroup), edge(B,C,refers_to), node(B,phenotype),
    nodes_distinct([B,C,A]).
e_path(A,B) ← node(A,gene), edge(A,C,codes_for), node(C,protein),
    edge(D,C,subsumes), node(D,protein), edge(D,E,interacts_with),
    node(E,protein), edge(B,E,refers_to), node(B,phenotype),
    nodes_distinct([B,E,D,C,A]).
e_path(A,B) ← node(A,gene), edge(A,C,participates_in),
    node(C,pathway), edge(D,C,participates_in), node(D,gene),
    edge(D,E,codes_for), node(E,protein), edge(B,E,refers_to),
    node(B,phenotype), nodes_distinct([B,E,D,C,A]).
e_path(A,B) ← node(A,gene), edge(A,C,is_found_in),
    node(C,cellularcomponent), edge(D,C,is_found_in),
    node(D,protein), edge(B,D,refers_to),
    node(B,phenotype), nodes_distinct([B,D,C,A]).
```

Table 10.6 shows four other explanations obtained for relationships between a gene (such as ROBO1) and a phenotype (such as AD). These explanations are all semantically meaningful. For instance, the first one indicates that gene A is related to phenotype B if A belongs to a group of homologous (i.e., evolutionarily related) genes that relate to B. The three other explanations are based on interaction of proteins: either an explicit one, by participation in the same pathway, or by being found in the same cellular component.

Such an explanation can then be used to query the database for a list of other genes connected to AD by the same type of pattern, and to rank them according to the probability of that connection, which may help the scientist to further examine the information obtained.

While the linear explanation used for illustration here could also be obtained using standard shortest-path algorithms, PEBL offers a more general framework for finding explanations where the structure is defined by background knowledge in the form of an arbitrary logic program.

Theory compression (Section 10.7) The most likely explanation for *path('ROBO1', 'AD')* is just a single proof and does not capture alternative proofs, not to mention the whole network of related and potentially relevant objects. Theory compression can be used here to automatically extract a suitable subgraph for illustration. By definition, the extracted subgraph aims at maximizing the probability of *path('ROBO1', 'AD')*, i.e., it contains the most relevant nodes and edges.

Looking at a small graph of, say 12 nodes, helps to give an overview of the most relevant connections between ROBO1 and AD. Such a look actually indicates that the association of AD to genomic context hsa10q24 is possibly due to the PLAU gene, which is suspected to be associated with late-onset Alzheimer disease. The life scientist could now add *path('ROBO1', 'hsa10q24')* as a negative example, in order to remove connections using the genomic context from the extracted graph.

Local pattern mining (Section 10.6) Given a number of genes he considers relevant for the problem at hand, our life scientist could now be interested in relationships these genes take part in with high probability. Local pattern mining offers a way to query ProbLog for such patterns or subgraphs of relationships without relying on predefined specific connections such as path.

Parameter estimation (Section 10.8) Imagine our life scientist got information on new entities and links between them, for example performing experiments or using information extraction techniques on a collection of texts. However, he does not know all the probabilities that should be attached to these new links, but only the probabilities of some of the links, of some specific paths, and of some pairs of entities being connected by some path. He could now use this knowledge as training examples for LeProbLog to automatically adjust the parameters of the new network to fit the available information.

10.10 Related Work in Statistical Relational Learning

In this section, we position ProbLog in the field of statistical relational learning [49] and probabilistic inductive logic programming [50]. In this context, its distinguishing features are that it is a probabilistic logic programming language based on Sato's distribution semantics [14], that it also can serve as a target language into which many of the other statistical relational learning formalisms can be compiled [51] and that several further approaches for learning ProbLog are being developed. Let us now discuss each of these aspects in turn.

First, ProbLog is closely related to some alternative formalisms such as PHA and ICL [7, 16], pD [6] and PRISM [8] as their semantics are all based on Sato's distribution semantics even though there exist also some subtle differences. However, ProbLog is – to the best of the authors' knowledge – the first implementation that tightly integrates Sato's original distribution semantics [14] in a state-of-the-art Prolog system without making additional restrictions (such as the exclusive explanation assumption made in PHA and PRISM). As ProbLog, both PRISM and the ICL implementation AILog2 use a two-step approach to inference, where proofs are collected in the first phase, and probabilities are calculated once all proofs are known. AILog2 is a meta-interpreter implemented in SWI-Prolog for didactical purposes, where the disjoint-sum-problem is tackled using a symbolic disjoining technique [16]. PRISM, built on top of B-Prolog, requires programs to be written such that alternative explanations for queries are mutually exclusive. PRISM uses a meta-interpreter to collect proofs in a hierarchical datastructure called explanation graph. As proofs are mutually exclusive, the explanation graph directly mirrors the sum-of-products structure of probability calculation [8]. ProbLog is the first probabilistic logic programming system using BDDs as a basic datastructure for probability calculation, a principle that receives increased interest in the probabilistic logic learning community, cf. for instance [52, 53].

Furthermore, as compared to SLPs [54], CLP($\mathscr{BN}$) [55], and BLPs [43], ProbLog is a much simpler and in a sense more primitive probabilistic programming language. Therefore, the relationship between probabilistic logic programming and ProbLog is, in a sense, analogous to that between logic programming and Prolog. From this perspective, it is our hope and goal to further develop ProbLog so that it can be used as a general purpose programming language with an efficient implementation for use in statistical relational learning [49] and probabilistic programming [50]. One important use of such a probabilistic programming language is as a target language in which other formalisms can be efficiently compiled. For instance, it has already been shown that CP-logic [56], a recent elegant probabilistic knowledge representation language based on a probabilistic extension of clausal logic, can be compiled into ProbLog [52] and it is well-known that SLPs [54] can be compiled into Sato's PRISM, which is closely related to ProbLog. Further evidence is provided in [51].

Another, important use of ProbLog is as a vehicle for developing learning and mining algorithms and tools [13, 39, 19, 31], an aspect that we have also discussed in the present paper. In the context of probabilistic representations [49, 50], one typically distinguishes two types of learning: parameter estimation and structure learning. In parameter estimation in the context of ProbLog and PRISM, one starts from a set of queries and the logical part of the program and the problem is to find good estimates of the parameter values, that is, the probabilities of the probabilistic facts in the program. In the present paper and [19], we have discussed a gradient descent approach to parameter learning for ProbLog in which the examples are ground facts together with their target probability. In [57], an approach to learning from interpretations based on an EM algorithm is introduced. There, each example specifies a possible world, that is, a set of ground facts together with their truth value. This setting closely corresponds to the standard setting for learning in statistical relational learning systems such as Markov Logic [58] and probabilistic relational models [59]. In structure learning, one also starts from queries but has to find the logical part of the program as well. Structure learning is therefore closely related to inductive logic programming. An initial approach to learning the structure, that is, the rules of a ProbLog program has recently been introduced in [60].

10.11 Conclusions

In this chapter, we provided a survey of the developments around ProbLog, a simple probabilistic extension of Prolog based on the distribution semantics. This combination of definite clause logic and probabilities leads to an expressive general framework supporting both inductive and probabilistic querying. Indeed, probabilistic explanation based learning, local pattern mining, theory compression and parameter estimation as presented in this chapter all share a common core: they all use the probabilistic inference techniques offered by ProbLog to score queries or examples. ProbLog has been motivated by the need to develop intelligent tools for supporting life scientists analyzing large biological networks involving uncertain data. All techniques presented here have been evaluated in the context of such a biological network; we refer to [3, 13, 31, 39, 19] for details.

Acknowledgements We would like to thank our co-workers Kate Revoredo, Bart Demoen, Ricardo Rocha and Theofrastos Mantadelis for their contributions to ProbLog. This work is partially supported by IQ (European Union Project IST-FET FP6-516169) and the GOA project 2008/08 Probabilistic Logic Learning. Angelika Kimmig and Bernd Gutmann are supported by the Research Foundation-Flanders (FWO-Vlaanderen).

References

1. Suciu, D.: Probabilistic databases. SIGACT News **39**(2) (2008) 111–124
2. Imielinski, T., Mannila, H.: A database perspective on knowledge discovery. Commun. ACM **39**(11) (1996) 58–64
3. De Raedt, L., Kimmig, A., Toivonen, H.: ProbLog: A probabilistic Prolog and its application in link discovery. In Veloso, M., ed.: IJCAI. (2007) 2462–2467
4. Dantsin, E.: Probabilistic logic programs and their semantics. In Voronkov, A., ed.: Proc. 1st Russian Conf. on Logic Programming. Volume 592 of LNCS. (1992) 152–164
5. Dalvi, N.N., Suciu, D.: Efficient query evaluation on probabilistic databases. In: VLDB. (2004) 864–875
6. Fuhr, N.: Probabilistic Datalog: Implementing logical information retrieval for advanced applications. Journal of the American Society for Information Science **51**(2) (2000) 95–110
7. Poole, D.: Probabilistic Horn abduction and Bayesian networks. Artificial Intelligence **64** (1993) 81–129
8. Sato, T., Kameya, Y.: Parameter learning of logic programs for symbolic-statistical modeling. J. Artif. Intell. Res. (JAIR) **15** (2001) 391–454
9. Poole, D.: Logic programming, abduction and probability. New Generation Computing **11** (1993) 377–400
10. Wrobel, S.: First order theory refinement. In De Raedt, L., ed.: Advances in Inductive Logic Programming. IOS Press, Amsterdam (1996) 14 – 33
11. Richards, B.L., Mooney, R.J.: Automated refinement of first-order horn-clause domain theories. Machine Learning **19**(2) (1995) 95–131
12. Sevon, P., Eronen, L., Hintsanen, P., Kulovesi, K., Toivonen, H.: Link discovery in graphs derived from biological databases. In: DILS. Volume 4075 of LNCS., Springer (2006) 35–49
13. Kimmig, A., De Raedt, L., Toivonen, H.: Probabilistic explanation based learning. In Kok, J.N., Koronacki, J., de Mantaras, R.L., Matwin, S., Mladenic, D., Skowron, A., eds.: 18th European Conference on Machine Learning (ECML). Volume 4701 of LNCS., Springer (2007) 176–187
14. Sato, T.: A statistical learning method for logic programs with distribution semantics. In Sterling, L., ed.: ICLP, MIT Press (1995) 715–729
15. Bryant, R.E.: Graph-based algorithms for boolean function manipulation. IEEE Trans. Computers **35**(8) (1986) 677–691
16. Poole, D.: Abducing through negation as failure: stable models within the independent choice logic. Journal of Logic Programming **44**(1-3) (2000) 5–35
17. Lloyd, J.W.: Foundations of Logic Programming. 2. edn. Springer, Berlin (1989)
18. Valiant, L.G.: The complexity of enumeration and reliability problems. SIAM Journal on Computing **8**(3) (1979) 410–421
19. Gutmann, B., Kimmig, A., De Raedt, L., Kersting, K.: Parameter learning in probabilistic databases: A least squares approach. In Daelemans, W., Goethals, B., Morik, K., eds.:

Proceedings of the European Conference on Machine Learning and Principles and Practice of Knowledge Discovery in Databases (ECML PKDD 2008), Part I. Volume 5211 of LNCS (Lecture Notes In Computer Science)., Antwerp, Belgium, Springer Berlin/Heidelberg (September 2008) 473–488

20. Kimmig, A., Santos Costa, V., Rocha, R., Demoen, B., De Raedt, L.: On the Efficient Execution of ProbLog Programs. In de la Banda, M.G., Pontelli, E., eds.: International Conference on Logic Programming. Number 5366 in LNCS, Springer (December 2008) 175–189

21. Santos Costa, V.: The life of a logic programming system. In de la Banda, M.G., Pontelli, E., eds.: Logic Programming, 24th International Conference, ICLP 2008, Udine, Italy, December 9-13 2008, Proceedings. Volume 5366 of Lecture Notes in Computer Science., Springer (2008) 1–6

22. Fredkin, E.: Trie Memory. Communications of the ACM **3** (1962) 490–499

23. Ramakrishnan, I.V., Rao, P., Sagonas, K., Swift, T., Warren, D.S.: Efficient Access Mechanisms for Tabled Logic Programs. Journal of Logic Programming **38**(1) (January 1999) 31–54

24. Mantadelis, T., Demoen, B., Janssens, G.: A simplified fast interface for the use of CUDD for binary decision diagrams (2008) http://people.cs.kuleuven.be/~theofrastos.mantadelis/tools/simplecudd.html.

25. Kimmig, A., Demoen, B., De Raedt, L., Santos Costa, V., Rocha, R.: On the implementation of the probabilistic logic programming language ProbLog. Theory and Practice of Logic Programming (TPLP) (2010) to appear; https://lirias.kuleuven.be/handle/123456789/259607.

26. Mitchell, T.M., Keller, R.M., Kedar-Cabelli, S.T.: Explanation-based generalization: A unifying view. Machine Learning **1**(1) (1986) 47–80

27. DeJong, G., Mooney, R.J.: Explanation-based learning: An alternative view. Machine Learning **1**(2) (1986) 145–176

28. Hirsh, H.: Explanation-based generalization in a logic-programming environment. In: IJCAI'87: Proceedings of the 10th international joint conference on Artificial intelligence, San Francisco, CA, USA, Morgan Kaufmann Publishers Inc. (1987) 221–227

29. Van Harmelen, F., Bundy, A.: Explanation-based generalisation = partial evaluation. Artificial Intelligence **36**(3) (1988) 401–412

30. Langley, P.: Unifying themes in empirical and explanation-based learning. In: Proceedings of the sixth international workshop on Machine learning, San Francisco, CA, USA, Morgan Kaufmann Publishers Inc. (1989) 2–4

31. Kimmig, A., De Raedt, L.: Local query mining in a probabilistic Prolog. In Boutilier, C., ed.: International Joint Conference on Artificial Intelligence. (2009) 1095–1100

32. Dehaspe, L., Toivonen, H., King, R.D.: Finding frequent substructures in chemical compounds. In Agrawal, R., Stolorz, P., Piatetsky-Shapiro, G., eds.: Proceedings of the 4th ACM-SIGKDD International Conference on Knowledge Discovery and Data Mining, AAAI Press (1998) 30–36

33. Tsur, S., Ullman, J.D., Abiteboul, S., Clifton, C., Motwani, R., Nestorov, S., Rosenthal, A.: Query flocks: A generalization of association-rule mining. In: SIGMOD Conference. (1998) 1–12

34. De Raedt, L., Ramon, J.: Condensed representations for inductive logic programming. In Dubois, D., Welty, C.A., Williams, M.A., eds.: Proceedings of the 9th International Conference on Principles and Practice of Knowledge Representation. AAAI Press (2004) 438–446

35. Esposito, F., Fanizzi, N., Ferilli, S., Semeraro, G.: Ideal refinement under object identity. In Langley, P., ed.: Proceedings of the 17th International Conference on Machine Learning, Morgan Kaufmann (2000) 263–270

36. Morishita, S., Sese, J.: Traversing itemset lattice with statistical metric pruning. In: Proceedings of the 19th ACM SIGACT-SIGMOD-SIGART Symposium on Principles of Database Systems, ACM Press (2000) 226–236

37. Mannila, H., Toivonen, H.: Levelwise search and borders of theories in knowledge discovery. Data Mining and Knowledge Discovery **1**(3) (1997) 241–258

38. Chui, C.K., Kao, B., Hung, E.: Mining frequent itemsets from uncertain data. In Zhou, Z.H., Li, H., Yang, Q., eds.: PAKDD. Volume 4426 of Lecture Notes in Computer Science., Springer (2007) 47–58

39. De Raedt, L., Kersting, K., Kimmig, A., Revoredo, K., Toivonen, H.: Compressing probabilistic Prolog programs. Machine Learning **70**(2-3) (2008) 151–168

40. Koppel, M., Feldman, R., Segre, A.M.: Bias-driven revision of logical domain theories. J. Artif. Intell. Res. (JAIR) **1** (1994) 159–208

41. Zelle, J., Mooney, R.: Inducing deterministic Prolog parsers from treebanks: A machine learning approach. In: Proceedings of the 12th National Conference on Artificial Intelligence (AAAI-94). (1994) 748–753

42. Paes, A., Revoredo, K., Zaverucha, G., Santos Costa, V.: Probabilistic first-order theory revision from examples. In Kramer, S., Pfahringer, B., eds.: ILP. Volume 3625 of Lecture Notes in Computer Science., Springer (2005) 295–311

43. Kersting, K., De Raedt, L.: Basic principles of learning bayesian logic programs. [50] 189–221

44. Gupta, R., Sarawagi, S.: Creating probabilistic databases from information extraction models. In: VLDB. (2006) 965–976

45. Chen, J., Muggleton, S., Santos, J.: Learning probabilistic logic models from probabilistic examples (extended abstract). In: ILP. (2007) 22–23

46. Gutmann, B., Kimmig, A., Kersting, K., De Raedt, L.: Parameter estimation in ProbLog from annotated queries. Technical Report CW 583, Department of Computer Science, Katholieke Universiteit Leuven, Belgium (April 2010)

47. Saul, L., Jaakkola, T., Jordan, M.: Mean field theory for sigmoid belief networks. JAIR **4** (1996) 61–76

48. Perez-Iratxeta, C., Bork, P., Andrade, M.: Association of genes to genetically inherited diseases using data mining. Nature Genetics **31** (2002) 316–319

49. Getoor, L., Taskar, B., eds.: Statistical Relational Learning. The MIT press (2007)

50. De Raedt, L., Frasconi, P., Kersting, K., Muggleton, S., eds.: Probabilistic Inductive Logic Programming — Theory and Applications. Volume 4911 of Lecture Notes in Artificial Intelligence. Springer (2008)

51. De Raedt, L., Demoen, B., Fierens, D., Gutmann, B., Janssens, G., Kimmig, A., Landwehr, N., Mantadelis, T., Meert, W., Rocha, R., Santos Costa, V., Thon, I., Vennekens, J.: Towards digesting the alphabet-soup of statistical relational learning. In Roy, D., Winn, J., McAllester, D., Mansinghka, V., Tenenbaum, J., eds.: Proceedings of the 1st Workshop on Probabilistic Programming: Universal Languages, Systems and Applications, Whistler, Canada (December 2008)

52. Riguzzi, F.: A top down interpreter for LPAD and CP-logic. In: AI*IA 2007: Artificial Intelligence and Human-Oriented Computing. Volume 4733 of LNCS. (2007)

53. Ishihata, M., Kameya, Y., Sato, T., ichi Minato, S.: Propositionalizing the EM algorithm by BDDs. In Železný, F., Lavrač, N., eds.: Proceedings of Inductive Logic Programming (ILP 2008), Late Breaking Papers, Prague, Czech Republic (September 2008) 44–49

54. Muggleton, S.: Stochastic logic programs. In De Raedt, L., ed.: ILP. (1995)

55. Santos Costa, V., Page, D., Cussens, J.: Clp(bn): Constraint logic programming for probabilistic knowledge. In: In Proceedings of the 19th Conference on Uncertainty in Artificial Intelligence (UAI03, Morgan Kaufmann (2003) 517–524

56. Vennekens, J., Verbaeten, S., Bruynooghe, M.: Logic programs with annotated disjunctions. In Demoen, B., Lifschitz, V., eds.: ICLP. Volume 3132 of LNCS., Springer, Heidelberg (2004) 431–445

57. Gutmann, B., Thon, I., De Raedt, L.: Learning the parameters of probabilistic logic programs from interpretations. Technical Report CW 584, Department of Computer Science, Katholieke Universiteit Leuven, Belgium (April 2010)

58. Domingos, P., Lowd, D.: Markov Logic: an interface layer for AI. Morgan & Claypool (2009)

59. Getoor, L., Friedman, N., Koller, D., Pfeffer, A.: Learning probabilistic relational models. In Džeroski, S., Lavrač, N., eds.: Relational Data Mining. Springer (2001) 307–335

60. De Raedt, L., Thon, I.: Probabilistic rule learning. Technical Report CW 580, Department of Computer Science, Katholieke Universiteit Leuven, Belgium (April 2010)

Part III
Inductive Databases:
Integration Approaches

Part III

Inductive Databases:
Integration Approaches

Chapter 11
Inductive Querying with Virtual Mining Views

Hendrik Blockeel, Toon Calders, Élisa Fromont, Bart Goethals, Adriana Prado, and Céline Robardet

Abstract In an inductive database, one can not only query the data stored in the database, but also the patterns that are implicitly present in these data. In this chapter, we present an inductive database system in which the query language is traditional SQL. More specifically, we present a system in which the user can query the collection of all possible patterns as if they were stored in traditional relational tables. We show how such tables, or mining views, can be developed for three popular data mining tasks, namely itemset mining, association rule discovery and decision tree learning. To illustrate the interactive and iterative capabilities of our system, we describe a complete data mining scenario that consists in extracting knowledge from real gene expression data, after a pre-processing phase.

Hendrik Blockeel
Katholieke Universiteit Leuven, Belgium
Leiden Institute of Advanced Computer Science, Universiteit Leiden, The Netherlands
e-mail: hendrik.blockeel@cs.kuleuven.be

Toon Calders
Technische Universiteit Eindhoven, The Netherlands
e-mail: t.calders@tue.nl

Élisa Fromont · Adriana Prado
Université de Lyon (Université Jean Monnet), CNRS, Laboratoire Hubert Curien, UMR5516, F-42023 Saint-Etienne, France
e-mail: {elisa.fromont, adriana.bechara.prado}@univ-st-etienne.fr

Bart Goethals
Universiteit Antwerpen, Belgium
e-mail: bart.goethals@ua.ac.be

Céline Robardet
Université de Lyon, INSA-Lyon, CNRS, LIRIS, UMR5205, F-69621, France
e-mail: celine.robardet@insa-lyon.fr

11.1 Introduction

Data mining is an interactive process in which different tasks may be performed sequentially. In addition, the output of those tasks may be repeatedly combined to be used as input for subsequent tasks. For example, one could (a) first learn a decision tree model from a given dataset and, subsequently, mine association rules which describe the misclassified tuples with respect to this model or (b) first look for an interesting association rule that describes a given dataset and then find all tuples that violate such rule.

In order to effectively support such a knowledge discovery process, the integration of data mining into database systems has become necessary. The concept of *Inductive Database Systems* has been proposed in [1] so as to achieve such integration. The idea behind this type of system is to give to the user the ability to query not only the data stored in the database, but also patterns that can be extracted from these data. Such database should be able to store and manage patterns as well as provide the user with the ability to query them.

In this chapter, we show how such an inductive database system can be implemented in practice, as studied in [2, 3, 4, 5, 6, 7]. To allow the users to query patterns as well as standard data, several researchers proposed extensions to the popular query language SQL as a natural way to express such mining queries [8, 9, 10, 11, 12, 13]. As opposed to those proposals, we present here an inductive database system in which the query language is traditional SQL. We propose a relational database model based on what we call *virtual mining views*. The mining views are relational tables that virtually contain the complete output of data mining tasks. For example, for the itemset mining task, there is a table called *Sets* virtually storing all itemsets. As far as the user is concerned, all itemsets are stored in table *Sets* and can be queried as any other relational table. In reality, however, table *Sets* is empty. Whenever a query is formulated selecting itemsets from this table, the database system triggers an itemset mining algorithm, such as Apriori [14], which computes the itemsets in the same way as normal views in databases are only computed at query time. The user does not notice the emptiness of the tables; he or she can simply assume their existence and query accordingly. Therefore, we prefer to name these special tables virtual mining views.

In this chapter, we show how such tables, or virtual mining views, can be developed for three popular data mining tasks, namely itemset mining, association rule discovery and decision tree learning. To make the model as generic as possible, the output of these tasks are represented by a unifying set of mining views. In Section 11.2, we present these mining views in detail.

Since the proposed mining views are empty, they need to be filled (materialized) by the system once a query is posed over them. The mining process itself needs to be performed by the system in order to answer such queries. Note that the user may impose certain constraints in his or her queries, asking for only a subset of all possible patterns. As an example, the user may query from the mining view *Sets* all frequent itemsets with a certain support. Therefore, the entire set of patterns does not always need to be stored in the mining views, but only those that satisfy the constrains imposed by the user. In [2], Calders et al. present an algorithm that extracts from a query a set of constraints relevant for association rules to be pushed into the mining algorithm. We have extended this constraint extraction algorithm to extract constraints from queries over decision trees. The reader can refer to [7] for the details on the algorithm.

All ideas presented here, from querying the mining views and extracting constraints from the queries to the actual execution of the data mining process itself and the materialization of the mining views, have been implemented into the well-known open source database system PostgreSQL[1]. Details of the implementation are given in Section 11.3.

We have therefore organized the rest of this chapter in the following way. The next section is dedicated to the virtual mining views framework. We also present how the 4 prototypical tasks described in Chapter 3 can be executed by SQL queries over the mining views. The implementation of the system along with an extended illustrative data mining scenario is presented in Section 11.3. Finally, the conclusions of this chapter are presented in Section 11.4, stressing the main contributions and pointing to related future work.

11.2 The Mining Views Framework

In this section, we present the mining views framework in detail. This framework consists of a set of relational tables, called mining views, which virtually represent the complete output of data mining tasks. In reality, the mining views are empty and the database system finds the required tuples only when they are queried by the user.

11.2.1 The Mining View Concepts

We assume to be working in a relational database which contains the table $T(A_1,\ldots,A_n)$, having only categorical attributes. We denote the domain of A_i by $dom(A_i)$, for all $i = 1 \ldots n$. A tuple of T is therefore an element of $dom(A_i) \times \ldots \times dom(A_n)$. The active domain of A_i of T, denoted by $adom(A_i, T)$, is defined as the set of values that are currently assigned to A_i, that is, $adom(A_i, T) := \{t.A_i \mid t \in T\}$.

In the mining views framework, the patterns extracted from table T are generically represented by what we call *concepts*. We denote a concept as a conjunction of attribute-value pairs that is definable over table T. For example,

$$(Outlook = \text{`Sunny'} \wedge Humidity = \text{`High'} \wedge Play = \text{`No'})$$

is a concept defined over the classical relational data table Playtennis [24], a sample of which is illustrated in Figure 11.1.

---
[1] http://www.postgresl.org/

Fig. 11.1 The data table
Playtennis.

	Playtennis				
Day	Outlook	Temperature	Humidity	Wind	Play
D1	Sunny	Hot	High	Weak	No
D2	Sunny	Hot	High	Strong	No
D3	Overcast	Hot	High	Weak	Yes
D4	Rain	Mild	High	Weak	Yes
...	...	...	...	...	...

To represent each concept as a database tuple, we use the symbol '?' as the *wild-card value* and assume it does not exist in the active domain of any attribute of T.

Definition 11.1. A concept over table T is a tuple $(c_1, \ldots, c_n)$ with $c_i \in adom(A_i) \cup \{'?'\}$, for all $i=1 \ldots n$.

Following Definition 11.1, the example concept above is represented by the tuple

$$('?', 'Sunny', '?', '?', 'High', '?', 'No').$$

We are now ready to introduce the mining view $T\_Concepts$. In the proposed framework, the mining view $T\_Concepts(cid, A_1, \ldots, A_n)$ virtually contains all concepts that are definable over table T. We assume that these concepts can be sorted in lexicographic order and that an identifier can unambiguously be given to each concept.

Definition 11.2. The mining view $T\_Concepts(cid, A_1, \ldots, A_n)$ contains one tuple $(cid, c_1, \ldots, c_n)$ for every concept defined over table T. The attribute cid uniquely identifies the concepts.

In fact, the mining view $T\_Concepts$ represents exactly a *data cube* [25] built from table T, with the difference that the wildcard value "ALL" introduced in [25] is replaced by the value '?'. By following the syntax introduced in [25], the mining view $T\_Concepts$ would be created with the SQL query shown in Figure 11.2 (consider adding the identifier cid after its creation).

```
1. create table T_Concepts
2. select A1, A2,..., An
3. from T
4. group by cube A1, A2,..., An
```

Fig. 11.2 The data cube that represents the contents of the mining view $T\_Concepts$.

11.2.2 Representing Patterns and Models as Sets of Concepts

We now explain how patterns extracted from the table Playtennis can be represented by the concepts in the mining view *Playtennis_Concepts*. In the remainder of this section, we refer to table Playtennis as T and use the concepts in Figure 11.3 for the illustrative examples.

11.2.2.1 Itemsets and Association Rules

As itemsets in a relational database are conjunctions of attribute-value pairs, they can be represented as concepts. Itemsets are represented in the proposed framework by the mining view:

$$T\_Sets(cid, supp, sz).$$

The view T_Sets contains a tuple for each itemset, where cid is the identifier of the itemset (concept), $supp$ is its support (the number of tuples satisfied by the concept), and sz is its size (the number of attribute-value pairs in which there are no wildcards).

Similarly, association rules are represented by the view:

$$T\_Rules(rid, cida, cidc, cid, conf).$$

T_Rules contains a tuple for each association rule that can be extracted from table T. We assume that a unique identifier, rid, can be given to each rule. The attribute rid is the rule identifier, $cida$ is the identifier of the concept representing its left hand side (referred to here as antecedent), $cidc$ is the identifier of the concept representing its right hand side (referred to here as consequent), cid is the identifier of the union of the last two, and $conf$ is the confidence of the rule.

Fig. 11.3 A sample of the mining view *Playtennis_Concepts*, which is used for the illustrative examples in Section 11.2.2.

Playtennis_Concepts

cid	Day	Outlook	Temperature	Humidity	Wind	Play
...	...	...	...	...	...	...
101	?	?	?	?	?	Yes
102	?	?	?	?	?	No
103	?	Sunny	?	High	?	?
104	?	Sunny	?	High	?	No
105	?	Sunny	?	Normal	?	Yes
106	?	Overcast	?	?	?	Yes
107	?	Rain	?	?	Strong	No
108	?	Rain	?	?	Weak	Yes
109	?	Rain	?	High	?	No
110	?	Rain	?	Normal	?	Yes
...	...	...	...	...	...	...

Fig. 11.4 Mining views for
representing itemsets and association rules. The attributes
cida, *cidc*, and *cid* refer to
concepts given in Figure 11.3.

T_Sets

cid	supp	sz
102	5	1
103	3	2
104	3	3
...	...	...

T_Rules

rid	cida	cidc	cid	conf
1	103	102	104	100%
...	...	...	...	...

Figure 11.4 shows the mining views *T_Sets* and *T_Rules*, and illustrates how
the rule "if outlook is sunny and humidity is high, you should not play tennis" is
represented in these views by using three of the concepts given in Figure 11.3.

In Figure 11.5, queries (A) and (B) are example mining queries over itemsets and
association rules, respectively. Query (A) asks for itemsets having support of at least
3 and size of at most 5, while query (B) asks for association rules having support
of at least 3 and confidence of at least 80%. Note that these two common data
mining tasks and the well known constraints "minimum support" and "minimum
confidence" can be expressed quite naturally with SQL queries over the mining
views.

(A)	(B)

`select C.*, S.supp, S.sz`
`from T_Concepts C, T_Sets S`
`where C.cid = S.cid`
`  and S.supp >= 3`
`  and S.sz <= 5`
`  and C.Outlook = 'Sunny'`
	`select Ante.*, Cons.*,`
	`       S.supp, R.conf`
	`from T_Sets S, T_Rules R,`
	`   T_Concepts Ante,`
	`   T_Concepts Cons`
	`where R.cid = S.cid`
	`  and Ante.cid = R.cida`
	`  and Cons.cid = R.cidc`
	`  and S.supp >= 3`
	`  and R.conf >= 80`

Fig. 11.5 Example queries over itemsets and association rules.

11.2.2.2 Decision Trees

A decision tree learner typically learns a single decision tree from a dataset. This
setting strongly contrasts with discovery of itemsets and association rules, which is
set-oriented: given certain constraints, the system finds all itemsets or association
rules that fit the constraints. In decision tree learning, given a set of (sometimes
implicit) constraints, one tries to find one tree that fulfills the constraints and, besides

that, optimizes some other criteria, which are again not specified explicitly but are a consequence of the algorithm used.

In the inductive databases context, we treat decision tree learning in a somewhat different way, which is more in line with the set-oriented approach. Here, a user would typically write a query asking for all trees that fulfill a certain set of constraints, or optimizes a particular condition. For example, the user might ask for the tree with the highest training set accuracy among all trees of size of at most 5. This leads to a much more declarative way of mining for decision trees, which can easily be integrated into the mining views framework. The set of all trees predicting a particular target attribute A_i from other attributes is represented by the view:

$$T\_Trees\_A_i(treeid, cid).$$

The mining view $T\_Trees\_A_i$ is such that, for every decision tree predicting a particular target attribute A_i, it contains as many tuples as the number of leaf nodes it has. We assume that a unique identifier, $treeid$, can be given to each decision tree. Each decision tree is represented by a set of concepts cid, where each concept represents one path from the root to a leaf node.

Additionally, a view representing several characteristics of a tree learned for one specific target attribute A_i is defined as:

$$T\_Treescharac\_A_i(treeid, acc, sz).$$

It contains a tuple for every decision tree in $T\_Trees\_A_i$, where $treeid$ is the decision tree identifier, acc is its corresponding accuracy, and sz is its size in number of nodes.

Figure 11.6 shows how a decision tree that predict the attribute *Play* of table T is represented in the mining views T_Trees_Play and $T\_Treescharac\_Play$ by using the concepts in Figure 11.3.

In Figure 11.7, we present some example mining queries over decision trees. Query (C) creates a table called "BestTrees" with all decision trees that predict the attribute *Play*, having maximal accuracy among all possible decision trees of size of

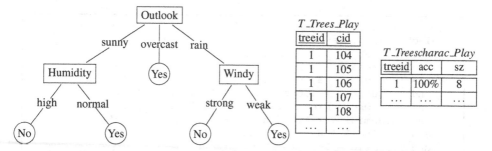

Fig. 11.6 Mining views representing a decision tree which predicts the attribute *Play*. Each attribute *cid* of view T_Trees_Play refers to a concept given in Figure 11.3.

at most 5. Query (D) asks for decision trees having a test on "Outlook=Sunny" and on "Wind=Weak", with a size of at most 5 and an accuracy of at least 80%.

```
(C)                                    (D)

                                      select T1.treeid,
                                              C1.*, C2.*
create table BestTrees as              from T_Trees_Play T1,
select T.treeid, C.*, D.*                   T_Trees_Play T2,
from T_Concepts C,                          T_Concepts C1,
     T_Trees_Play T,                        T_Concepts C2,
     T_Treescharac_Play D                   T_Treescharac_Play D
where T.cid = C.cid                    where C1.Outlook = 'Sunny'
  and T.treeid = D.treeid                and C2.Wind = 'Weak'
  and D.sz <= 5                          and T1.cid = C1.cid
  and D.acc =                            and T2.cid = C2.cid
    (select max(acc)                     and T1.treeid = T2.treeid
     from T_Treescharac_Play             and T1.treeid = D.treeid
     where sz <= 5)                      and D.sz <= 5
                                         and D.acc >= 80
```

Fig. 11.7 Example queries over decision trees.

Prediction. In order to classify a new tuple using a learned decision tree, one simply searches for the concept in this tree (path) that is satisfied by the new tuple. More generally, if we have a test set S, all predictions of the tuples in S are obtained by equi-joining S with the semantic representation of the decision tree given by its concepts. We join S to the concepts of the tree by using a variant of the equi-join that requires that either the values are equal, or there is a wildcard value.

Consider the table BestTrees created after the execution of query (C), in Figure 11.7. Figure 11.8 shows a query that predicts the attribute *Play* for all unclassified tuples in an example table Test_Set(*Day,Outlook,Temperature, Humidity,Wind*) by using the tree in table BestTrees that has identification number 1.

```
(E)
select S.*, T.Play
from Test_Set S,
     BestTrees T
where (S.Day = T.Day or T.Day = '?')
  and (S.Outlook = T.Outlook or T.Outlook = '?')
  and (S.Temperature = T.Temperature or T.Temperature = '?')
  and (S.Humidity = T.Humidity or T.Humidity = '?')
  and (S.Wind = T.Wind or T.Wind = '?')
  and T.treeid = 1
```

Fig. 11.8 An example prediction query.

11.2.3 Putting It All Together

For every data table $T(A_1, \ldots, A_n)$ in the database, with T having only categorical attributes, the virtual mining views framework consists of a set of relational tables, called virtual mining views, which virtually contain the complete output of data mining tasks executed over T. These mining views are the following:

- $T\_Concepts(cid, A_1, \ldots, A_n)$.
- $T\_Sets(cid, supp, sz)$.
- $T\_Rules(rid, cida, cidc, cid, conf)$.
- $T\_Trees\_A_i(treeid, cid)$, for all $i=1 \ldots n$.
- $T\_Treescharac\_A_i(treeid, acc, sz)$, for all $i=1 \ldots n$.

As shown in the examples given in this section, in order to retrieve patterns over table T, the user simply needs to write SQL queries over the proposed mining views. The semantics of these queries is the same as that of queries over traditional relational tables. For more example queries over the mining views, we refer the reader to [7].

Another important thing to note is that if the user wants to mine itemsets, association rules, or learn a decision tree from only a portion of table T, he or she should first create a new table T' from T, applying the appropriate selections and (or) projections. Then, the mining views associated with T', which are automatically created, will represent the patterns extracted from that corresponding portion of the data.

11.2.4 Mining Views vs. Data Mining Tasks

We now present how the 4 prototypical tasks described in Chapter 3 can be executed by SQL queries over the mining views.

11.2.4.1 Discretization task: Discretize attribute Temperature into 3 intervals. The discretized attribute should be used in the subsequent tasks

Since the data mining query language is SQL, our approach does not offer any new operator for pre-processing tasks. The discretization task can thus be performed by creating a new table called "MyPlaytennis" with the SQL CASE query introduced in Chapter 3 (when presenting the MINE RULE operator).

11.2.4.2 Area task: Find all intra-tuple itemsets with relative support of at least 20%, size of at least 2, and area, that is, absolut support × size, of at least 10.

The area task can be performed with an SQL query involving the mining views *MyPlaytennis_Concepts* and *MyPlaytennis_Sets*, which are created automatically after the creation of table MyPlaytennis for the discretization task. The query is shown below. Notice that the property area can be constrained quite naturally in our framework (see line 6), due to the flexibility of ad hoc querying.

```
1. select C.*, S.supp, S.sz,
            S.supp * S.sz as area
2. from MyPlaytennis_Sets S,
        MyPlaytennis_Concepts C
3. where C.cid = S.cid
4.   and S.supp >= 3
5.   and S.sz >= 2
6.   and S.supp * S.sz >= 10
```

11.2.4.3 Right hand side task: Find all intra-tuple association rules with relative support of at least 20%, confidence of at most 80%, size of at most 3, and a singleton right hand size.

Since the next task (lift task) requires a post-processing query over the results output by this one, it is necessary to store these results so that they can be further queried. The SQL query to perform the right hand side task is the following:

```
1. create table MyRules as
2. select Ant.Day as DayA, ... ,Ant.Play as PlayA,
            Con.Day as DayC, ..., Con.Play as PlayC,
            R.conf, SCon.supp/14 as suppC
3. from MyPlaytennis_Sets S, MyPlaytennis_Rules R,
        MyPlaytennis_Concepts Ant,
        MyPlaytennis_Concepts Con,
        MyPlaytennis_Sets SCon
4. where R.cid = S.cid
5.   and Ant.cid = R.cida
6.   and Con.cid = R.cidc
7.   and S.supp >= 3
8.   and R.conf >= 80
9.   and S.sz <= 3
10.  and SCon.cid = R.cidc
11.  and SCon.sz = 1
```

The query above creates a new table called "MyRules". We also store in this table the confidence of the rules along with the relative supports of their consequents, since they are necessary to perform the lift task (the number 14, which is used to compute the relative supports of the consequents, refers to the total number of tuples in table MyPlaytennis). Observe that the mining views framework does not restrain the user from any format in which the rules are to be stored, thanks again to the flexibility of ad hoc querying.

11.2.4.4 Lift task: Find, from the result of the right hand side task, rules with attribute Play as consequent that have a lift greater than 1.

In order to perform the lift task, one needs to query table MyRules, created for the previous task. The query in question is the one depicted below:

```
1. select M.*, (M.conf/100)/M.suppC as lift
2. from MyRules M
3. where M.PlayC <> '?'
4.     and (M.conf/100)/M.suppC >=1
```

Note that the two constraints required by the lift task can be expressed quite naturally in our framework. In line 3, we assure that the rules in the result have the attribute *Play* as consequent, i.e., it is not a wildcard value. In line 4, we compute the property lift of the rules.

11.2.5 Conclusions

Observe that the mining views framework is able to perform all data mining tasks described in Chapter 3 without any type of pre- or post-processing, as opposed to the other proposals. Also note that the choice of the schema for representing item-sets and association rules implicitly determines the complexity of the queries a user needs to write. For instance, by adding the attributes *sz* and *supp* to the mining views *T_Sets*, the area constraint can be expressed quite naturally in our framework. Without these attributes, one could still obtain their values. Nevertheless, it would imply that the user would have to write more complicated queries.

The addition of the attribute *cid* in the mining view *T_Rules* can be justified by the same argument. Indeed, one of the 3 concept identifiers for an association rule, *cid*, *cida* or *cidc* is redundant, as it can be determined from the other two. However, this redundancy eases query writing. Still with regard to the mining view *T_Rules*, while the query for association rule mining seems to be more complex than the queries for the same purpose in other data mining query languages (e.g., in MSQL), one could easily turn it into a view definition so that association rules can be mined with simple queries over that database view.

It is also important to notice that some types of tasks are not easily expressed with the mining views. For example, if the tuples over which the data mining tasks are to be executed come from different tables in the database, a new table containing these tuples should be created before the mining can start. In DMQL, MINE RULE, SPQL the relevant set of tuples can be specified in the query itself. In the case of DMX, this can be done while training the model. Another example is the extraction of inter-tuple patterns, which are possible to be performed with DMQL, MINE RULE, SPQL, and DMX. To mine inter-tuple patterns in the mining views framework, one would need to first pre-process the dataset that is to be mined, by changing its representation: the relevant attributes of a group of tuples should be added to a single tuple of a new table. Constraints on the corresponding groups of tuples being considered, which are allowed to be specified in the proposals mentioned above, can be specified in a post-processing step over the results. Our proposal is more related to MSQL and SIQL, as they also only allow the extraction of intra-tuple patterns over a single relation.

Some data mining tasks that can be performed in SIQL and DMX, such as clustering, cannot currently be executed with the proposed mining views. On the other hand, note that one could always extend the framework by defining new mining views that represent clusterings, as studied in [7]. In fact, one difference between our approach and those presented in Chapter 3 is the fact that to extend the formalism, it is necessary to define new mining views or simply add new attributes to the existing ones, whereas in other formalisms one would need to extend the language itself.

To finalize, although the mining views do not give the user the ability to express every type of query the user can think of (similarly to any relational database), the set of mining tasks that can be executed by the system is consistent and large enough to cover several steps in a knowledge discovery process.

We now list how the mining views overcome the drawbacks found in at least one of the proposals surveyed in chapter 3:

Satisfaction of the closure principle.   Since, in the proposed framework, the data mining query language is standard SQL, the closure principle is clearly satisfied.

Flexibility to specify different kinds of patterns.   The mining views framework provides a very clear separation between the patterns it currently represents, which in turn can be queried in a very declarative way (SQL queries). In addition to itemsets, association rules and decision trees, the flexibility of ad hoc querying allows the user to think of new types of patterns which may be derived from those currently available. For example, in [7] we show how frequent closed itemsets [26] can be extracted from a given table T with an SQL query over the available mining views $T\_Concepts$ and T_Sets.

Flexibility to specify ad hoc constraints.   The mining views framework is meant to offer exactly this flexibility: by virtue of a full-fledged query language that allows of ad hoc querying, the user can think of new constraints that were not considered at the time of implementation. An example is the constraint lift, which could be computed by the framework for the execution of the lift task.

Intuitive way of representing mining results. In the mining views framework, patterns are all represented as sets of concepts, which makes the framework as generic as possible, not to mention that the patterns are easily interpretable.

Support for post-processing of mining results. Again, thanks to the flexibility of ad hoc querying, post-processing of mining results is clearly feasible in the mining views framework.

11.3 An Illustrative Scenario

One of the main advantages of our system is the flexibility of ad hoc querying, that is, the user can iteratively specify new types of constraints and query the patterns in combination with the data themselves. In this section, we illustrate this feature with a complete data mining scenario that consists in extracting knowledge from real gene expression data, after an extensive pre-processing phase. Differently to the scenario presented in [6], here we do not learn a classifier, but mine for non-redundant correct association rules.

We begin by presenting how the implementation of our inductive database system was realized. Next, the aforementioned scenario is presented.

11.3.1 Implementation

Our inductive database system was developed into the well-known open source database system PostgreSQL[2], which is written in C language. Every time a data table is created into our system, its mining views are automatically created. Accordingly, if this data table is removed from the system, its mining views are deleted as well.

The main steps of the system are illustrated in Figure 11.9. When the user writes a query, PostgreSQL generates a data structure representing its corresponding relational algebra expression. A call to our Mining Extension was added to PostgreSQL's source code after the generation of this data structure. In the Mining Extension, which was implemented in C language, we process the relational algebra structure. If it refers to one or more mining views, we then extract the constraints (as described in detail in [7]), trigger the data mining algorithms and materialize the virtual mining views with the obtained mining results. Just after the materialization (i.e., upon return from the miningExtension() call), the work-flow of the database system continues and the query is executed as if the patterns or models were there all the time. We refer the reader to [5, 7] for more details on the implementation and efficiency evaluation of the system.

---

[2] http://www.postgresl.org/

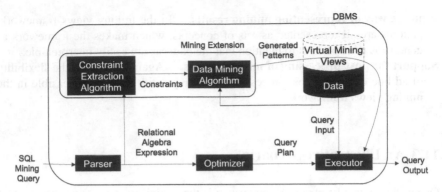

Fig. 11.9 The proposed inductive database system implemented into PostgreSQL.

Additionally, we adapted the web-based administration tool PhpPgAdmin[3] so as to have a user-friendly interface to the system.

11.3.2 Scenario

The scenario presented in this section consists in extracting knowledge from the gene expression data which resulted from a biological experimentation concerning the transcription of *Plasmodium Falciparum* [27] during its reproduction cycle (IDC) within the human blood cells.

The Plasmodium Falciparum is a parasite that causes human malaria. The data gather the expression profiles of 472 genes of this parasite in 46 different biological samples.[4] Each gene is known to belong to a specific biological function. Each sample in turn corresponds to a time point (hour) of the IDC, which lasts for 48 hours. During this period, the merozoite (initial stage of the parasite) evolves to 3 different identified stages: Ring, Trophozoite, and Schizont. In addition, due to reproduction, one merozoite leads to up to 32 new ones during each cycle, after which a new developmental cycle is started. Figure 11.10 shows the percentage of parasites (y-axis) that are at the Ring (black curve), Trophozoite (light gray curve), or Schizont (dark gray curve) stage, at every time point of the IDC (x-axis).

These data were stored into 3 different tables in our system, as illustrated in Figure 11.11. They are the following:

- GeneFunctions(function_id, function): represents the biological functions. There are in total 12 different functional groups.
- Samples(sample_name, stage): represents the samples themselves. Two data points are missing, namely the 23rd and 29th hours. We added to this table the

---

[3] http://phppgadmin.sourceforge.net/

[4] The data is available at http://malaria.ucsf.edu/SupplementalData.php

Fig. 11.10 Major developmental stages of Plasmodium Falciparum parasite (Figure from [27]). The three curves, in different levels of gray, represent the percentage of parasites (y-axis) that are at the Ring (black), Trophozoite (light gray), or Schizont stage (dark gray), at every time point of the IDC (x-axis).

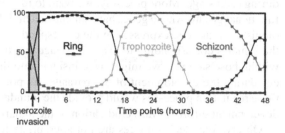

attribute called *stage*, the values of which are based on the curves illustrated in Figure 11.10: this new attribute discriminates the samples having at least 75% of the parasites in the Ring (stage=1), Trophozoite (stage=2) or Schizont (stage=3) stage. Samples that contain less than 75% of any parasite stage were assigned to stage 4, a "non-identified" stage. Thus, for our scenario, stage 1 corresponds to time points between 1 and 16 hours; stage 2 corresponds to time points between 18 and 28 hours; and stage 3 gathers time points between 32 and 43 hours.

- Plasmodium(gene_id, function_id, tp_1, tp_2,..., tp_48): represents, for each of the genes, its corresponding function and its expression profile. As proposed in [27], we take the logarithm to the base 2 of the raw expression values.

Plasmodium

gene_id	function_id	tp_1	tp_2	...	tp_48
1	12	-0.13	0.12	...	0.11
2	12	0.24	0.48	...	-0.03
...	...	...	...	...	...
472	5	1.2	0.86	...	1.15

GeneFunctions

function_id	function
1	Actin_myosin_mobility
2	Cytoplasmic_translation_machinery
...	...
12	Transcription_machinery

Samples

sample_name	stage
tp_1	1
tp_2	1
...	...
tp_48	4

Fig. 11.11 The Plasmodium data.

Having presented the data, we are now ready to describe the goal of our scenario. In gene expression analysis, a gene is said to be highly expressed, according to a biological sample, if there are many RNA transcripts in the considered sample. These RNA transcripts can be translated into proteins, which can, in turn, influence the expression of other genes. In other words, it can make other genes also highly expressed. This process is called *gene regulation* [27].

In this context, analogously to what the biologists have studied in [27], we want to characterize the parasite's different stages by identifying the genes that are active during each stage. More precisely, we want to identify, for each different stage, the functional groups whose genes have an unusual high level of expression or, as the biologists say, are overexpressed in the corresponding set of samples. By considering the samples corresponding to a specific stage and the genes that are overexpressed within those samples, we might have insights into the regulation processes that occur during the development of the parasite. As pointed out in [27], understanding these regulation processes would provide the foundation for future drug and vaccine development efforts toward eradication of the malaria.

Observe that decision trees are not appropriate for the analysis we want to perform; they are most suited for predicting, which is not our intention here. Therefore, in our scenario we mine for association rules. A couple of pre-processing steps have to be performed initially, such as the discretization of the expression values. These steps are described in detail in the first 3 subsequent subsections. The remaining subsections show how the desired rules can be extracted from the data.

11.3.2.1 Step 1: Pre-processing 1

Since our intention is to characterize the parasite's stages by means of the functional groups and not of the individual genes themselves, we first create a view on the data that groups the genes by the function they belong to. The corresponding pre-process query is shown below:[5]

```
1. create view PlasmodiumAvg as
2. select G.function,
3.            avg(p.tp_1) as tp_1,
4.            ...,
5.            avg(p.tp_48) as tp_48
6. from Plasmodium P, GeneFunctions G
7. where P.function_id = G.function_id
8. group by G.function
```

The view called "PlasmodiumAvg" calculates, for every different functional group, the average expression profile (arithmetic mean) over all time points (see lines from 2 to 5).

11.3.2.2 Step 2: Pre-processing 2

Since we want the functional groups as components of the desired rules (antecedents and/or consequents), it is therefore necessary to transpose the view PlasmodiumAvg, which was created in the previous step. In other words, we need a new view in which

---

[5] For the sake of readability, ellipsis were added to some of the SQL queries presented in this section and in the following ones, which represent sequences of attribute names, attribute values, clauses etc.

the gene functional groups are the columns and the expression profiles are the rows. To this end, we use the PostgreSQL function called *crosstab*[6]. As *crosstab* requires the data to be listed down the page (not across the page), we first create a view on PlasmodiumAvg, called "PlasmodiumAvgTemp", which lists data in such format. The corresponding queries are shown below.

```
 1. create view PlasmodiumAvgTemp as
 2. select function as tid, `tp_1' as item,
                         tp_1 as val
 3. from PlasmodiumAvg
 4. union
 5. ...
 6. union
 7. select function as tid, `tp_48' as item,
                         tp_48 as val
 8. from PlasmodiumAvg
 9. create view PlasmodiumTranspose as
10. select * from crosstab
11. (`select item, tid, val from PlasmodiumAvgTemp
          order by item',
      `select distinct tid from PlasmodiumAvgTemp
          order by item')
12. as (sample_name text, Actin_myosin_mobility real,
                      ...,
                 Transcription_machinery real)
```

11.3.2.3 Step 3: Pre-processing 3

Having created the transposed view PlasmodiumTranspose, the third and last pre-processing step is to discretize the gene expression values so as to encode the expression property of each functional group of genes.

In gene expression data analysis, a gene is considered to be overexpressed if its expression value is high with respect to its expression profile. One approach to identify the level of expression of a gene is the method called $x\%$ *cut-off*, which was proven to be successful in [28]: a gene is considered overexpressed if its expression value is among the $x\%$ highest values of its expression profile, and underexpressed otherwise. With $x=50$, a gene is tagged as overexpressed if its expression value is above the median value of its profile.

As in this scenario the data are log transformed (very high expression values are deemphasized), the distribution of the data is symmetrical and, therefore, median expression values are very similar to mean values. As computing the mean value is straightforward in SQL and as we are not dealing with genes independently, but with

---

[6] We refer the reader to http://www.postgresql.org/docs/current/static/ tablefunc.html for more details on the *crosstab* function.

groups of genes, we use a slight adaptation of the *50% cut-off* method: we encode
the overexpression property by comparing it to the mean value observed for each
group, rather than the median. We first create a view, called "PlasmodiumTrans-
poseAvg", which calculates, for every group of genes, its mean expression value.
This computation is performed by the following query:

```
1. create view PlasmodiumTransposeAvg as
2. select avg(Actin_myosin_mobility) as avg_Actin_mm,
3. ...
4. avg(Transcription_machinery) as avg_Tran_m
5. from PlasmodiumTranspose
```

Afterwards, we create the new table named "PlasmodiumSamples" applying the
aforementioned discretization rule. The query that performs this discretization step
is shown below. Notice that the attribute *stage* is also added to the new table PlasmodiumSamples (see line 2).

```
 1. create table PlasmodiumSamples as
 2. select P.sample_name, S.stage,
 3. case when P.Actin_myosin_mobility > avg_Actin_mm
 4.     then 'overexpressed'
 5. else
 6.     'underexpressed'
 7. end as Actin_myosin_mobility,
 8. ...
 9. case when P.Transcription_machinery > avg_Tran_m
10.     then 'overexpressed'
11. else
12.     'underexpressed'
13. end as Transcription_machinery
14. from PlasmodiumTranspose P,
            PlasmodiumTransposeAvg,
            Samples S
15. where P.sample_name = S.sample_name
16. order by S.stage
```

11.3.2.4 Step 2: Mining over Association Rules

After creating the table PlasmodiumSamples, in this new step, we search for the
desired rules. The corresponding query is shown below:

```
 1. create table RulesStage as
 2. select R.rid, S.sz, S.supp, R.conf,
            CAnt.stage as stage_antecedent
            CCon.*
 3. from PlasmodiumSamples_Sets S,
PlasmodiumSamples_Sets SAnt,
            PlasmodiumSamples_Concepts CAnt,
            PlasmodiumSamples_Concepts CCon,
            PlasmodiumSamples_Rules R
 4. where R.cid = S.cid
 5.    and CAnt.cid = R.cida
 6.    and CCon.cid = R.cidc
 7.    and S.supp >= 10
 8.    and R.conf = 100
 9.    and R.cida = SAnt.cid
10.    and SAnt.sz = 1
11.    and CAnt.stage <> '?'
12. order by Ant.stage
```

As we want to characterize the parasite's stages themselves by means of the gene functions, we look for rules having only the attribute *stage* as the antecedent (see lines 9, 10 and 11) and gene function(s) in the consequent. Additionally, since we want to characterize the stages without any uncertainty, we only look for correct association rules, that is, rules with a confidence of 100% (see line 8). Finally, as the shortest stage is composed of 10 time points in total (not considering the dummy stage), we set 10 as the minimum support (line 7). The 381 resultant rules are eventually stored in the table called "RulesStage" (see line 1).

11.3.2.5 Step 3: Post-processing

The previous query has generated many redundant rules [29]: for each different antecedent, all rules have the same support and 100% confidence. Notice, however, that as we are looking for all gene groups that are overexpressed according to a given stage, it suffices to analyze, for each different stage (antecedent of the rules), only the rule that has maximal consequent. Given this, all one has to do is to select, for each different stage, the longest rule. The corresponding query is presented below. The sub-query, in lines from 3 to 5, computes, for every antecedent (stage), the maximal consequent size.

```
1. select R.*
2. from RulesStage R,
3. (select max(sz) as max_sz,
        stage_antecedent
4. from RulesStage
5. group by stage_antecedent) R1
6. where R.sz = R1.max_sz
7.   and R.stage_antecedent = R1.stage_antecedent
```

The 3 rules output by the last query are presented in Figure 11.12. As shown in Figure 11.13, they are consistent with the conclusion drawn in the corresponding biological article [27]. Each graph in Figure 11.13, from B to M, corresponds to the average expression profile of the genes of a specific functional group (the names of the functions are shown at the bottom of the figure). The functions are ordered, from left to right, with respect to the time point when there is a peak in their expression profiles (the peak value is shown in parentheses) and they are assigned to the parasite's stage during which this peak occurs (the name of the stages are presented at the top of the figure). Observe that, according to Figure 11.13, the functions *Early ring transcripts* and *Transcription machinery* are related to the early Ring and Ring stages, which is in fact indicated by the first extracted rule. The *Glycolytic pathway*, *Ribonucleotide synthesis*, *Deoxynucleotide synthesis*, *DNA replication*, and *Proteasome* are related to the early Trophozoite and the Trophozoite stages, which is also consistent with the second extracted rule. Finally, *Plastid genome*, *Merozoite Invasion*, and *Actin myosin mobility* have been associated to the Schizont stage by the biologists, which is indeed consistent with the third extracted rule.

antecedent (stage)	consequent	
	overexpressed	underexpressed
Ring	Early ring transcripts Transcription machinery	Deoxynucleotide synthesis DNA replication machine Plastid genome TCA cycle
Trophozoite	Glycolytic pathway Ribonucleotide synthesis Deoxynucleotide synthesis DNA replication Proteasome	Actin myosin motors Early ring transcripts Merozoite invasion
Schizont	Plastid genome Merozoite invasion Actin myosin mobility	Cytoplasmic translation machinery Ribonucleotide synthesis Transcription machinery

Fig. 11.12 Correct association rules with maximum consequent.

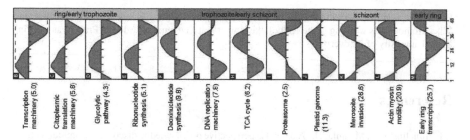

Fig. 11.13 The temporal ordering of functional groups of genes (an adapted figure from [27]). Each graph, from B to M, corresponds to the average expression profile of the genes of a specific functional group. The biologists of [27] have assigned each functional group to the parasite's stage in which it achieves its highest expression value.

11.4 Conclusions and Future Work

In this chapter, we described an inductive database system in which the query language is SQL. More specifically, we presented a system in which the user can query the collection of all possible patterns as if they were stored in traditional relational tables. The development of the proposed system was motivated by the need to (a) provide an intuitive framework that covers different kinds of patterns in a generic way and, at the same time, allows of (b) ad hoc querying, (c) definition of meaningful operations and (d) querying of mining results.

As for future work, we identify the following three directions:

- Currently, the mining views are in fact empty and only materialized upon request. Therefore, inspired by the work of Harinarayan et al. [30], the first direction for further research is to investigate which mining views (or which parts of them) could actually be materialized in advance. This would speed up query evaluation.
- Our system deals with intra-tuple patterns only. To mine inter-tuple patterns, one would need to first pre-process the dataset that is to be mined, by changing its representation. Although this is not a fundamental problem, this pre-processing step may be laborious. For example, in the context of market basket analysis, a table would need to be created in which each transaction is represented as a tuple with as many boolean attributes as are the possible items that can be bought by a customer. An interesting direction for future work would then be to investigate how inter-tuple patterns can be integrated into the system.
- Finally, the prototype developed so far covers only itemset mining, association rules and decision trees. An obvious direction for further work is to extend it with other models, taking into account the exhaustiveness nature of the queries the users are allowed to write.

Acknowledgements This work has been partially supported by the projects IQ (IST-FET FP6-516169) 2005/8, GOA 2003/8 "Inductive Knowledge bases", FWO "Foundations for inductive databases", and BINGO2 (ANR-07-MDCO 014-02). When this research was performed, Hendrik

Blockeel was a post-doctoral fellow of the Research Foundation - Flanders (FWO-Vlaanderen), Élisa Fromont was working at the Katholieke Universteit Leuven, and Adriana Prado was working at the University of Antwerp.

References

1. Imielinski, T., Mannila, H.: A database perspective on knowledge discovery. Communications of the ACM **39** (1996) 58–64
2. Calders, T., Goethals, B., Prado, A.: Integrating pattern mining in relational databases. In: Proc. ECML-PKDD. (2006) 454–461
3. Fromont, E., Blockeel, H., Struyf, J.: Integrating decision tree learning into inductive databases. In: ECML-PKDD Workshop KDID (Revised selected papers). (2007) 81–96
4. Blockeel, H., Calders, T., Fromont, E., Goethals, B., Prado, A.: Mining views: Database views for data mining. In: ECML-PKDD Workshop CMILE. (2007)
5. Blockeel, H., Calders, T., Fromont, E., Goethals, B., Prado, A.: Mining views: Database views for data mining. In: Proc. IEEE ICDE. (2008)
6. Blockeel, H., Calders, T., Fromont, E., Goethals, B., Prado, A.: An inductive database prototype based on virtual mining views. In: Proc. ACM SIGKDD. (2008)
7. Prado, A.: An Inductive Database System Based on Virtual Mining Views. PhD thesis, University of Antwerp, Belgium (December 2009)
8. Han, J., Fu, Y., Wang, W., Koperski, K., Zaiane, O.: DMQL: A data mining query language for relational databases. In: ACM SIGMOD Workshop DMKD. (1996)
9. Imielinski, T., Virmani, A.: Msql: A query language for database mining. Data Mining Knowledge Discovery **3**(4) (1999) 373–408
10. Meo, R., Psaila, G., Ceri, S.: An extension to sql for mining association rules. Data Mining and Knowledge Discovery **2**(2) (1998) 195–224
11. Wicker, J., Richter, L., Kessler, K., Kramer, S.: Sinbad and siql: An inductive databse and query language in the relational model. In: Proc. ECML-PKDD. (2008) 690–694
12. Bonchi, F., Giannotti, F., Lucchese, C., Orlando, S., Perego, R., Trasarti, R.: A constraint-based querying system for exploratory pattern discovery information systems. Information System (2008) Accepted for publication.
13. Tang, Z.H., MacLennan, J.: Data Mining with SQL Server 2005. John Wiley & Sons (2005)
14. Agrawal, R., Srikant, R.: Fast algorithms for mining association rules. In: Proc. VLDB. (1994) 487–499
15. Botta, M., Boulicaut, J.F., Masson, C., Meo, R.: Query languages supporting descriptive rule mining: A comparative study. In: Database Support for Data Mining Applications. (2004) 24–51
16. Han, J., Kamber, M.: Data Mining - Concepts and Techniques, 1st ed. Morgan Kaufmann (2000)
17. Han, J., Chiang, J.Y., Chee, S., Chen, J., Chen, Q., Cheng, S., Gong, W., Kamber, M., Koperski, K., Liu, G., Lu, Y., Stefanovic, N., Winstone, L., Xia, B.B., Zaiane, O.R., Zhang, S., Zhu, H.: Dbminer: a system for data mining in relational databases and data warehouses. In: Proc. CASCON. (1997) 8–12
18. Srikant, R., Agrawal, R.: Mining generalized association rules. Future Generation Computer Systems **13**(2–3) (1997) 161–180
19. Meo, R., Psaila, G., Ceri, S.: A tightly-coupled architecture for data mining. In: Proc. IEEE ICDE. (1998) 316–323
20. Mannila, H., Toivonen, H.: Levelwise search and borders of theories in knowledge discovery. Data Mining and Knowledge Discovery **1**(3) (1997) 241–258
21. Ng, R., Lakshmanan, L.V.S., Han, J., Pang, A.: Exploratory mining and pruning optimizations of constrained associations rules. In: Proc. ACM SIGMOD. (1998) 13–24

22. Pei, J., Han, J., Lakshmanan, L.V.S.: Mining frequent itemsets with convertible constraints. In: Proc. IEEE ICDE. (2001) 433–442

23. Bistarelli, S., Bonchi, F.: Interestingness is not a dichotomy: Introducing softness in constrained pattern mining. In: Proc. PKDD. (2005) 22–33

24. Mitchell, T.M.: Machine Learning. McGraw-Hill, New York (1997)

25. Gray, J., Chaudhuri, S., Bosworth, A., Layman, A., Reichart, D., Venkatrao, M.: Data cube: A relational aggregation operator generalizing group-by, cross-tab, and sub-total. Data Mining and Knowledge Discovery (1996) 152–159

26. Pasquier, N., Bastide, Y., Taouil, R., Lakhal, L.: Discovering frequent closed itemsets for association rules. In: Proc. ICDT. (1999) 398–416

27. Bozdech, Z., Llinás, M., Pulliam, B.L., Wong, E.D., Zhu, J., DeRisi, J.L.: The transcriptome of the intraerythrocytic developmental cycle of plasmodium falciparum. PLoS Biology 1(1) (2003) 1–16

28. Becquet, C., Blachon, S., Jeudy, B., Boulicaut, J.F., Gandrillon, O.: Strong association rule mining for large-scale gene-expression data analysis: a case study on human SAGE data. Genome Biology 12 (2002)

29. Zaki, M.J.: Generating non-redundant association rules. In: Proc. ACM SIGKDD. (2000) 34–43

30. Harinarayan, V., Rajaraman, A., Ullman, J.D.: Implementing data cubes efficiently. In: Proc. ACM SIGMOD. (1996) 205–216

Chapter 12
SINDBAD and SiQL: Overview, Applications and Future Developments

Jörg Wicker, Lothar Richter, and Stefan Kramer

Abstract The chapter gives an overview of the current state of the SINDBAD system and planned extensions. Following an introduction to the system and its query language SiQL, we present application scenarios from the areas of gene expression/regulation and small molecules. Next, we describe a web service interface to SINDBAD that enables new possibilities for inductive databases (distributing tasks over multiple servers, language and platform independence, ...). Finally, we discuss future plans for the system, in particular, to make the system more 'declarative' by the use of signatures, to integrate the useful concept of mining views into the system, and to support specific pattern domains like graphs and strings.

12.1 Introduction

Many of the recent proposals for inductive databases and constraint-based data mining focus on single pattern domains (such as itemsets or molecular fragments) or single tasks, such as pattern discovery or decision tree induction [20, 5, 7, 19, 13]. Although the closure property is fulfilled by many of those approaches, the possibilities of combining various techniques in multi-step and compositional data mining are rather limited.

In this chapter, we give an overview of the SINDBAD project that explores a different avenue. SINDBAD stands for structured inductive database development, with structured in the sense of SQL – structured query language. The project aims at the development of a prototype of an inductive database system that supports the most basic preprocessing and data mining operations, such that they can be combined more or less arbitrarily. One explicit goal of the project is to support

Jörg Wicker · Lothar Richter · Stefan Kramer
Technische Universität München, Institut für Informatik I12, Boltzmannstr. 3,
D-85748 Garching b. München, Germany
e-mail: {joerg.wicker,lothar.richter,stefan.kramer}@in.tum.de

the complete knowledge discovery process, from preprocessing to post-processing. Since it is at the moment far from clear what the requirements of a full-fledged inductive database will be, it is our belief that we can only find out by building prototype systems.

The research described here follows ideas worked out at the Dagstuhl perspectives workshop "Data Mining: The Next Generation" [1], where a system of types and signatures of data manipulation and mining operators was proposed to support compositionality in the knowledge discovery process. At the workshop, the idea of using the data as the bridge between the various operators was explicitly articulated. In this work, the main idea was to use the simplest possible signature (mapping tables onto tables) as a starting point for the exploration of more complex scenarios.

We started out with concrete scenarios for multi-step, compositional data mining and then identified the building blocks necessary for supporting them. The SINDBAD project started in 2004 as a students project of a group of six students. Since then, several diploma theses and follow-up student projects continued the development of the system.

For the development, various paradigms could have been adopted. In SINDBAD, we chose the relational model, as it possesses several desirable properties. First, closure can easily be achieved. Second, it allows handling collections of tuples conveniently and in a declarative manner. Third, the technology scales up well, and highly optimized implementations are available. Fourth, systems supporting (variants of) SQL are well-known and established, making it easier to get users acquainted with new querying facilities. Thus, we took the same approach as Meo *et al.* [20] and devised an extension of SQL, but for the most basic pre-processing and data mining techniques, discretization, feature selection, pattern discovery, clustering and classification. Similar approaches have been taken by Imielinski and Virmani [16] and Han *et al.* [15].

For a comprehensive discussion of these query languages and the current lack of preprocessing and post-processing primitives, we refer the reader to a survey by Boulicaut and Masson [17]. Another, more recent approach based on the relational model by Blockeel *et al.* [2, 3, 4] focuses on the automatic generation of mining views, in which the relevant parts are materialized on demand when a model is queried. The focus of the work is on the storage model and evaluation logic of data mining results. In contrast to mining views, SINDBAD uses an own query language, SiQL, to execute the mining operations. Mining views only use standard SQL queries to generate models and apply them (for details see section 12.5.2). SINDBAD also differs from related work [5] in – among other things – the support of pre-processing features. Also, it should be viewed as a prototype to be used for the exploration of concepts and requirements on such systems.

This chapter is organized as follows: After sketching the main ideas of the inductive query language, we present details of the SINDBAD implementation. Subsequently, we show how the query language can be used in three typical multi-step data mining scenarios [22, 25]. After this, we present an extension of SINDBAD, SINDBAD SAILS which implements a Web Service interface to SINDBAD [24]. In the subsequent section, we discuss the plans for future developments and extensions of SINDBAD, before we come to our conclusions.

12.2 SiQL

SiQL (structured inductive database query language), the query language of the SINDBAD system, is a straightforward extension of SQL. Instead of just adding complicated data mining operators to SQL, we focused on incorporating small, but extensible and adjustable operators that can be combined to build more complex functions. The query language supports the knowledge discovery process by a successive transformation of data. As each pre-processing and data mining operator returns a table, the queries can be nested arbitrarily, and the kind of compositionality needed in multi-step data mining can be achieved easily. As most queries return a table, these queries can be arbitrarily nested with SQL queries. For example, instead of table names, the query can be given a SELECT statement. Also, the result of a mining operation can be directly used in a SQL query.

The mining operators were designed in analogy to relational algebra and SQL: For instance, we made heavy use of the extend-add-as operator which adds the results of a data mining operation in terms of new columns to a relation (see below). Also, we devised a feature-select clause in analogy to the select clause. It selects certain features which fulfill a given condition, for example an information gain above a certain threshold or the best features with respect to the correlation coefficient to a given column.

12.2.1 Preliminaries

For every relation, we assume that an attribute heading as well as non-deletable and non-mutable tuple identifiers are given. Since in many of the envisaged applications rows and columns should be interchangeable, we included an operator for transposing a table. Table transposition is only possible if all attributes are of the same type. If a table is transposed, the tuple identifiers become attributes, and vice versa. If tables are joined, the new tuple identifiers are concatenations of the tuple identifiers of the tuples from the joined tables.[1]

Most of the query operators below can be parametrized. Parameters can be either passed directly, or set in a so-called configure clause (see Table 12.1). For the sake of simplicity, we did not include the parameters in the following definitions in Backus-Naur Form (BNF), and assume the parameters are set in a configure clause.

---

[1] For a more formal discussion of the operators, we have to refer to a previous paper [18].

Table 12.1 Main parameters to be set in a configure clause.

```
<configure-clause>        ::=
              configure <group-expression-value>;

<group-expression-value> ::=
              knn_k=<integer>                            |
              kmed_k=<integer>                           |
              apriori_minSupport=<float>                 |
              discretization_<disc-method-value>         |
              sampling_method=<sampling-method-value>    |
              sampling_percentage=<double>               |
              sampling_class_column=<string>

<disc-method-value>       ::=
              numofintervals=<integer>            |
              method=(frequency|width|manual)     |
              classColumn=<string>

<sampling-method-value>  ::=
              holdout            |
              leave-one-out
```

12.2.2 Main Ideas

Adopting the relational model, queries are mappings from relations onto a relation. We designed an extension of SQL (a superset of SQL) to support different kinds of preprocessing and data mining operations. Since every operator returns a table, queries can be arbitrarily nested. If we would like to work with more complex data, e.g., chemical compounds or substructures, we might handle tables of SMILES or SMARTS strings [6]. The mining operators were designed in analogy to relational algebra and SQL: For instance, we made heavy use of the extend-add-as operator and devised a feature-select clause in analogy to the select clause.

The results of mining operations applied to tables are again tables. For instance, the discretization and feature selection operators return modified tables. More importantly, the classification from a nearest-neighbor query can be added to a table as a new attribute. Similarly, clustering results (cluster membership) can simply be added to a table as a new attribute. The frequent itemsets are stored in a table, each itemset in one row, the columns in the same schema as the input relation. The values are Boolean giving information if the item is present in the itemset.

Since one goal of the project was to explore the power of compositionality in data mining, we chose the most basic building blocks and implemented one fundamental technique per category. For discretization, we implemented: equal-frequency/equal width, for feature selection: a filter approach based on information gain or variance, for pattern discovery: APriori, for clustering: k-Medoids and for classification: k-Nearest Neighbor, for model learning: a propositional FOIL [21] variant. The resulting rule set is stored in a table. The columns representing literals, each row one rule. This makes it possible to execute queries on the results of the mining operations. The goal is to support the whole knowledge discovery process, including pre-processing steps as discretization and feature selection. However, it is not our ambition to re-implement every technique ourselves, but to make the system extensible by design.

External tools can easily be integrated by declaring wrappers for exporting and importing tables as plug-ins. Still, every analysis step can be performed via queries from a command line interface. For instance, once a wrapper for molecular data is written and declared as a plug-in, we might run an external graph mining tool and import the results, e.g., a table of frequent or significant molecular substructures, or a table of the occurrences of substructures in small molecules.

12.2.3 The *extend add as* Query

We adopted the extend operator to add the results of the various data mining operations as new attributes to a relation. The extend operator, one of the simplest extensions of the original relational algebra proposal, adds computational capabilities to the algebra [6]. It computes a function for each tuple and adds the result as the value of a new attribute. Although the same functionality can be achieved with SQL, the introduction of the extend query makes it easier to execute complex operations on the data. The most general form of an extend clause is given as follows:

```
<extend-clause> ::=  extend <relation>
                         add <function>
                           as <att>
```

As an example, consider we want to add a new attribute gmwt to a table p, defined as the attribute weight multiplied by 454 [6]:

```
extend p add (weight*454) as gmwt
```

In SQL, extending a table by computed attributes can easily be achieved by the first part of a select statement (SELECT AS). All the data mining operations would then be treated in the same way as aggregate functions (e.g., SELECT ..., KMEDOIDS(*) AS ... FROM ...). Somewhat related, but conceptually different, is the ALTER TABLE operator in today's SQL systems that *changes the structure* of existing tables (ALTER TABLE ... ADD ...).

In SINDBAD, by contrast, the extend operator is modified in several ways and used directly in the query language. The complete syntax of the new operators in BNF is shown in Table 12.2. The operators support a variety of pre-processing and data mining operations.

Now we are going to explain some of the extension functions in more detail. *kmedoid* provides distance-based clustering, which can come in two flavors. If combined with *membership*, the new attribute values are the identifiers (integers greater than or equal to one) of the clusters the respective example falls into, whereas in

Table 12.2 The extend clause was adapted for clustering, sampling and k-Nearest Neighbor prediction. The last two clauses are variants of k-medoids and k-NN that might be useful in practice: The k-medoid clause returns a relation with the medoids only. The k-NN clause retrieves the closest instances from a relation for a given instance. Thus, the results of k-medoids or k-NN can be stored directly in an own table without adding it to the input relation.

```
<extend-clause> ::=
    extend <relation> add
    (
        kmedoid membership as <att>                                    |
        kmedoid centers as <att>                                      |
        knn prediction of <att> from <relation> as <att>|
        sample membership as <att>                                    |
        distances from <relation> [as <prefix-att>]      |
        covered by <relation> [as <prefix-att>]          |
        external <external-program> [<relation>]
                                    [as <prefix-att>]
    )

<kmedoid-clause> ::= kmedoid relation <relation>

<knn-clause> ::= <singleton-relation> knns from <relation>
```

combination with *centers* the value of the attribute indicates whether it is a medoid or not (one for centers, zero otherwise). Another, less space-intensive way is to use the *k-medoid* clause from Table 12.2, only returning a table of medoids. Even simpler, one could only return a table with the keys of the medoids. Another possibility (not implemented yet) is to return both cluster membership and centers to facilitate an easy reconstruction of clusters.[2]

A simple prediction method is included by *knn prediction of*. The class identified by the first attribute in the clause is predicted on the basis of training examples (the relation specified after the *from* keyword), and the resulting prediction is stored in the new attribute specified following *as*.

Particularly useful for testing purposes is the *sample membership* operation, which allows the user to split the set of examples into test and training set, simply indicated by zero or one values of the added attribute. Cross-validation is currently not supported, but will be integrated into one of the next versions of SINDBAD.

If distances to certain examples (one to many) are desired, they can be easily created by the *distances from* operation, which adds the distances from the examples in the given relation as new attributes, either with attribute names generated from the examples' identifiers or with a specified name prefix.

---

[2] Note that the user perspective need not coincide with the implementation perspective. We might use a very compact representation of clusters internally and present them to the user in a seemingly space-intensive way. Further, the main idea of SINDBAD is to transform data successively, and not to create too many extra tables containing results in various forms.

Table 12.3 The feature select clause, reminiscent of the select clause in SQL.

```
<feature-select-clause> :: =
                feature select <conditions-on-tuples>
                from <relation>
                where <fs-condition>

<fs-condition> ::= ((variance | infogain <att>)
                   (( <|>|=|<=|>=) <real> |
                    in top <integer>)            |
                   <attribute-condition-expression>)
```

To use the frequent itemsets generated by the APriori algorithm (see below), the *covered by* operation was included, that maps the occurrence of itemsets back to the examples.

The genuine extensibility of the system comes into play with the *external* keyword. This is not merely an operator transforming the data, but rather indicates an external plug-in to the system, whose results are used as input for the new values of the attribute.

12.2.4 The *feature select* Query

In Table 12.3, several variants of feature selection are offered, which is an indispensable step in the knowledge discovery process. Feature selection can be done according to various criteria. These criteria are specified in the *<fs-condition>*. Feature selection can be done either by applying hard thresholds for variance or information gain, or by relative thresholds (*in top*). Alternatively, simple string matching over the attributes' names can be applied, where the keyword *attribute* is used to refer to attribute names (see Table 12.4).

Table 12.4 Various other operators, for discretization, pattern discovery, table transposition and projection on another table's attributes.

```
<disc-clause> ::=          discretize (* | <att-list>)
                           in <relation>

<pattern-disc-clause> ::=  frequent itemsets
                           in <relation>

<transpose-clause> ::=     transpose <relation>

<project-onto-clause> ::=  project <relation>
                           onto <relation> attributes
```

In a way, the feature-select clause resembles the select clause "rotated by 90 degrees". However, in the feature-select clause, we can apply criteria for attributes to be included, and need not specify explicit lists of attributes.[3]

12.2.5 Parsing and Executing SiQL Queries

The SINDBAD prototype is implemented in Java. For parsing the queries, we used the lexical analyzer generator JFlex[4] and the parser generator Cup[5]. The implementation supports arbitrarily nested queries. In the future, we are planning to integrate a full-fledged analysis of parse trees, opening possibilities for query optimization. The system is built on top of PostgreSQL[6], an open source relational database management system. The queries are analyzed, the SQL parts of the query are redirected to PostgreSQL, the SiQL queries are handled by the Java implementation. Most of the inductive queries are broken down and translated into a larger number of less complex non-inductive queries. The implementation of data mining features as PostgreSQL functions seems to be critical for performance.

12.3 Example Applications

In this section, we will highlight some of the main features of SINDBAD in three real-world applications [22, 25]. In the first application, we test it on the gene expression data from Golub *et al.* [14], which contains the expression levels of genes from two different types of leukemia. In the second application, the task is to predict gene regulation dependent on the presence of binding sites and the state of regulators [12]. The third application is to predict anti-HIV activity for more than 40,000 small molecules [19].

12.3.1 Gene Expression Analysis

We aim at finding a classifier that predicts the cancer type either acute myeloid leukemia (AML) or acute lymphoblastic leukemia (ALL) based on gene expression monitoring by DNA microarrays. Table 12.5 shows how the AML/ALL gene expression dataset is analyzed step by step. The input relation contains attributes

---

[3] In principle, it would be desirable to support arbitrary Boolean expressions (analogously to the select clause [6], pp. 973-976), composed of syntactic criteria regarding the attribute name as well as criteria regarding the variance or information gain of the attribute.

[4] http://jflex.de/

[5] http://www2.cs.tum.edu/projects/cup/

[6] http://www.postgresql.org/

stating the expression levels of the genes (that is, one attribute per gene) and one class attribute, which gives the actual tumor subtype (AML or ALL) of the cell. Table 12.6 shows the input and output of the system without displaying the actual relations.

First, the dataset is loaded, discretized and divided into a training and a test set (queries (10) to (13)). Note that the discretization and labeling as training or test example is done in the second query. The `sample membership` statement conceptually splits a set of examples into two subsets, simply indicated by an additional attribute containing either the value zero or one. The following two queries split the table into two tables based on the previously added information. Queries (14) perform class-sensitive feature selection. As a result, we reduce the dataset to the fifty genes with maximal information gain with respect to the tumor subtype to be predicted. Since the test set should have the same attributes as the training set, we project the former onto the attributes of the latter in query (15).

Next, we query for frequent itemsets, that is, co-expressed genes. The co-expressed genes are used to transform the data, because individual genes are usually only predictive in conjunction with other genes. In the following queries, one new attribute per frequent itemset is added to training (17) and test table (18), which specifies which gene occurs in which frequent item set. Then, it uses feature selection to remove the original expression attributes. In this way, each example is represented only by attributes indicating co-expression with other genes. Finally, query (19) induces a k-nearest neighbor classifier on the training table and applies it to the examples in the test table. The predictions are added to the test table as values of the new attribute `predicted_tumor_subtype`. More generally, the k-nn clause adds the values of a predicted attribute to a given test set on the basis of a target attribute of a given training set:

```
extend <testset>
  add knn prediction
    of <targetatt>
    from <trainset>
  as <predictatt>
```

12.3.2 Gene Regulation Prediction

In the following, we briefly demonstrate multi-relational clustering and classification on gene regulation data [12]. Gene expression is the complex process of conversion of genetic information into resulting proteins. It mainly consists of two steps: transcription and translation. Transcription is the copying of a DNA-template in mRNA mediated by special proteins, so called transcription factors. Translation is the protein formation based on the information coded in the mRNA.

The data used here in this work reflects regulatory dependencies involved in the step of transcription. The task on this data is to learn a model that predicts the level of gene expression, i.e., to learn under which experimental conditions a gene is up-

Table 12.5 Sample run of SINDBAD on leukemia gene expression dataset

```
(10) create table expression_profiles as
        import ALLAML.arff;

(11) create table train_test_expression_profiles as
        extend (discretize * in expression_profiles)
        add sample membership as test_flag;

(12) create table train_expression_profiles as
        select * from
            train_test_expression_profiles
        where test_flag = true;

(13) create table test_expression_profiles as
        select * from
            train_test_expression_profiles
        where test_flag = false;

(14) create table reduced_train_expression_profiles as
        feature select * from
            train_expression_profiles
        where infogain tumor_subtype in top 50;

(15) create table reduced_test_expression_profiles as
        project test_expression_profiles onto
            reduced_train_expression_profiles attributes;

(16) create table coexpressed_genes as
        frequent itemsets in reduced_train_expression_profiles;

(17) create table train_set as
        feature select * from
          (extend reduced_train_expression_profiles
            add covered by coexpressed_genes as fp)
        where attribute like 'fp%' or
            attribute = 'tumor_subtype';

(18) create table test_set as
        feature select * from
          (extend reduced_test_expression_profiles
            add covered by coexpressed_genes as fp)
        where attribute like 'fp%' or
            attribute = 'tumor_subtype';

(19) create table classified_test_expression_profiles as
        extend test_set
          add knn prediction of tumor_subtype
          from train_set
        as predicted_tumor_subtype;
```

Table 12.6 Sample outputs of SINDBAD models and patterns from leukemia gene expression dataset

```
(1) select * from coexpressed_genes;

    sindbadrownames | gene_id72 | gene_id77 | gene_id716 | ...
    ----------------+-----------+-----------+------------+----
          itemset1  |         f |         t |          t | ...
          itemset2  |         t |         f |          f | ...
            ...     |       ... |       ... |        ... | ...

(2) select predicted_tumor_subtype from
        classified_test_expression_profiles;

    predicted_tumor_subtype
    --------------------------
    ALL
    AML
    AML
    ...
```

or down-regulated. The expression level of a gene depends on certain experimental conditions and properties of the genes such as the presence of transcription factor binding sites, functional categorizations, and protein-protein interactions.

The data is represented in five relations (see Table 12.7). The main table stores the gene identifiers and their expression level, as well as an identifier of the experimental condition. The experimental conditions are given in two separate relations, information about the genes and their interactions to each other.

Table 12.7 Relational schema of gene regulation data. The relation gene is the main table and connects genes with experimental setups and expression levels. The fun_cat relation gives the functional category membership of a gene according to the FunCat database. The third relation, has_tfbs indicates occurrence of transcription factor binding sites in respective genes, whereas in the regulators table experimental conditions and activated regulators are given. The last table p_p_interaction gives the gene product interaction data.

```
    gene(             fun_cat(            p_p_interaction(
        gene_id,          gene_id,            gene1_id,
        cond_id,          fun_cat_id)         gene2_id)
        level)
    has_tfbs(          regulators(
        gene_id,          cond_id,
        yaac3_01,         yb1005w,
        yacc1_01,         ycl067c,
        yacs1_07,         ydl214c,
        yacs2_01,         ydr277c,
        ...)              ...)
```

Given this input, we can compute the similarity of gene-condition pairs using multi-relational distance measures. The results of k-medoids clustering is shown in Table 12.8.

The results of k-nearest neighbor classification is shown in Table 12.9. The target attribute in this case is the increase or decrease in expression level. The class attribute is set to $+1$ if the expression is above a certain threshold, and -1 if it is below. A gene in an experiment is represented by the functional category membership of a gene according to FunCat, occurrence of certain transcription factor binding sites, the experimental conditions causing the gene to over- or under-express, activated regulators and gene product interactions.

K-nearest neighbor is configured for $k = 10$ and the "majority wins" strategy for prediction. This is a good example for the advantage of the support of multi-relational distance measures over simple propositional distance measures. Multi-relational distances make it possible to analyze complex data with algorithms designed for propositional data in an easy and transparent way without further modifications. Six multi-relational distance measures are currently implemented: single linkage, complete linkage, average linkage, sum of minimum distances, Hausdorff distance and matching distance. Which is used in the query can be set using a `configure` clause. The connections between the tables are read from the database schema. Each connection is represented by key constraints in PostgreSQL.

Table 12.8 k-Medoids for gene regulation prediction. The resulting table shows in column 2 the gene identifiers, in column 3 the experimental conditions, followed by the change of expression level and the cluster membership in columns 3 and 4.

```
(20)  configure kmedoids_k = 5;

(21)  configure multirelational_recursion_depth = 3;

(22)  configure multirelational_exclude_tables = '';

(23)  configure distance_between_instances = 'euclidean';

(24)  configure distance_between_instance_sets ='single_linkage';

(25)  extend gene add k medoid membership of gene;

(26)  show table gene;

row|gene_id|cond_id                  |level|cluster|
1   |YAL003W|2.5mM DTT 120 m dtt-1    |-1   |2      |
2   |YAL005C|2.5mM DTT 180 m dtt-1    |-1   |3      |
3   |YAL005C|1.5 mM diamide (20 m)    |+1   |5      |
4   |YAL005C|1.5 mM diamide (60 m)    |+1   |1      |
5   |YAL005C|aa starv 0.5 h           |-1   |2      |
...|...     |...                      |...  |...    |
```

Table 12.9 k-nearest neighbor for gene regulation prediction. This resulting table, column 2 and 3 are the same as in Table 12.8 followed by the predicted class label in column 4.

```
(30)  configure KNearestNeighbour_K = 10;

(31)  extend gene_test add knn prediction
         of level from gene_train;

(32)  show table gene_test;

row|gene_id|cond_id                       |class|
 1  |YBL064C|aa starv 1 h                  |+1   |
 2  |YDL170W|YPD 3 d ypd-2                 |-1   |
 3  |YER126C|Heat shock 40 minutes hs -1  |-1   |
 4  |YJL109C|dtt 240 min dtt-2             |+1   |
 5  |YKL180W|Nitrogen Depletion 1 d        |+1   |
...|...    |...                           |...  |
```

12.3.3 Structure-Activity Relationships

In the last application, we predict the anti-HIV activity of small molecules using the NCI Developmental Therapeutics Program HIV data [19]. Here, the AIDS antiviral screen data of the National Cancer Institute[7] is used. The data is a collection of about 43,000 chemical structures, which are labeled according to how effectively they protect human CEM cells from HIV-1 infection [23]. We search the data for rules describing a compound's activity against HIV. Hence, the data is prepared and randomly split into test and training set. We chose a representation where each attribute in the training and test relation specifies whether or not a chemical substructure occurs in a substance. The attributes are named f1 to f688, each of them representing a chemical substructure occurring in the antiviral screen data. The numbers refer to the order of their detection by the tree mining algorithm which searched these frequent subtrees in the data set.

An additional attribute gives the target label, that is, the compound's effectiveness in protecting against HIV. In Table 12.10, a protocol of the analysis steps is shown. In the first few queries the datasets are prepared and the FOIL rule induction algorithm [21] is configured (40)-(46). In the main step, rules are learned (47) and displayed (48). The sample rule refers to substructures 683, 262, 219, and 165 to predict a compound as active. Finally, the rule set is applied to a test set, adding its predictions as an additional attribute (49).

---

[7] http://dtp.nci.nih.gov/docs/aids/aids_screen.html

Table 12.10 Rule learning applied to the NCI HIV data.

```
(40) configure sampling_method = 'holdout';

(41) configure sampling_percentage = '0.25';

(42) configure sampling_classcolumn =
        'activity';

(43) create table hiv_train_test as
        extend hiv
        add sample membership as test_flag;

(44) create table testset as
        select * from hiv
        where  test_flag = false;

(45) create table trainset as
        select * from hiv
        where  test_flag = true;

(46) configure foil_mdl = 'true';

(47) create table hiv_rules as learn rules
        for activity in  trainset;

(48) show table hiv_rules;

(activity = true  <-  f1 = true AND
 f3 = true AND f4 =  true AND
 f165 = true)
...

(49) extend testset add
        model prediction of
        hiv_rules
        as learned_activity;
```

Table 12.11 Learned rule from the NCI HIV data set in the PostgreSQL table.

```
(1) select * from hiv_rules;

 activity | f1  | f2  | f3  | f4  | ...
----------+-----+-----+-----+-----+----
    true  | t   |     | t   | t   | ...
    ...   | ... | ... | ... | ... | ...
```

12.4 A Web Service Interface for SINDBAD

To show the benefits of using an inductive database for service-oriented knowledge discovery, in this section, we present a Web Service interface to SINDBAD [24]. Using this interface, all features of SINDBAD can be made available on a dedicated server. In this way, SINDBAD data mining services can be started from arbitrary clients. It is possible to distribute tasks over multiple servers to decrease the load on each machine. Another effect is the availability of features of SINDBAD in many different programming languages and on many different platforms. Thus, machine learning and data mining methods can be combined easily with native programming language constructs, such as conditional statements or loops, without having to install specialized libraries or software packages.

12.4.1 Web Services

A Service-Oriented Architecture (SOA) is a design paradigm for distributed computational resources, described by their capabilities and typically made accessible on the Internet [9]. Encapsulating functionality in an SOA, parts of a software system can be reused regardless of specific requirements on the underlying system, programming language or location of the provided service.

One possible implementation of an SOA is a Web Service. Web Services are offered on the Internet and can be accessed by the Simple Object Access Protocol (SOAP). The specification of a Web Service is split into three parts:

1. SOAP (Simple Object Access Protocol), an XML-based message format for the communication and embedding into transport protocols,
2. WSDL (Web Service Description Language), an XML-based description language to describe the Web Service, its interfaces and parameters, and
3. UDDI (Universal Description, Discovery and Integration Protocol) (optional), the directory service for Web Services, specifying the standardized directory structure for administration and search for Web Service meta-data.

12.4.2 Motivation

Building a Web Service on top of an inductive database offers many advantages: First of all, it is possible to run the (in most cases) computationally intensive operations on separate systems and distribute work that can be done simultaneously. As the computations are carried out on the server, the hardware requirements on the client are not very high. While this feature can be achieved by other implementations than a Web Service interface, common implementations in most cases require certain packages or software on the client machines. When using Web Services, this does not necessarily apply. In some cases it is beneficial to install packages to handle the access to the service. This depends on the used programming language on the clients. However, as the implementation of the service and the client are completely independent, it is up to the user which language, packages or software is used.

The distinction between the implementation of the data mining and preprocessing algorithms on the server running SINDBAD and the implementation of the user code on the client side makes it easier to use the data mining algorithms. The users do not need to know the details of the algorithms: They just need to submit the data in the right format and send it to the server.

The advantage of inductive databases compared to other possible implementations is due to the status of patterns and models in such systems. Just as regular data items, patterns and models are viewed as *first-class objects* in inductive databases. Taking advantage of a service-oriented architecture, it is possible to transfer data, patterns and models from one inductive database to another. In this way, distributing data mining tasks and integrating methods and results from multiple servers becomes an easy task.

Finally, the use of data mining and machine learning features in programming languages has not received much attention so far. Whereas machine learning is considered important in the context of reasoning, or more generally, artificial intelligence systems [8], the use of inductive queries in regular computer programs has not yet been discussed in the literature. The approach differs from R and MATLAB implementations and interfaces, and older libraries like MLC++, in its additional layers of abstraction (Web Service and SiQL). This level of abstraction in terms of well-defined interfaces can also be useful in workflow systems like KNIME[8], where components of workflows could be replaced as long as the interface is the same. Using SINDBAD SAILS, it is easily possible to use basic machine learning and data mining in (almost) arbitrary programming languages, without the need to install specialized software.

12.4.3 Features

A SINDBAD SAILS call can be split in several steps. Data mining methods are translated into SiQL queries and executed by the underlying SINDBAD database. The generated queries are passed to SINDBAD via command line, the output of SINDBAD is parsed for thrown exceptions and problems.Methods for uploading the data on and downloading results from the server are provided. The input for the algorithms is either uploaded from a given URL or a result of a former query on the database is used. The data is processed by the preprocessing and data mining algorithms and the results are stored on the server. They can be downloaded or used in further computations. Additionally, each intermediate result can be obtained from the server.

---

[8] http://www.knime.org

A sample call in Java is shown in Table 12.12. To connect to the Web Service, the Axis package of Apache is used[9]. First, a connection to the Web Service is established. Then the task is set to frequent itemset mining, which is done by the APriori algorithm. Finally, an URL of the data is sent to the service for download. With this URL, the minimum support and the level of detail for the results is passed to SINDBAD. If any errors occur during the execution of the APriori algorithm, an exception is thrown, containing a detailed error message from SINDBAD. The example is given in Java, but similar programs can be written in almost any other programming language.

Table 12.12 Simple Java example program using the Web Service interface of SINDBAD. In the example, the Axis package of Apache is used to establish a connection to the Web Service.

```
import org.apache.axis.client.Call;
import org.apache.axis.client.Service;
import javax.xml.namespace.QName;
public class SoapClient
{
    public static void main(String[] args) throws Exception
    {
        String endpoint = "http://sindbad.in.tum.de/soap";
        Service service = new Service();
        Call call = (Call) service.createCall();
        call.setTargetEndpointAddress(
                        new java.net.URL(endpoint) );
        call.setOperationName("frequentItemsets");
        Object[] returnset =
            (Object[]) call.invoke(
                new Object[]{
                    'http://wwwkramer.in.tum.de/soybean.arff',
                    '0.5',
                    'low'} );
    }
}
```

12.5 Future Developments

12.5.1 Types and Signatures

As mentioned above, the Dagstuhl perspectives workshop [1] introduced the concept of signatures of KDD operations. The signature of an operator prescribes the types of its inputs and its outputs. One obvious use of signatures in an inductive database

---

[9] http://ws.apache.org/axis/

is to prevent data mining operations being applied to data not suitable for them (for instance, itemset data cannot be discretized). Signatures can be used to help the user avoid wrong combinations of operations. To structure the data processing and mining operations conceptually, signatures can be organized in a hierarchy. The hierarchy starts with a general declaration at the top level, which becomes more specific on the lower levels. For instance, a signature may describe a mapping from data and patterns onto patterns, which is specialized for itemsets on a lower level.

At the workshop, three possible ways to organize data processing and mining operations were discussed:

Generic base operations   Mappings from data and patterns to data and patterns
Type of data   Pattern domains, for instance, items, strings, graphs, ...
Type of operation   For instance, clustering, classification, ...

All these concepts are conceivable to be included in SINDBAD. A hierarchy could be built with signatures for generic base operations on the top level. These could ensure that queries are only applied to the right type of input. Thus, data mining algorithms could only be applied to data, and patterns could only be used in post-processing in combination with data. On a lower level, signatures for the type of operation could provide better control over the specific data mining algorithms. For instance, the input data for the APriori algorithm need to be a relation holding itemsets.

12.5.2 Integration of Mining Views

Blockeel *et al.* presented an approach to integrate constraint-based data mining into a relational database [3]. Using the concept of *mining views*, they propose a schema to consistently represent a large number of models in a relational database. The approach is implemented in PostgreSQL using a virtual *Concepts* table. Using this table, constraints can be formulated using SQL queries. Association rule mining and decision tree learning have already been implemented within this framework. The implementation provides fast querying over the data and patterns.

Considering the implementation and its concepts, it appears to be feasible and reasonable to include mining views into SINDBAD. Nevertheless, there are still some open questions to be addressed. For instance, it is not clear whether it is preferable to include new query primitives for these functions or if everything should be accessed via SQL as proposed by Blockeel and co-authors. Introducing new query primitives could improve the user interface to this structure. The SQL queries to access the mining views tend to become quite complex. Also, good interoperability between the existing algorithms in SINDBAD and mining views needed to be elaborated.

12.5.3 String Mining

Although SINDBAD is intended as a general-purpose tool, it is still desirable to offer extensions for specific data types and pattern domains like strings (sequences) and graphs. These extensions are particularly useful in bioinformatics and cheminformatics applications. In preliminary work, we have already developed a basic concept for the integration of string mining algorithms and data structures into SINDBAD. As a use case, we focused on DNA and protein sequences. As it turns out, this requires only minor modifications for some of the operators, while it takes major changes for others. Minor modifications are needed, for instance, for the distance-based methods in SINDBAD. Here, only the definition of the distance measures needs to be adapted. For pattern mining, the query language is more or less the same as before, but implementations of suitable algorithms [10, 11] have to be incorporated. Major changes are required to handle operations that are specific to certain pattern domains, for instance, alignments (pairwise or multiple) or sequence profiles. Nevertheless, there are still open questions regarding such an effort. First of all, we have to develop an efficient way to store and access very large sequence data in our approach to inductive databases. For this purpose, efficient index structures for strings and sequences like suffix arrays could be taken into account, as done by Fischer *et al.* [10, 11]. Second, it has to be figured out whether and how string mining could be fit into the infrastructure of a relational database system.

12.6 Conclusion

The chapter gave an overview of the SINDBAD system, summarized a few recent use cases [22, 25], gave an overview of a web service based architecture for SINDBAD [24], and discussed possible extensions. The query language of the systems, SiQL, is an extension of SQL for inductive databases in the tradition of Imielinski and Virmani [16], Han *et al.* [15] and Meo *et al.* [20]. For a detailed comparison with commercial systems like MS SQL Server 2005, we refer to a previous article [18]. In short, SINDBAD focuses of the successive transformation of data, whereas SQL Server focuses almost exclusively on prediction. One of the main purposes of the SINDBAD system is to elucidate the requirements for inductive database systems in the relational model. Moreover, SINDBAD should be viewed as an integration effort, to provide various basic building blocks that can be plugged together almost arbitrarily in complex application scenarios. As the project progresses, we hope to be able to provide the research prototype as an open source implementation and come up with a (tentative) list of requirements on inductive database systems.

References

1. R. Agrawal, T. Bollinger, C.W. Clifton, S. Dzeroski, J.-C. Freytag, J. Gehrke, J. Hipp, D.A. Keim, S. Kramer, H.-P. Kriegel, B. Liu, H. Mannila, R. Meo, S. Morishita, R.T. Ng, J. Pei, P. Raghavan, R. Ramakrishnan, M. Spiliopoulou, J. Srivastava, V. Torra, and A. Tuzhilin. Data mining: The next generation. *Report based on a Dagstuhl perspectives workshop organized by R. Agrawal, J-C. Freytag, and R. Ramakrishnan*, 2005.
2. H. Blockeel, T. Calders, É. Fromont, B. Goethals, and A. Prado. Mining views: Database views for data mining. In *Proceedings of the International Workshop on Constrained-Bawsed Mining andLearning*, 2007.
3. H. Blockeel, T. Calders, É. Fromont, B. Goethals, and A. Prado. Mining views: Database views for data mining. In *Proceedings of the IEEE International Conference on Data Engineering*, 2008.
4. H. Blockeel, T. Calders, E. Fromont, B. Goethals, A. Prado, and C. Robardet. An inductive database prototype based on virtual mining views. In *KDD '08: Proceeding of the 14th ACM SIGKDD international conference on Knowledge discovery and data mining*, pages 1061–1064, New York, NY, USA, 2008. ACM.
5. M. Botta, Boulicaut J.-F., C. Masson, and R. Meo. Query languages supporting descriptive rule mining: A comparative study. In *Database Support for Data Mining Applications*, pages 24–51, 2004.
6. C. J. Date. *An Introduction to Database Systems*. Addison Wesley, 4th edition, 1986.
7. L. De Raedt and S. Kramer. The levelwise version space algorithm and its application to molecular fragment finding. In *Proc. 17th International Joint Conference on Artificial Intelligence (IJCAI 2001, Seattle, USA)*, pages 853–862. Morgan Kaufmann, San Francisco, CA, USA, 2001.
8. P. Domingos. Structured machine learning: Ten problems for the next ten years. In *Proceedings of Seventeenth International Conference on Inductive Logic Programming*, Corvallis, Oregon, 2007. Springer.
9. C. Ferris, D. Booth, M. Champion, H. Haas, D. Orchard, E. Newcomer, and F. McCabe. Web services architecture. W3C note, W3C, 2004. http://www.w3.org/TR/2004/NOTE-ws-arch-20040211/.
10. J. Fischer, V. Heun, and S. Kramer. Fast frequent string mining using suffix arrays. In *Proceedings of the Fifth IEEE International Conference on Data Mining*. IEEE Computer Society Press, 2005.
11. J. Fischer, V. Heun, and S. Kramer. Optimal string mining under frequency constraints. In *Proceedings of the 10th European Conference on Principles and Practice of Knowledge Discovery in Databases (PKDD 2006)*, pages 139–150, 2006.
12. S. Fröhler and S. Kramer. Inductive logic programming for gene regulation prediction. *Machine Learning*, 70(2-3):225–240, 2008.
13. M. Garofalakis, D. Hyun, R. Rastogi, and K. Shim. Efficient algorithms for constructing decision trees with constraints. In *KDD '00: Proceedings of the Sixth ACM SIGKDD International Conference on Knowledge Discovery and Data Mining*, pages 335–339, New York, NY, USA, 2000. ACM.
14. T.R. Golub, D.K. Slonim, P. Tamayo, P. Huard, M. Gaasenbeek, J.P. Mesirov, H. Coller, M.L. Loh, J.R. Downing, M.A. Caligiuri, C.D. Bloomfield, and E.S. Lander. Molecular classification of cancer: class discovery and class prediction by gene expression monitoring. *Science*, 286(5439):531–7, 1999.
15. J. Han, Y. Fu, W. Wang, K. Koperski, and O. Zaiane. DMQL: A data mining query language for relational databases. In *SIGMOD'96 Workshop on Research Issues in Data Mining and Knowledge Discovery (DMKD'96)*, Montreal, Canada, 1996.
16. T. Imielinski and A. Virmani. MSQL: A query language for database mining. *Data Min. Knowl. Discov*, 3(4):373–408, 1999.
17. Boulicaut J.-F. and C. Masson. Data mining query languages. In O. Maimon and L. Rokach, editors, *The Data Mining and Knowledge Discovery Handbook*, pages 715–727. Springer, 2005.

18. S. Kramer, V. Aufschild, A. Hapfelmeier, A. Jarasch, K. Kessler, S. Reckow, J. Wicker, and L. Richter. Inductive databases in the relational model: The data as the bridge. In Francesco Bonchi and Jean-François Boulicaut, editors, *Proceedings of the Fourth International Workshop on Knowledge Discovery in Inductive Databases (KDID 2005)*, volume 3933 of *Lecture Notes in Computer Science*, pages 124–138. Springer, 2005.

19. S. Kramer, L. De De Raedt, and C. Helma. Molecular feature mining in HIV data. In *Proceedings of the Seventh ACM SIGKDD International Conference on Knowledge Discovery and Data Mining (KDD-01)*, pages 136–143, 2001.

20. R. Meo, G. Psaila, and S. Ceri. An extension to SQL for mining association rules. *Data Mining and Knowledge Discovery*, 2(2):195–224, 1998.

21. J. R. Quinlan. Learning logical definitions from relations. *Machine Learning*, 5:239, 1990.

22. L. Richter, J. Wicker, K. Kessler, and S. Kramer. An inductive database and query language in the relational model. In *Proceedings of the 10th International Conference on Extending Database Technology (EDBT 2008)*, pages 740–744. ACM Press, 2008.

23. O.S. Weislow, R. Kiser, D.L. Fine, J.P. Bader, R.H. Shoemakerand, and M.R. Boyd. New soluble formazan assay for HIV-1 cytopathic effects: application to high flux screening of synthetic and natural products for aids antiviral activity. *Journal of the National Cancer Institute*, 81:577–586, 1989.

24. J. Wicker, C. Brosdau, L. Richter, and S. Kramer. SINDBAD SAILS: A service architecture for inductive learning schemes. In Nada Lavrač, Joost Kok, Jeroen de Bruin, and Vid Podpečan, editors, *Proceedings of the First Workshop on Third Generation Data Mining: Towards Service-Oriented Knowledge Discovery*, 2008.

25. J. Wicker, L. Richter, K. Kessler, and S. Kramer. SINDBAD and SiQL: An inductive database and query language in the relational model. In Walter Daelemans, Bart Goethals, and Katharina Morik, editors, *Machine Learning and Knowledge Discovery in Databases, European Conference, ECML/PKDD 2008, Antwerp, Belgium, September 15-19, 2008, Proceedings, Part II*, pages 690–694. Springer, 2008.

Chapter 13
Patterns on Queries

Arno Siebes and Diyah Puspitaningrum

Abstract One of the most important features of any database system is that it supports *queries*. For example, in relational databases one can construct new tables from the stored tables using relational algebra. For an inductive database, it is reasonable to assume that the stored tables have been modelled. The problem we study in this chapter is: do the models available on the stored tables help to model the table constructed by a query? To focus the discussion, we concentrate on one type of modelling, i.e., computing frequent item sets. This chapter is based on results reported in two earlier papers [12, 13]. Unifying the approaches advocated by those papers as well as comparing them is the main contribution of this chapter.

13.1 Introduction

By far the most successful type of DBMS is relational. In a relational database, the data is stored in tables and a query constructs a new table from these stored tables using, e.g., relational algebra [5]. While querying an inductive relational database, the user will, in general, not only be interested in the table that the query yields, but also -if not more- in particular models induced from that result-table. Since inductive databases have models as first-class citizens -meaning they can be stored and queried- it is reasonable to assume that the original, stored, tables are already modelled. Hence, a natural question is: does knowing a model on the original tables help in inducing a model on the result of a query?

Slightly more formally, let M_{DB} be the model we induced from database DB and let Q be a query on DB. Does knowing M_{DB} help in inducing a model M_Q on $Q(DB)$, i.e., on the result of Q when applied to DB. For example, if M_{DB} is a classifier and

Arno Siebes · Diyah Puspitaningrum
Department Of Information and Computing Sciences, Universiteit Utrecht, The Netherlands
e-mail: {arno,diyah}@cs.uu.nl

Q selects a subset of DB, does knowing M_{DB} help the induction of a new classifier M_Q on the subset $Q(DB)$?

This formulation is only slightly more formal as the term "help" is a non-technical and, thus, ill-defined concept. In this chapter, we will formalise "help" in two different ways. Firstly, in the sense that we can compute M_Q directly from M_{DB} *without* consulting either DB or $Q(DB)$. While this is clearly the most elegant way to formalise "help", it puts such stringent requirements on the class of models we consider that the answer to our question becomes *no* for many interesting model-classes; we'll exhibit one in this chapter.

Hence, secondly, we interpret "help", far less ambitiously, as meaning "speeding-up" the computation of M_Q. That is, let $\mathscr{A}lg$ be the algorithm used to induce M_{DB} from DB, i.e., $\mathscr{A}lg(DB) = M_{DB}$. We want to transform $\mathscr{A}lg$ into an algorithm $\mathscr{A}lg^*$, which takes M_{DB} as extra input such that

$$\mathscr{A}lg^*(Q(DB), M_{DB}) \approx \mathscr{A}lg(Q(DB))$$

Note that we do not ask for exactly the same model, approximately the same answer is acceptable if the speed-up is considerable. In fact, for many application areas, such as marketing, a *good enough model* rather than the *best model* is all that is required.

The problem as stated is not only relevant in the context of inductive databases, but also in existing data mining practice. In the data mining literature, the usual assumption is that we are given some database that has to be mined. In practice, however, this assumption is usually not met. Rather, the construction of the mining database is often one of the hardest parts of the KDD process [9]. The data often resides in a data warehouse or in multiple databases, and the mining database is constructed from these underlying databases.

From most perspectives, it is not very interesting to know whether one mines a specially constructed database or an original database. For example, if the goal is to build the best possible classifier on that data set, the origins of the database are of no importance whatsoever.

It makes a difference, however, if the underlying databases have already been modelled. Then, like with inductive databases, one would hope that knowing such models would help in modelling the specially constructed mining database. For example, if we have constructed a classifier on a database of customers, one would hope that this would help in developing a classifier for the female customers only.

In other words, the problem occurs both in the context of inductive databases and in the everyday practice of data miners. Hence, it is a relevant problem, but isn't it trivial? After all, if M_{DB} is a good model on DB, it is almost always also a good model on a random subset of DB; almost always, because a random subset may be highly untypical. The problem is, however, *not* trivial because queries in general do *not* compute a random subset. Rather, queries construct a very specific result.

For the usual "project-select-join" queries, there is not even a natural way in which the query-result can be seen as subset of the original database. Even if Q is just a "select"-query, the result is usually not random and M_{DB} can even be highly misleading on $Q(DB)$. This is nicely illustrated by the well-known example of *Simpson's Paradox* on Berkeley's admission data [3]. Overall, 44% of the male applicants were admitted, while only 35% of the females were admitted. Four of the six departments, however, have a bias that is in favour of female applicants. While the overall model may be adequate for certain purposes, it is woefully inadequate for a query that selects a single department.

In other words, we do address a relevant and non-trivial problem. Addressing the problem, in either sense of "help", for all possible model classes and/or algorithms is, unfortunately, too daunting a task for this chapter. In the sense of "direct construction" it would require a discussion of all possible model classes, which is too large a set to consider (and would result in a rather boring discussion). In the "speed-up and approximation" sense it would require either a transformation of all possible induction algorithms or a generic transformation that would transform any such algorithm to one with the required properties. The former would, again, be far too long, while a generic transformation is unlikely to exist.

Therefore we restrict ourselves to one type of model, i.e., frequent item sets [1] and one induction algorithm, i.e., our own KRIMP algorithm [14]. The structure of this chapter is as follows. In the next section, Section 13.2, we introduce our data, models -that is code-tables-, and the KRIMP algorithm. Next, in Section 13.3 we investigate the "direct computation" interpretation of "help" in the context of frequent item set mining. This is followed in Section 13.4 by the introduction of a transformed variant of KRIMP for the "speed-up" interpretation of "help". In Section 13.5, we discuss and compare these two approaches. The chapter ends with conclusions and prospects for further research.

13.2 Preliminaries

In this section we give a brief introduction to the data, models, and algorithms as used in this chapter.

13.2.1 Data

In this chapter we restrict ourselves to databases with categorical data only, the biggest impact being that we do not consider real-valued attributes. Moreover, rather than using the standard representation for relational databases, we represent them as transaction databases familiar from item set mining. After briefly introducing such databases, we will briefly discuss how a (categorical) relational database can be transformed into such a transaction database. Moreover, for each relational algebra operator, we will briefly discuss how they should be interpreted in the transaction setting.

13.2.1.1 Transaction Databases

The problem of frequent item set mining [1] can be described as follows. The basis is a set of items $\mathscr{I}$, e.g., the items for sale in a store; $|\mathscr{I}| = n$. A transaction t is a set of items, i.e., $t \in \mathscr{P}(\mathscr{I})$ in which $P(X)$ denotes the power set of X. For example, t represents the set of items a client bought at the store. A table (normally called a database) over $\mathscr{I}$ is simply a bag of transactions, e.g., the different sale transactions in the store on a given day.

A transaction database is a set of transaction tables that is related through the familiar key-foreign key mechanism known from the relational model [5]. Without loss of generality we assume that there is at most one key-foreign key relation between any two tables. That is, we assume that the *join* between two tables is unambiguous without explicit key-foreign key identification.

An item set $I \subset \mathscr{I}$ occurs in a transaction $t \in T$ iff $I \subseteq t$. The *support* of I in T, denoted by $sup_T(I)$ is the number of transactions in the table in which t occurs. The problem of frequent item set mining is: given a threshold *min-sup*, determine all item sets I such that $sup_T(I) \geq min\text{-}sup$. These *frequent item sets* represent, e.g., sets of items customers buy together often enough.

Based on the A Priori property,

$$I \subseteq J \Rightarrow sup_T(I) \geq sup_T(J),$$

reasonably efficient frequent item set miners exist.

13.2.1.2 Relational Databases as Transaction Databases

Transforming a relational database into a transaction database is straight-forward. Let T be a table in the relational database DB, having (non-key) attributes $A_1, \ldots, A_k$. Let the (finite!) domain of A_i be $D_i = \{d_{i,1}, \ldots d_{i,m_i}\}$. Then we define the set of items $\mathscr{I}_{T,i} = \{A_i = d_{i,1}, \ldots, A_i = d_{i,m_i}\}$. Moreover, define $\mathscr{I}_T = \bigcup_{i \in \{1, \ldots, k\}} \mathscr{I}_{T,i}$ and, obviously, $\mathscr{I} = \bigcup_{T \in DB} \mathscr{I}_T$.

The "transactified" table T' is then defined over the items in $\mathscr{I}_T$. The "transactified" version $t' \in T'$ of a $t \in T$ is given by:

$$\text{``} A_i = d''_{i,j} \in t' \Leftrightarrow t.A_i = d_{i,j}$$

The keys and foreign keys of T are simply copied in T'.

Note that this is not the most efficient way to encode a relational database as a transaction database. However, the efficiency of this encoding is irrelevant in this chapter. Moreover, while being inefficient, it is the most intuitive encoding; which is far more important for the purposes of this chapter.

From now on, we assume that all our databases are transaction databases.

13.2.1.3 Relational Algebra on Transaction Databases

To investigate models on the results of queries, we have to make our query language precise. Since we focus on relational databases -albeit in their "transactified" form- a relational query language is the obvious choice. Of these query languages, relational algebra is the most suited. More precisely, we focus on the usual "select-project-join" queries. That is, on the selection operator σ, the projection operator π, and the (equi-)join operator $\bowtie$; see [5].

We interpret these operators on transactions in the intuitive way. That is, σ selects those transactions that satisfy the selection predicate. The projection π returns that part of each transaction that is specified by the projection predicate. That is, we do not take the original relational representation into account. More in particular, this means that we, e.g., project on $A_i = d_{i,j}$ rather than on A_i. The former is more natural in the transaction context and the latter can easily be simulated by the former.

Finally the join is computed using key-foreign key relations only. That is, $\bowtie$ itself does not have items -attribute-value pairs- in its predicate. The reason is that such further selections can easily be accomplished using σ

Two final remarks on the queries in this chapter are the following, Firstly, as usual in the database literature, we use *bag* semantics. That is, we do allow duplicates tuples in tables and query results.

Secondly, as mentioned in the introduction, the mining database is constructed from *DB* using queries. Given the compositionality of the relational algebra, we may assume, again without loss of generality, that the analysis database is constructed using one query Q. That is, the analysis database is $Q(DB)$, for some relational algebra expression Q. Since DB is fixed, we will often simply write Q for $Q(DB)$; that is, we will use Q to denote both the query and its result.

13.2.2 Models

In this paper we consider two different types of models. The first is simply the set of all frequent item sets. The second are the models as computed by our KRIMP algorithm [14]. Since this later kind of model is less well-known, we provide a brief review of thses models.

The models computed by KRIMP consist of two components. First a constant -the same for all possible models- component, the COVER algorithm. Second a variable -database dependent- component, a *code table*.

Given a prefix code $\mathscr{C}$ a code table CT over $\mathscr{I}$ and $\mathscr{C}$ is a two-column table containing item sets and codes such that:

- each $I \in \mathscr{P}(\mathscr{I})$ and each $C \in \mathscr{C}$ occurs at most once in CT
- all the singleton item sets occur in CT
- The item sets in the code table are ordered descending on 1) item set length and 2) support size and 3) lexicographically.

Slightly abusing notation we say $I \in CT$ and $C \in CT$.

To encode a database with a code table, each transaction is partitioned into item sets in the code table:

COVER(CT, t)
If there exists $I \in CT$ such that $I \subseteq t$
   Then $Res := \{I\}$ where I is the first such element
       If $t \setminus I \neq \emptyset$
          Then $Res := Res \cup$ COVER$(CT, t \setminus I)$
   Else Fail
Return Res

D can be encoded by CT using COVER in the obvious way:

- compute the cover of each transaction $t \in D$
- replace each $I \in$ COVER(CT, t) by its code and concatenate these codes

Decoding is similarly easy because C is a prefix code:

- determine the codes in the code string
- take the union of the item sets that belong to these codes

Defined in this way, not all code tables are equally satisfying as a model of a given database DB. For, CT may assign very long codes to things that occur very often in DB, while it may assign very short codes to rare things. This is clearly unsatisfactory. We want the encoding to be optimal given the item sets in the code table.

The *usage* of an $I \in CT$ while coding DB is defined by:

$$usage(I) = |\{t \in DB | I \in \text{COVER}(CT, t)\}|$$

Usage yields a probability distribution on the $I \in CT$:

$$\mathbb{P}(I) = \frac{usage(I)}{\sum_{J \in CT} usage(J)}$$

A Shannon code, which always exists [7], for CT is a prefix code with:

$$length(code(I)) = -\log(\mathbb{P}(I))$$

Such a code is optimal in the sense that the more often a code is used, the shorter its length is. From now on we assume that the code tables we consider have such Shannon-codes for database DB.

13.2.3 Algorithms

To induce the frequent item sets used in Section 13.3 we simply use one of the well-known frequent item set miners. For the code tables used in Section 13.4 we use our

KRIMP algorithm, since this is not as well-known, we provide a brief introduction here. For a more detailed description please refer to [14].

13.2.3.1 MDL for Code Tables

Even if all code tables we consider have Shannon optimal codes, not all such code tables are equally good models for *DB*. For example, there is one that contains the singleton item sets only. This is a model that specifies nothing about the correlation between the various items. To determine the best code table, we use the Minimum Description Length principle (MDL).

MDL [10] embraces the slogan *Induction by Compression*. It can be roughly described as follows.

Given a set of models[1] $\mathcal{H}$, the best model $H \in \mathcal{H}$ is the one that minimises

$$L(H) + L(D|H)$$

in which

- $L(H)$ is the length, in bits, of the description of H, and
- $L(D|H)$ is the length, in bits, of the description of the data when encoded with H.

One can paraphrase this by: the smaller $L(H) + L(D|H)$, the better H models D. In our terminology we want the code table that compresses *DB* best.

We already know how to compute the size of the compressed database. Simply encode *DB* and add the lengths of all the codes, which are Shannon optimal. That is,

$$L(DB|CT) = - \sum_{I \in CT : freq(I) \neq 0} usage(I) \log (\mathbb{P}(I))$$

Note that the stipulation $freq(I) \neq 0$ is only there because we require that all singleton item sets are present in *CT*. All other item sets are only present in *CT* if they are actually used.

Similarly, we already know the size in bits of the second column of *CT*, it is simply the sum of the sizes of all codes in *DB*. So, we only have to determine the size in bits of the first column, i.e,. of all the item sets in *CT*.

To determine that size we encode those item sets with the code table for *DB* that consists of the singleton item sets only.

- this means we can reconstruct D up to the actual label of the $i \in \mathcal{I}$.

This is actually a good feature. It means, among other things, that the model we find does not depend on the actual language used to describe the data.

The size of the left-hand column is the sum of these encoded sizes, The size of *CT*, denoted by $L(CT)$ is simply the sum of the sizes of the two columns. Hence, for a given database *DB* we have:

---

[1] MDL-theorists tend to talk about *hypothesis* in this context, hence the $\mathcal{H}$; see [10] for the details.

$$\mathscr{L}(CT,DB) = L(CT) + L(DB|CT)$$

Note that we omit the size of COVER as it is the same for all databases and code tables. That is, it is just an additive constant, which does not influence the search for the optimal model.

13.2.3.2 KRIMP

Unfortunately, finding the best code table is too expensive. Therefore we use a heuristic algorithm called KRIMP. KRIMP starts with a valid code table (only the collection of singletons) and a sorted list of candidates (frequent item sets). These candidates are assumed to be sorted descending on 1) support size, 2) item set length and 3) lexicographically. Each candidate item set is considered by inserting it at the right position in CT and calculating the new total compressed size. A candidate is only kept in the code table iff the resulting total size is smaller than it was before adding the candidate. If it is kept, all other elements of CT are reconsidered to see if they still positively contribute to compression. The whole process is illustrated in Figure 13.1; see [14]. If we assume a fixed minimum support threshold for a database, KRIMP has only one essential parameter: the database. For, given the database and the (fixed) minimum support threshold, the candidate list is also specified. Hence, we will simply write CT_{DB} and KRIMP(DB), to denote the code table induced by KRIMP from DB. Similarly CT_Q and KRIMP(Q) denote the code table induced by KRIMP from the result of applying query Q to DB.

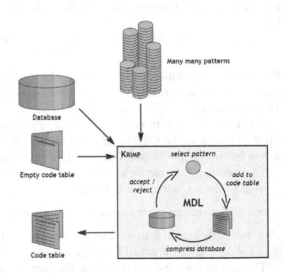

Fig. 13.1 KRIMP in action

13.3 Frequent Item Set Mining

The goal of this section is to investigate whether we can determine the set of frequent item sets on $Q(DB)$ without consulting $Q(DB)$. Rather, we are given the frequent item sets on DB and the query Q and from that only we should determine the frequent item sets on $Q(DB)$. That is, we want to *lift* the relational operators to sets of frequent item sets.

13.3.1 Selection

The relational algebra operator σ (select) is a mapping:

$$\sigma : \mathscr{B}(D) \to \mathscr{B}(D)$$

in which $\mathscr{B}(D)$ denotes all possible bags over domain D.

Lifting means that we are looking for an operator $\sigma_{(D,\mathscr{A}lg)}$ that makes the diagram in Figure 13.2 commute: Such diagrams are well-known in , e.g., category theory [2] and the standard interpretation is:

$$\mathscr{A}lg \circ \sigma = \sigma_{(D,\mathscr{A}lg)} \circ \mathscr{A}lg$$

In other words, first inducing the model using algorithm $\mathscr{A}lg$ followed by the application of the *lifted* selection operator $\sigma_{(D,\mathscr{A}lg)}$ yields the same result as first applying the *standard* selection operator σ followed by induction with algorithm $\mathscr{A}lg$.

In fact, we are willing to settle for commutation of the diagram in a loose sense: That is, if we are able to give reasonable support bounds for those item sets whose support we can not determine exactly, we are satisfied.

For frequent item sets the three basic selections are $\sigma_{I=0}$, $\sigma_{I=1}$, and $\sigma_{I_1=I_2}$. More complicated selections can be made by conjunctions of these basic comparisons. We look at the different basic selections in turn.

First consider $\sigma_{I=0}$. If it is applied to a table, all transactions in which I occurs are removed from that table. Hence, all item sets that contain I get a support of zero in the resulting table. For those item sets in which I doesn't occur, we have to compute which part of their support consists of transactions in which I does occur and subtract that number. Hence, for support for item sets J, we have:

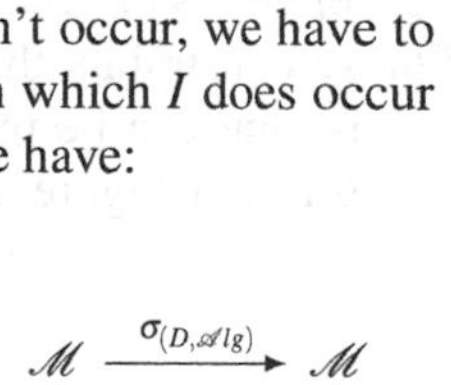

Fig. 13.2 Lifting the selection operator

$$
\begin{array}{ccc}
\mathscr{M} & \xrightarrow{\sigma_{(D,\mathscr{A}lg)}} & \mathscr{M} \\
\uparrow{\scriptstyle \mathscr{A}lg} & & \uparrow{\scriptstyle \mathscr{A}lg} \\
\mathscr{B}(D) & \xrightarrow{\ \sigma\ } & \mathscr{B}(D)
\end{array}
$$

$$sup_{\sigma_{I=0}(T)}(J) = \begin{cases} 0 & \text{if } I \in J, \\ sup_T(J) - sup_T(J \cup \{I\}) & \text{otherwise.} \end{cases}$$

If we apply $\sigma_{I=1}$ to the table, all transactions in which I doesn't occur are removed from the table. In other words, the support of item sets that contain I doesn't change. For those item sets that do not contain I, the support is given by those transactions that *also* contained I. Hence, we have:

$$sup_{\sigma_{I=1}(T)}(J) = \begin{cases} sup_T(J) & \text{if } I \in J, \\ sup_T(J \cup \{I\}) & \text{otherwise.} \end{cases}$$

If we apply $\sigma_{I_1=I_2}$ to the table, the only transactions that remain are those that either contain both I_1 and I_2 or neither. In other words, for frequent item sets that contain both, the support remains the same. For all others, the support changes. For those item sets J that contain just one of the I_i the support will be the support of $J \cup \{I_1, I_2\}$. For those that contain neither of the I_i, we have to correct for those transactions that contain one of the I_i in their support. If we denote this by $sup_T(J \neg I_1 \neg I_2)$ (a support that can be easily computed) We have:

$$sup_{\sigma_{I_1=I_2}(T)}(J) = \begin{cases} sup_T(J \cup \{I_1, I_2\}) & \text{if } \{I_1, I_2\} \cap J \neq \emptyset, \\ sup_T(J \neg I_1 \neg I_2) & \text{otherwise.} \end{cases}$$

Clearly, we can also "lift" conjunctions of the basic selections, simply processing one at the time. So, in principle, we can lift all selections for frequent item sets. But only in principle, because we need the support of item sets that are *not necessarily frequent*. Frequent item sets are a lossy model (not all aspects of the data distribution are modelled) and that can have its repercussions: in general the lifting will *not* be commutative. In our loose sense of "commutativity", the situation is slightly better. For, we can give reasonable bounds for the resulting supports; for those supports we do not know are bounded (from above) by *min-sup*.

We haven't mentioned constraints [11] so far. Constraints in frequent item set mining are the pre-dominant way to select a subset of the frequent item sets. In general the constraints studied do not correspond to selections on the database. The exception is the class of *succinct anti-monotone constraints* introduced in [11]. For these constraints there is such a selection (that is what succinct means) and the constraint can be pushed into the algorithm. This means we get the commutative diagram in Figure 13.3. Note that in this case we know that the diagonal arrow

Fig. 13.3 Lifting selections
for succinct constraints

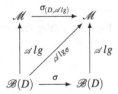

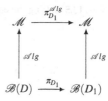

Fig. 13.4 Lifting projections

makes the bottom right triangle commute in the strict sense of the word. For the upper left triangle, as well as the square, our previous analysis remains true.

13.3.2 Project

For the projection operator π, we have a new domain D_1 such that $D = D_1 \times D_2$. Projection on D_1 has thus as signature:

$$\pi_{D_1} : \mathscr{B}(D) \to \mathscr{B}(D_1)$$

Hence, we try to find an operator $\pi_{D_1}^{\mathscr{A}lg}$ that makes the diagram in Figure 13.4 commute. Note that D_1 is spanned by the set of variables (or items) we project on.

We project on a set of items $\mathscr{J} \subseteq \mathscr{I}$, let $J \subseteq \mathscr{I}$ be a frequent item set. There are three cases to consider:

1. if $J \subseteq \mathscr{J}$, then all transactions in the support of J will simply remain in the table, hence J will remain frequent.
2. if $J \cap \mathscr{J} \neq \emptyset$, then $J \cap \mathscr{J}$ is also frequent and will remain in the set of frequent item sets.
3. if $J \cap \mathscr{J} = \emptyset$, then its support will vanish.

In other words, if $\mathscr{F}$ denotes the set of all frequent item sets, then:

$$\pi_{\mathscr{J}}(\mathscr{F}) = \{J \in \mathscr{F} | J \subseteq \mathscr{J}\}$$

Clearly, this method of lifting will make the diagram commute in the strict sense if we use absolute minimal frequency. In other words, for projections, frequent item sets do capture enough of the underlying data distribution to allow lifting.

13.3.3 EquiJoin

The equijoin has as signature:

$$\bowtie: \mathscr{B}(D_1) \times \mathscr{B}(D_2) \to \mathscr{B}(D_1 \bowtie D_2)$$

Fig. 13.5 Lifting the equijoin

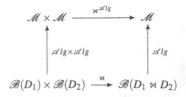

Hence, the diagram we want to make commute is given in Figure 13.5. The join can be computed, though not very efficiently, starting with the Cartesian product of the two tables. Since in extreme cases, the equi-join equals the Cartesian product, we discuss that operator.

Let J_1 be a frequent item set for the first table and J_2 for the second. The frequency of the pair on the Cartesian product of the two tables is simply given by:

$$sup_{T_1 \times T_2}(J_1, J_2) = sup_{T_1}(J_1) \times sup_{T_2}(J_2)$$

While this is easy to compute, it means again that in general we will not be able to compute all frequent item sets on the Cartesian product without consulting the database. Even if we set the minimal frequency to the product of the two minimal frequencies, the combination of an infrequent item set on one database with a frequent one on the other may turn out to be frequent.

In other words, we cannot even make the diagram commute in the approximate sense of the word. For, the bound is given by $\max\{|T_1| \times (\text{min-sup} - 1), |T_2| \times (\text{min-sup} - 1)\}$, which is hardly a reasonable bound.

Given that the number of joins possible in a database is limited and known beforehand, we may make our lives slightly easier. That is, we may allow ourselves to do some pre-computations.

Assume that we compute the tables $T_1^2 = \pi_{T_1}(T_1 \bowtie T_2)$ and $T_2^1 = \pi_{T_2}(T_1 \bowtie T_2)$ and their frequent item sets, say $\mathscr{F}_1^2$ and $\mathscr{F}_2^1$, off-line. Are those sets enough to lift the join? For the extreme case, the Cartesian product, the answer is clearly: *yes*. By "blowing" up the original tables we add enough information to compute the support of any item set in the join *iff* that item set exceeds the minimal support.

Unfortunately, the same is not true for the join in general. Since we cannot see from either $\mathscr{F}_1^2$ or $\mathscr{F}_2^1$ which combinations of frequent item sets will actually occur in $(T_1 \bowtie T_2)$. That is, we can only compute a superset of the frequent item sets on the join.

Hence, the only way to lift the join is to compute and store the frequent item sets on all possible joins. While this is doable given the limited number of possible joins, this can hardly count as lifting.

13.3.4 Discussion

The fact that lifting the relational algebra operators to sets of frequent item sets is only partially possible should hardly come as a surprise: the *min-sup* constraint makes this into an inherently lossy model. For models that do try to capture the complete distribution, such as Baysian networks, one would expect far better results; see [12] for a discussion of lifting for such networks.

13.4 Transforming KRIMP

Recall from the Introduction that the problem we investigate in this Section is that we want to transform an induction algorithm $\mathscr{A}lg$ into an algorithm $\mathscr{A}lg^*$ that takes at least two inputs, i.e, both Q and $\mathscr{M}_{DB}$, such that:

1. $\mathscr{A}lg^*$ gives a reasonable approximation of $\mathscr{A}lg$ when applied to Q, i.e.,

$$\mathscr{A}lg^*(Q,\mathscr{M}_{DB}) \approx \mathscr{M}_Q$$

2. $\mathscr{A}lg^*(Q,\mathscr{M}_{DB})$ is simpler to compute than $\mathscr{M}_Q$.

The second criterion is easy to formalise: the runtime of $\mathscr{A}lg^*$ should be shorter than that of $\mathscr{A}lg$. The first one is harder. What do we mean that one model is an approximation of another? Moreover, what does it mean that it is a *reasonable* approximation?

Before we discuss how KRIMP can be transformed and provide experimental evidence that our approach works, we first formalise this notion of approximation.

13.4.1 Model Approximation

The answer to the question of how to formalise that one model approximates another depends very much on the goal. If $\mathscr{A}lg$ induces classifiers, approximation should probably be defined in terms of prediction accuracy, e.g., on the Area Under the ROC-curve (AUC).

KRIMP computes code tables. Hence, the quick approximating algorithm we are looking for, KRIMP* in the notation used above, also has to compute code tables. So, one way to define the notion of approximation is by comparing the resulting code tables. Let CT_{KRIMP} be the code table computed by KRIMP and similarly, let CT_{KRIMP^*} denote the code table computed by KRIMP* on the same data set. The more similar CT_{KRIMP^*} is to CT_{KRIMP}, the better KRIMP* approximates KRIMP.

While this is intuitively a good way to proceed, it is far from obvious how to compare two code tables. Fortunately, we do not need to compare code tables directly. KRIMP is based on MDL and MDL offers another way to compare models, i.e., by their *compression-rate*. Note that using MDL to define "approximation" has the advantage that we can formalise our problem for a larger class of algorithms than just KRIMP. It is formalised for all algorithms that are based on MDL. MDL is quickly becoming a popular formalism in data mining research, see, e.g., [8] for an overview of other applications of MDL in data mining.

What we are interested in is comparing two algorithms on the same data set, i.e., on $Q(DB)$. Slightly abusing notation, we will write $\mathscr{L}(\mathscr{A}lg(Q))$ for $L(\mathscr{A}lg(Q)) + L(Q(DB)|\mathscr{A}lg(Q))$, similarly, we will write $\mathscr{L}(\mathscr{A}lg^*(Q,\mathscr{M}_{DB}))$. Then, we are interested in comparing $\mathscr{L}(\mathscr{A}lg^*(Q,\mathscr{M}_{DB}))$ to $\mathscr{L}(\mathscr{A}lg(Q))$. The closer the former is to the latter, the better the approximation is.

Just taking the difference of the two, however, can be quite misleading. Take, e.g., two databases db_1 and db_2 sampled from the same underlying distribution, such that db_1 is far bigger than db_2. Moreover, fix a model H. Then necessarily $L(db_1|H)$ is bigger than $L(db_2|H)$. In other words, big absolute numbers do not necessarily mean very much. We have to *normalise* the difference to get a feeling for how good the approximation is. Therefore we define the asymmetric dissimilarity measure (ADM) as follows [15].

Definition 13.1. Let H_1 and H_2 be two models for a dataset D. The asymmetric dissimilarity measure $ADM(H_1,H_2)$ is defined by:

$$ADM(H_1,H_2) = \frac{|\mathscr{L}(H_1) - \mathscr{L}(H_2)|}{\mathscr{L}(H_2)}$$

Note that this dissimilarity measure is related to the Normalised Compression Distance [4]. The reason why we use this asymmetric version is that we have a "gold standard". We want to know how far our approximate result $\mathscr{A}lg^*(Q,\mathscr{M}_{DB})$ deviates from the optimal result $\mathscr{A}lg(Q)$.

The remaining question is, of course, what ADM scores indicate a good approximation? In a previous paper [15], we took two random samples from data sets, say D_1 and D_2. Code tables CT_1 and CT_2 were induced from D_1 and D_2 respectively. Next we tested how well CT_i compressed D_j. For the four data sets also used in this paper, *Iris, Led7, Pima* and, *PageBlocks*, the "other" code table compressed 16% to 18% worse than the "own" code table; the figures for other data sets are in the same ball-park. In other words, an ADM score of 0.2 is in-line with the "natural variation" in a data set. If it gets much higher, it shows that the two code tables are rather different.

Clearly, $ADM(\mathscr{A}lg^*(Q,\mathscr{M}_{DB}),\mathscr{A}lg(Q))$ does not only depend on $\mathscr{A}lg^*$ and on $\mathscr{A}lg$, but also very much on Q. We do not seek a low ADM on one particular Q, rather we want to have a reasonable approximation on all possible queries. Requiring that the ADM is equally small on all possible queries seems too strong a requirement. Some queries might result in a very untypical subset of DB, the ADM is probably higher on the result of such queries than it is on queries that result in more typical subsets. Hence, it is more reasonable to require that the ADM is small most of the time. This is formalised through the notion of an (ε,δ)-approximation

Definition 13.2. Let *DB* be a database and let *Q* be a random query on *DB*. Moreover, let $\mathscr{A}lg_1$ and $\mathscr{A}lg_2$ be two data mining algorithms on *DB*. Let $\varepsilon \in \mathbb{R}$ be the threshold for the maximal acceptable ADM score and $\delta \in \mathbb{R}$ be the error tolerance for this maximum. $\mathscr{A}lg_1$ is an (ε, δ)-approximation of $\mathscr{A}lg_2$ iff

$$\mathbb{P}(ADM(\mathscr{A}lg_1(Q), \mathscr{A}lg_2(Q)) > \varepsilon) < \delta$$

13.4.2 Transforming KRIMP

Given that KRIMP results in a code table, there is only one sensible way in which KRIMP(*DB*) can be re-used to compute KRIMP(*Q*): provide KRIMP only with the item sets in CT_{DB} as candidates. While we change nothing to the algorithm, we'll use the notation KRIMP* to indicate that KRIMP got only code table elements as candidates. So, e.g., KRIMP*(*Q*) is the code table that KRIMP induces from *Q*(*DB*) using the item sets in CT_{DB} only.

Given our general problem statement, we now have to show that KRIMP* satisfies our two requirements for a transformed algorithm. That is, we have to show for a random database *DB*:

- For reasonable values for ε and δ, KRIMP* is an (ε, δ)-approximation of KRIMP, i.e, for a random query *Q* on *DB*:

$$\mathbb{P}(ADM(\text{KRIMP}^*(Q), \text{KRIMP}(Q)) > \varepsilon) < \delta$$

Or in MDL-terminology:

$$\mathbb{P}\left(\frac{|\mathscr{L}(\text{KRIMP}^*(Q)) - \mathscr{L}(\text{KRIMP}(Q))|}{\mathscr{L}(\text{KRIMP}(Q))} > \varepsilon\right) < \delta$$

- Moreover, we have to show that it is faster to compute KRIMP*(*Q*) than it is to compute KRIMP(*Q*).

Neither of these two properties can be formally proven, if only because KRIMP and thus KRIMP* are both heuristic algorithms. Rather, we report on extensive tests of these two requirements.

13.4.3 The Experiments

In this subsection, we describe our experimental set-up. First we briefly describe the data sets we used. Next we discuss the queries used for testing. Finally we describe how the tests were performed.

13.4.3.1 The Data Sets

To test our hypothesis that KRIMP* is a good and fast approximation of KRIMP, we have performed extensive tests mostly on 6 well-known UCI [6] data sets and one data set from the KDDcup 2004.

In particular, we have used the data sets *connect, adult, chessBig, letRecog, PenDigits* and *mushroom* from UCI. These data sets were chosen because they are well suited for KRIMP. Some of the other data sets in the UCI repository are simply too small for KRIMP to perform well. MDL needs a reasonable amount of data to be able to function. Some other data sets are very dense. While KRIMP performs well on these very dense data sets, choosing them would have turned our extensive testing prohibitively time-consuming.

Since all these data sets are single table data sets, they do not allow testing with queries involving joins. To test such queries, we used tables from the "Hepatitis Medical Analysis"[2] of the KDDcup 2004. From this relational database we selected the tables *bio* and *hemat*. The former contains biopsy results, while the latter contains results on hematological analysis. The original tables have been converted to item set data and rows with missing data have been removed.

13.4.3.2 The Queries

To test our hypothesis, we need to consider randomly generated queries. On first sight this appears a daunting task. Firstly, because the set of all possible queries is very large. How do we determine a representative set of queries? Secondly, many of the generated queries will have no or very few results. If the query has no results, the hypothesis is vacuously true. If the result is very small, MDL (and thus KRIMP) doesn't perform very well.

To overcome these problems, we restrict ourselves to queries that are built by using selections (σ), projections (π), and joins ($\bowtie$) only. The rationale for this choice is twofold. Firstly, simple queries will have, in general, larger results than more complex queries. Secondly, we have seen in Section 13.3 that lifting these operators is already a problem.

13.4.3.3 The Experiments

The experiments preformed for each of the queries on each of the data sets were generated as follows.

Projection: The projection queries were generated by randomly choosing a set X of n items, for $n \in \{1, 3, 5, 7, 9\}$. The generated query is then $\pi_{\overline{X}}$. That is, the elements of X are projected out of each of the transactions. For example, $\pi_{\overline{\{I_1, I_3\}}}(\{I_1, I_2, I_3\}) = \{I_2\}$. For this case, the code table elements generated on

---

[2] http://lisp.vse.cz/challenge/

the complete data set were projected in the same way. For each value of n, 10 random sets X were generated on each data set.

As an aside, note that the rationale for limiting X to maximally 9 elements is that for larger values too many result sets became too small for meaningful results.

Selection: The random selection queries were again generated by randomly choosing a set X of n items, with $n \in \{1,2,3,4\}$. Next for each random item I_i a random value v_i (0 or 1) in its domain D_i was chosen. Finally, for each I_i in X a random $\theta_i \in \{=, \neq\}$ was chosen. The generated query is thus $\sigma(\bigwedge_{I_i \in X} I_i \theta_i v_i)$. As in the previous case, we performed 10 random experiments on each of the data sets for each of the values of n.

Project-Select: The random project-select queries generated are essentially combinations of the simple projection and selection queries as explained above. The only difference is that we used $n \in \{1,3\}$ for the projection and $n \in \{1,2\}$ for the selections. That is we select on 1 or 2 items and we project away either 1 or 3 items. The size of the results is, of course, again the rationale for this choice. For each of the four combinations, we performed 100 random experiments on each of the data sets: first we chose randomly the selection (10 times for each selection), for each such selection we performed 10 random projections.

Project-Select-Join: Since we only use one "multi-relational" data set and there is only one possible way to join the *bio* and *hemat* tables, we could not do random tests for the join operator. However, in combination with projections and selections, we can perform random tests. These tests consist of randomly generated project-select queries on the join of *bio* and *hemat*. In this two-table case, KRIMP* got as input all pairs $(\mathscr{I}_1, \mathscr{I}_2)$ in which $\mathscr{I}_1$ is an item set in the code table of the "blown-up" version of *bio*, and $\mathscr{I}_2$ is an item set in the code table of the "blown-up" version of *hemat*. Again we select on 1 or 2 items and we project away either 1 or 3 items. And, again, we performed again 100 random experiments on the database for each of the four combinations; as above.

13.4.4 The Results

In this subsection we give an overview of the results of the experiments described in the previous section. Each test query is briefly discussed in its own subsubsection.

13.4.4.1 Projection Queries

In Figure 13.6 the results of the random projection queries on the *letRecog* data set are visualised. The marks in the picture denote the averages over the 10 experiments, while the error bars denote the standard deviation. Note that, while not statistically significant, the average ADM grows with the number of attributes projected away. This makes sense, since the more attributes are projected away, the smaller the result

Fig. 13.6 Projection results
on *letRecog*

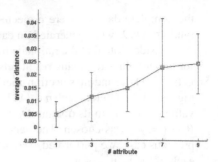

set becomes. On the other data sets, KRIMP* performs similarly. Since this is also
clear from the project-select query results, we do not provide all details here.

13.4.4.2 Selection Queries

The results of the random selection queries on the *penDigits* data set are visualised
in Figure 13.7. For the same reason as above, it makes sense that the average ADM
grows with the number of attributes selected on. Note, however, that the ADM averages for selection queries seem much larger than those for projection queries. These
numbers are, however, not representative for the results on the other data sets. It
turned out that *penDigits* is actually too small and sparse to test KRIMP* seriously.
In the remainder of our results section, we do not report further results on *penDigits*. The reason why we report on it here is to illustrate that even on rather small
and sparse data sets KRIMP* still performs reasonably well. On all other data sets
KRIMP* performs far better, as will become clear next.

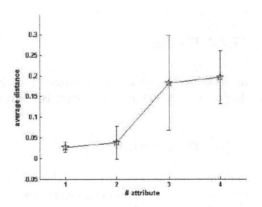

Fig. 13.7 Selection results on
penDigits

Table 13.1 The results of Project-Select Queries

ADM ± STD		connect	adult	chessBig	letRecog	mushroom
Select 1	**Project out 1**	0.1 ± 0.01	0.1 ± 0.01	0.04 ± 0.01	0.1 ± 0.01	0.3 ± 0.02
	Project out 3	0.1 ± 0.02	0.1 ± 0.01	0.04 ± 0.03	0.1 ± 0.01	0.3 ± 0.16
Select 2	**Project out 1**	0.2 ± 0.01	0.1 ± 0.01	0.1 ± 0.03	0.04 ± 0.01	0.2 ± 0.04
	Project out 3	0.2 ± 0.02	0.1 ± 0.01	0.1 ± 0.03	0.04 ± 0.01	0.2 ± 0.05

13.4.4.3 Project-Select Queries

The results of the projection-select queries are given in Table 13.1. All numbers are the average ADM score ± the standard deviation for the 100 random experiments. All the ADM numbers are rather small, only for mushroom do they get above 0.2.

Two important observations can be made from this table. Firstly, as for the projection and selection queries reported on above, the ADM scores get only slightly worse when the query results get smaller: "Select 2, Project out 3" has slightly worse ADM scores than "Select 1, Project out 1". Secondly, even more importantly, combining algebra operators only degrades the ADM scores slightly. This can be seen if we compare the results for "Project out 3" on *letRecog* in Figure 13.6 with the "Select 1, Project out 3" and "Select 2, Project out 3" queries in Table 13.1 on the same data set. These results are very comparable, the combination effect is small and mostly due to the smaller result sets. While not shown here, the same observation holds for the other data sets.

To give insight in the distribution of the ADM scores of the "Select 2, Project out 3" queries on the *connect* data set are given in Figure 13.8. From this figure we see that if we choose $\varepsilon = 0.2$, $\delta = 0.08$. In other words, KRIMP* is a pretty good approximation of KRIMP. Almost always the approximation is less than 20% worse than the optimal result. The remaining question is, of course, how much faster is KRIMP*? This is illustrated in Table 13.2.

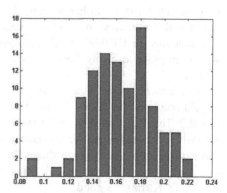

Fig. 13.8 Histogram of 100 Project-Select Queries on *connect*

Table 13.2 Relative number of candidates for KRIMP*

Relative #candidates		connect	adult	chessBig	letRecog	mushroom
Select 1	Project out 1	0.01 ± 0.001	0.01 ± 0.002	0.21 ± 0.012	0.01 ± 0.001	0.01 ± 0.001
	Project out 3	0.01 ± 0.001	0.01 ± 0.004	0.26 ± 0.031	0.02 ± 0.004	0.01 ± 0.001
Select 2	Project out 1	0.01 ± 0.001	0.03 ± 0.003	0.76 ± 0.056	0.02 ± 0.002	0.03 ± 0.002
	Project out 3	0.01 ± 0.002	0.03 ± 0.008	0.96 ± 0.125	0.02 ± 0.004	0.03 ± 0.003

Table 13.2 gives the average number of candidates KRIMP* has to consider relative to those that the full KRIMP run has to consider. Since, both KRIMP* and KRIMP are linear in the number of candidates, this table shows that the speed-up is considerable; a factor of 100 is often attained; except for *chessBig* were the query results get small and, thus, have few frequent item sets. The experiments are those that are reported on in Table 13.1.

13.4.4.4 Select-Project-Join Queries

The results for the select-project-join queries are very much in line with the results reported on above. In fact, they are even better. Since the join leads to rather large results, the ADM score is almost always zero: in only 15 of the 400 experiments the score is non-zero (average of non-zero values is 1%). The speed-up is also in line with the numbers reported above, a factor of 100 is again often attained.

13.4.5 Discussion

As noted in the previous section, the speed-up of KRIMP* is easily seen. The number of candidates that KRIMP* has to consider is often a factor 100 smaller than those that the full KRIMP run has to consider. Given that the algorithm is linear in the number of candidates, this means a speed-up by a factor 100. In fact, one should also note that for KRIMP*, we do not have to run a frequent item set miner. In other words, in practice, using KRIMP* is even faster than suggested by the Speed-up scores.

But, how about the other goal: how good is the approximation? That is, how should one interpret ADM scores? Except for some outliers, ADM scores are below 0.2. That is, a full-fledged KRIMP run compresses the data set 20% better than KRIMP*. As noted when we introduced the ADM score, this about as good as one can expect, such a percentage shows the natural variation in the data. Hence, given that the average ADM scores are often much lower we conclude that the approximation by KRIMP* is good.

In other words, the experiments verify our hypothesis: KRIMP* gives a fast and good approximation of KRIMP. The experiments show this for simple "project-select-join" queries, but as noticed with the results of the "project-select" queries, the effect of combining algebra operators is small. If the result set is large enough, the approximation is good.

13.5 Comparing the two Approaches

In this chapter, we introduced two ways in which the models present in an inductive database DB help in computing the models on the results of a query Q on the data in that database. The first, if applicable, gives results without consulting $Q(DB)$. The result is computed directly from the models M_T induced on the tables used by Q. For the relational algebra we formalised this by lifting the relational algebra operators to the set of all models.

The second approach does allow access to $Q(DB)$. The induction algorithm $\mathscr{A}lg$ is transformed into an algorithm $\mathscr{A}lg^*$ that takes at least two inputs, i.e, both Q and $\mathscr{M}_{DB}$, such that:

1. $\mathscr{A}lg^*$ gives a reasonable approximation of $\mathscr{A}lg$ when applied to Q, i.e.,

$$\mathscr{A}lg^*(Q, \mathscr{M}_{DB}) \approx \mathscr{M}_Q$$

2. $\mathscr{A}lg^*(Q, \mathscr{M}_{DB})$ is simpler to compute than $\mathscr{M}_Q$.

The first requirement was formalised using MDL into the requirement:

$$\mathbb{P}\left(\frac{|\mathscr{L}(\mathscr{A}lg^*(Q)) - \mathscr{L}(\mathscr{A}lg(Q))|}{\mathscr{L}(\mathscr{A}lg(Q))} > \varepsilon \right) < \delta$$

for reasonably small ε and δ. The second requirement was simply interpreted as a significant speed-up in computation.

Clearly, when applicable, the first approach is to be preferred above the second approach. Firstly because it doesn't even require the computation of $Q(DB)$, and is, hence, likely to be much faster. Secondly, because an algebraic structure on the set of all models opens up many more possible applications.

In this chapter, we investigated both approaches on item sets. More precisely, we investigated lifting the relational algebra operators to sets of frequent item sets. Moreover, we transformed our KRIMP algorithm to investigate the second approach.

As noted already in Section 13.3, lifting the relational algebra operators to sets of frequent item sets has its problems. Only for the projection it works well. For the selection operator we get a reasonable approximation. Reasonable in the sense that we can put a bound on the error of the approximated support; an upper bound that is determined by the minimal support threshold. Since this bound is an upperbound, this means that we may declare too many item sets to be frequent. If we declare an item set to be infrequent, it is infrequent on the result of the selection.

The join operator, unfortunately, can not be lifted at all. Not even if we provide extra information by giving access to the frequent item sets on the "blown-up" version of the underlying tables. In that case, we again only have an upperbound on the support. That is, again, we declare too many item sets to be frequent. In the case of the join, however, there is no bound on the error. For, if I_1 has a high support on $T_1^2 = \pi_{T_1}(T_1 \bowtie T_2)$, say n_1, while I_2 has a high support on $T_2^1 = \pi_{T_2}(T_1 \bowtie T_2)$, say n_2, then the computed upperbound on the support of (I_1, I_2) on $T_1 \bowtie T_2$ will be $n_1 \times n_2$, while there may be no transaction in $T_1 \bowtie T_2$ which actually supports this pair! Again, if we declare an item set to be infrequent on the join, it is infrequent.

Again as noted before, the reason for this failure is that sets of frequent item sets are an inherently lossy model. As our analysis above shows, this loss of information makes us overestimate the support of item sets on $Q(DB)$, in the case of the join with an unbounded error.

The transformation of KRIMP proved to be far more successful. The algorithm KRIMP*, which is simply KRIMP with a restricted set of candidates proved in the experiments to be much faster and provide models which approximate the true model very well. Given the lack of success for frequent item sets, this is a surprising result.

For, from earlier research [15] we know that the code tables produced by KRIMP determine the support of all item sets rather accurately. More precisely, in that paper we showed that these code tables can be used to generate a new code table. The support of an arbitrary frequent item set in this generated database, say DB_{gen}, is almost always almost equal to the support of that item set in the original database, say DB_{orig}. As usual, this sentence is probably more clear in its mathematical formulation:

$$\mathbb{P}\left(|sup_{DB_{orig}}(I) - sup_{DB_{gen}}(I)| > \varepsilon\right) < \delta$$

This surprise raises two immediate questions:

1. Why does transforming KRIMP work and
2. Can we transform frequent item set mining?

The reason that transforming KRIMP work is firstly exactly the fact that it determines the support of all item sets so well. Given a code table, which KRIMP* produces, we know the support of these item sets. Clearly, as for the set of frequent item sets, this means that we will overestimate the support of item sets on the result of a query. However, different from the lifting approach, we do allow access to the query result and, hence, the overestimation can be corrected. This is the second reason why transforming KRIMP works.

This reasoning makes the question "Can we transform item set mining?" all the more relevant. Unfortunately, the answer to this question is *probably not*. This can be easily seen from the join. The input for the transformed item set miner would be the joined tables as well as the Cartesian product of the sets of frequent item sets on the "blown-up" individual tables. This set of candidate frequent item sets will be **prohibitively large**, far larger than the final set of item sets that is frequent on the join. Hence, checking all these candidates will be more expensive than computing only the frequent ones efficiently.

Pruning the set of candidates while searching for the frequent ones requires a data structure that stores all candidates. Whenever, we can prune, a set of candidates has to be physically deleted from this data structure. The normal item set miners do not even generate most of these pruned candidates. In this approach we would first generate and then delete them. In other words, it is highly unlikely that this approach will have a performance similar to the best item set miners. Let alone that it will be significantly more efficient than these algorithms, as is required by the transformation approach.

In turn, this reasoning points to the third reason why transforming KRIMP works. The code tables KRIMP produces are small, far smaller than the set of frequent item sets. Hence, checking the support of all candidates suggested by KRIMP is not detrimental for the efficiency of KRIMP*.

From this discussion we can derive the following succinct all-encompassing reason why transforming KRIMP works. KRIMP produces, relatively, small code tables that capture the support of all item sets rather well, such that checking the set of all suggested candidates is rather cheap.

Note that the comparison of the two approaches for a single case, i.e., that of item sets does not imply at all that the second approach is inherently superior to the first one. In fact, we already argued at the start of this section that the first approach, if applicable, is to be preferred above the second one. Moreover, in [12] we argued that the first approach is applicable for the discovery of Bayesian networks from data. In other words, the first approach is a viable approach.

A conclusion we can, tentatively, draw from the discussion in this section is that for either approach to work, the models should capture the data distribution well.

13.6 Conclusions and Prospects for Further Research

In this chapter we introduced a problem that has received little attention in the literature on inductive databases or in the literature on data mining in general. This question is: does knowing models on the database help in inducing models on the result of a query on that database?

We gave two approaches to solve this problem, induced by two interpretations of "help". The first, more elegant, one produces results without access to the result of the query. The second one does allow access to this result.

We investigated both approaches for item set mining. It turned out that the first approach is not applicable to frequent item set mining, while the second one produced good experimental results for our KRIMP algorithm. In Section 13.5 we discussed this failure and success. The final tentative conclusion of this discussion is: for either approach to work, the models should capture the data distribution well.

This conclusion points directly to other classes of models that may be good candidates for either approach, i.e., those models that capture a detailed picture of the data distribution. One example are Bayesian networks already discussed in [12]. Just as interesting, if not even more, are models based on bagging or boosting or similar approaches. Such models do not concentrate all effort on the overall data distribution, but also take small selections with their own distribution into account. Hence, for such models one would expect that, e.g., lifting the selection operator should be relatively straight forward.

This is an example for a much broader research agenda: For which classes of models and algorithms do the approaches work? Clearly, we have only scratched the surface of this topic. Another, similarly broad, area for further research is: Are there other, better, ways to formalise "help"?

References

1. Rakesh Agrawal, Heikki Mannila, Ramakrishnan Srikant, Hannu Toivonen, and A. Inkeri Verkamo. Fast discovery of association rules. In *Advances in Knowledge Discovery and Data Mining*, pages 307–328. AAAI, 1996.
2. Andrea Asperti and Giuseppe Longo. *Categories, Types, and Structures*. MIT Press, 1991.
3. P.J. Bickel, E.A. Hammel, and J.W. O'Connell. Sex bias in graduate admissions: Data from berkeley. *Science*, 187(4175):398–404, 1975.
4. Rudi Cilibrasi and Paul Vitanyi. Automatic meaning discovery using google. In *IEEE Transactions on Knowledge and Data Engineering*, volume 19, pages 370–383. 2007.
5. E.F. Codd. A relational model of data for large shared data banks. *Communications of the ACM*, 13(6):377–387, 1970.
6. Frans Coenen. The LUCS-KDD discretised/normalised ARM and CARM data library: http://www.csc.liv.ac.uk/~frans/ KDD/Software/LUCS_KDD_DN/. 2003.
7. T.M. Cover and J.A. Thomas. *Elements of Information Theory, 2nd ed.* John Wiley and Sons, 2006.
8. C. Faloutsos and V. Megalooikonomou. On data mining, compression and kolmogorov complexity. In *Data Mining and Knowledge Discovery*, volume 15, pages 3–20. Springer Verlag, 2007.
9. Usama M. Fayyad, Gregory Piatetsky-Shapiro, and Padhraic Smyth. From data mining to knowledge discovery: An overview. 1996.
10. Peter D. Grünwald. Minimum description length tutorial. In P.D. Grünwald and I.J. Myung, editors, *Advances in Minimum Description Length*. MIT Press, 2005.
11. Raymond T. Ng, Laks V. S. Lakshmanan, Jiawei Han, and Alex Pang. Exploratory mining and pruning optimizations of constrained associations rules. In *Proc. ACM SIGMOD conference*, 1998.
12. Arno Siebes. Data mining in inductive databases. In Francesco Bonchi and Jean-François Boulicaut, editors, *Knowledge Discovery in Inductive Databases, 4th International Workshop, KDID 2005, Revised Selected and Invited Papers*, volume 3933 of *Lecture Notes in Computer Science*, pages 1–23. Springer, 2005.
13. Arno Siebes and Diyah Puspitaningrum. Mining databases to mine queries faster. In Wray L. Buntine, Marko Grobelnik, Dunja Mladenic, and John Shawe-Taylor, editors, *Proceedings ECML PKDD 2009, Part II*, volume 5782 of *Lecture Notes in Computer Science*, pages 382–397. Springer, 2009.
14. Arno Siebes, Jilles Vreeken, and Matthijs van Leeuwen. Item sets that compress. In *Proceedings of the SIAM Conference on Data Mining*, pages 393–404, 2006.
15. Jilles Vreeken, Matthijs van Leeuwen, and Arno Siebes. Preserving privacy through data generation. In *Proceedings of the IEEE International Conference on Data Mining*, pages 685–690, 2007.

Chapter 14
Experiment Databases

Joaquin Vanschoren and Hendrik Blockeel

Abstract Next to running machine learning algorithms based on inductive queries, much can be learned by immediately querying the combined results of many prior studies. Indeed, all around the globe, thousands of machine learning experiments are being executed on a daily basis, generating a constant stream of empirical information on machine learning techniques. While the information contained in these experiments might have many uses beyond their original intent, results are typically described very concisely in papers and discarded afterwards. If we properly store and organize these results in central databases, they can be immediately reused for further analysis, thus boosting future research. In this chapter, we propose the use of *experiment databases*: databases designed to collect all the necessary details of these experiments, and to intelligently organize them in online repositories to enable fast and thorough analysis of a myriad of collected results. They constitute an additional, queriable source of empirical meta-data based on principled descriptions of algorithm executions, without reimplementing the algorithms in an inductive database. As such, they engender a very dynamic, *collaborative* approach to experimentation, in which experiments can be freely shared, linked together, and immediately reused by researchers all over the world. They can be set up for personal use, to share results within a lab or to create open, community-wide repositories. Here, we provide a high-level overview of their design, and use an existing experiment database to answer various interesting research questions about machine learning algorithms and to verify a number of recent studies.

Joaquin Vanschoren · Hendrik Blockeel
Department of Computer Science, Katholieke Univeristeit Leuven, Leuven, Belgium
e-mail: firstname.lastname@cs.kuleuven.be

14.1 Introduction

"Study the past", Confucius said, "if you would divine the future". This applies to machine learning and data mining as well: when developing new machine learning algorithms, we wish to know which techniques have been successful (or not) on certain problems in the past, and when analyzing new datasets, we assess the potential of certain machine learning algorithms, parameter settings and preprocessing steps based on prior experience with similar problems.

Since machine learning algorithms are typically heuristic in nature, much of this information is extracted from experiments. Much like in many other empirical sciences, we collect *empirical evidence* of the behavior of machine learning algorithms by observing their performance on different datasets. If we have a hypothesis about how algorithms will perform under certain conditions, we test this by running controlled experiments, hopefully discovering empirical laws that contribute to a better understanding of learning approaches. Additionally, exploratory studies also probe many algorithms to study their behavior or to assess their utility on new datasets.

As such, all around the globe, thousands of machine learning experiments are being executed on a daily basis, generating a constant stream of empirical information on machine learning techniques. Unfortunately, most of these experiments are interpreted with a single focus of interest, described only concisely in papers and discarded afterwards, while they probably have many uses beyond their original intent. If we properly store and organize these results, they can be immediately reused by other researchers and accelerate future research. But in order to make this possible, we need a system that can store descriptions of data mining run, including the learners and datasets used, and the models produced.

In this chapter, we present *experiment databases* (ExpDBs): databases designed to collect all necessary details of machine learning experiments, and to intelligently organize them in online repositories to enable fast and thorough analysis of a myriad of collected results. They engender a much more dynamic, *collaborative* approach to experimentation, in which experiments can be freely shared, linked together, and immediately reused by researchers all over the world, simply by querying them. As we shall see, the use of such public repositories is common practice in many other scientific disciplines, and by developing similar repositories for machine learning, we similarly aim to create an "open scientific culture where as much information as possible is moved out of people's heads and labs, onto the network and into tools that can help us structure and filter the information" [26].

ExpDBs thus constitute an additional, queriable source of empirical meta-data, generated by many different researchers. They are a kind of inductive databases in that they store models which can be queried afterwards; however, they differ from regular inductive databases in a number of ways.

First, an inductive database (IDB) stores a *single* dataset, together with models that may have been produced from that dataset by running inductive queries, and with properties of those models. An experiment database (ExpDB), on the other hand, stores *multiple* datasets, multiple learners, and multiple models resulting from running those learners on those datasets.

Second, rather than storing the datasets, learners, and models themselves, an ExpDB may in practice store only descriptions (in terms of predefined properties) of them. In a regular IDB, this would not make sense, as the model itself is what the user is interested in.

Finally, in an IDB, one typically queries the data, or the set of models stored in the database (as in the virtual mining views approach, see Chapter 11), to get a model as a result. In an ExpDB, one typically queries the datasets, models, and experimental results in order to find possible relationships between their properties.

In the following sections, we discuss the main benefits of experiment databases in Sect. 14.2 and present related work in other scientific disciplines in Sect. 14.3. Next, we provide a high-level overview of their design in Sect. 14.4. Finally, we illustrate their use in Sect. 14.5 by querying an existing experiment database to answer various interesting questions about machine learning algorithms and to verify a number of recent studies.

This chapter is based on prior work on experiment databases [6, 40, 42, 43].

14.2 Motivation

Thousands of machine learning research papers contain extensive experimental evaluations of learning algorithms. However, it is not always straightforward to interpret these published results and use them as stepping stones for further research: they often lack the details needed to reproduce or reuse them, and it is often difficult to see how generally valid they are.

14.2.1 Reproducibility and Reuse

Indeed, while much care and effort goes into machine learning studies, they are usually conducted with a single focus of interest and summarize the empirical results accordingly. The individual experiments are usually not made publicly available, thus making it impossible to reuse them for further or broader investigation. Moreover, because of space restrictions imposed on publications, it is often practically infeasible to publish all details of the experimental setup, making it, in turn, very hard for other researchers to reproduce the experiments and verify if the results are interpreted correctly. This lack of reproducibility has been warned against repeatedly [21, 34, 29, 17], and some conferences have started to require that all submitted research be fully reproducible [24], adding notices to the ensuing publications stating whether or not the results could be verified.

14.2.2 Generalizability and Interpretation

A second issue is that of generalizability: in order to ensure that results are generally valid, the empirical evaluation must cover many different conditions such as various parameter settings and various kinds of datasets, e.g., differing in size, skewness, noisiness or with or without being preprocessed with basic techniques such as feature selection. Unfortunately, many studies limit themselves to algorithm benchmarking, often exploring only a small set of different conditions. It has long been recognized that such studies are in fact only 'case studies' [1], and should be interpreted with caution.

A number of studies have illustrated that sometimes, overly general conclusions can be drawn. In time series analysis research, for instance, it has been shown that many studies were biased toward the datasets being used, leading to contradictory results [21]. Furthermore, Perlich et al. [30] describe how the relative performance of logistic regression and decision trees depends strongly on the *size* of dataset samples, which is often not taken into account. Finally, it has been shown that the relative performance of lazy learning and rule induction is easily dominated by the effects of parameter optimization, data sampling and feature selection [19]. These studies underline that there are good reasons to thoroughly explore different conditions, or at least to clearly state under which conditions certain conclusions may or may not hold.

14.2.3 Experiment Databases

The idea of (inductive) databases that log and organize all the details of one's machine learning experiments, providing a full and fair account of conducted research, was first proposed by one us (Blockeel) [5] as an elegant way to remedy the low reproducibility and generalizability of many machine learning experiments. Still, this work did not present details on how to construct such a database.

Blockeel and Vanschoren [6] provided the first implementation of an experiment database for supervised classification, and further work details how to query this database to gain insight into the performance of learning algorithms [39, 43].

14.2.3.1 Collaborative Experimentation

However, given the amount of effort invested in empirical assessment, and the potential value of machine learning results beyond the summarized descriptions found in most papers, it would be even more useful to employ such databases to create searchable, community-wide repositories, complete with tools to automatically *publish* experimental results online. Such repositories would be a tremendously valuable source of unambiguous information on all known algorithms for further investigation, verification and comparison.

It engenders a more dynamic, *collaborative* form of experimentation, in which as many experiments as possible are reused from previous studies, and in return, any additional experiments are again shared with the community [40]. The experiment databases discussed in this chapter allow exactly this: they offer a formal experiment description language (see Sect. 14.4) to import large numbers of experiments directly from data mining tools, performed by many different researchers, and make them immediately available to everyone. They can be set up for personal use, to share results within a lab and to create open, community-wide repositories.

14.2.3.2 Automatic Organization

Most importantly, they also make it easy to *reuse* all stored experiments by automatically organizing them. Every new experiment is broken down to its components (such as the algorithm, parameter settings and dataset used), and its results are related to the exact configuration of those components. It then only takes a query (e.g in SQL) to ask for all results under specific conditions. For instance, requesting the parameter settings of an algorithm and its performance results allows to track the general effect of each parameter. Additionally requesting the dataset size allows to highlight what influence that may have on those parameters. As will be illustrated in Sect. 14.5, such queries allow to quickly peruse the results under different conditions, enabling fast and thorough analysis of large numbers of collected results. The expressiveness of database query languages warrants that many kinds of hypothesis can be tested by writing only one or perhaps a few queries, and the returned results can be interpreted unambiguously, as all conditions under which they are valid are stated in the query itself.

As such, instead of setting up new experiments for each question one may be interested in, often a laborious procedure involving the manual collection of datasets and algorithms and the manual organization of results, one could simply write a query to retrieve the results of hundreds of algorithms on perhaps thousands of datasets, thus obtaining much more detailed results in a matter of seconds.

14.2.3.3 Meta-learning

Experiment databases also serve as a great platform for meta-learning studies [38, 41], i.e. to search for useful patterns in algorithm behavior. To this end, it is helpful to link the empirical results to known properties of datasets [25, 31], as well as properties of algorithms, such as the type of model used, or whether they produce high bias or variance error [20]. As such, all *empirical* results, past and present, are immediately linked to all known *theoretical* properties of algorithms and datasets, providing new grounds for deeper analysis.

Previous meta-learning projects, especially the StatLog [25] and METAL [8] projects, also collected large numbers of machine learning experiments with the goal of using this meta-data to discover patterns in learning behavior, but these repositories were not developed to ensure reproducibility, were not open to new results, nor facilitated thorough querying.

14.2.3.4 e-Sciences

As will be discussed in Sect. 14.3, many scientific fields have developed online infrastructures to share and combine empirical results from all over the world, thus enabling ever larger studies and speeding up research. In the resulting *deluge* of combined experimental results, machine learning techniques have proven very successful, discovering useful patterns and speeding up scientific progress. Still, in an apparent contradiction, machine learning experiments themselves are currently not being documented and organized well enough to engender the same automatic discovery of insightful patterns that may speed up the design of better algorithms or the selection of algorithms to analyze new collections of data. We aim to solve this contradiction.

14.2.4 Overview of Benefits

We can summarize the benefits of sharing machine learning experiments and storing them in public databases as follows:

Reproducibility   The database stores all details of the experimental setup, thus attaining the scientific goal of truly reproducible research.

Reference   All experiments, including algorithms and datasets, are automatically organized in one resource, creating a useful 'map' of all known approaches, their properties, and results on how well they fared on previous problems. This also includes *negative results*, which usually do not get published in the literature. As such, we get a detailed overview of how algorithms from many studies perform relative to one another, and many aspects of learning behavior, that may only be known to some experts, can be instantly explored by writing a query.

Visibility   It adds visibility to (better) algorithms that the user may not have been aware of.

Reuse   It saves time and energy, as previous experiments can be readily reused. Especially when benchmarking new algorithms on commonly used datasets, there is no need to run older algorithms over and over again, as their evaluations are likely to be available online. This would also improve the quality of many algorithm comparisons, because the original authors probably know best how to tune their algorithms, and because one can also easily take the stored dataset properties into account to find out how they affect the relative performance of algorithms.

Larger studies   It enables larger and more generalizable studies. Studies covering many algorithms, parameter settings and datasets are hugely expensive to run, but could become much more feasible if a large portion of the necessary experiments are available online. Even if many experiments are missing, one can use the existing experiments to get a first idea, and run additional experiments to fill in the blanks. And even when all the experiments have yet to be run, the automatic storage and organization of experimental results markedly simplify conducting such large scale experimentation and thorough analysis thereof.

Integration   The formalized descriptions of experiments also allow the integration of such databases in data mining tools, for instance, to automatically log and share every experiment in a study or to reuse past experiments to speed up the analysis of new problems.

14.3 Related Work

The idea of sharing empirical results is certainly not new: it is an intrinsic aspect of many sciences, especially *e-Sciences*: computationally intensive sciences, which use the internet as a global, user-driven collaborative workspace.

14.3.1 e-Sciences

In all these scientific fields, both the need for reproducibility and the recognition of the potential value of empirical results beyond the summarized descriptions found in most papers, has led to the creation of online, public infrastructures for experiment exchange. Although these infrastructures have evolved somewhat differently in each field, they do share the same three components:

A formal representation language   To enable a free exchange of experimental data, a standard and formal representation language needs to be agreed upon. Such a language may also contain guidelines about the information necessary to ensure reproducibility.

Ontologies   Defining a coherent and unambiguous description language is not straightforward. It requires a careful analysis of the concepts used within a domain and their relationships. This is formally represented in *ontologies* [12]: machine manipulable models of a domain providing a controlled vocabulary, clearly describing the interpretation of each concept.

A searchable repository   To reuse experimental data, we need to locate it first. Experiment repositories therefore still need to organize all data to make it easily retrievable.

Bioinformatics. Expression levels of thousands of genes, recorded to pinpoint their functions, are collected through high-throughput screening experiments called *DNA-microarrays*. To allow verification and reuse of the obtained data in further studies, *microarray databases* [35] were created to collect all such experiments. Experiment submission is even a condition for publication in several journals [4].

To support the sharing of these results, a set of guidelines was drawn up regarding the required Minimal Information About a Microarray Experiment (MIAME [9]). Moreover, a MicroArray Gene Expression Markup Language (MAGE-ML) was conceived so that data could be exchanged uniformly, and an ontology (MAGE-MO) was designed [35] to provide a controlled core vocabulary, in addition to more specific ontologies, such as the Gene Ontology [2]. Their success has instigated similar approaches in related fields, such as proteomics [44] and mass spectrometry data analysis. One remaining drawback is that experiment description is still partially performed manually. Still, some projects are automating the process further. The Robot Scientist [23] stores all experiments automatically, including all physical aspects of their execution and the hypotheses under study. It has autonomously made several novel scientific discoveries.

Astronomy. A similar evolution has taken place in the field of astronomy. Astronomical observations from telescopes all over the world are collected in so-called *Virtual Observatories* [36]. This provides astronomers with an unprecedented catalog - a World-Wide Telescope - to study the evolving universe. An extensive list of different protocols supports the automatic sharing of observations, such as XML formats for tabular information (VOTable) [27] and astronomical image data (FITS, including meta-data on how the image was produced), as well an Astronomical Data Query Language (ADQL) [45] and informal ontologies [13]. The data is stored in databases all over the world and is queried for by a variety of portals [32], now seen as indispensable to analyze the constant flood of data.

Physics. Various subfields of physics also share their experimental results in common repositories. Low-energy nuclear reaction data can be expressed using the Evaluated Nuclear Data File (ENDF) format and collected into searchable ENDF libraries.[1] In high-energy particle physics, the HEPDATA[2] website scans the literature and downloads the experimental details directly from the machines performing the experiments. Finally, XML-formats and databases have been proposed for high-energy nuclear physics as well [10].

14.3.2 Extension to Machine Learning

We will use the same three components to develop a similar infrastructure for the exchange of machine learning experiments. While different kinds of machine learning experiments exist, we can similarly express their structure and vocabulary to describe, share and organize them in a uniform fashion.

Moreover, experiments in machine learning should be much easier to manipulate. First, compared to the *in vitro* experiments in bioinformatics, the exchange of the *in silico* experiments in machine learning can be automated completely. Indeed, a great

---

[1] http://www.nndc.bnl.gov/exfor/endf00.jsp

[2] http://durpdg.dur.ac.uk/hepdata/

deal of experimentation is performed through data mining workbenches and smaller software tools. As such, experiments could be exported at the click of a button.

Second, in contrast to scientific equipment, we can store datasets and algorithms into the database as well, cross-linked with existing repositories for datasets [3] and machine learning algorithms [34]. As such, all information necessary to reproduce the stored experiments can be found easily.

14.4 A Pilot Experiment Database

In this section, we provide a high-level outline of how we designed our current experiment database, which, although built to extend easily to other tasks, is focused on supervised classification. A detailed discussion is outside the scope of this chapter: we will only highlight the most important aspects of its design and how it can be used in data mining research. All further details, including detailed design guidelines, database models, ontologies and XML definitions, can be found on the ExpDB website: http://expdb.cs.kuleuven.be

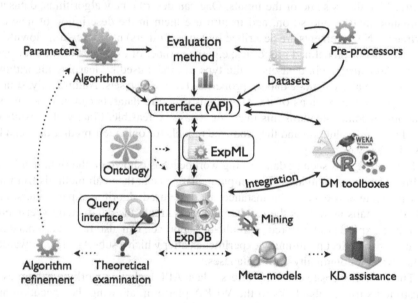

Fig. 14.1 An online infrastructure for experiment exchange.

14.4.1 Conceptual Framework

An overview of how experiment databases are used in practice is shown in Fig. 14.1. The five boxed components include the three components also used in e-Sciences: an ontology of domain concepts involved in running data mining experiments, a formal experiment description language (ExpML) and an experiment database to store and organize all experiments (ExpDB). In addition, two interfaces are defined: an application programming interface (API) to automatically export experiments from data mining software tools, and a query interface to browse the results of all stored experiments. Each is briefly discussed below.

14.4.1.1 Software Interface

First, to facilitate the automatic exchange of data mining experiments, an application programming interface (API) is provided that builds uniform, manipulable experiment instances (java objects) out of all necessary details and exports them as descriptions in ExpML language or directly stores them in a database. The top of Fig. 14.1 shows some of the inputs. One can describe new algorithms, datasets, evaluation metrics and so on, and in turn use them in the description of new experiments. New elements are described by (among others) name, version, download url and a list of predefined properties, e.g. the number of examples or the skewness of the class attribute in datasets or the type of model used by learning algorithms. The API can also calculate dataset properties for new datasets. Additionally, source code, executable versions of the algorithms or entire datasets can also be stored, although in some situations this may not always be feasible. Finally, the results of the algorithm evaluation and the produced models (or only their predictions) can be described as well.

Software agents such as data mining workbenches (shown on the right hand side in Fig. 14.1) or custom algorithm implementations can then call methods from the API to create new experiment instances, add the used algorithms, parameters, and all other details as well as the results, and then stream the completed experiments to online ExpDBs to be stored. A multi-tier approach can also be used: a personal database can collect preliminary experiments, after which a subset can be forwarded to lab-wide or community-wide databases.

The ExpDB website currently offers a Java API, including working examples to illustrate its use. It also links to the WEKA platform, allowing the execution and automatic storage of experiments on WEKA algorithms. Further extensions to other platforms, such as KNIME and Rapidminer are also planned.

This approach is quite different from other, more recent proposals for experiment databases, such as MLComp.[3] They require algorithms to be uploaded into the system, and scripts to be written that interface with the algorithm execution system.

---

[3] http://mlcomp.org

14.4.1.2 The Exposé Ontology

The vocabulary and structure of the ExpML files and database model is provided by an ontology of data mining experimentation, called Exposé. It provides a formal domain model that can be adapted and extended on a conceptual level, thus fostering collaboration between many researchers. Moreover, any conceptual extensions to the domain model can be translated consistently into updated or new ExpML definitions and database models, thus keeping them up to date with recent developments.

Exposé is built using concepts from several other data mining ontologies. First, OntoDM [28] (See Chap. 2) is a general ontology for data mining which tries to relate various data mining subfields. It provides the top-level classes for Exposé, which also facilitates the extension of Exposé to other subfields covered by OntoDM. Second, EXPO [33] models scientific experiments in general, and provides the top-level classes for the parts involving experimental designs and setups. Finally, DMOP [16] models the internal structure of learning algorithms, providing detailed concepts for general algorithm definitions. Exposé unites these three ontologies and adds many more concepts regarding specific types of experiments, evaluation techniques, evaluation metrics, learning algorithms and their specific configurations in experiments. In future work, we also wish to extend it to cover preprocessing techniques in more depth, for instance using the KD ontology [46] and DMWF ontology [22], which model the wider KD process. The full OWL-DL^4 description can be found online.

Exposé defines various kinds of experiments, such as 'learner evaluations', which apply a learning algorithm with fixed parameter settings on a static dataset, and evaluate it using a specific performance estimation method (e.g., 10-fold cross validation) and a range of evaluation metrics (e.g., predictive accuracy). As shown at the top of Fig. 14.2, experiments are described as workflows, with datasets as inputs and evaluations or models as outputs, and can contain sub-workflows of preprocessing techniques. Algorithms can also be workflows, with participants (components) such as kernels, distance functions or base-learners fulfilling a certain role. The top of Fig. 14.3 clarifies the structure of workflows: they can have several inputs and outputs, and consist of participants (operators), which in turn can also have multiple in- and outputs. Exposé also differentiates between general algorithms (e.g., 'decision trees'), versioned implementations (e.g., weka.J48) and applications (weka.J48 with fixed parameters). Finally, the context of sets of experiments can also be described, including conclusions, the employed experimental designs, and the papers in which they are used so they can be easily looked up afterwards.

---

^{4} http://www.w3.org/TR/owl-guide/

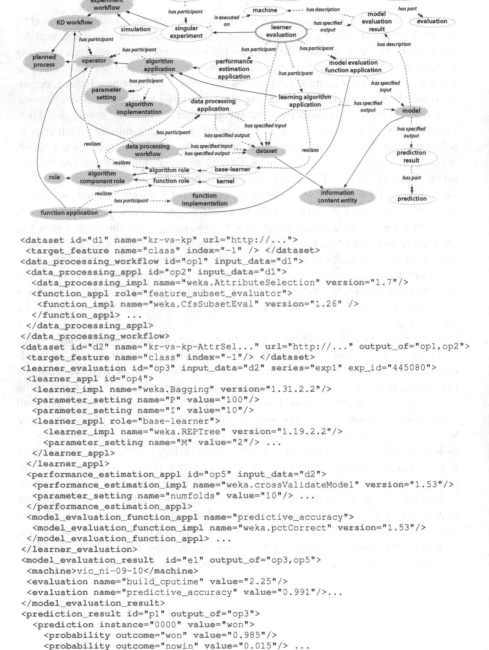

```
<dataset id="d1" name="kr-vs-kp" url="http://...">
 <target_feature name="class" index="-1" /> </dataset>
<data_processing_workflow id="op1" input_data="d1">
 <data_processing_appl id="op2" input_data="d1">
  <data_processing_impl name="weka.AttributeSelection" version="1.7"/>
  <function_appl role="feature_subset_evaluator">
   <function_impl name="weka.CfsSubsetEval" version="1.26" />
  </function_appl> ...
 </data_processing_appl>
</data_processing_workflow>
<dataset id="d2" name="kr-vs-kp-AttrSel..." url="http://..." output_of="op1,op2">
 <target_feature name="class" index="-1"/> </dataset>
<learner_evaluation id="op3" input_data="d2" series="exp1" exp_id="445080">
 <learner_appl id="op4">
  <learner_impl name="weka.Bagging" version="1.31.2.2"/>
  <parameter_setting name="P" value="100"/>
  <parameter_setting name="I" value="10"/>
  <learner_appl role="base-learner">
    <learner_impl name="weka.REPTree" version="1.19.2.2"/>
    <parameter_setting name="M" value="2"/> ...
  </learner_appl>
 </learner_appl>
 <performance_estimation_appl id="op5" input_data="d2">
  <performance_estimation_impl name="weka.crossValidateModel" version="1.53"/>
  <parameter_setting name="numfolds" value="10"/> ...
 </performance_estimation_appl>
 <model_evaluation_function_appl name="predictive_accuracy">
  <model_evaluation_function_impl name="weka.pctCorrect" version="1.53"/>
 </model_evaluation_function_appl> ...
</learner_evaluation>
<model_evaluation_result  id="e1" output_of="op3,op5">
 <machine>vic_ni-09-10</machine>
 <evaluation name="build_cputime" value="2.25"/>
 <evaluation name="predictive_accuracy" value="0.991"/>...
</model_evaluation_result>
<prediction_result id="p1" output_of="op3">
  <prediction instance="0000" value="won">
    <probability outcome="won" value="0.985"/>
    <probability outcome="nowin" value="0.015"/> ...
</prediction_result>
```

Fig. 14.2 Experimental workflows in Exposé (top) and ExpML (below).

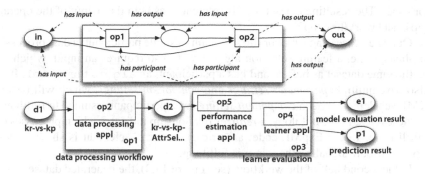

Fig. 14.3 Structure of a workflow (top) and an example (below, also see Fig. 14.2)

14.4.1.3 The ExpML Experiment Markup Language

Using the Exposé ontology as our core vocabulary, we can define a formal markup language for describing experiments, called ExpML. It is complementary to PMML[5], which allows to exchange predictive models, but not detailed experimental setups nor evaluations. It is derived by translating the ontological classes and properties (relationships) to XML elements and syntax. This translation process is especially useful because it allows ontological extensions (e.g. to new machine learning tasks) to be translated into updated ExpML definitions.

Because ontological relationships are more expressive than XML syntax, different relationships between these concepts need to be translated quite differently. Table 14.1 provides a short overview of these relationships and their XML equivalent. Figure 14.2 illustrates this process, showing a real experiment (experiment 445080 in our experiment database) expressed in ExpML. The structure of this particular experiment is shown at the bottom of Fig. 14.3. We assign an id to each operator and in- or output (e.g. datasets). For each operator, we state its inputs in an `input data` attribute, and for each output, we state the operator that generated that output in an `output of` attribute. As shown in the ExpML code, a dataset with `id='d1'` is used as the input of workflow 'op1' and data processing opera-

Table 14.1 Translating ontological properties to XML syntax.

Ontological property	XML syntax
has-part, pas-participant	target: subelement of source
has-description	(required) attribute
has-quality	subelement called `property`
is-concretization-of	`implementation_of` attribute
has-component	target: subelement of source with role attribute
has-specified-input	input given id, referenced in `input_data` attribute
has-specified-output	source given id, referenced in `output_of` attribute

---

[5] See `http://www.dmg.org/pmml-v3-2.html`

tor 'op2'. The resulting dataset 'd2' is references as both the output of the operator 'op1' and workflow 'op2'.

Our data processing sub-workflow contains a single participant: a data processing application, i.e. a feature selection algorithm. It also requires an input, which will be the same dataset as before, and has a participant: an `algorithm impl`. It can also have multiple *parameter settings* and *component settings*, which will become XML subelements. Each *component setting* has a participant assumed to fulfill each of these roles. In the ontology, a *realizes* relationship indicates which processes can fulfill them. In the ExpML code, a `function appl` element is shows, with a `role` attribute signaling the role it is fulfilling.

In the second half of the workflow (see Figure 14.3), the generated dataset serves as the input for a learner evaluation, which will in turn produce a model evaluation result and a prediction result. The evaluation consists of a learning algorithm application complete with parameter and component settings (in this case including a base learner application with its own parameter settings), the performance estimation technique (10-fold cross-validation) and a list of evaluation functions to assess the produced models, each pointing to their precise implementations.

The output of the experiment is shown next, consisting of all evaluation results (also stating the machine used in order to interpret cpu time) and all predictions, including the probabilities for each class. Although omitted for brevity, evaluation error margins are stored as well. Storing predictions is especially useful if we want to apply new evaluation metrics afterwards without rerunning all prior experiments.

14.4.1.4 The Experiment Databases (ExpDBs)

Finally, all submitted ExpML descriptions are interpreted and automatically stored into (relational) databases. The database model, also based on Exposé, is very fine-grained, so that queries can be written about any aspect of the experimental setup, evaluation results, or properties of involved components (e.g., dataset size). A working implementation in MySQL, containing over 650,000 experiments, can be queried online.

14.4.1.5 Query Interfaces

The database can be accessed through two query interfaces: an online interface on the homepage itself and an open-source desktop application. Both allow to launch queries written in SQL (many examples of SQL queries are supplied, including the ones used in Sect. 14.5), or composed in a graphical query interface, and can show the results in tables or graphical plots. The desktop application offers a wider range of plots, including self-organizing maps.

14.4.2 Using the Database

The three arrows emanating from the ExpDB at the bottom of Fig. 14.1 show different ways to tap into the stored information:

Querying This allows a researcher to formulate questions about the stored experiments, and immediately get all results of interest. Such queries could, for instance, be aimed at discovering ways in which an algorithm can be improved (e.g., see Sect. 14.5.2.1), after which that algorithm can be refined and tested again, thus completing the algorithm development cycle.

Mining A second use is to automatically look for patterns in algorithm performance by mining the stored results and theoretical meta-data. The insights provided by such *meta-models* can then be used to design better algorithms or to assist in knowledge discovery applications [8].

Integration Data mining toolboxes could also interface with ExpDBs directly, for instance to download the results of experiments that have been run before by a different user of that toolbox.

14.4.3 Populating the Database

The current database is populated with very diverse experiments to test algorithms under different conditions. First, we entered 54 classification algorithms from the WEKA platform together with all their parameters, 45 implementations of evaluation measures, 87 datasets from the UCI repository [3], 56 data characteristics calculated for each dataset, and two data preprocessors: correlation-based feature selection [14], and a subsampling procedure.

Next, three series of experiments were performed, in which a number of algorithms were explored more thoroughly than others:

- The first series simply ran all algorithms with default parameter settings.
- The second series varied each parameter, with at least 20 different values, of a selection of popular algorithms: SMO (a support vector machine (SVM) trainer), MultilayerPerceptron, J48 (C4.5), 1R, RandomForest, Bagging and Boosting. Moreover, different SVM kernels were used with their own parameter ranges, and all learners were used as base-learners for ensemble learners. We used a one-factor-at-a-time design to vary multiple parameters: each parameter (including the choice of base-learner or kernel) is varied in turn while keeping all others at default.
- Finally, the third series of experiments used a random sampling design to uniformly cover the entire parameter space (with at least 1000 settings) of an even smaller selection of algorithms: J48, Bagging and 1R.

All parameter settings were run on all datasets, and repeated 20 times with different random seeds for all algorithms that have them. In all cases, all 45 evaluation metrics were calculated in a 10-fold cross-validation procedure, with the same folds for each dataset. A large portion was additionally evaluated with a bias-variance analysis.

Quality Control. It is worth noting that, in the collaborative context, some form of quality control must be implemented to avoid *contamination* by bad (perhaps even fraudulent) ExpML descriptions. One solution, used in several repositories in bio-informatics, is to attach a trustworthiness value to the source of certain results. Experiments submitted from a trusted tool may be labeled very trustworthy, while custom submissions might get a lower value until the results are verified. Alternatively, if the database system can automatically run the algorithms in question, it could rerun all submitted experiments to verify the results.

14.5 Learning from the Past

In this section, we use the existing experiment database to illustrate how easily the results of previously stored experiments can be exploited for the discovery of new insights into a wide range of research questions, as well as to verify a number of recent studies. These illustrations can also be found in Vanschoren et al. [43]. Similar to Van Someren [37], we distinguish between three types of studies, increasingly making use of the available meta-level descriptions, and offering increasingly generalizable results:

1. Model-level analysis. These studies evaluate the produced models through a range of performance measures, but typically consider only individual datasets and algorithms. They typically try to identify HOW a specific algorithm performs, either on average or under specific conditions.
2. Data-level analysis. These studies investigate how known or measured data properties, not individual datasets, affect the performance of specific algorithms. They identify WHEN (on which kinds of data) an algorithm can be expected to behave a certain way.
3. Method-level analysis. These studies don't look at individual algorithms, but take general properties of the algorithms (eg. their bias-variance profile) into account to identify WHY an algorithm behaves a certain way.

14.5.1 Model-level Analysis

In the first type of study, we are interested in how individual algorithms perform on specific datasets. This type of study is typically used to benchmark, compare or rank algorithms, but also to investigate how specific parameter settings affect performance.

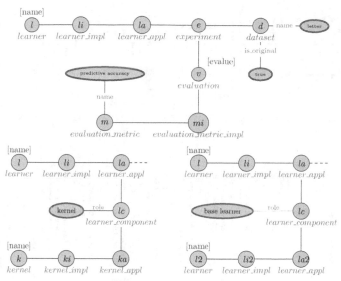

Fig. 14.4 A graph representation of our query. The top shows the main query, and below are two subqueries selecting the used kernels (left) and the base-learners of an ensemble method (right).

14.5.1.1 Comparing Algorithms

To compare the performance of all algorithms on one specific dataset, we write a query that simply selects the name of the algorithm used and the predictive accuracy recorded in all stored experiments on, for instance, the dataset 'letter'. A graph representation of this query is shown in Fig. 14.4. It joins the tables (nodes in the graph) of the learning algorithm, dataset, and evaluation based on the experiment in which they are used. It also selects the algorithm name and its evaluation (in brackets), and adds constraints (in ellipses) on the dataset name and the evaluation metric used. `is_original` indicates that the dataset is not preprocessed. For more detail, we can also select the kernel in the case of a SVM and the base-learner in the case of an ensemble. This is done in the subqueries shown in the bottom of Fig. 14.4. We order the results by their performance and plot the results in Fig. 14.5.

Since the returned results are always as general as the query allows, we now have a complete overview of how each algorithm performed. Next to their optimal performance, it is also immediately clear how much variance is caused by suboptimal parameter settings (at least for those algorithms whose parameters were varied). For instance, when looking at SVMs, it is clear that especially the RBF-kernel is of great use here (indeed, RBF kernels are popular in letter recognition problems), while the polynomial kernel is much less interesting. However, there is still much variation in the performance of the SVM's, so it might be interesting to investigate this in more detail. Also, while most algorithms vary smoothly as their parameters are altered, there seem to be large jumps in the performances of SVMs and RandomForests, which are, in all likelihood, caused by parameters that heavily af-

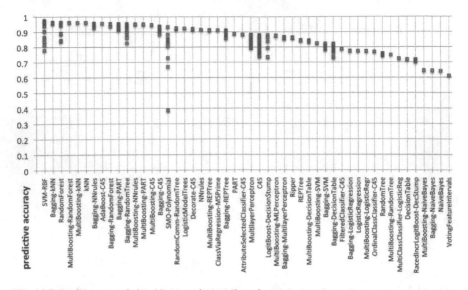

Fig. 14.5 Performance of algorithms on dataset 'letter'.

fect their performance. Moreover, when looking at bagging and boosting, it is clear that some base-learners are much more interesting than others. For instance, it appears that while bagging and boosting do give an extra edge to the nearest neighbor and logistic regression algorithms, the effect is rather limited. Conversely, bagging RandomTree seems to be hugely profitable, but this does not hold for boosting. It also seems more rewarding to fine-tune RandomForests, MultiLayerPerceptrons and SVMs than to bag or boost their default setting. Still, this is only one dataset, further querying is needed. Given the generality of the results, each query is likely to highlight things we were not expecting, providing interesting cases for further study.

14.5.1.2 Investigating Parameter Effects

First, we examine the effect of the parameters of the RBF kernel. Based on the first query, we can focus on the SVM's results by adding a constraint. Then we simply ask for the value of the parameter we are interested in. By selecting the value of the gamma parameter and plotting the results, we obtain Fig. 14.6. We constrain the datasets to a selection with the same default accuracy (10%).

On the 'mfeat_morphological' (and 'pendigits') dataset, performance increases when increasing gamma up to a point, after which it slowly declines. The other curves show that the effect on accuracy on other datasets is very different: performance is high for low gamma values, but quickly drops down to the default accuracy for higher values. Looking at the number of attributes in each dataset (shown in parentheses) we can observe some correlation.

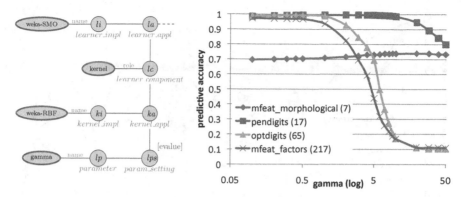

Fig. 14.6 The effect of parameter gamma of the RBF-kernel in SVMs on a number of different datasets, with their number of attributes shown in parentheses, and the accompanying query graph.

A possible explanation for this lies in the fact that this SVM implementation normalizes all attributes into the interval $[0,1]$. Therefore, the maximal squared distance between two examples, $\sum (a_i - b_i)^2$ for every attribute i, is equal to the number of attributes. Since the RBF-kernel computes $e^{(-\gamma * \sum (a_i - b_i)^2)}$, the kernel value will go to zero very quickly for large gamma-values and a large number of attributes, making the non-zero neighborhood around a support vector very small. Consequently, the SVM will overfit these support vectors, resulting in low accuracies. This suggests that the RBF kernel should take the number of attributes into account to make the default gamma value more suitable across a range of datasets. It also illustrates how the experiment database allows the investigation of algorithms in detail and assist their development.

14.5.1.3 General Comparisons

By simply dropping the constraints on the datasets used, the query will return the results over a large number of different problems. Furthermore, to compare algorithms over a range of performance metrics, instead of only considering predictive accuracy, we can use a normalization technique used by Caruana and Niculescu-Mizil [11]: normalize all performance metrics between the baseline performance and the best observed performance over all algorithms on each dataset. Using the aggregation functions of SQL, we can do this normalization on the fly, as part of the query.

We can now perform a very general comparison of supervised learning algorithms. We select all algorithms whose parameters were varied (see Sect. 14.4.3) and, though only as a point of comparison, logistic regression, nearest neighbors (kNN), naive Bayes and RandomTree with their default parameter settings. As for the performance metrics, we selected predictive accuracy, F-measure, precision and recall, the last three of which were averaged over all classes. We then queried for

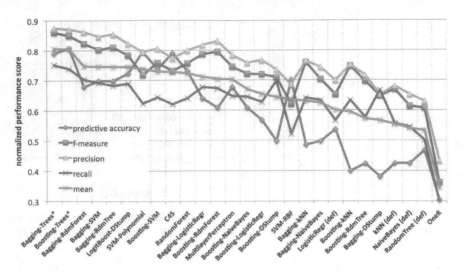

Fig. 14.7 Ranking of algorithms over all datasets on different performance metrics.

the maximal (normalized) performance of each algorithm for each metric on each dataset, averaged each of these scores over all datasets, and finally ranked all classifiers by the average of predictive accuracy, precision and recall.[6] The results of this query are shown in Fig. 14.7.

The overall best performing algorithms are mostly bagged and boosted ensembles. Especially bagged and boosted trees perform very well, in agreement with the previous results [11]. In Fig. 14.7 these are grouped as Trees* since they perform very similarly, and include C4.5, PART, Ripper, NaiveBayesTree, REPTree and similar tree-based learners. Another shared conclusion is that boosting full trees performs dramatically better than boosting stumps (see Boosting-DStump) or boosting random trees. While C45 seems to perform slightly better than RandomForests on predictive accuracy, this is only the case for multi-class datasets. When constraining the results to binary datasets (not shown here), RandomForests do outperform C45 on all metrics.

Furthermore, since this study contains many more algorithms, we can make a number of additional observations. For instance, the bagged versions of most strong learners (SVM, C45, RandomForest, etc.) seem to improve primarily on precision and recall, while the original base-learners (with optimized parameters) perform better on predictive accuracy. Apparently, tuning the parameters of these strong learners has a much larger effect on accuracy than on the other metrics, for which it is better to employ bagging than parameter tuning, at least on multi-class datasets.

---

[6] Since all algorithms were evaluated over all of the datasets (with 10-fold cross-validation), we could not optimize their parameters on a separate calibration set for this comparison. To limit the effect of overfitting, we only included a limited set of parameter settings, all of which fairly close to the default setting. Nevertheless, these results should be interpreted with caution as they might be overly optimistic.

14.5.2 Data-level Analysis

While the queries in the previous section allow the examination of the behavior of learning algorithms to a high degree of detail, they give no indication of exactly *when* (on which kind of datasets) certain behavior is to be expected. In order to obtain results that generalize over different datasets, we need to look at the properties of each individual dataset, and investigate how they affect learning performance.

14.5.2.1 Data Property Effects

In a first such study, we examine what causes the 'performance jumps' that we noticed with the RandomForest algorithm in Fig. 14.5. Querying for the effects of the number of trees in the forest and the dataset size yields Fig. 14.8.

This shows that predictive accuracy increases with the number of trees, usually leveling off between 33 and 101 trees.[7] One dataset, *monks problems 2*, is a notable exception: obtaining less than 50% accuracy on a binary problem, it actually performs worse as more trees are included. We also see that on large datasets, the accuracies for a given forest size vary less since the trees become more stable on large datasets, thus causing clear performance jumps on very large datasets. However, for very small datasets, the benefit of using more trees is overpowered by the randomization occurring in the trees (the algorithm considers K random features at each node).

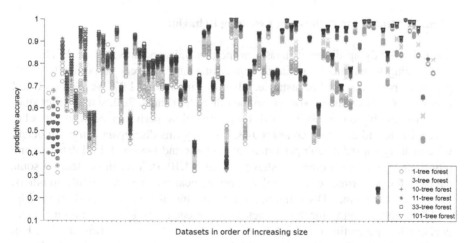

Fig. 14.8 The effect of dataset size and the number of trees for random forests. The dataset names are omitted since they are too small to be printed legibly.

---

[7] We used a geometric progression (1,3,11,33,101) of the number of trees, choosing for odd numbers to break ties while voting.

Fig. 14.9 Learning curves on
the Letter-dataset.

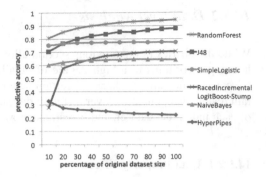

14.5.2.2 Preprocessing Effects

We can also study the effect of preprocessing methods. For instance, to investigate if
the results in Fig. 2 are also valid on smaller samples, we can query for the results on
downsampled versions of the dataset, yielding a learning curve for each algorithm,
as shown in Fig. 14.9. This provides further evidence that the ranking of algorithms
depends on the size of the dataset sample [30]. While logistic regression is initially
stronger than J48, the latter keeps on improving when given more data. Also note
that RacedIncrementalLogitBoost has a particularly steep learning curve, crossing
two other curves, and that the performance of the HyperPipes algorithm actually
worsens given more data, which suggests it was 'lucky' on the smaller samples.

14.5.2.3 Mining for Patterns in Learning Behavior

Instead of studying different dataset properties independently, we could also use
data mining techniques to relate the effect of many different properties to an algorithm's performance. For instance, when looking at Fig. 14.7, we see that OneR
performs obviously much worse than the other algorithms. Still, some earlier studies, most notably one by Holte [18], found very little performance differences between OneR and the more complex J48. To study this discrepancy in more detail,
we can query for the default performance of OneR and J48 on all UCI datasets, and
plot them against each other, as shown in Fig. 14.10(a). This shows that on some
datasets, the performances are similar (crossing near the diagonal), while on others,
J48 is the clear winner. Discretizing these results into three classes as shown in Fig.
14.10(a), and querying for the characteristics of each dataset, we can train a meta-
decision tree predicting on which kinds of datasets J48 has the advantage (see Fig.
14.10(b)). From this we learn that a high number of class values often leads to a
large win of J48 over OneR. Indeed, the original study [18] only had one dataset
with that many classes.

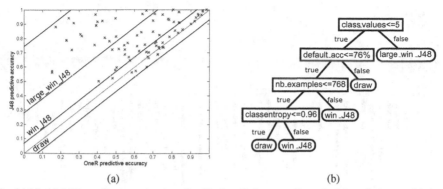

(a)   (b)

Fig. 14.10 (a) J48's performance against OneR's for all datasets, discretized into 3 classes. (b) A meta-decision tree predicting algorithm superiority based on data characteristics.

14.5.3 Method level analysis

While the results in the previous section are clearly more generalizable towards the datasets used, they don't explain *why* algorithms behave a certain way. They only consider individual algorithms and thus do not generalize over different techniques. Hence, we need to include algorithm properties in our queries as well.

14.5.3.1 Bias-variance Profiles

One very interesting property of an algorithm is its bias-variance profile [20]. Since the database contains a large number of bias-variance decomposition experiments, we can give a realistic, numerical assessment of how capable each algorithm is in reducing bias and variance error. In Fig. 14.11 we show, for each algorithm, the proportion of the total error that can be attributed to bias error, using default parameter settings and averaged over all datasets.

The algorithms are ordered from large bias (low variance), to low bias (high variance). NaiveBayes is, as expected, one of the algorithms with the strongest variance management (it avoids overfitting), but poor bias management (the ability to model complex target concepts). RandomTree, on the other hand, has very good bias management, but generates more variance error. When looking at ensemble methods, it shows that bagging reduces variance, as it causes REPTree to shift significantly to the left. Conversely, boosting reduces bias, shifting DecisionStump to the right in AdaBoost and LogitBoost.

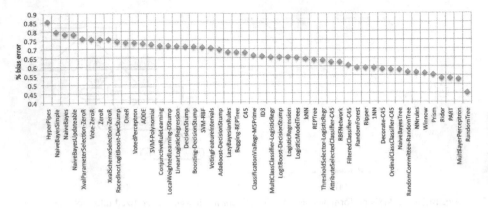

Fig. 14.11 Average percentage of bias-related error for each algorithm over all datasets.

14.5.3.2 Bias-variance Effects

As a final study, we investigate the claim by Brain and Webb [7] that on large datasets, the bias-component of the error becomes the most important factor, and that we should use algorithms with high bias management to tackle them. To verify this, we look for a connection between the dataset size and the proportion of bias error in the total error of a number of algorithms, using the previous figure to select algorithms with very different bias-variance profiles. By plotting the percentage of bias error generated on each dataset against the size of that dataset we obtain Fig. 14.12. Datasets or similar size are grouped for legibility. It shows that bias error is of varying significance on small datasets, but steadily increases in importance on larger datasets, for all algorithms. This validates the previous study on a much larger set of datasets. In this case (on UCI datasets), bias becomes the most important factor on datasets larger than 50000 examples, no matter which algorithm is used. As such, it is indeed advisable to look to algorithms with good bias management when dealing with large datasets, as variance becomes a less important factor.

14.6 Conclusions

Experiment databases are databases specifically designed to collect all the details on large numbers of past experiments, possibly performed by many different researchers, and make them immediately available to everyone. They ensure that experiments are repeatable and automatically organize them so that they can be easily reused in future studies.

In this chapter, we have provided a high-level overview of their design. Similar to experiment repositories actively used in e-Sciences, it consists of an ontologi-

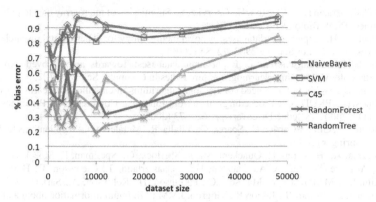

Fig. 14.12 Average percentage of bias-related error in algorithms vs dataset size.

cal domain model, which is in turn used to create a formal experiment description language and a detailed database model.

We also used an existing experiment database to illustrate how easily the results of previously stored experiments can be exploited to gain new insight: we performed elaborate algorithms comparisons, investigated the effects of algorithm parameters and data properties, suggested algorithm improvements, built meta-models of algorithm performance, and studied the bias-variance profiles of learning algorithms, each time by writing only a single query.

Experiment databases offer the possibility to truly unite the results of many individual machine learning studies, enhance cooperation, and facilitate large-scale empirical studies. As such, we are confident that they can contribute greatly to the vigor of machine learning research.

Our database is available online at http://expdb.cs.kuleuven.be

Acknowledgements Hendrik Blockeel was a Postdoctoral Fellow of the Fund for Scientific Research - Flanders (Belgium) (F.W.O.-Vlaanderen) at the time of this work, and this research is further supported by GOA 2003/08 "Inductive Knowledge Bases" and F.W.O.-Vlaanderen G.0108.06 "Foundations of Inductive Databases for Data Mining".

References

1. Aha, D.: Generalizing from case studies: A case study. Proceedings of the Ninth International Conference on Machine Learning pp. 1–10 (1992)
2. Ashburner, M., Ball, C.A., Blake, J.A., Botstein, D., Butler, H., Cherry, J.M., Davis, A.P., Dolinski, K., Dwight, S.S., Eppig, J.T., Harris, M.A., Hill, D.P., Issel-Tarver, L., Kasarskis, A., Lewis, S., Matese, J.C., Richardson, J.E., Ringwald, M., Rubin, G.M., Sherlock, G.: Gene ontology: tool for the unification of biology. nature genetics **25**, 25–29 (2000)
3. Asuncion, A., Newman, D.: UCI machine learning repository. University of California, School of Information and Computer Science (2007)

4. Ball, C., Brazma, A., Causton, H., Chervitz, S.: Submission of microarray data to public repositories. PLoS Biology **2**(9), e317 (2004)
5. Blockeel, H.: Experiment databases: A novel methodology for experimental research. Lecture Notes in Computer Science **3933**, 72–85 (2006)
6. Blockeel, H., Vanschoren, J.: Experiment databases: Towards an improved experimental methodology in machine learning. Lecture Notes in Computer Science **4702**, 6–17 (2007)
7. Brain, D., Webb, G.: The need for low bias algorithms in classification learning from large data sets. PKDD '02: Proceedings of the 6th European Conference on Principles of Data Mining and Knowledge Discovery pp. 62—73 (2002)
8. Brazdil, P., Giraud-Carrier, C., Soares, C., Vilalta, R.: Metalearning: Applications to data mining. Springer (2009)
9. Brazma, A., Hingamp, P., Quackenbush, J., Sherlock, G., Spellman, P., Stoeckert, C., Aach, J., Ansorge, W., Ball, C.A., Causton, H.C., Gaasterland, T., Glenisson, P., Holstege, F.C., Kim, I.F., Markowitz, V., Matese, J.C., Parkinson, H., Robinson, A., Sarkans, U., Schulze-Kremer, S., Stewart, J., Taylor, R., Vingron, J.V.M.: Minimum information about a microarray experiment. nature genetics **29**, 365 – 371 (2001)
10. Brown, D., Vogt, R., Beck, B., Pruet, J.: High energy nuclear database: a testbed for nuclear data information technology. International Conference on Nuclear Data for Science and Technology p. Article 250 (2007)
11. Caruana, R., Niculescu-Mizil, A.: An empirical comparison of supervised learning algorithms. Proceedings of the 23rd International Conference on Machine Learning (ICML'06) pp. 161–168 (2006)
12. Chandrasekaran, B., Josephson, J.: What are ontologies, and why do we need them? IEEE Intelligent systems **14**(1), 20–26 (1999)
13. Derriere, S., Preite-Martinez, A., Richard, A.: UCDs and ontologies. ASP Conference Series **351**, 449 (2006)
14. Hall, M.: Correlation-based feature selection for machine learning. Ph.D dissertation Hamilton, NZ: Waikato University, Department of Computer Science (1998)
15. Hall, M., Frank, E., Holmes, G., Pfahringer, B., Reutemann, P., Witten, I.: The weka data mining software: An update. SIGKDD Explorations **11**(1), 10–18 (2009)
16. Hilario, M., Kalousis, A., Nguyen, P., Woznica, A.: A data mining ontology for algorithm selection and meta-mining. Proceedings of the ECML/PKDD09 Workshop on 3rd generation Data Mining (SoKD-09) pp. 76–87 (2009)
17. Hirsh, H.: Data mining research: Current status and future opportunities. Statistical Analysis and Data Mining **1**(2), 104–107 (2008)
18. Holte, R.: Very simple classification rules perform well on most commonly used datasets. Machine Learning **11**, 63–91 (1993)
19. Hoste, V., Daelemans, W.: Comparing learning approaches to coreference resolution. there is more to it than bias. Proceedings of the Workshop on Meta-Learning (ICML-2005) pp. 20–27 (2005)
20. Kalousis, A., Hilario, M.: Building algorithm profiles for prior model selection in knowledge discovery systems. Engineering Intelligent Systems **8**(2) (2000)
21. Keogh, E., Kasetty, S.: On the need for time series data mining benchmarks: A survey and empirical demonstration. Data Mining and Knowledge Discovery **7**(4), 349–371 (2003)
22. Kietz, J., Serban, F., Bernstein, A., Fischer, S.: Towards cooperative planning of data mining workflows. Proceedings of the Third Generation Data Mining Workshop at the 2009 European Conference on Machine Learning (ECML 2009) pp. 1–12 (2009)
23. King, R., Rowland, J., Oliver, S., Young, M., Aubrey, W., Byrne, E., Liakata, M., Markham, M., Pir, P., Soldatova, L., Sparkes, A., Whelan, K., Clare, A.: The automation of science. Science **324**(3)(5923), 85–89 (2009)
24. Manolescu, I., Afanasiev, L., Arion, A., Dittrich, J., Manegold, S., Polyzotis, N., Schnaitter, K., Senellart, P., Zoupanos, S.: The repeatability experiment of SIGMOD 2008. ACM SIGMOD Record **37**(1) (2008)
25. Michie, D., Spiegelhalter, D., Taylor, C.: Machine learning, neural and statistical classification. Ellis Horwood (1994)

26. Nielsen, M.: The future of science: Building a better collective memory. APS Physics **17**(10) (2008)
27. Ochsenbein, F., Williams, R., Davenhall, C., Durand, D., Fernique, P., Hanisch, R., Giaretta, D., McGlynn, T., Szalay, A., Wicenec, A.: Votable: tabular data for the virtual observatory. Toward an International Virtual Observatory. Springer pp. 118–123 (2004)
28. Panov, P., Soldatova, L., Džeroski, S.: Towards an ontology of data mining investigations. Discovery Science (DS09). Lecture Notes in Artificial Intelligence **5808**, 257–271 (2009)
29. Pedersen, T.: Empiricism is not a matter of faith. Computational Linguistics **34**, 465–470 (2008)
30. Perlich, C., Provost, F., Simonoff, J.: Tree induction vs. logistic regression: A learning-curve analysis. The Journal of Machine Learning Research **4**, 211–255 (2003)
31. Pfahringer, B., Bensusan, H., Giraud-Carrier, C.: Meta-learning by landmarking various learning algorithms. Proceedings of the Seventeenth International Conference on Machine Learning pp. 743–750 (2000)
32. Schaaff, A.: Data in astronomy: From the pipeline to the virtual observatory. Lecture Notes in Computer Science **4832**, 52–62 (2007)
33. Soldatova, L., King, R.: An ontology of scientific experiments. Journal of the Royal Society Interface **3**(11), 795–803 (2006)
34. Sonnenburg, S., Braun, M., Ong, C., Bengio, S., Bottou, L., Holmes, G., LeCun, Y., Muller, K., Pereira, F., Rasmussen, C.E., Ratsch, G., Scholkopf, B., Smola, A., Vincent, P., Weston, J., Williamson, R.: The need for open source software in machine learning. Journal of Machine Learning Research **8**, 2443–2466 (2007)
35. Stoeckert, C., Causton, H., Ball, C.: Microarray databases: standards and ontologies. nature genetics **32**, 469–473 (2002)
36. Szalay, A., Gray, J.: The world-wide telescope. Science **293**, 2037–2040 (2001)
37. Van Someren, M.: Model class selection and construction: Beyond the procrustean approach to machine learning applications. Lecture Notes in Computer Science **2049**, 196–217 (2001)
38. Vanschoren, J., Van Assche, A., Vens, C., Blockeel, H.: Meta-learning from experiment databases: An illustration. Proceedings of the 16th Annual Machine Learning Conference of Belgium and The Netherlands (Benelearn07) pp. 120–127 (2007)
39. Vanschoren, J., Blockeel, H.: Investigating classifier learning behavior with experiment databases. Data Analysis, Machine Learning and Applications: 31st Annual Conference of the Gesellschaft für Klassifikation pp. 421–428 (2008)
40. Vanschoren, J., Blockeel, H.: A community-based platform for machine learning experimentation. Lecture Notes in Artificial Intelligence **5782**, 750–754 (2009)
41. Vanschoren, J., Blockeel, H., Pfahringer, B.: Experiment databases: Creating a new platform for meta-learning research. Proceedings of the ICML/UAI/COLT Joint Planning to Learn Workshop (PlanLearn08) pp. 10–15 (2008)
42. Vanschoren, J., Blockeel, H., Pfahringer, B., Holmes, G.: Organizing the world's machine learning information. Communications in Computer and Information Science **17**, 693–708 (2008)
43. Vanschoren, J., Pfahringer, B., Holmes, G.: Learning from the past with experiment databases. Lecture Notes in Artificial Intelligence **5351**, 485–492 (2008)
44. Vizcaino, J.A., Cote, R., Reisinger, F., Foster, J.M., Mueller, M., Rameseder, J., Hermjakob, H., Martens, L.: A guide to the proteomics identifications database proteomics data repository. Proteomics **9**(18), 4276–4283 (2009)
45. Yasuda, N., Mizumoto, Y., Ohishi, M., amd T Budavári, W.O., Haridas, V., Li, N., Malik, T., Szalay, A., Hill, M., Linde, T., Mann, B., Page, C.: Astronomical data query language: Simple query protocol for the virtual observatory. ASP Conference Proceedings **314**, 293 (2004)
46. Žáková, M., Kremen, P., Železný, F., Lavrač, N.: Planning to learn with a knowledge discovery ontology. Second planning to learn workshop at the joint ICML/COLT/UAI Conference pp. 29–34 (2008)

Part IV
Applications

Chapter 15
Predicting Gene Function using Predictive Clustering Trees

Celine Vens, Leander Schietgat, Jan Struyf, Hendrik Blockeel, Dragi Kocev, and Sašo Džeroski

Abstract In this chapter, we show how the predictive clustering tree framework can be used to predict the functions of genes. The gene function prediction task is an example of a hierarchical multi-label classification (HMC) task: genes may have multiple functions and these functions are organized in a hierarchy. The hierarchy of functions can be such that each function has at most one parent (tree structure) or such that functions may have multiple parents (DAG structure).

We present three predictive clustering tree approaches for the induction of decision trees for HMC, as well as an empirical study of their use in functional genomics. We show that the predictive performance of the best of the approaches outperforms C4.5H, a state-of-the-art decision tree system used in functional genomics, while yielding equally interpretable results.

By upgrading our method to an ensemble learner, the predictive performances outperform those of a recently proposed statistical learning method. The ensemble method also scales better and is easier to use. Our evaluation makes use of precision-recall-curves. We argue that this is a better evaluation criterion than previously used criteria.

Celine Vens · Leander Schietgat · Jan Struyf · Hendrik Blockeel
Department of Computer Science, Katholieke Universiteit Leuven,
Celestijnenlaan 200A, 3001 Leuven, Belgium
e-mail: {Celine.Vens, Leander.Schietgat, Jan.Struyf, Hendrik. Blockeel}@cs.kuleuven.be

Dragi Kocev · Sašo Džeroski
Department of Knowledge Technologies, Jožef Stefan Institute,
Jamova cesta 39, 1000 Ljubljana, Slovenia
e-mail: {Dragi.Kocev, Saso.Dzeroski}@ijs.si

15.1 Introduction

The completion of several genome projects in the past decade has generated the full genome sequence of many organisms. Identifying genes in the sequences and assigning biological functions to them has now become a key challenge in modern biology. This last step is often guided by automatic discovery processes, which interact with the laboratory experiments.

More precisely, biologists have a set of possible functions that genes may have, and these functions are organized in a hierarchy (see Fig. 15.1 for an example). It is known that a single gene may have multiple functions. Machine learning techniques are used to predict these gene functions. Afterwards, the predictions with highest confidence can be tested in the lab.

There are two characteristics of the function prediction task that distinguish it from common machine learning problems: (1) a single gene may have multiple functions; and (2) the functions are organized in a hierarchy: a gene that is related to some function is automatically related to all its parent functions (this is called the hierarchy constraint). This particular problem setting is known as hierarchical multi-label classification (HMC).

Several methods can be distinguished that handle HMC tasks. A first approach transforms an HMC task into a separate binary classification task for each class in the hierarchy and applies a known classification algorithm. We refer to it as the SC (single-label classification) approach. This technique has several disadvantages. First, it is *inefficient*, because the learner has to be run $|C|$ times, with $|C|$ the number of classes, which can be hundreds or thousands in this application. Second, from the knowledge discovery point of view, the learned models *identify features relevant for one class*, rather than identifying features with high overall relevance. Finally, *the hierarchy constraint is not taken into account*, i.e., it is not automatically imposed that an instance belonging to a class should belong to all its superclasses.

A second approach is to adapt the SC method, so that this last issue is dealt with. Some authors have proposed to hierarchically combine the class-wise models in the prediction stage, so that a classifier constructed for a class c can only predict positive if the classifier for the parent class of c has predicted positive [4]. In addition, one can also take the hierarchy constraint into account during training by restrict-

```
1 METABOLISM
1.1 amino acid metabolism
1.1.3 assimilation of ammonia, metabolism of the glutamate group
1.1.3.1 metabolism of glutamine
1.1.3.1.1 biosynthesis of glutamine
1.1.3.1.2 degradation of glutamine
...
1.2 nitrogen, sulfur, and selenium metabolism
...
2 ENERGY
2.1 glycolysis and gluconeogenesis
...
```

Fig. 15.1 A small part of the hierarchical FunCat classification scheme [34].

ing the training set for the classifier for class c to those instances belonging to the parent class of c [11, 12]. This approach is called the HSC (hierarchical single-label classification) approach throughout the text.

A third approach is to develop learners that learn a single multi-label model that predicts all the classes of an example at once [16, 7]. In this way, the hierarchy constraint can be taken into account and features can be identified that are relevant to all classes. We call this the HMC approach.

In this work, we do not only consider tree structured class hierarchies, such as the example shown in Fig. 15.1, but also support more complex hierarchies structured as directed acyclic graphs (DAGs), where classes may have multiple parents. The latter occurs for example in the widely used Gene Ontology classification scheme [2].

Given our target application of functional genomics, we focus on decision tree methods, because they yield models that are interpretable for domain experts. Decision trees are well-known classifiers, which can handle large datasets, and produce accurate results. In Chapter 7 decision trees have been placed in the predictive clustering tree (PCT) context. We show how the three HMC approaches outlined above can be set in the PCT framework.

An experimental comparison shows that the approach that learns a single model (the HMC approach) outperforms the other approaches on all fronts: predictive performance, model size, and induction time. We show that the results obtained by this method also outperform previously published results for predicting gene functions in *S. cerevisiae* (baker's or brewer's yeast) and *A. thaliana*. Moreover, we show that by upgrading our method to an ensemble technique, classification accuracy improves further. Throughout these comparisons, we use precision-recall curves to evaluate predictive performance, which are better suited for this type of problems than commonly used measures such as accuracy, precision and ROC curves.

The text is organized as follows. We start by discussing previous work on HMC approaches in gene function prediction in Section 15.2. Section 15.3 presents the three PCT approaches for HMC in detail. In Section 15.4, we describe the precision-recall based performance measures. Section 15.5 presents the classification schemes and datasets used in the empirical study described in Section 15.6 and Section 15.7. Finally, we conclude in Section 15.8.

15.2 Related Work

A number of HMC approaches have been proposed in the area of functional genomics. Several approaches predict functions of unannotated genes based on known functions of genes that are nearby in a functional association network or protein-protein interaction network [46, 13, 29, 15, 35, 30, 45]. These approaches are based on label propagation, whereas the focus of this work is on learning global predictive models.

Deng et al. [20] predict gene functions with Markov random fields using protein interaction data. They learn a model for each gene function separately and ignore the hierarchical relationships between the functions. Lanckriet et al. [32] represent the data by means of a kernel function and construct support vector machines for each gene function separately. They only predict top-level classes in the hierarchy. Lee et al. [33] have combined the Markov random field approach of [20] with the SVM approach of [32] by computing diffusion kernels and using them in kernel logistic regression.

Obozinski et al. [36] present a two-step approach in which SVMs are first learned independently for each gene function separately (allowing violations of the hierarchy constraint) and are then reconciled to enforce the hierarchy constraint. Barutcuoglu et al. [4] have proposed a similar approach where unthresholded support vector machines are learned for each gene function separately (allowing violations of the hierarchy constraint) and then combined using a Bayesian network so that the predictions are consistent with the hierarchical relationships. Guan et al. [27] extend this method to an ensemble framework that is based on three classifiers: a classifier that learns a single support vector machine for each gene function, the Bayesian corrected combination of support vector machines mentioned above, and a classifier that constructs a single support vector machine per gene function and per data source and forms a Naive Bayes combination over the data sources. Valentini and Re [48] also propose a hierarchical ensemble method that uses probabilistic support vector machines as base learners and combines the predictions by propagating the *weighted true path rule* both top-down and bottom-up through the hierarchy, which ensures consistency with the hierarchy constraint.

Rousu et al. [41] present a more direct approach that does not require a second step to make sure that the hierarchy constraint is satisfied. Their approach is based on a large margin method for structured output prediction [44, 47]. Such work defines a joint feature map $\Psi(x, y)$ over the input space X and the output space Y. In the context of HMC, the output space Y is the set of all possible subtrees of the class hierarchy. Next, it applies SVM based techniques to learn the weights w of the discriminant function $F(x, y) = \langle w, \Psi(x, y) \rangle$, with $\langle \cdot, \cdot \rangle$ the dot product. The discriminant function is then used to classify a (new) instance x as $\text{argmax}_{y \in Y} F(x, y)$. There are two main challenges that must be tackled when applying this approach to a structured output prediction problem: (a) defining Ψ, and (b) finding an efficient way to compute the argmax function (the range of this function is Y, which is of size exponential in the number of classes). Rousu et al. [41] describe a suitable Ψ and propose an efficient method based on dynamic programming to compute the argmax. Astikainen et al. [3] extend this work by applying two kernels for structured output to the prediction of enzymatic reactions.

If a domain expert is interested in knowledge that can provide insight in the biology behind the predictions, a disadvantage of using support vector machines is the lack of interpretability: it is very hard to find out why a support vector machine assigns certain classes to an example, especially if a non-linear kernel is used.

Clare [16] presents an HMC decision tree method in the context of predicting gene functions of *S. cerevisiae*. She adapts the well-known decision tree algorithm

C4.5 [39] to cope with the issues introduced by the HMC task. First, where C4.5 normally uses class entropy for choosing the best split, her version uses the sum of entropies of the class variables. Second, she extends the method to predict classes on several levels of the hierarchy, assigning a larger cost to misclassifications higher up in the hierarchy. The resulting tree is transformed into a set of rules, and the best rules are selected, based on a significance test on a validation set. Note that this last step violates the hierarchy constraint, since rules predicting a class can be dropped while rules predicting its subclasses are kept. The non-hierarchical version of her method was later used to predict gene functions for *A. thaliana* [17]. Here the annotations are considered one level at the time, which also results in violations of the hierarchy constraint.

Geurts et al. [25] recently presented a decision tree based approach related to predictive clustering trees. They start from a different definition of variance and then kernelize this variance function. The result is a decision tree induction system that can be applied to structured output prediction using a method similar to the large margin methods mentioned above [47, 44]. Therefore, this system could also be used for HMC after defining a suitable kernel. To this end, an approach similar to that of Rousu et al. [41] could be used.

Blockeel et al. [7, 5] proposed the idea of using predictive clustering trees [6] for HMC tasks. This work [7] presents the first thorough empirical comparison between an HMC and SC decision tree method in the context of tree shaped class hierarchies. Vens et al. [49] extend the algorithm towards hierarchies structured as DAGs and show that learning one decision tree for predicting all classes simultaneously, outperforms learning one tree per class (even if those trees are built taking into account the hierarchy). In Schietgat et al. [42], the predictive performance of the HMC method and ensembles thereof is compared to results reported in the biomedical literature. The latter two articles form the basis for this chapter.

15.3 Predictive Clustering Tree Approaches for HMC

We start this section by defining the HMC task more formally (Section 15.3.1). Next, we instantiate three decision tree algorithms for HMC tasks in the PCT framework: an HMC algorithm (Section 15.3.2), an SC algorithm (Section 15.3.3), and an HSC algorithm (Section 15.3.4).

15.3.1 Formal Task Description

We define the task of hierarchical multi-label classification as follows:

Given:

- an instance space X,

- a class hierarchy $(C, \leq_h)$, where C is a set of classes and $\leq_h$ is a partial order representing the superclass relationship (for all $c_1, c_2 \in C$: $c_1 \leq_h c_2$ if and only if c_1 is a superclass of c_2),
- a set T of examples (x_k, S_k) with $x_k \in X$ and $S_k \subseteq C$ such that $c \in S_k \Rightarrow \forall c' \leq_h c$: $c' \in S_k$, and
- a quality criterion q (which typically rewards models with high predictive accuracy and low complexity).

Find: a function $f : X \to 2^C$ (where 2^C is the power set of C) such that f maximizes q and $c \in f(x) \Rightarrow \forall c' \leq_h c : c' \in f(x)$. We call this last condition the hierarchy constraint.

In our work, the function f is represented with predictive clustering trees.

15.3.2 Clus-HMC: An HMC Decision Tree Learner

The approach that we present is based on decision trees and is set in the predictive clustering tree (PCT) framework [6], see Chapter 7. This framework views a decision tree as a hierarchy of clusters: the top-node corresponds to one cluster containing all training examples, which is recursively partitioned into smaller clusters while moving down the tree. PCTs can be applied to both clustering and prediction tasks. The PCT framework is implemented in the CLUS system, which is available at `http://dtai.cs.kuleuven.be/clus`.

Before explaining the approach in more detail, we show an example of a (partial) predictive clustering tree predicting the functions of *S. cerevisiae* using homology data from Clare [16] (Fig. 15.2). The homology features are based on a sequence similarity search for each gene in yeast against all the genes in SwissProt. The functions are taken from the FunCat classification scheme [34]. Each internal node of the tree contains a test on one of the features in the dataset. Here, the attributes are binary and have been obtained after preprocessing the relational data with a frequent pattern miner. The root node, for instance, tests whether there exists a SwissProt protein that has a high similarity (e-value $< 1.0 \cdot 10^{-8}$) with the gene under consideration G, is classified into the rhizobiaceae group and has references to the database Interpro. In order to predict the functions of a new gene, the gene is routed down the tree according to the outcome of the tests. When a leaf node is reached, the gene is assigned the functions that are stored in it. Only the most specific functions are shown in the figure. In the rest of this section, we explain how the PCT is constructed. A detailed explanation is given in Vens et al. [49].

PCTs [6] are explained in Chapter 7 and can be constructed with a standard "top-down induction of decision trees" (TDIDT) algorithm, similar to CART [10] or C4.5 [39]. The algorithm (see Fig. 7.1) takes as input a set of training instances (i.e., the genes and their annotations). It searches for the best acceptable test that can be put in a node. If such a test can be found then the algorithm creates a new internal node and calls itself recursively to construct a subtree for each subset (cluster) in

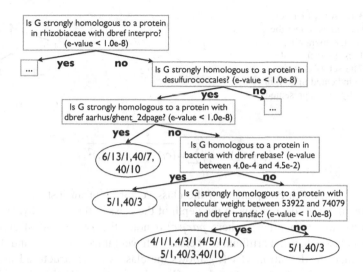

Fig. 15.2 Example of a predictive clustering tree, where the functions of a gene G are predicted, based on homology data.

the partition induced by the test on the training instances. To select the best test, the algorithm scores the tests by the reduction in variance (which is to be defined further) they induce on the instances. Maximizing variance reduction maximizes cluster homogeneity and improves predictive performance. If no acceptable test can be found, that is, if no test significantly reduces variance, then the algorithm creates a leaf and labels it with a representative case, or prototype, of the given instances.

To apply PCTs to the task of hierarchical multi-label classification, the variance and prototype are instantiated as follows [49].

First, the set of labels of each example is represented as a vector with binary components; the i'th component of the vector is 1 if the example belongs to class c_i and 0 otherwise. It is easily checked that the arithmetic mean of a set of such vectors contains as i'th component the proportion of examples of the set belonging to class c_i. We define the variance of a set of examples as the average squared distance between each example's class vector v_k and the set's mean class vector $\bar{v}$, i.e.,

$$Var(S) = \frac{\sum_k d(v_k, \bar{v})^2}{|S|}.$$

In HMC applications, it is generally considered more important to avoid making mistakes for terms at higher levels of the hierarchy than for terms at lower levels. For example in gene function prediction, predicting an "energy" gene function (i.e. FunCat class 1, see Fig. 15.1) while the gene is involved in "metabolism" (FunCat class 2) is worse than predicting "biosynthesis of glutamine" (FunCat class 1.1.3.1.1) instead of "degradation of glutamine" (FunCat class 1.1.3.1.2). To that aim, we use a weighted Euclidean distance

Fig. 15.3 (a) A toy hierarchy.
Class label names reflect the
position in the hierarchy,
e.g., '2/1' is a subclass of
'2'. (b) The set of classes
{1,2,2/2}, indicated in bold in
the hierarchy, and represented
as a vector.

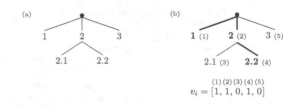

$$d(v_1,v_2) = \sqrt{\sum_i w(c_i) \cdot (v_{1,i} - v_{2,i})^2},$$

where $v_{k,i}$ is the i'th component of the class vector v_k of an instance x_k, and the class weights $w(c)$ decrease with the depth of the class in the hierarchy. We choose $w(c) = w_0 \cdot \text{avg } w(par_j(c))$, where $par_j(c)$ denotes the j'th parent of class c (the top-level classes have an artificial root class with weight $w(root) = 1$) and $0 < w_0 < 1$. Note that our definition of $w(c)$ allows the classes to be structured in a DAG, as is the case with the Gene Ontology. Consider for example the class hierarchy shown in Fig. 15.3, and two examples (x_1, S_1) and (x_2, S_2) with $S_1 = \{1, 2, 2/2\}$ and $S_2 = \{2\}$. Using a vector representation with consecutive components representing membership of class 1, 2, 2/1, 2/2 and 3, in that order,

$$d([1,1,0,1,0],[0,1,0,0,0]) = \sqrt{w_0 + w_0^2}.$$

The heuristic for choosing the best test for a node of the tree is then maximization of the variance reduction as discussed before, with the above definition of variance.

Second, a classification tree stores in a leaf the majority class for that leaf; this class will be the tree's prediction for examples arriving in the leaf. But in our case, since an example may have multiple classes, the notion of "majority class" does not apply in a straightforward manner. Instead, the mean $\bar{v}$ of the class vectors of the examples in that leaf is stored. Recall that $\bar{v}_i$ is the proportion of examples in the leaf belonging to c_i. An example arriving in the leaf can therefore be predicted to belong to class c_i if $\bar{v}_i$ is above some threshold t_i, which can be chosen by a domain expert. To ensure that the predictions fulfil the hierarchy constraint (whenever a class is predicted its superclasses are also predicted), it suffices to choose $t_i \leq t_j$ whenever c_i is a superclass of c_j. The PCT that is shown in Fig. 15.2 has a threshold of $t_i = 0.4$ for all i.

We call the resulting instantiation of the PCT algorithm in the CLUS system CLUS-HMC.

15.3.3 Clus-SC: Learning a Separate Tree for Each Class

The second approach that we consider builds a separate tree for each class in the hierarchy. Each of these trees is a single-label binary classification tree. Assume that the tree learner takes as input a set of examples labeled positive or negative. To construct the tree for class c with such a learner, we label the class c examples positive and all the other examples negative. The resulting tree predicts the probability that a new instance belongs to c. We refer to this method as single-label classification (SC).

In order to classify a new instance, SC thresholds the predictions of the different single-label trees, similar to CLUS-HMC. Note, however, that this does not guarantee that the hierarchy constraint holds, even if the thresholds are chosen such that $t_i \leq t_j$ whenever $c_i \leq_h c_j$.

The class-wise trees can be constructed with any classification tree induction algorithm. Note that CLUS-HMC reduces to a single-label binary classification tree learner when applied to such data; its class vector then reduces to a single component and its heuristic reduces to CART's Gini index [10]. We can therefore use the same induction algorithm (CLUS-HMC) for both the HMC and SC approaches. This makes the results easier to interpret. It has been confirmed [7] that on binary classification tasks, CLUS-HMC performs comparably to state-of-the-art decision tree learners. We call the SC approach with CLUS-HMC as decision tree learner CLUS-SC.

15.3.4 Clus-HSC: Learning a Separate Tree for Each Hierarchy Edge

Building a separate decision tree for each class has several disadvantages, such as the possibility of violating the hierarchy constraint. In order to deal with this issue, the CLUS-SC algorithm can be adapted as follows.

For a non top-level class c in a tree structured hierarchy, it holds that an instance can only belong to c if it belongs to c's parent $par(c)$. An alternative approach to learning a tree that directly predicts c, is therefore to learn a tree that predicts c given that the instance belongs to $par(c)$. Learning such a tree requires fewer training instances: only the instances belonging to $par(c)$ are relevant. The subset of these instances that also belong to c become the positive instances and the other instances (those belonging to $par(c)$ but not to c) the negative instances. The resulting tree predicts the conditional probability $P(c \mid par(c))$. W.r.t. the top-level classes, the approach is identical to CLUS-SC, i.e., all training instances are used.

To make predictions for a new instance, we use the product rule $P(c) = P(c \mid par(c)) \cdot P(par(c))$ (for non top-level classes). This rule applies the trees recursively, starting from the tree for a top-level class. For example, to compute the probability that the instance belongs to class 2.2, we first use the tree for class 2 to predict $P(2)$ and next the tree for class 2.2 to predict $P(2.2 \mid 2)$. The resulting probability is then $P(2.2) = P(2.2 \mid 2) \cdot P(2)$. For DAG structured hierarchies, the product rule can be applied for each parent class separately, and will yield a valid estimate of $P(c)$ based on that parent. To obtain an estimate of $P(c)$ based on all parent classes, we aggregate over the parent-wise estimates. In order to fulfil the hierarchy constraint, we use as aggregate function the minimum of the parent-wise estimates, i.e., $P(c) = \min_j P(c \mid par_j(c)) \cdot P(par_j(c))$.

Again, these probabilities are thresholded to obtain the predicted set of classes. As with CLUS-HMC, to ensure that this set fulfills the hierarchy constraint, it suffices to choose a threshold $t_i \le t_j$ whenever $c_i \le_h c_j$. We call the resulting algorithm CLUS-HSC (hierarchical single-label classification).

15.3.5 Ensembles of Predictive Clustering Trees

Ensemble methods are learning methods that construct a set of classifiers for a given prediction task and classify new examples by combining the predictions of each classifier. In this chapter, we consider bagging, an ensemble learning technique that has primarily been used in the context of decision trees.

Bagging [8] is an ensemble method where the different classifiers are constructed by making bootstrap replicates of the training set and using each of these replicates to construct one classifier. Each bootstrap sample is obtained by randomly sampling training instances, with replacement, from the original training set, until an equal number of instances is obtained. The individual predictions given by each classifier can be combined by taking the average (for numeric targets) or the majority vote (for nominal targets). Breiman [8] has shown that bagging can give substantial gains in predictive performance of decision tree learners. Also in the case of learning PCTs for predicting multiple targets at once, decision tree methods benefit from the application of bagging [31]. However, it is clear that, by using bagging on top of the PCT algorithm, the learning time of the model increases significantly, resulting in a clear trade-off between predictive performance and efficiency to be considered by the user.

The algorithm for bagging the PCTs takes an extra parameter k as input that denotes the number of trees in the ensemble. In order to make predictions, the average of all class vectors predicted by the k trees in the ensemble is computed, and then the threshold is applied as before. This ensures that the hierarchy constraint holds.

In the experiments, we will use bagged CLUS-HMC trees. We call the resulting instantiation of the bagging algorithm around the CLUS-HMC algorithm CLUS-HMC-ENS.

15.4 Evaluation Measure

We will report our predictive performance results with precision-recall curves. Precision is the probability that a positive prediction is correct, and recall is the probability that a positive instance is predicted positive. Remember that every leaf in the tree contains a vector $\bar{v}$ with for each class the probability that the instance has this class. When decreasing CLUS-HMC's prediction threshold t_i from 1 to 0, an increasing number of instances is predicted as belonging to class c_i, causing the recall for c_i to increase whereas precision may increase or decrease (with normally a tendency to decrease). Thus, a tree with specified threshold has a single precision and recall, and by varying the threshold a precision-recall curve (PR curve) is obtained. Such curves allow us to evaluate the predictive performance of a model regardless of t. In the end, a domain expert can choose a threshold according to the point on the curve which is most interesting to him.

Our decision to conduct a precision-recall based evaluation is motivated by the following three observations: (1) precision-recall evaluation was used in earlier approaches to gene function prediction [20, 15], (2) it allows one to simultaneously compare classifiers for different classification thresholds, and (3) it suits the characteristics of typical HMC datasets, in which many classes are infrequent (i.e., typically only a few genes have a particular function). Viewed as a binary classification task for each class, this implies that for most classes the number of negative instances by far exceeds the number of positive instances. We are more interested in recognizing the positive instances (i.e., that a gene has a given function), rather than correctly predicting the negative ones (i.e., that a gene does not have a particular function). ROC curves [38] are less suited for this task, exactly because they reward a learner if it correctly predicts negative instances (giving rise to a low false positive rate). This can present an overly optimistic view of the algorithm's performance [19].

Although a PR curve helps in understanding the predictive behavior of the model, a single performance score is more useful to compare models. A score often used to this end is the area between the PR curve and the recall axis, the so-called "area under the PR curve" (AUPRC). The closer the AUPRC is to 1.0, the better the model is.

With hundreds of classes, each of which has its own PR curve, there is the question of how to evaluate the overall performance of a system. We can construct a single "average" PR curve for all classes together by transforming the multi-label problem into a binary single-label one, i.e., by counting instance-class-couples instead of instances [49]. An instance-class couple is (predicted) positive if the instance has (is predicted to have) that class, it is (predicted) negative otherwise. The definition of precision and recall is then as before. We call the corresponding area the "area under the average PR curve" (AU($\overline{\text{PRC}}$)).

15.5 Datasets

Gene functions are categorized into ontologies for several reasons. First, they provide biologists with a controlled vocabulary; second, they reflect biological interdependences; and third, they ease the use of computational analysis. In this work, we consider two such ontologies: the Functional Catalogue and the Gene Ontology

The MIPS Functional Catalogue (FunCat, http://mips.gsf.de/projects/funcat) [34] is a tree structured vocabulary with functional descrip-

Table 15.1 Saccharomyces cerevisiae data set properties: number of instances $|D|$, number of attributes $|A|$.

Data set	$|D|$	$|A|$	Data set	$|D|$	$|A|$
D_1 Sequence [16] (seq)	3932	478	D_7 DeRisi et al. [21] (derisi)	3733	63
D_2 Phenotype [16] (pheno)	1592	69	D_8 Eisen et al. [22] (eisen)	2425	79
D_3 Secondary structure [16] (struc)	3851	19628	D_9 Gasch et al. [24] (gasch1)	3773	173
D_4 Homology search [16] (hom)	3867	47034	D_{10} Gasch et al. [23] (gasch2)	3788	52
D_5 Spellman et al. [43] (cellcycle)	3766	77	D_{11} Chu et al. [14] (spo)	3711	80
D_6 Roth et al. [40] (church)	3764	27	D_{12} All microarray [16] (expr)	3788	551

tions of gene products, consisting of 28 main categories. A small part of it is shown in Fig.15.1.

The structure of the Gene Ontology (GO, http://www.geneontology.org) [2] scheme differs substantially from FunCat, as it is not strictly hierarchical but organized as directed acyclic graphs, i.e. it allows more than one parent term per child. Another difference of the GO architecture is that it is organized as three separate ontologies: biological process, molecular function, and cellular localization. As can be seen in Table 15.3, GO has much more terms than FunCat.

Next to using two different classification schemes, we predict gene functions of two organisms: *Saccharomyces cerevisiae* and *Arabidopsis thaliana*, two of biology's classic model organisms. We use datasets described in [4], [16], and [17], with different sources of data that highlight different aspects of gene function. All datasets are available at the following webpage: http://dtai.cs.kuleuven.be/clus/hmc-ens.

15.5.1 Saccharomyces cerevisiae datasets

The first dataset we use (D_0) was described by Barutcuoglu et al. [4] and is a combination of different data sources. The input feature vector for a gene consists of pairwise interaction information, membership to colocalization locale, possession of transcription factor binding sites and results from microarray experiments, yielding a dataset with in total 5930 features. The 3465 genes are annotated with function terms from a subset of 105 nodes from the Gene Ontology's *biological process* hierarchy.

We also use the 12 yeast datasets ($D_1 - D_{12}$) from [16] (Table 15.1). The datasets describe different aspects of the genes in the yeast genome. They include five types of bioinformatics data: sequence statistics, phenotype, secondary structure, homology, and expression. The different sources of data highlight different aspects of gene function. The genes are annotated with functions from the FunCat classification schemes. Only annotations from the first four levels are given.

Table 15.2 Arabidopsis thaliana data set properties: number of instances $|D|$, number of attributes $|A|$.

Data set	$\|D\|$	$\|A\|$	Data set	$\|D\|$	$\|A\|$
D_{13} Sequence (seq)	3719	4450	D_{14} Expression (exprindiv)	3496	1251
D_{15} SCOP superfamily (scop)	3097	2003	D_{16} Secondary structure (struc)	3719	14804
D_{17} InterProScan data (interpro)	3719	2815	D_{18} Homology search (hom)	3473	72870

D_1 (seq) records sequence statistics that depend on the amino acid sequence of the protein for which the gene codes. These include amino acid frequency ratios, sequence length, molecular weight and hydrophobicity.

D_2 (pheno) contains phenotype data, which represents the growth or lack of growth of knock-out mutants that are missing the gene in question. The gene is removed or disabled and the resulting organism is grown with a variety of media to determine what the modified organism might be sensitive or resistant to.

D_3 (struc) stores features computed from the secondary structure of the yeast proteins. The secondary structure is not known for all yeast genes; however, it can be predicted from the protein sequence with reasonable accuracy, using Prof [37]. Due to the relational nature of secondary structure data, Clare performed a preprocessing step of relational frequent pattern mining; D_3 includes the constructed patterns as binary attributes.

D_4 (hom) includes for each yeast gene, information from other, homologous genes. Homology is usually determined by sequence similarity; here, PSI-BLAST [1] was used to compare yeast genes both with other yeast genes and with all genes indexed in SwissProt v39. This provided for each yeast gene a list of homologous genes. For each of these, various properties were extracted (keywords, sequence length, names of databases they are listed in, ...). Clare preprocessed this data in a similar way as the secondary structure data to produce binary attributes.

$D_5, \ldots, D_{12}$. Many microarray datasets exist for yeast and several of these were used [16]. Attributes for these datasets are real valued, representing fold changes in expression levels.

15.5.2 Arabidopsis thaliana datasets

We use six datasets from [17] (Table 15.2), originating from different sources: sequence statistics, expression, predicted SCOP class, predicted secondary structure, InterPro and homology. Each dataset comes in two versions: with annotations from the FunCat classification scheme and from the Gene Ontology's *molecular function* hierarchy. Again, only annotations for the first four levels are given. We use the manual annotations for both schemes.

D_{13} (seq) records sequence statistics in exactly the same way as for *S. cerevisiae*. D_{14} (exprindiv) contains 43 experiments from NASC's Affymetrix service "Affy-

Table 15.3 Properties of the two classification schemes for the updated yeast datasets. $|C|$ is the average number of classes actually used in the data sets (out of the total number of classes defined by the scheme). $|S|$ is the average number of labels per example, with between parentheses the average number counting only the most specific classes of an example.

	FunCat	GO
Scheme version	2.1 (2007/01/09)	1.2 (2007/04/11)
Yeast annotations	2007/03/16	2007/04/07
Total classes	1362	22960
Data set average $|C|$	492 (6 levels)	3997 (14 levels)
Data set average $|S|$	8.8 (3.2 most spec.)	35.0 (5.0 most spec.)

watch" (http://affymetrix.arabidopsis.info/AffyWatch.html), taking the signal, detection call and detection p-values. D_{15} (scop) consists of SCOP superfamily class predictions made by the Superfamily server [26]. D_{16} (struc) was obtained in the same way as for *S. cerevisiae*. D_{17} (interpro) includes features from several motif or signature finding databases, like PROSITE, PRINTS, Pfam, ProDom, SMART and TIGRFAMs, calculated using the EBI's stand-alone Inter-ProScan package [51]. To obtain features, the relational data was mined in the same manner as the structure data. D_{18} (hom) was obtained in the same way as for *S. cerevisiae*, but now using SWISSPROT v41.

15.6 Comparison of Clus-HMC/SC/HSC

In order to compare the three PCT approaches for HMC tasks, we use the 12 yeast data sets D_1 to D_{12} from Clare [16], but with new and updated class labels. We construct two versions of each data set. The input attributes are identical in both versions, but the classes are taken from the two different classification schemes FunCat and GO (we use GO's "is-a" relationship between terms). GO has an order of magnitude more classes than FunCat for our data sets: the FunCat datasets have 1362 classes on average, spread over 6 levels, while the GO datasets have 3997 classes, spread over 14 levels, see Table 15.3. The 24 resulting datasets can be found on the following webpage: http://dtai.cs.kuleuven.be/clus/hmcdatasets.html.

CLUS-HMC was run as follows. For the weights used in the weighted Euclidean distance in the variance calculation, w_0 was set to 0.75. The minimal number of examples a leaf has to cover was set to 5. The F-test stopping criterion takes a "significance level" parameter s, which was optimized as follows: for each out of 6 available values for s, CLUS-HMC was run on 2/3 of the training set and its PR curve for the remaining 1/3 validation set was constructed. The s parameter yielding the largest area under this average validation PR curve was then used to train the model on the complete training set. The results for CLUS-SC and CLUS-HSC were

Table 15.4 Predictive performance (AU($\overline{\text{PRC}}$)) of CLUS-HMC, CLUS-SC and CLUS-HSC.

Data set	FunCat labels			GO labels		
	HMC	HSC	SC	HMC	HSC	SC
seq	0.211	0.091	0.095	0.386	0.282	0.197
pheno	0.160	0.152	0.149	0.337	0.416	0.316
struc	0.181	0.118	0.114	0.358	0.353	0.228
hom	0.254	0.155	0.153	0.401	0.353	0.252
cellcycle	0.172	0.111	0.106	0.357	0.371	0.252
church	0.170	0.131	0.128	0.348	0.397	0.289
derisi	0.175	0.094	0.089	0.355	0.349	0.218
eisen	0.204	0.127	0.132	0.380	0.365	0.270
gasch1	0.205	0.106	0.104	0.371	0.351	0.239
gasch2	0.195	0.121	0.119	0.365	0.378	0.267
spo	0.186	0.103	0.098	0.352	0.371	0.213
expr	0.210	0.127	0.123	0.368	0.351	0.249
Average:	0.194	0.120	0.118	0.365	0.361	0.249

obtained in the same way as for CLUS-HMC, but with a separate run for each class (including separate optimization of s for each class).

Each algorithm was trained on 2/3 of each data set and tested on the remaining 1/3.

Table 15.4 shows the AU($\overline{\text{PRC}}$) of the three decision tree algorithms. Table 15.5 shows summarizing Wilcoxon outcomes comparing the AU($\overline{\text{PRC}}$) of CLUS-HMC to CLUS-SC and CLUS-HSC[1]. We see that CLUS-HMC performs better than CLUS-SC and CLUS-HSC, both for FunCat and GO. We see also that CLUS-HSC performs better than CLUS-SC on FunCat and on GO.

Table 15.6 shows the average number of leaves in the trees. We see that the SC trees are smaller than the HMC trees, because they each model only one class. Nevertheless, the total size of all SC trees is on average a factor 398 (FunCat) and 1392 (GO) larger than the corresponding HMC tree. This difference is bigger for GO than for FunCat because GO has an order of magnitude more classes and therefore also an order of magnitude more SC trees. Comparing HMC to HSC yields similar conclusions.

Observe that the HSC trees are smaller than the SC trees. We see two reasons for this. First, HSC trees encode less knowledge than SC ones because they are conditioned on their parent class. That is, if a given feature subset is relevant to all

---

[1] Given a pair of methods X and Y, the input to the Wilcoxon test is the test set performance (AUPRC) of the two methods on the 12 data sets. The null-hypothesis is that the median of the performance difference $Z_i = Y_i - X_i$ is zero. Briefly, the test orders the Z_i values by absolute value and then assigns them integer ranks such that the smallest $|Z_i|$ is ranked 1. It then computes the rank sum of the positive (W^+) and negative (W^-) Z_i. If $W^+ > W^-$, then Y is better than X because the distribution of Z is skewed to the right. Let $S = \min(W^+, W^-)$. The p-value of the test is the probability of obtaining a sum of ranks (W^+ or W^-) smaller than or equal to S, given that the null-hypothesis is true. In the results, we report the p-value together with W^+ and W^-.

Table 15.5 Comparison of the AU($\overline{\text{PRC}}$) of CLUS-HMC, CLUS-SC and CLUS-HSC. A '$\oplus$' ('$\ominus$') means that the first method performs better (worse) than the second method according to the Wilcoxon signed rank test. The table indicates the rank sums and corresponding p-values. Differences significant at the 0.01 level are indicated in bold.

	HMC vs. SC		HMC vs. HSC		HSC vs. SC	
	Score	p	Score	p	Score	p
FunCat	$\oplus$**78/0**	**$4.9 \cdot 10^{-4}$**	$\oplus$**78/0**	**$4.9 \cdot 10^{-4}$**	$\oplus$62/16	$7.7 \cdot 10^{-2}$
GO	$\oplus$**78/0**	**$4.9 \cdot 10^{-4}$**	$\oplus$43/35	$7.9 \cdot 10^{-1}$	$\oplus$**78/0**	**$4.9 \cdot 10^{-4}$**

classes in a sub-lattice of the hierarchy, then CLUS-SC must include this subset in each tree of the sub-lattice, while CLUS-HSC only needs them in the trees for the sub-lattice's most general border. Second, HSC trees use fewer training examples than SC trees, and tree size typically grows with training set size.

We also measure the total induction time for all methods. CLUS-HMC requires on average 3.3 (FunCat) and 24.4 (GO) minutes to build a tree. CLUS-SC is a factor 58.6 (FunCat) and 129.0 (GO) slower than CLUS-HMC. CLUS-HSC is faster than CLUS-SC, but still a factor 6.3 (FunCat) and 55.9 (GO) slower than CLUS-HMC.

Table 15.6 Average tree size (number of tree leaves) for FunCat and GO datasets. For CLUS-SC and CLUS-HSC we report both the total number of leaves in the collection of trees, and the average number of leaves per tree.

	CLUS-HMC	CLUS-SC		CLUS-HSC	
		Total	Average	Total	Average
FunCat	19.8	7878	15.9	3628	7.3
GO	22.2	30908	7.6	16988	3.0

15.7 Comparison of (Ensembles of) CLUS-HMC to State-of-the-art Methods

15.7.1 Comparison of CLUS-HMC to Decision Tree based Approaches

The previous section clearly showed the superiority of CLUS-HMC over CLUS-HSC and CLUS-SC. We now investigate how this method performs compared to state-of-the-art decision tree methods for functional genomics. As explained in Section 15.2, Clare [16] has presented an adaptation of the C4.5 decision tree algorithm towards HMC tasks. We compare our results to the results reported by Clare and

Fig. 15.4 Left: Average precision/recall over all classes for C4.5H, CLUS-HMC and CLUS-HMC-ENS on D_4 with FunCat annotations. Right: Precision-recall curve for class 29 on D_4 with FunCat annotations.

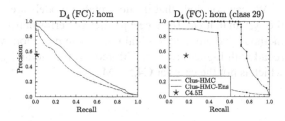

King [18] on *S. cerevisiae* (D_1 to D_{12}), and by Clare et al. [17] on *A. thaliana* D_{13} to D_{18}. The datasets that we use in this evaluation are exactly those datasets that are used in the mentioned articles. For the 18 datasets that are annotated with FunCat classes, we will compare to the hierarchical extension of C4.5 [18], which we will refer to as C4.5H. For the 6 datasets with GO annotations, we will use the non-hierarchical version [17], as C4.5H cannot handle hierarchies structured as a DAG. We refer to this system as C4.5M. For CLUS-HMC, all parameters were set as in the previous experiment.

For evaluating their systems, Clare et al. [17] report average precision. Indeed, as the biological experiments required to validate the learned rules are costly, it is important to avoid false positives. However, precision is always traded off by recall: a classifier that predicts one example positive, but misses 1000 other positive examples may have a precision of 1, although it can hardly be called a good classifier. Therefore, we also computed the average recall of the models obtained by C4.5H/M. These models were presented as rules derived from the trees, which enables us to plot only one point in PR space.

For each of the datasets these PR points are plotted against the average PR curves for CLUS-HMC. As we are comparing curves with points, we speak of a "win" for CLUS-HMC when its curve is above C4.5H/M's point, and of a "loss" when it is below the point. Under the null hypothesis that both systems perform equally well, we expect as many wins as losses. We observed that only in one case out of 24, C4.5H/M outperforms CLUS-HMC. For all other cases there is a clear win for CLUS-HMC. Representative PR curves can be found in Fig. 15.4 (left) and 15.5. For each of these datasets, we also compared the precision of C4.5H/M and CLUS-HMC, at the recall obtained by C4.5H/M. The results can be found in Fig. 15.6. The average gain in precision w.r.t. C4.5H/M is 0.209 for CLUS-HMC. Note that these figures also contain the results for the ensemble version of CLUS-HMC (see further).

Every leaf of a decision tree corresponds to an *if ... then ...* rule. When comparing the interpretability and precision/recall of these individual rules, CLUS-HMC also performs well. For instance, take FunCat class 29, with a prior frequency of 3%. Figure 15.4 (right) shows the PR evaluation for the algorithms for this class using homology dataset D_4. The PR point for C4.5H corresponds to one rule, shown in Fig. 15.7. This rule has a precision/recall of 0.55/0.17. CLUS-HMC's most precise rule for 29 is shown in Fig. 15.8. This rule has a precision/recall of 0.90/0.26.

We can conclude that if interpretable models are to be obtained, CLUS-HMC is the system that yields the best predictive performance. Compared with other existing methods, we are able to obtain the same precision with higher recall, or the same recall with higher precision. Moreover, the hierarchy constraint is always fulfilled, which is not the case for C4.5H/M.

15.7.2 Comparison of Ensembles of CLUS-HMC to an SVM based Approach

As explained in Sect. 15.3.5, we have extended CLUS-HMC to an ensemble induction algorithm (referred to as CLUS-HMC-ENS) in order to increase its predictive performance. More precisely, we built a bagging procedure around the PCT induction algorithm, each bag containing 50 trees in all experiments. As can be seen in Figures 15.4, 15.5, and 15.6, the improvement in predictive performance that is obtained by using ensembles carries over to the HMC setting.

We now compare CLUS-HMC-ENS to Bayesian-corrected SVMs [4]. This method was discussed in Sect. 15.2, and we refer to it as BSVM.

Barutcuoglu et al. [4] have used one dataset ($\mathbf{D_0}$) to evaluate their method. It is a combination of different data sources. The input feature vector for each *S. cerevisiae* gene consists of pairwise interaction information, membership to colocalization locale, possession of transcription factor binding sites, and results from microarray experiments. The genes are annotated with function terms from a subset of 105 nodes from the Gene Ontology's *biological process* hierarchy. They report classwise area under the ROC convex hull (AUROC) for these 105 functions. Although we have argued that precision-recall based evaluation is more suited for HMC problems, we adopt the same evaluation metric for this comparison. We also use the same evaluation method, which is based on out-of-bag estimates [9].

Fig. 15.9 compares the classwise out-of-bag AUROC estimates for CLUS-HMC-ENS and BSVM outputs. CLUS-HMC-ENS scores better on 73 of the 105 functions, while BSVM scores better on the remaining 32 cases. According to the (two-sided) Wilcoxon signed rank test [50], the performance of CLUS-HMC-ENS is significantly better ($p = 4.37 \cdot 10^{-5}$).

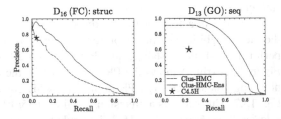

Fig. 15.5 Left: Average precision/recall over all classes for C4.5H, CLUS-HMC and CLUS-HMC-ENS on D_{16} with FunCat annotations. Right: Average curve for C4.5M, CLUS-HMC and CLUS-HMC-ENS on D_{13} with GO annotations.

Fig. 15.6 Precision of the C4.5H/M, CLUS-HMC and CLUS-HMC-ENS algorithms, at the recall obtained by C4.5H/M on FunCat (FC) and Gene Ontology (GO) annotations. The dark grey surface represents the gain in precision obtained by CLUS-HMC, the light grey surface represents the gain for CLUS-HMC-ENS. D_{14}(FC) was not included, since C4.5H did not find significant rules.

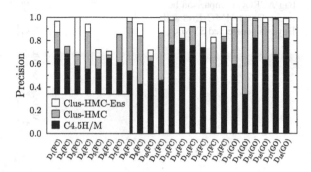

Moreover, CLUS-HMC-ENS is faster than BSVM. Run times are compared for one of the previously used datasets having annotations from Gene Ontology's com-

---

if	the ORF is NOT homologous to another yeast protein ($e \geq 0.73$) and homologous to a protein in rhodospirillaceae ($e < 1.0 \cdot 10^{-8}$)
and	NOT homologous to another yeast protein ($5.0 \cdot 10^{-4} < e < 3.3 \cdot 10^{-2}$) and homologous to a protein in anabaena ($e \geq 1.1$)
and	homologous to another yeast protein ($2.0 \cdot 10^{-7} < e < 5.0 \cdot 10^{-4}$) and homologous to a protein in beta_subdivision ($e < 1.0 \cdot 10^{-8}$)
and	NOT homologous to a protein in sinorhizobium with keyword transmembrane ($e \geq 1.1$)
and	NOT homologous to a protein in entomopoxvirinae with dbref pir ($e \geq 1.1$)
and	NOT homologous to a protein in t4-like_phages with molecular weight between 1485 and 38502 ($4.5 \cdot 10^{-2} < e < 1.1$)
and	NOT homologous to a protein in chroococcales with dbref prints ($1.0 \cdot 10^{-8} < e < 4.0 \cdot 10^{-4}$)
and	NOT homologous to a protein with sequence length between 344 and 483 and dbref tigr ($e < 1.0 \cdot 10^{-8}$)
and	homologous to a protein in beta_subdivision with sequence length between 16 and 344 ($e < 1.0 \cdot 10^{-8}$)
then	class 29/0/0/0 "transposable elements, viral and plasmid proteins"

Fig. 15.7 Rule found by C4.5H on the D_4 homology dataset.

---

if	the ORF is NOT homologous to a protein in rhizobiaceae_group with dbref interpro ($e < 1.0 \cdot 10^{-8}$)
and	NOT homologous to a protein in desulfurococcales ($e < 1.0 \cdot 10^{-8}$)
and	homologous to a protein in ascomycota with dbref transfac ($e < 1.0 \cdot 10^{-8}$)
and	homologous to a protein in viridiplantae with sequence length ≥ 970 ($e < 1.0 \cdot 10^{-8}$)
and	homologous to a protein in rhizobium with keyword plasmid ($1.0 \cdot 10^{-8} < e < 4.0 \cdot 10^{-4}$)
and	homologous to a protein in nicotiana with dbref interpro ($e < 1.0 \cdot 10^{-8}$)
then	class 29/0/0/0 "transposable elements, viral and plasmid proteins"

Fig. 15.8 Rule found by CLUS-HMC on the D_4 homology dataset.

Fig. 15.9 Class-wise out-of-bag AUROC comparison between CLUS-HMC-ENS and Bayesian-corrected SVMs.

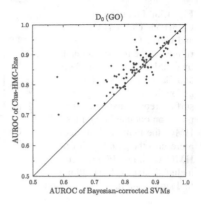

plete *biological process* hierarchy (in particular, we used D_{16} from Sect. 15.7.1, which is annotated with 629 classes). Run on a cluster of AMD Opteron processors (1.8 - 2.4GHz, $\geq$2GB RAM), CLUS-HMC-ENS required 34.8 hours, while SVM-light [28], which is the first step of BSVM, required 190.5 hours for learning the models (i.e., CLUS-HMC-ENS is faster by a factor 5.5 in this case).

15.8 Conclusions

An important task in functional genomics is to assign a set of functions to genes. These functions are typically organized in a hierarchy: if a gene has a particular function, it automatically has its superfunctions. This setting where instances can have multiple classes and where these classes are organized in a hierarchy is called hierarchical multi-label classification (HMC) in machine learning.

In this chapter, we have presented three instantiations of the predictive clustering tree framework for HMC: (1) an algorithm that learns a single tree that predicts all classes at once (CLUS-HMC), (2) an algorithm that learns a separate decision tree for each class (CLUS-SC), and (3) an algorithm that learns and applies such single-label decision trees in a hierarchical way (CLUS-HSC). The three algorithms are designed for problems where the class hierarchy is either structured as a tree or as a directed acyclic graph (DAG).

An evaluation of these approaches on functional genomics datasets shows that CLUS-HMC outperforms the other approaches on all fronts: predictive performance, model size, and induction times. We also show that CLUS-HMC outperforms a known decision tree learner (C4.5H). Moreover, it is possible to maximize predictive performance by constructing an ensemble of CLUS-HMC-trees. We show that the latter outperforms an approach based on SVMs, while still being efficient and easy to use.

Our evaluation makes use of precision-recall curves, which give the domain expert more insight into the relation between precision and recall. We argued that

PR-based evaluation measures are best suited for HMC problems, since they do not reward the negative predictions, i.e., predicting an example not to have particular labels (like ROC curves do).

We conclude that predictive clustering tree based methods are currently the most efficient, easy-to-use, and flexible approach to gene function prediction, flexible in the sense that they cover the spectrum from highly interpretable to highly accurate models.

Acknowledgements Part of the research presented in this chapter was conducted within the project IQ (*Inductive Queries for mining patterns and models*) funded by the European Commission of the EU under contract number FP6-IST 516169.

Celine Vens is a postdoctoral fellow of the Research Foundation - Flanders (FWO-Vlaanderen). Leander Schietgat is supported by a PhD grant of the Institute for the Promotion of Innovation through Science and Technology in Flanders (IWT-Vlaanderen) and the ERC Starting Grant 240186: Mining Graphs and Networks: a Theory-based approach. Dragi Kocev and Sašo Džeroski are currently supported by the project PHAGOSYS *Systems biology of phagosome formation and maturation - modulation by intracellular pathogens* funded by the European Commission of the EU under contract number FP7-HEALTH 223451. For a complete list of agencies, grants and institutions currently supporting Sašo Džeroski, please consult the Acknowledgements chapter of this volume.

The authors would like to thank Amanda Clare and Zafer Barutcuoglu for providing them with the datasets and the anonimous reviewers for providing many useful suggestions. This research was conducted utilizing high performance computational resources provided by K.U.Leuven, http://ludit.kuleuven.be/hpc.

References

1. Altschul, S., Madden, T., Schaffer, A., Zhang, J., Zhang, Z., Miller, W., Lipman, D.: Gapped BLAST and PSI-BLAST: A new generation of protein database search programs. *Nucleic Acids Research* **25**: 3389–3402 (1997)
2. Ashburner, M., Ball, C., Blake, J., Botstein, D., Butler, H., Cherry, J., Davis, A., Dolinski, K., Dwight, S., Eppig, J., Harris, M., Hill, D., Issel-Tarver, L., Kasarskis, A., Lewis, S., Matese, J., Richardson, J., Ringwald, M., Rubin, G., Sherlock, G.: Gene Ontology: Tool for the unification of biology. The Gene Ontology Consortium. *Nature Genetics* **25**(1): 25–29 (2000)
3. Astikainen, K., L., H., Pitkanen, E., S., S., Rousu, J.: Towards structured output prediction of enzyme function. *BMC Proceedings* **2**(Suppl 4): S2 (2008)
4. Barutcuoglu, Z., Schapire, R., Troyanskaya, O.: Hierarchical multi-label prediction of gene function. *Bioinformatics* **22**(7): 830–836 (2006).
5. Blockeel, H., Bruynooghe, M., Džeroski, S., Ramon, J., Struyf, J.: Hierarchical multi-classification. In: *Proc. Wshp on Multi-RelationalData Mining*, pp. 21–35. ACM SIGKDD (2002)
6. Blockeel, H., De Raedt, L., Ramon, J.: Top-down induction of clustering trees. In: *Proc. of the 15th Intl Conf. on Machine Learning*, pp. 55–63. Morgan Kaufmann (1998)
7. Blockeel, H., Schietgat, L., Struyf, J., Džeroski, S., Clare, A.: Decision trees for hierarchical multilabel classification: A case study in functional genomics. In: *Proc. of the 10th European Conf. on Principles and Practices of Knowledge Discovery in Databases*, LNCS, vol. 4213, pp. 18–29. Springer (2006)
8. Breiman, L.: Bagging predictors. Machine Learning **24**(2): 123–140 (1996)
9. Breiman, L.: Out-of-bag estimation. *Technical Report*, Statistics Department, University of California (1996)

10. Breiman, L., Friedman, J., Olshen, R., Stone, C.: *Classification and Regression Trees.* Wadsworth, Belmont (1984)
11. Cesa-Bianchi, N., Gentile, C., Zaniboni, L.: Incremental algorithms for hierarchical classification. *Journal of Machine Learning Research* **7**: 31–54 (2006)
12. Cesa-Bianchi, N., Valentini, G.: Hierarchical cost-sensitive algorithms for genome-wide gene function prediction. In *Proc. 3rd Intl Wshp on Machine Learning in Systems Biology, JMLR: Workshop and Conference Proceedings* **8**: 14–29 (2010)
13. Chen, Y., Xu, D.: Global protein function annotation through mining genome-scale data in yeast saccharomyces cerevisiae. *Nucleic Acids Research* **32**(21): 6414–6424 (2004)
14. Chu, S., DeRisi, J., Eisen, M., Mulholland, J., Botstein, D., Brown, P., Herskowitz, I.: The transcriptional program of sporulation in budding yeast. *Science* **282**: 699–705 (1998)
15. Chua, H., Sung, W., Wong, L.: Exploiting indirect neighbours and topological weight to predict protein function from protein-protein interactions. *Bioinformatics* **22**(13): 1623–1630 (2006)
16. Clare, A.: Machine Learning and Data Mining for Yeast Functional Genomics. Ph.D. thesis, University of Wales, Aberystwyth (2003)
17. Clare, A., Karwath, A., Ougham, H., King, R.D.: Functional bioinformatics for Arabidopsis thaliana. *Bioinformatics* **22**(9): 1130–1136 (2006)
18. Clare, A., King, R.D.: Predicting gene function in Saccharomyces cerevisiae. *Bioinformatics* **19**(Suppl. 2): 42–49 (2003).
19. Davis, J., Goadrich, M.: The relationship between precision-recall and ROC curves. In *Proc. of the 23rd Intl Conf. on Machine Learning*, pp. 233–240. ACM Press (2006)
20. Deng, M., Zhang, K., Mehta, S., Chen, T., Sun, F.: Prediction of protein function using protein-protein interaction data. In *Proc. of the IEEE Computer Society Bioinformatics Conf.*, pp. 197–206. IEEE Computer Society Press (2002)
21. DeRisi, J., Iyer, V., Brown, P.: Exploring the metabolic and genetic control of gene expression on a genomic scale. *Science* **278**: 680–686 (1997)
22. Eisen, M., Spellman, P., Brown, P., Botstein, D.: Cluster analysis and display of genome-wide expression patterns. In *Proc. National Academy of Sciences of USA* **95**(14): 14863–14868 (1998)
23. Gasch, A., Huang, M., Metzner, S., Botstein, D., Elledge, S., Brown, P.: Genomic expression responses to DNA-damaging agents and the regulatory role of the yeast ATR homolog Mec1p. *Molecular Biology of the Cell* **12**(10): 2987–3000 (2001)
24. Gasch, A., Spellman, P., Kao, C., Carmel-Harel, O., Eisen, M., Storz, G., Botstein, D., Brown, P.: Genomic expression program in the response of yeast cells to environmental changes. *Molecular Biology of the Cell* **11**: 4241–4257 (2000)
25. Geurts, P., Wehenkel, L., d'Alché Buc, F.: Kernelizing the output of tree-based methods. In *Proc. of the 23rd Intl Conf. on Machine learning*, pp. 345–352. ACM Press (2006).
26. Gough, J., Karplus, K., Hughey, R., Chothia, C.: Assignment of homology to genome sequences using a library of hidden markov models that represent all proteins of known structure. *Molecular Biology* **313**(4): 903–919 (2001)
27. Guan, Y., Myers, C., Hess, D., Barutcuoglu, Z., Caudy, A., Troyanskaya, O.: Predicting gene function in a hierarchical context with an ensemble of classifiers. *Genome Biology* **9**(Suppl 1): S3 (2008)
28. Joachims, T.: Making large-scale SVM learning practical. In: B. Scholkopf, C. Burges, A. Smola (eds.) *Advances in Kernel Methods – Support Vector Learning.* MIT Press (1999)
29. Karaoz, U., Murali, T., Letovsky, S., Zheng, Y., Ding, C., Cantor, C., Kasif, S.: Whole-genome annotation by using evidence integration in functional-linkage networks. *Proc. National Academy of Sciences of USA* **101**(9): 2888–2893 (2004)
30. Kim, W., Krumpelman, C., Marcotte, E.: Inferring mouse gene functions from genomic-scale data using a combined functional network/classification strategy. *Genome Biology* **9**(Suppl 1): S5 (2008)
31. Kocev, D., Vens, C., Struyf, J., Džeroski, S.: Ensembles of multi-objective decision trees. In: *Proc. of the 18th European Conf. on Machine Learning, LNCS*, vol. 4701, pp. 624–631. Springer (2007)

32. Lanckriet, G.R., Deng, M., Cristianini, N., Jordan, M.I., Noble, W.S.: Kernel-based data fusion and its application to protein function prediction in yeast. In *Proc. of the Pacific Symposium on Biocomputing*, pp. 300–311. World Scientific Press (2004)
33. Lee, H., Tu, Z., Deng, M., Sun, F., Chen, T.: Diffusion kernel-based logistic regression models for protein function prediction. *OMICS* **10**(1): 40–55 (2006)
34. Mewes, H., Heumann, K., Kaps, A., Mayer, K., Pfeiffer, F., Stocker, S., Frishman, D.: MIPS: A database for protein sequences and complete genomes. *Nucleic Acids Research* **27**: 44–48 (1999)
35. Mostafavi, S., Ray, D., Warde-Farley, D., Grouios, C., Morris, Q.: GeneMANIA: a real-time multiple association network integration algorithm for predicting gene function. *Genome Biology* **9**(Suppl 1): S4 (2008)
36. Obozinski, G., Lanckriet, G., Grant, C., Jordan, M., Noble, W.: Consistent probabilistic outputs for protein function prediction. *Genome Biology* **9**(Suppl 1): S6 (2008)
37. Ouali, M., King, R.: Cascaded multiple classifiers for secondary structure prediction. *Protein Science* **9**(6): 1162–76 (2000)
38. Provost, F., Fawcett, T.: Analysis and visualization of classifier performance: comparison under imprecise class and cost distributions. In *Proc. of the Third Intl Conf. on Knowledge Discovery and Data Mining*, pp. 43–48. AAAI Press (1998)
39. Quinlan, J.: *C4.5: Programs for Machine Learning*. Morgan Kaufmann (1993)
40. Roth, F., Hughes, J., Estep, P., Church, G.: Fining DNA regulatory motifs within unaligned noncoding sequences clustered by whole-genome mRNA quantitation. *Nature Biotechnology* **16**: 939–945 (1998)
41. Rousu, J., Saunders, C., Szedmak, S., Shawe-Taylor, J.: Kernel-based learning of hierarchical multilabel classification models. *Journal of Machine Learning Research* **7**: 1601–1626 (2006)
42. Schietgat, L., Vens, C., Struyf, J., Blockeel, H., Kocev, D., Džeroski, S.: Predicting gene function using hierarchical multi-label decision tree ensembles. *BMC Bioinformatics* **11**:2 (2010)
43. Spellman, P., Sherlock, G., Zhang, M., Iyer, V., Anders, K., Eisen, M., Brown, P., Botstein, D., Futcher, B.: Comprehensive identification of cell cycle-regulated genes of the yeast *Saccharomyces cerevisiae* by microarray hybridization. *Molecular Biology of the Cell* **9**: 3273–3297 (1998)
44. Taskar, B., Guestrin, C., Koller, D.: Max-margin Markov networks. *Advances in Neural Information Processing Systems 16*. MIT Press (2003)
45. Tian, W., Zhang, L., Tasan, M., Gibbons, F., King, O., Park, J., Wunderlich, Z., Cherry, J., Roth, F.: Combining guilt-by-association and guilt-by-profiling to predict saccharomyces cerevisiae gene function. *Genome Biology* **9**(Suppl 1): S7 (2008)
46. Troyanskaya, O., Dolinski, K., Owen, A., Altman, R., D., B.: A bayesian framework for combining heterogeneous data sources for gene function prediction (in saccharomyces cerevisiae). *Proc. National Academy of Sciences of USA* **100**(14): 8348–8353 (2003)
47. Tsochantaridis, I., Joachims, T., Hofmann, T., Altun, Y.: Large margin methods for structured and interdependent output variables. *Journal of Machine Learning Research* **6**: 1453–1484 (2005)
48. Valentini, G., Re, M.: Weighted true path rule: a multilabel hierarchical algorithm for gene function prediction. In *Proc. of the 1st Intl Wshp on Learning from Multi-Label Data*, pp. 133–146. ECML/PKDD (2009)
49. Vens, C., Struyf, J., Schietgat, L., Džeroski, S., Blockeel, H.: Decision trees for hierarchical multi-label classification. *Machine Learning* **73**(2): 185–214 (2008)
50. Wilcoxon, F.: Individual comparisons by ranking methods. *Biometrics* **1**: 80–83 (1945)
51. Zdobnov, E., Apweiler, R.: Interproscan - an integration platform for the signature-recognition methods in interpro. *Bioinformatics* **17**(9): 847–848 (2001)

Chapter 16
Analyzing Gene Expression Data with Predictive Clustering Trees

Ivica Slavkov and Sašo Džeroski

Abstract In this work we investigate the application of predictive clustering trees (PCTs) for analysing gene expression data. PCTs provide a flexible approach for both predictive and descriptive analysis, both often used on gene expression data. To begin with, we use gene expression data for building predictive models for associated clinical data, where we compare single-target with multi-target models. Related to this, random forests of PCTs (single and multi-target) are used to assess the importance of individual genes w.r.t. the clinical parameters. For a more descriptive analysis, we perform a so-called constrained clustering of expression data. Also, we extend the descriptive analysis to take into account a temporal component, by using PCTs for finding descriptions of short time series of gene expression data.

16.1 Introduction

Central to our interest is gene expression data, which in recent years is widespread in medical studies. Gene expression data record the activity of each gene in the cell. A tissue sample is taken from a patient and the overall gene expression levels are measured, most often by using microarrays.

In a typical gene expression dataset, each data instance is a single patient and each data attribute (feature) is the expression level of a gene. To both of these dimensions (instance and attribute) one can relate additional information. Patients can be related to a clinical record containing multiple clinical parameters, while each gene can be annotated with different descriptions of its function. This allows for many practically relevant and biologically interesting data analysis scenarios.

Ivica Slavkov · Sašo Džeroski
Department of Knowledge Technologies, Jožef Stefan Institute
Jamova cesta 39, 1000 Ljubljana, Slovenia
e-mail: {ivica.slavkov,saso.dzeroski}@ijs.si

We can divide the analysis scenarios into predictive and descriptive. The predictive scenarios include building predictive models from gene expression data w.r.t. individual clinical parameters. Also, for the purpose of biomarker discovery, individual gene importance is assessed, related to the clinical parameters. The descriptive analysis of gene expression data aims at discovering common descriptions of patient or gene groups with similar expression profiles.

In our work, we consider the application of predictive clustering trees (PCTs) [1] for both descriptive and predictive analysis of gene expression data. PCTs are a part of the predictive clustering framework, which unifies predictive modelling and clustering [1]. PCTs have been developed to work with different types of data, like multiple numeric or nominal attributes [15], hierarchies [16] and also time series data [3]. Taking into consideration this generality of PCTs, we present several application of PCTs for gene expression data analysis.

The predictive scenarios involve building PCTs from gene expression data w.r.t. the related clinical records. This includes both building PCTs related to individual clinical parameters (single-target) and building multi-target PCTs, which take into account all of the clinical parameters at the same time.

Besides building PCTs for predicting the clinical parameters, it is biologically relevant to evaluate the individual importance of genes with relation to them. For this purpose we use random forests (RFs) of PCTs. We also compare single-target with multi-target feature importance.

For descriptive analysis, we use PCTs to perform constrained clustering on the gene expression profiles. Constrained clustering discovers compact groups of gene expression data, which are not only similar but also completely explained (covered) by their descriptions, i.e., constraints. This is unlike classical clustering, where first the gene expression data is grouped by the similarity of the expression values and then the cluster descriptions are derived.

As descriptions we consider information related to both patient and gene dimensions of the gene expression data. The patient clinical data is used as a starting point for constrained clustering of the gene expression profiles, or to simulate itemset-constrained clustering [11]. As a result we get clusters of patients with similar gene expression profiles, described by either individual clinical parameters or a combination thereof. When performing constrained clustering on the gene dimension of the expression data, we use gene functional annotation from the Gene Ontology [4]. In this case, instead of clusters of patients, we obtain clusters of genes having similar expression which are explained by the genes sharing similar function, similar location or by being involved in similar biological processes.

The remainder of this chapter is organised as follows. We first present in Section 16.2 an overview of all the gene expression datasets used in our work. In Section 16.3, we discuss and compare single with multi-target PCTs for predicting patient clinical data from gene expression data. We use random forests of multi-target PCTs for evaluating individual gene importance in Section 16.4. The descriptive scenarios of analysis of gene expression data are covered in Section 16.5 and Section 16.6. Finally, in Section 17.7, we summarise and present the conclusions.

16.2 Datasets

In this section, we give a brief overview of all the datasets used in our different analysis scenarios. The datasets contain gene expression levels measurements. They originate from studies investigating different diseases in humans or response of model organisms to different kinds of stress. The datasets have also been generated by using different microarray platforms. Below we give a description of each dataset separately, as well as of the additional information related to patients and genes.

16.2.1 Liver cancer dataset

We use a liver cancer dataset previously used in [11]. It contains expression levels measured by ATAC-PCR for 213 patients and 1993 genes. The expression data also has patient clinical information related to it.

The patients clinical data contains the patients diagnosis (tumor presence) and other information related to the liver status like cirrhosis, hepatitis infection and the general status of liver function (abnormal). The patient data also contains more general patient information like the patients age and gender.

16.2.2 Huntington's disease dataset

Huntington's disease is an autosomal dominant neurodegenerative disorder characterised by progressive motor impairment, cognitive decline, and various psychiatric symptoms. Its typical age of onset is in the third to fifth decade.

The dataset contains microarray data and basic patient records. The microarray data was obtained by using the Affymetrix HG.U133A chip, measuring the expression levels for 54.675 probes for 27 patients [12]. The patient records consist of three attributes: huntington disease status (presymptomatic (9 patients), symptomatic (5 patients) and controls (13)), age and gender.

16.2.3 Neuroblastoma dataset

Neuroblastoma is the most common extracranial solid tumour of childhood. A large proportion (88%) of neuroblastoma patients are 5 years or younger.

Gene expression was measured for 63 primary neuroblastomas. The Affymetrix U95Av2 microarrays were used, which measure the expression levels for a total of 12625 probes (genes). These data are included in the 68 patients analysed by Schramm et al. [10].

The patient data include information about the clinical course of the disease, in particular whether the tumour re-occurs after a prolonged period of time (relapse) or not (no event). Additional clinical information includes data about the possible chromosomal aberrations present in a patient. In particular, amplification (multiple copies) of the MYCN gene, as well as deletions (losses) in the 1p chromosomal region.

16.2.4 Yeast time series expression data

We use the time-series expression data from the study conducted by Gasch et al. [5]. The dataset records the changes in expression levels of yeast (*Saccharomyces cerevisiae*) genes, under diverse environmental stresses. The gene expression levels of around 5000 genes are measured at different time points using microarrays. Various sudden changes in environmental conditions are tested, ranging from heat shock to amino acid starvation for a prolonged period of time.

The datasets consist of time series of gene expression levels and gene descriptions [13]. As gene descriptions we use annotations for each yeast gene from the Gene Ontology [4] (version June, 2009). To limit the number of features, we set a minimum frequency threshold: each included GO term must appear in the annotations for at least 50 of the 5000 genes.

16.3 Predicting Multiple Clinical Parameters

Medical studies involving microarray data usually have one clinical parameter of interest. For example, it is important to identify a gene expression profile specific to patients having cancer, as compared to healthy individuals. Although for each patient there is additional clinical data (e.g., age), this information is usually not of primary interest and is not directly used in the predictive modelling process.

In this section we aim to demonstrate the advantage of building predictive models for multiple clinical parameters at the same time, as compared to building models for single clinical parameters separately. To build models for multiple target parameters, we use predictive clustering trees (PCTs). We consider the Huntington disease and Neuroblastoma datasets.

16.3.1 Huntington disease progress

In the case of the Huntington disease (HD), a useful gene expression profile is one that distinguishes between either of the following: a subject has HD or is healthy (control); the stage of the disease of HD patients (whether the patient is presymp-

Table 16.1 Predictive accuracy estimates of single- and multi-target PCTs constructed for the HD dataset, derived by 10-fold cross-validation

Targets	HD	Stage	{HD, Stage}
Accuracy	51%	41%	74%, 74%

tomatic or symptomatic). First, we constructed single target PCTs for each of these two possible targets individually. The targets are Huntington disease vs. control as a target attribute and "Stage" (with three possible values: presymtpomatic, symptomatic and controls).

For comparison, we then constructed a single predictive model (PCT) by considering both of the targets simultaneously. The class labels that are output from the multi-target model can be used to predict whether a patient has Huntington disease and also to determine the stage of progression of the disease. We evaluated the models predictive performance with leave-one-out cross validation and give a comparison below.

Table 16.1 gives the classification accuracy of the constructed predictive models. It shows that the multi-target model outperforms the single-target ones in terms of predictive accuracy. From the application point of view, the multi-target model performs better because the two classes are correlated. When constructing the models, the algorithm is constrained to prefer putting presymtpomatic and symptomatic patients in one cluster and controls in the other.

16.3.2 Neuroblastoma recurrence

We perform a similar analysis with the Neuroblastoma dataset and its associated clinical data. For this data, we have only one real target of interest from the clinical record, namely Neuroblastoma (NB) status. This is a clinical parameter which indicates whether NB patients have a re-occurrence of the tumour or not. It is important to have a predictive model for this, so that the course of therapy for a patient can be decided.

In our experiments, we considered additional clinical parameters, namely whether a patient has multiple copies of the MYCN gene or deletions on the 1p chromosome. Both parameters are related to tumour aggressiveness and have been previously used for predictive modelling [8], but only as descriptive attributes and not as targets for prediction.

We constructed a single-target PCT only for NB status, as it was the only parameter of interest. We compared this model to multi-target models by considering three different combinations of NB status with the other two parameters. Namely, we combine NB status with MYCN amplification or 1p deletion, as well as with both parameters.

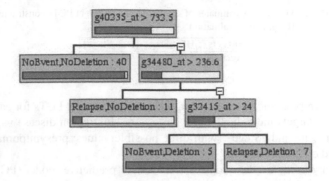

Fig. 16.1 As any decision tree model, a PCT can be easily interpreted. The first node of the tree, with attribute 40235_at (TNK2, "tyrosine kinase, non-receptor, 2"), splits the samples into two groups. In the first group, there are patients without event and with no deletion of the 1p chromosome region. The remaining group is split by a node containing 34480_at (CDH16, "cadherin 16, KSP-cadherin") of the PCT that essentially distinguishes between patients that have/do not have a 1p deletion. The last node containing g32415_at (IFNA5, "interferon, alpha 5") further differentiates between the patients with 1p deletion that had a relapse or are without event.

An illustrative example is presented in Figure 16.1. The accuracy results are summarised in Table 16.2. As in the case with Huntington's disease, it can be seen that including 1p deletion and MYCN amplification on the target side during modelling drastically improves the predictive performance of the PCTs.

Table 16.2 Predictive accuracy estimates of single- and multi-target PCTs constructed for the NB dataset derived by 10-fold cross validation

Targets	NB	{NB, 1p}	{NB, MYCN}	{NB, 1p, MYCN}
Accuracy	74.6%	90.5%	84.1%	74.6%

16.4 Evaluating Gene Importance with Ensembles of PCTs

When examining the relation of gene expression data to clinical parameters, besides building predictive models (Section 16.3), it is also biologically relevant to determine the individual importance of each gene w.r.t the clinical parameter(s) of interest. From a biological perspective, this is the initial step in the process of biomarker discovery. In machine learning terminology, this is equivalent to the process of feature ranking and selection.

Typically, the ranking is produced with respect to a single target variable. Following the intuition from Section 16.3, we investigate feature ranking by considering multiple targets simultaneously. In particular, to determine the feature importance we use Random Forests [2] (RFs) of PCTs [6].

Below, we first describe the methodology used for feature ranking. In particular, we describe feature ranking with random forests and then discuss how this methodology can be extended for multi-target feature ranking. We then describe the application of multi-target ranking with RFs of PCTs to the Neuroblastoma dataset.

16.4.1 Feature ranking with multi-target Random Forests

Typical feature ranking methods consider the relation between each feature and the target separately. They rank the features based on the strength of their relation to the target. More recent methods also consider interactions among the features themselves. One such method is based on the ensemble learning approach of random forests.

Random forests [2] is an ensemble learning method. It learns base classifiers on bootstrap replicates of the dataset, by using a randomised decision tree algorithm. For each bootstrap replicate, there is a corresponding out-of-bag (OOB) dataset. This OOB dataset contains only the instances from the original data that do not appear in the bootstrap replicate. These OOB datasets are used to determine the feature importance as proposed in [2].

Table 16.3 The algorithm for feature ranking via random forests. I is the set of the training instances, F is the number of features, k is the number of trees in the forest, and f_{sub} is the size of the feature subset that is considered at each node during tree construction.

procedure Induce_RF(I, k, f_{sub}) **returns** feature importance vector Imp	**procedure** Update_Imp(E_{OOB}, T, I)
1: $Imp = \emptyset$	1: $Err_{OOB} = $ Evaluate(T, I_{OOB})
2: $Forest = \emptyset$	2: **for** $j = 1$ **to** F **do**
3: **for** $i = 1$ **to** k **do**	3: $I_j = $ Randomize(I_{OOB}, j)
4: $I_i = $ Bootstrap_sample(I)	4: $Err_j = $ Evaluate(T, I_j)
5: $T_i = PCT(I_i, f_{sub})$	5: $Imp_j = Imp_j + (Err_j - Err_{OOB})/Err_{OOB}$
6: $I_{OOB} = I \setminus I_i$	6: **return**
7: Update_Imp(I_{OOB}, T, Imp)	**procedure** Average(Imp, k)
8: $Imp = $ Average(Imp, k)	1: $Imp^T = \emptyset$
9: **return** Imp	2: **for** $l = 1$ **to** $size(Imp)$ **do**
	3: $Imp_l^T = Imp_l/k$
	4: **return** Imp^T

First, for each bootstrap replicate, a random tree is built and its predictive performance (e.g., misclassification rate) is measured on the corresponding OOB dataset. Then, for each feature in turn, the values for the instances in the OOB dataset are randomly permuted. Again, the predictive performance of the random trees are calculated for the permuted OOB data. Finally, the importance of a feature is computed as the average increase of the error rate of the permuted OOB datasets compared to the error rate of the original OOB datasets. The rationale behind this is that if a feature is important for the target concept(s) it should have an increased error rate when its values are randomly permuted in the OOB dataset. The full procedure is described in Table 16.3.

The above methodology for feature rankings, works for single-target rankings by using a single-target decision trees for the random forests. We have extended it in [6] to work for multiple targets by using multi-target PCTs as base classifiers in the random forests. When working with multiple targets, the feature importance is simply calculated as an average percent increase of the error rate over all targets.

16.4.2 Gene importance in Neuroblastoma

We apply the proposed method for feature ranking with multiple targets to the Neuroblastoma microarray dataset. For targets, we consider the NB status, 1p deletion and MYCN amplification. We produce ranked lists of genes with respect only to NB status and also by considering all of the target variables simultaneously.

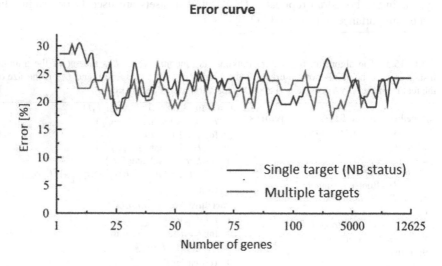

Fig. 16.2 In this figure we compare two error curves constructed from ranked gene lists by using a single-target (NB status) and multi-target RF of PCTs. Each point of the curves represents an error estimate of a predictive model built with the corresponding set of genes. The genes in the gene sets are incrementally added from the previously generated ranked lists of genes.

We compare the ranked lists by using the so-called error curves, as proposed in [14]. These error curves give us an intuition of how the error of predictive models changes, as we incrementally add more and more genes from the previously generated ranked list of genes. A ranked list of genes is better than another one if the corresponding error curve drops to a lower error and/or drops faster.

In Figure 16.2, we present a comparison between the error curves of the single-target and multi-target ranked lists of genes. When considering multiple targets simultaneously, the highly ranked genes can be used to construct better predictive models than the highly ranked genes in the single-target scenario. Besides the boost in predictive performance, it is important to note that the same set of genes produced by the multi-target approach can be used for predicting all of the different clinical variables instead of having a different set of genes for each one.

16.5 Constrained Clustering of Gene Expression Data

In Sections 16.3 and Section 16.4 our main focus was on using PCTs for predictive modelling of gene expression data with relation to clinical data. Considering that PCTs unify prediction and clustering, we now turn to investigate the clustering aspect of PCTs and its application to the descriptive analysis of gene expression data.

We use PCTs to perform constrained clustering, where only clusters that can be described by using a given vocabulary are considered. We consider the direct use of PCTs to cluster gene expression data and also the combination of frequent pattern mining and PCTs. In both cases, we consider cluster descriptions (constraints) related to the patient dimension of the data, i.e., the clinical data.

16.5.1 Predictive clustering of gene expression profiles

We first consider a simple scenario of constrained clustering of gene expression data by using the ability for multi-target modelling of PCTs. As targets for constructing PCTs, we use the gene expression data of the liver cancer dataset and as descriptive vocabulary the related patient record.

As a result we get a PCT, which can be considered as hierarchical clustering of the gene expression data. In Figure 16.3, we present a PCT of the gene expression profiles of the liver cancer dataset. Each leaf of the PCT is a cluster of patients who share a similar gene expression profile. The cluster descriptions can be obtained by following the path from the root of the tree to the corresponding leaf (cluster). It should be noted that these clusters are non-overlapping, unlike the clusters obtained in the following section by simulation of the so-called itemset constrained clustering.

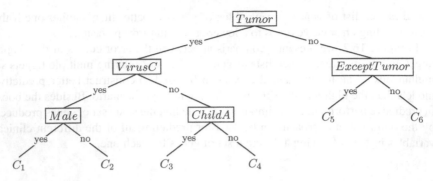

Fig. 16.3 Constrained clustering of the gene expression profiles of the liver cancer dataset by using a single PCT

16.5.2 Itemset constrained clustering

The basic intuition behind itemset constrained clustering (IC-clustering) is to use itemsets as constraints when performing clustering of the gene expression data. In our specific application, the itemsets are produced from the patients clinical data. Instead of using a dedicated algorithm, we decompose the IC-clustering algorithm into several steps. These steps are executed with already known modelling approaches (e.g., PCTs) and their corresponding implementations, which in the end results in an overall IC-clustering simulation.

In short, the IC-clustering algorithm can be described as in Table 16.4. As input it requires minimum cluster size, which translates into a frequency constraint on the itemsets. After searching the space of possible clusters described by the itemsets, it outputs a list of N itemsets sorted by interclass variance of the clusters. Essentially, the IC-clustering algorithm can be decomposed into a frequent itemsets mining algorithm and a cluster evaluation algorithm.

Considering Table 16.4, we simulate IC-clustering in two steps. In the first step, we find frequent itemsets in the associated patient clinical data. By specifying the minimum support of the frequent itemsets, we also simulate the parameter C (minimum cluster size) in the IC-clustering algorithm. Between step one and two, there is an intermediary step, where we modify the gene expression dataset by including the produced frequent itemsets as patient features. This means that for each itemset we

Table 16.4 A simplified description of the IC-clustering algorithm

Input: Minimum cluster size C

1. Search for a feature itemset that splits the tuples into two clusters

2. Compute the interclass variance between the clusters

Output: List of the top N itemsets sorted by interclass variance

Fig. 16.4 A sample PCT stub
is presented in this figure.
In its single decision node
(rectangle), it contains a
description "{tumour, man}"
which is true only for the
"yes" branch of the PCT stub.
The instances which fall into
this leaf represent the actual
cluster of interest and their
expression values are used
to calculate the interclass
variance.

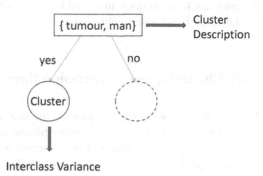

add a binary feature, which is true for those patients (instances) which are described
by all of the clinical features contained in the frequent itemset. For example, a frequent itemset {male, has tumour}, will be true only for those patients (instances) of
the dataset which are male and have tumour, but not for those that are only male or
only have tumour.

On this modified dataset, we proceed with step two of the IC-clustering simulation, by inducing a beam of PCT stubs with width N. A PCT stub contains only a
single node, i.e., performs only a single split of the data according to some feature
value. An illustrative example can be seen on Figure 16.4.

The single decision nodes of the PCT stubs contain descriptive attributes that are
chosen from the set of frequent itemset features. To calculate the interclass variance
of the clusters, we use the multi-target ability of the PCTs, with all of the genes
expression values as targets. We use the beam-search version of PCTs, as described
by Struyf and Džeroski [7], which at all times keeps a set of partially built candidate
trees.

The results from the simulation of the IC-clustering can be seen in Table 16.5.
The specific dataset that was used was the liver cancer dataset, from the original
publication of IC-clustering by Morishita et al.[11]. The clusters and their clinical
descriptions are identical to the ones published in [11]. As noted by Morishita et al.,
IC-clustering (its simulation in this case) reveal interesting compact clusters with

Table 16.5 Results from the simulation of IC-clustering

Itemset constraint	Cluster size	Interclass variance
{tumour}	107	3126.9
{except tumour, normal liver function}	88	2534.7
{except tumor, HBV-}	88	2397.3
{tumor, man}	86	2181.5
{except tumor, HBV-, normal liver function}	74	2098.9
{except tumor, man}	83	2037.87
{except tumor, no cirrhosis}	68	1979.74
...	...	...
{tumor, not over 65 years old}	55	1587.7

descriptions like {tumour, man} which are overlooked by conventional clustering methods, such as k-means clustering.

16.6 Clustering gene expression time series data

In this section, we continue exploring the use of PCTs for descriptive analysis of gene expression data. Instead of using patient data, here we consider descriptions related to the gene dimension of the expression data, more specifically gene ontology descriptions. We analyse the previously described yeast time-course expression dataset. Clustering the temporal gene expression profiles requires for PCTs to be adapted to work with time-series data on the target side. This was first done by Džeroski et al. [3] and we discuss it in more detail in Section 16.6.1. The specific analysis scenario, the results and their interpretation are given in Section 16.6.2.

16.6.1 PCTs for clustering short time-series

In order for PCTs to handle time-course data on the target side, three things have to be adapted. These are the distance measure used for calculating the difference between the time-series, the cluster centroid (prototype) calculation, and the cluster quality estimation.

16.6.1.1 Qualitative distance measure

For our application (i.e., clustering short time course gene expression data), the differences in scale and size are not of great importance; only the shape of the time series matters. Namely, we are interested in grouping together time-course profiles of genes that react in the same way to a given condition, regardless of the intensity of the up- or down-regulation.

For that reason, we use the qualitative distance measure(QDM) proposed by Todorovski et al. [17]. It is based on a qualitative comparison of the shape of the time series. Consider two time series X and Y. Then choose a pair of time points i and j; observe the qualitative change of the value of X and Y at these points. There are three possibilities: increase ($X_i > X_j$), no-change ($X_i \approx X_j$), and decrease ($X_i < X_j$). d_{qual} is obtained by summing the difference in qualitative change observed for X and Y for all pairs of time points, i.e.,

$$d_{\text{qual}}(X,Y) = \sum_{i=1}^{n-1} \sum_{j=i+1}^{n} \frac{2 \cdot Diff(q(X_i,X_j), q(Y_i,Y_j))}{N \cdot (N-1)}, \tag{16.1}$$

with $Diff(q_1, q_2)$ a function that defines the difference between different qualitative changes (16.6). Roughly speaking, d_{qual} counts the number of disagreements in the direction of change of X and Y.

Table 16.6 The definition of $Diff(q_1, q_2)$.

$Diff(q_1, q_2)$	increase	no-change	decrease
increase	0	0.5	1
no-change	0.5	0	0.5
decrease	1	0.5	0

16.6.1.2 Computing the cluster centroid and variance

The definition of the cenroid c of a cluster C is directly related to the calculation of the cluster variance. The variance of a cluster C can be defined based on a distance measure as:

$$Var(C) = \frac{1}{|C|} \sum_{X \in C} d^2(X, c) \tag{16.2}$$

To cluster time series, d should be a distance measure defined on time series, such as the previously defined QDM.

The centroid c can be computed as $\text{argmin}_q \sum_{X \in C} d^2(X, q)$. The centroid c can be either calculated from the time-series in the cluster or it can be one of the time series from the cluster, which minimises the variance. Because there is no closed form for the centroid for the QDM distance, we choose the second option for calculating the cluster centroid.

For computational reasons, we re-define and approximate the calculation of the variance by means of sampling as:

$$Var(C) = \frac{1}{2|C|m} \sum_{X \in C} \left(\sum_{Y \in \text{sample}(C,m)} d^2(X, Y) \right), \tag{16.3}$$

with sample(C, m) a random sample without replacement of m elements from C, where $|C| \geq m$.

The PCT induction algorithm places cluster centroids in its leaves, which can be inspected by the domain expert and used as predictions. For these centroids, we use the representation discussed above.

16.6.1.3 Estimating cluster centroid error

Although in this application we use PCTs for descriptive analysis, we can still make predictions with them just like with regular decision trees [9]. They sort each test instance into a leaf and assign as prediction the label of that leaf. PCTs label their leaves with the training set centroids of the corresponding clusters.

We use the predictive performance of PCTs as a way to assess how well the cluster centroid approximates the time-series in that cluster. The error measure we use is the root mean squared error (RMSE), which is defined as:

$$\text{RMSE}(I, T) = \sqrt{\frac{1}{|I|} \sum_{X \in I} d^2(T(X), \text{series}(X))}, \qquad (16.4)$$

with I the set of test instances, T the PCT that is being tested, $T(X)$ the time series predicted by T for instance X, series(X) the actual series of X, and d the QDM distance.

16.6.2 Explained groups of yeast time-course gene expression profiles

We perform constrained clustering with Gene Ontology terms as descriptions (constraints) and with time-series of the expression data on the target side. The whole process of generating compact clusters of genes, with GO terms as descriptions, begins with inducing PCTs and then discerning the descriptions by following just the positive (yes) branches of the PCT. This is graphically illustrated in Figure 16.5 and described in more detail by Slavkov et al. in [13].

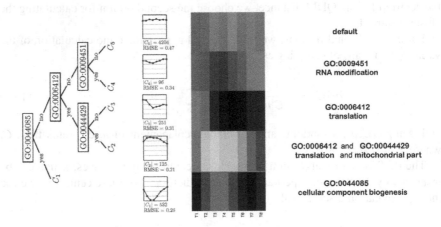

Fig. 16.5 Using a PCT to cluster time series of gene expression data, describing response to stress in yeast (diamide exposure).

On the left side of Figure 16.5, we present a sample PCT. For practical purposes, we show a small tree with just 5 leaves, obtained when yeast is exposed to diamide.

We also show the cluster centroids for each of the leaves, and their related cluster sizes and the root mean squared error (RMSE). We use the RMSE as an estimate of how well the cluster centroids represent all of the instances in the cluster, i.e., as an estimate of cluster quality. Visually, the cluster prototypes can be represent by using a heatmap. Each row in the heatmap represents a cluster prototype: the more intense the colours, the larger the up- or down-regulation of the genes contained in that cluster.

The associated cluster descriptions in Figure 16.5, can be easily obtained from the PCT by following the path from the root of the tree to a leaf. For example, if we want to derive the description of cluster C_2, we begin from the root GO term "GO:004408", we follow the "no" branch, obtaining the description "GO:0044085 = no". We then add the "GO:0006412 = yes" and "GO:0044429 = yes" by following the "yes" branches ending up at cluster C_2. So, the final description of cluster C_2 is the following conjunction: "GO:0044085 = no AND GO:0006412 = yes AND GO:0044429 = yes". This can be interpreted as follows: genes that are annotated by both "GO:0006412" and "GO:0044429", but not by "GO:0044085" are contained in cluster C_2 and have a temporal profile represented by the prototype of cluster C_2. It should be noted here that for our application only the positive branches of the tree are semantically meaningful in a biological context.

In Figure 16.5, clusters C_1 to C_4 show significant temporal changes in gene expression and have a relatively low error. C_1 includes genes with an immediate response to stress, while C_3 and C_4 include down-regulated genes with a short time lag in response. C_2 includes genes that are up-regulated under diamide exposure. All cluster prototypes show that the changes of gene expression levels are transient, i.e., after the initial stress response and cell adaptation, the cell continues with its regular function. The size of C_5 indicates that the bulk of genes fall into this cluster. It includes genes that do not have a coordinated stress response and major changes in gene expression.

The application of PCTs for descriptive analysis of time course data is investigated in detail in [13]. The results of the descriptive analysis of yeast exposed to different environmental stresses are consistent with previously published work [5].

16.7 Conclusions

In recent years, gene expression data has become common in almost all medical studies. Its advantage is that it measures the level of activity (least approximately) of all known genes in the human genome. This data, together with all of the additional knowledge about gene functions and specific disease mechanisms allows for a plethora of possible data analysis scenarios.

We roughly divided these scenarios into predictive and descriptive and considered some which are common for gene expression data analysis. We considered the analysis scenarios in the context of the predictive clustering framework [1]. More specifically, we used predictive clustering trees (PCTs). We demonstrated that PCTs are general enough to be used in all of the considered scenarios.

The predictive scenarios were investigated in Section 16.3 and Section 16.4. First, we considered an application scenario of building PCTs for diagnosis from gene expression data. We considered two specific instances of the scenarios for two different diseases: Huntington's disease and Neuroblastoma. In both, we compared single-target with multi-target PCTs, thus utilising the whole clinical information available while constructing the models. The accuracy estimates show a distinct advantage of using multiple clinical parameters as targets when constructing the PCTs.

In the predictive scenarios context, we also extended the use of PCTs towards evaluating individual feature importance w.r.t multiple targets. We performed experiments with the Neuroblastoma gene expression dataset, where on the target side we considered three parameters NB status, 1p deletion and MYCN amplification. Our results show that the multi-target approach is beneficial as compared to the single-target variable approach. The produced ranked list of genes is more accurate in terms of predictive performance and it can be applied to each of the target variables separately.

Descriptive scenarios of analysis involved constrained clustering of the gene expression profiles, by using constraints both on the patient and gene dimension of the gene expression datasets. In Section 16.5 we investigated the use of patient clinical data as constraints for clustering the gene expression data. We first considered the use of a single PCT for constrained clustering, where as a result we got a hierarchy of clusters of patients described by individual clinical parameters. Also, we performed simulation of itemset-constrained clustering [11], where the clusters of patients were described by frequent patterns of the patient clinical parameters.

Section 16.6 investigates the use of PCTs for constrained clustering of gene expression time-course data. PCTs had to be adapted so they can handle the temporal aspect of the data, by using a qualitative distance measure. The constraints used for the clustering were on the gene dimension of the time series data and composed of gene ontology terms. The detailed results [13] were consistent with previous knowledge about yeast stress response [5], which demonstrates the utility of PCTs for time-series analysis.

Further work would include the use of PCTs in other domains, for example analysis of the human immune response to infections with *M. leprae* and *M. tuberculosis*. Also, predictive clustering rules (PCRs) have been developed by Ženko et al. in [18], which have been adapted for multi-target modelling, rule ensembles learning and time-series analysis. Applying them to the descriptive analysis scenarios can prove to be more helpful, as PCRs (unlike PCTs) output clusters which can be overlapping and are thus similar to IC-clustering.

In summary, all previously described scenarios demonstrated that PCTs are flexible enough to be used for various gene expression data analysis tasks. Taking into account the generality of the predictive clustering framework, they can be used to perform the most common predictive and descriptive gene expression analysis tasks. They also open up the possibility of additional scenarios which are not immediately obvious, like multi-target predictive models which take into account the whole clinical data of a patient.

Acknowledgements The authors would like to acknowledge Dragi Kocev for his work on multi-target feature ranking. Part of the research presented in this chapter was conducted within the projects E.E.T. Pipeline (*European Embryonal Tumour Pipeline*) and IQ (Inductive Queries for mining patterns and models), funded by the European Commission of the EU under contract numbers FP6-IST 516169 and FP6-LSHC 037260, respectively. Ivica Slavkov and Sašo Džeroski are currently supported by the project PHAGOSYS *Systems biology of phagosome formation and maturation - modulation by intracellular pathogens* funded by the European Commission of the EU under contract number FP7-HEALTH 223451. For a complete list of agencies, grants and institutions currently supporting Sašo Džeroski, please consult the Acknowledgements chapter of this volume.

References

1. H. Blockeel, L. De Raedt, and J. Ramon. Top-down induction of clustering trees. In Proc. *15th Int'l Conf. on Machine Learning*, pages 55–63. Morgan Kaufman, 1998.
2. L. Breiman. Random forests. *Machine Learning*, 45(1):5–32, 2001.
3. S. Džeroski, V. Gjorgjioski, I. Slavkov, and J. Struyf. Analysis of time series data with predictive clustering trees. In *5th Int'l Workshop on Knowledge Discovery in Inductive Databases: Revised Selected and Invited Papers*, pages 63–80, Springer Berlin, 2007.
4. Ashburner, M., Ball, C., Blake, J., Botstein, D., Butler, H., Cherry, J., Davis, A., Dolinski, K., Dwight, S., Eppig, J., Harris, M., Hill, D., Issel-Tarver, L., Kasarskis, A., Lewis, S., Matese, J., Richardson, J., Ringwald, M., Rubin, G., Sherlock, G.: Gene Ontology: Tool for the unification of biology. The Gene Ontology Consortium. *Nature Genetics* 25(1): 25–29, 2000
5. A. Gasch, P. Spellman, C. Kao, O. Carmel-Harel, M. Eisen, G. Storz, D. Botstein, and P. Brown. Genomic expression program in the response of yeast cells to environmental changes. *Molecular Biology of the Cell*, 11:4241–4257, 2000.
6. D. Kocev, I. Slavkov, and S. Džeroski. More is better: ranking with multiple targets for biomarker discovery. In Proc. *2nd Int'l Wsp on Machine Learning in Systems Biology*, page 133, University of Liege 2008.
7. D. Kocev, J. Struyf, and S. Džeroski. Beam search induction and similarity constraints for predictive clustering trees. In *5th Int'l Workshop on Knowledge Discovery in Inductive Databases: Revised Selected and Invited Papers*, pages 134–151. Springer, Berlin 2007.
8. J. M. Maris. The biologic basis for neuroblastoma heterogeneity and risk stratification. *Current Opinion in Pediatrics*, 17(1):7–13, 2005.
9. J.R. Quinlan. *C4.5: Programs for Machine Learning*. Morgan Kaufmann, San Mateo, CA 1993.
10. A. Schramm, J. H. Schulte, L. Klein-Hitpass, W. Havers, H. Sieverts, B. Berwanger, H. Christiansen, P. Warnat, B. Brors, J. Eils, R. Eils, and A. Eggert. Prediction of clinical outcome and biological characterization of neuroblastoma by expression profiling. *Oncogene*, 7902–7912, 2005.
11. J. Sese, Y. Kurokawa, M. Monden, K. Kato, and S. Morishita. Constrained clusters of gene expression profiles with pathological features. *Bioinformatics*, 20:3137–3145, 2004.
12. I. Slavkov, S. Džeroski, B. Peterlin, and L. Lovrečić. Analysis of huntington's disease gene expression profiles using constrained clustering. *Informatica Medica Slovenica*, 11(2):43–51, 2006.
13. I. Slavkov, V. Gjorgjioski, J. Struyf, and S. Džeroski. Finding explained groups of time-course gene expression profiles with predictive clustering trees. *Molecular bioSystems*, 6(7):729–740, 2010.

14. I. Slavkov, B. Ženko, and S. Džeroski. Evaluation method for feature rankings and their aggregations for biomarker discover. In *Proc. 3rd Intl Wshp on Machine Learning in Systems Biology, JMLR: Workshop and Conference Proceedings* **8**: 122–135 (2010)
15. J. Struyf and S. Džeroski. Constraint based induction of multi-objective regression trees. In *4th Int'l Workshop on Knowledge Discovery in Inductive Databases: Revised Selected and Invited Papers*, pages 222–233. Springer, Berlin 2006.
16. J. Struyf, S. Dzeroski, H. Blockeel, and A. Clare. Hierarchical multi-classification with predictive clustering trees in functional genomics. In *12th Portuguese Conference on Artificial Intelligence*, pages 272–283. Springer 2005.
17. L. Todorovski, B. Cestnik, M. Kline, N. Lavrač, and S. Džeroski. Qualitative clustering of short time-series: A case study of firms reputation data. In *Proc. Wshp on Integration and Collaboration Aspects of Data Mining, Decision Support and Meta-Learning*, pages 141–149, ECML/PKDD 2002.
18. B. Ženko, S. Džeroski, and J. Struyf. Learning predictive clustering rules. In *4th Int'l Workshop on Knowledge Discovery in Inductive Databases: Revised Selected and Invited Papers*, pages 234–250. Springer, Berlin 2005.

Chapter 17
Using a Solver Over the String Pattern Domain to Analyze Gene Promoter Sequences

Christophe Rigotti, Ieva Mitašiūnaitė, Jérémy Besson, Laurène Meyniel,
Jean-François Boulicaut, and Olivier Gandrillon

Abstract This chapter illustrates how inductive querying techniques can be used to support knowledge discovery from genomic data. More precisely, it presents a data mining scenario to discover putative transcription factor binding sites in gene promoter sequences. We do not provide technical details about the used constraint-based data mining algorithms that have been previously described. Our contribution is to provide an abstract description of the scenario, its concrete instantiation and also a typical execution on real data. Our main extraction algorithm is a complete solver dedicated to the string pattern domain: it computes string patterns that satisfy a given conjunction of primitive constraints. We also discuss the processing steps necessary to turn it into a useful tool. In particular, we introduce a parameter tuning strategy, an appropriate measure to rank the patterns, and the post-processing approaches that can be and have been applied.

17.1 Introduction

Understanding the regulation of gene expression remains one of the major challenges in molecular biology. One of the elements through which the regulation

Christophe Rigotti · Laurène Meyniel · Jean-François Boulicaut
Laboratoire LIRIS CNRS UMR 5205, INSA-Lyon, 69621 Villeurbanne, France
e-mail: christophe.rigotti@insa-lyon.fr, laurene.meyniel@wanadoo.fr,
jean-francois.boulicaut@insa-lyon.fr

Ieva Mitašiūnaitė · Jérémy Besson
Faculty of Mathematics and Informatics, Vilnius University, Lithuania
e-mail: ieva.mitasiunaite@gmail.com, contact.jeremy.besson@gmail.com

Olivier Gandrillon
Centre de Génétique Moléculaire et Cellulaire CNRS UMR 5534,
Université Claude Bernard Lyon I, 69622 Villeurbanne, France
e-mail: gandrillon@cgmc.univ-lyon1.fr

works is the initiation of the transcription by the interaction between short DNA sequences (called *gene promoters*) and multiple activator and repressor proteins called Transcription Factors (TFs). These gene promoter elements are located in sequences called promoter sequences, that are DNA sequences close to the sequences that encode the genes. In fact, on a promoter sequence various compounds can bind, having then an impact on the activation/repression of the gene associated to this promoter sequence, but among these compounds, the TFs are known to play a very important role. Therefore, many researchers are working on TFs and Transcription Factor Binding Sites (TFBSs). These are subsequences of the promoter sequences where the TFs are likely to bind. In practice, identifying patterns corresponding to putative TFBSs help the biologists to understand which TFs are involved in the regulation of the different genes.

In this study, we report our contribution to gene promoter sequence analysis and TFBS discovery by means of generic constraint-based data mining techniques over strings. Indeed, we consider that the promoter sequences are sequences of nucleotides represented by the symbols A, C, G and T (i.e., a data sequence is a string over the alphabet $\{A, C, G, T\}$ and a pattern is a substring in such sequences). Contrary to many approaches that support motif discovery in promoter sequences, we do not take into account domain knowledge about that quite specific type of strings. Instead, we use a generic solver over the string pattern domain.

The recent advances in constraint-based mining (see [2] and [7] for an overview), and more generally the current developments in the domain of inductive querying (i.e., the vision proposed in [10]), lead to the design of many mining tools based on the constraint paradigm. We have now at hand scalable complete solvers, in particular over the string domain, that can be used to find substring patterns in sequences. However, this is far from being sufficient to tackle a real application. In this chapter, we present all the necessary processing, beyond the pattern extraction, that is needed to support knowledge discovery from a biological perspective, hopefully leading to the discovery of new putative TFBSs. First, we describe the corresponding data mining abstract scenario, and then we give its concrete instantiation. Finally, we illustrate its execution by means of a typical case study. We also give technical details about aspects that are important to run the scenario in practice. This includes, in particular, the tuning of the parameters in the early exploratory mining stage, the ranking of the patterns using a measure adapted to the domain, and the designed pattern post-processing technique to exhibit putative TFBSs.

Methodological and technical details about the method and the algorithms can be found in several papers. The *Marguerite* solver over the string pattern domain has been described in details in [14, 15]. A concrete instance of the scenario is described in the journal publication [16]. This is also where our measure of interest, the so-called TZI measure, is studied in depth. Our parameter tuning method has been introduced in [1]. Last by not least, the Ph.D. thesis [13] considers all these issues in detail.

The rest of the chapter is organized as follows. In Section 17.2, we present the scenario both at an abstract and instantiated level. Then, in Section 17.3, we describe the kind of patterns and the constraints that are handled by the solver. The parameter tuning strategy is discussed in Section 17.4 and the dedicated measure to rank the patterns is introduced in Section 17.5. Then, a typical example of a real execution of the scenario is presented in Section 17.6. Finally, we conclude with a short summary in Section 17.7.

17.2 A Promoter Sequence Analysis Scenario

Let us present the scenario which has been designed and used in our case study. First, we describe it in abstract terms and then we explain how it has been instantiated into a concrete scenario.

17.2.1 A generic scenario

This abstract view describes the main steps of the general process that has been studied. It can be decomposed as a workflow containing the following sequence of operations:

- Use the results of SAGE experiments [21] to select two groups of genes, one group corresponding to genes active in a context (called the positive context), and the second group corresponding to genes active in an opposite context (called the negative context). These positive vs. negative issues are application dependent. Notice that SAGE is one technology for recording gene expression values in biological samples and that other popular approaches could be used, e.g., microarrays.
- Retrieve from a gene database the promoter sequences of the selected genes. Construct two sets D^+ and D^- of promoter sequences: one for the genes active in the positive context (D^+), and the other for the genes active in the negative context (D^-).
- Perform a differential extraction of substrings between datasets D^+ and D^-, to find substrings frequent in D^+ and not frequent in D^-.
- Compute for each extracted substring a dedicated interestingness measure.
- Select some of the patterns, according to their ranking on the measure value and/or to their support in D^+ and/or support in D^-.
- Perform a complementary post-processing:
  - Cluster the set of selected patterns (pairwise alignment).
  - In each cluster, perform a multiple alignment of the patterns in the cluster, to obtain a consensus motif (centroid) for each cluster.
  - Search these consensus motifs in a database of known TF binding sites (e.g., $Transfac^{®}$ database [12]), to look for their corresponding TFs and the known functions of these TFs (if any).

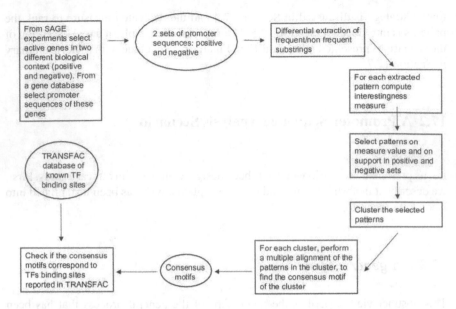

Fig. 17.1 Workflow of the abstract scenario.

The workflow of the whole process is depicted in Figure 17.1. Notice that numerous efforts have given rise to a variety of computational methods to discover putative TFBSs in sets of promoters of co-regulated genes (see [16] for an overview). Among them two families can be distinguished: statistical or stochastic approaches, and combinatorial approaches [20]. Concerning the family of statistical and stochastic approaches, a recent review of the most widely used algorithms exhibits rather limited results [19]. The scenario presented in this chapter uses a combinatorial approach, and its main originality w.r.t. the other combinatorial algorithms, which allow to extract patterns from several datasets (e.g., SPEXS [3] or DRIM [9]), is that the maximal support threshold is set explicitly. This is particularly interesting, when there is a clear semantic cut between a positive and negative datasets, i.e., the negative dataset has an opposite biological sense (presence/absence of a mutation; addition or not of a given drug, etc.), and does not just represent random background.

17.2.2 Instantiation of the abstract scenario

We focus the search on putative TFBSs that could be used to regulate the transcription of the genes associated to promoter sequences of the positive dataset (D^+) while they are not likely to have an important impact on the regulation of the genes associated to the other set. To collect the sets D^+ and D^-, the method starts with a classical operation used in molecular biology: the search for differentially expressed genes[1], using SAGE experiments. This allows to obtain two groups of genes from which we derive two sets of associated promoter sequences using a promoter database. To look for putative TFBSs regulating the overexpressed genes, we choose the first set (the promoters of the overexpressed genes) to be used as a positive set, and the second set as a negative one[2]. The promoter sequences are sequences of compounds called bases. There are four different bases, commonly represented by the symbols A, C, G and T, and a sequence is simply represented by a string over the alphabet $\{A,C,G,T\}$. Then the method consists in finding patterns that are substrings occurring in at least α_{min} promoters from the positive set and in at most α_{max} promoters from the negative set, where the parameter α_{min} (resp. α_{max}) is supposed to be a large (resp. small) threshold value. Typical sizes of the promoter sequences are about a few thousands of symbols, and the positive and negative datasets contain each a few tens of such sequences.

We consider two kinds of patterns: Exact Matching Patterns (EMPs) and Soft Matching Patterns (SMPs). Both are strings of bases, but they differ in the way their *supports* are defined. The support of an EMP in a dataset is the number of sequences of the dataset that contain at least one exact occurrence of this EMP. Let α_{dist} be a given threshold, termed the *soft matching* threshold, then the support of a SMP is the number of sequences containing at least one soft occurrence of the pattern, where a soft occurrence is a part of the sequence different from the pattern in at most α_{dist} positions (i.e., the Hamming distance between this part of the sequence and the pattern is at most α_{dist}). Both SMPs and EMPs are necessary: SMPs allow to gather the degenerated TFBSs while EMPs are dedicated to pick out the conserved ones.

The two kinds of patterns are extracted using a solver over the string pattern domain called *Marguerite* (see Section 17.3). This tool performs a differential extraction of patterns between the two sets of sequences D^+ and D^-. To run an extraction, the user has to set the four following constraints: L the length of the patterns, α_{min} their minimal support in D^+, α_{max} their maximal support in D^-, and α_{dist} the soft matching threshold (for SMPs). *Marguerite* is complete in the sense that it finds all possible patterns satisfying the constraints according to the user setting. In the case of SMPs, the solver enforces an additional constraint: the patterns must have at-least one exact occurrence in D^+. This additional constraint enables to focus on SMPs that appear at-least one time in a non-degenerated way. Concerning the use of the solver, setting four parameters is not an easy task, so we developed and used a dedicated parameter tuning tool (see Section 17.4).

In order to assess the significance of a pattern we used the notion of Twilight Zone (TZ) [11] to build a Twilight Zone Indicator (TZI). A twilight zone is a zone in a parameter space, where we are likely to obtain patterns produced by the random background. For a pattern ϕ of length L, the indicator $TZI(\phi)$ is an estimate of the

---

[1] It consists in comparing two biological situations, Sit_1 and Sit_2, in order to obtain two groups of genes: one that is up-regulated, and the other one that is down-regulated, when going from Sit_1 to Sit_2.

[2] Notice that if we exchange the positive and negative datasets, then we could find putative TFBSs regulating the underexpressed genes.

minimum number of patterns of length L, due to the random background, that are likely to be extracted together with ϕ, in the most stringent conditions (i.e., with the strongest constraints, that still lead to the extraction of ϕ). The computation of the TZI is detailed in Section 17.5. It is based on the same hypothesis made in [11]: the data sequences are composed of independent and uniformly distributed nucleotides, and the possible overlap of the occurrences of the patterns is considered to have a limited impact on the number of extracted patterns. In addition, we suppose that the positive and the negative datasets are independent.

During the next step, the biologist browses and ranks the patterns (according to the TZI measure, and to the support of the patterns in D^+ and D^-) and then he/she selects some promising ones.

On these selected patterns, the following post-processing is applied (see Section 17.6.3). First the similar patterns are grouped by performing a hierarchical clustering. Then, for each cluster we compute the average of the TZI of the patterns in the cluster, and in each cluster, the patterns are aligned with a multiple alignment tool (*MultAlin* [5]) to build a *consensus pattern* of the cluster. Finally, the consensus patterns are checked w.r.t. the *Transfac*® [12] database, to find out if they are known TFBSs, close to some known TFBSs or unknown.

17.3 The *Marguerite* Solver

We introduce the solver *Marguerite* which supports inductive querying on strings. It has been used in the scenario described in this chapter. We define more precisely the patterns and constraints handled by this solver. More details can be found in [14, 15].

Let Σ be a finite alphabet (in the scenario $\Sigma = \{A, C, G, T\}$), then a string ϕ over Σ is a finite sequence of symbols from Σ. The language of patterns $\mathscr{L}$ is Σ^*, i.e, the set of all strings over Σ. A string dataset D is a multi-set[3] of strings from Σ^*. The length of a string ϕ is denoted $|\phi|$. A substring ϕ' of ϕ is a sequence of contiguous symbols in ϕ.

An *exact occurrence* of a pattern ϕ is simply a substring of a string in D that is equal to ϕ. The *exact support* of ϕ, denoted $supp_E(\phi, D)$, is the number of strings in D that contain at least one exact occurrence of ϕ. Notice that multiple occurrences of a pattern in the same string do not change its support.

Let α_{dist} be a positive integer, then an (α_{dist})-*soft occurrence* of a pattern ϕ is a substring ϕ' of a string in D, having the same length as ϕ and such that $hamming(\phi, \phi') \leq \alpha_{dist}$, where $hamming(\phi, \phi')$ is the Hamming distance between ϕ and ϕ' (i.e., the number of positions where ϕ and ϕ' are different). The (α_{dist})-*soft support* of ϕ is the number of strings in D that contain at least one (α_{dist})-soft occurrence of ϕ. It is denoted $supp_S(\phi, D, \alpha_{dist})$.

---

[3] The dataset may contain several times the same string.

Example 17.1. If $D = \{atgcaaac, acttggac, gatagata, tgtgtgtg, gtcaactg\}$, then we have $supp_E(gac, D) = 1$ since only string *acttggac* contains *gac*, and we also have $supp_S(gac, D, 1) = 3$ because *acttggac*, *gatagata* and *gtcaactg* contain some 1-soft occurrences of *gac*.

Definition 17.1 (Frequency constraints). In the case of the exact support, given a threshold value f, the minimal (resp. maximal) frequency constraint is $MinFr(\phi, D, f) \equiv supp_E(\phi, D) \geq f$ (resp. $MaxFr(\phi, D, f) \equiv supp_E(\phi, D) \leq f$). For the (α_{dist})-soft support, the constraints are defined as $MinFr(\phi, D, f) \equiv supp_S(\phi, D, \alpha_{dist}) \geq f \wedge supp_E(\phi, D) \geq 1$ and $MaxFr(\phi, D, f) \equiv supp_S(\phi, D, \alpha_{dist}) \leq f$.

Notice that, in the case of the soft support, our definition of $MinFr$ enforces the presence of at least one exact occurrence, in order to discard patterns that only occur as degenerated instances.

The generic conjunction of constraints handled by *Marguerite* is:

$\mathscr{C} \equiv MinFr(\phi, D^+, \alpha_{min}) \wedge MaxFr(\phi, D^-, \alpha_{max}) \wedge |\phi| = L$, where D^+ and D^- are string databases, α_{min} and α_{max} are frequency thresholds, and L is a user defined pattern length.

The algorithms used by *Marguerite* [14, 15] are based on the generic algorithm FAVST [6], designed for the efficient extraction of strings under combination of constraints, taking advantage of the so-called Version Space Tree (VST) [8] data structure. *Marguerite* extends FAVST to degenerated patterns discovery through similarity and soft-support constraints. It is implemented in $C/C++$ and can be used to compute both Exact Matching Patterns (EMPs) and Soft Matching Patterns (SMPs) in a complete way (i.e., all patterns satisfying the constraints are outputted).

17.4 Tuning the Extraction Parameters

In an exploratory data mining task based on pattern extraction, one of the most commonly used parameter tuning strategies, in the early exploration stage, is to run a few experiments with different settings, and to simply count the number of patterns that are obtained. Then, using some domain knowledge, the user tries to guess some potentially interesting parameter settings. After that stage, the user enters a more iterative process, in which she/he also looks at the patterns themselves and at their scores (according to various quality measures), and uses her/his knowledge of the domain to focus on some patterns and/or to change the parameters by some *local* variations of their values.

To support this early exploratory stage, so that the user can guess promising initial parameter settings, we decided to probe the parameter space in a more systematic way, so that it could be possible to provide graphics that depict the extraction landscape, i.e., the number of patterns that will be obtained for a wide range of parameter values. This idea is very simple, and many (if not all) of the practitioners have one day written their own script/code to run such sets of experiments. However, in many cases, the cost of running real extractions for hundreds of different parameter settings is clearly prohibitive.

Instead of running real experiments, a second way is to develop an analytical model, that estimates the number of patterns satisfying the constraint $\mathscr{C}$, with respect to the distribution of the symbols and the structure (number of strings and size) of the datasets, and with respect to the values of the parameters used in $\mathscr{C}$. In this approach, an important effort has to be made on the design of the model, and in most cases this is a non-trivial task. For instance, to the best of our knowledge, in the literature there is no analytical model of the number of patterns satisfying $\mathscr{C} \equiv MinFr(\phi, D^+, \alpha_{min}) \wedge MaxFr(\phi, D^-, \alpha_{max}) \wedge |\phi| = L$ when soft-occurrences are used to handle degenerated patterns (even in the simple case where $\alpha_{dist} = 1$). Designing an analytical model to handle this case is certainly not straightforward, in particular because of the specific symbol distribution that has to be incorporated in the model.

We developed a third approach based on the following key remark. When a pattern ϕ is given, together with the distribution of the symbols, the structure of the datasets and the values of the parameters in $\mathscr{C}$, we can compute $P(\phi\ sat.\ \mathscr{C})$ the probability that ϕ satisfies $\mathscr{C}$ in this dataset. In most cases, designing a function to compute $P(\phi\ sat.\ \mathscr{C})$ is rather easy in comparison to the effort needed to exhibit an analytical model that estimates the number of patterns satisfying the constraint $\mathscr{C}$. Having at hand a function to compute $P(\phi\ sat.\ \mathscr{C})$, the next step is then to estimate the total number of patterns that will be extracted, but without having to compute $P(\phi\ sat.\ \mathscr{C})$ for all patterns in the pattern space. Therefore, we propose a simple pattern space sampling approach, that leads to a fast and accurate estimate of the number of patterns that will be extracted. Finally, we can compute such an estimate for a large number of points in the parameter space and provide views of the whole extraction landscape.

To determine $P(\phi\ sat.\ \mathscr{C})$, we first compute the different frequencies of occurrence of the symbols. We consider that all occurrences of the symbols are independent, and then, for a given pattern ϕ we can easily compute the probability that ϕ occurs in a string of a given length. If we suppose that all strings in the dataset have the same length, the probability to appear in each string is the same, and we can use a binomial law to obtain the probability for this pattern to satisfy the constraint $MinFr(\phi, D^+, \alpha_{min})$ and the probability to satisfy the constraint $MaxFr(\phi, D^-, \alpha_{max})$. Finally, if we suppose that the datasets D^+ and D^- are independent, we can multiply these two probabilities to obtain $P(\phi\ sat.\ \mathscr{C})$.

Let $S_\mathscr{C}$ be the set of patterns in $\mathscr{L}$ that satisfy the constraint $\mathscr{C} \equiv MinFr(\phi, D^+, \alpha_{min}) \wedge MaxFr(\phi, D^-, \alpha_{max}) \wedge |\phi| = L$, using $P(\phi\ sat.\ \mathscr{C})$ we can estimate $|S_\mathscr{C}|$ by sampling the pattern space as follows. Let us associate to each pattern ϕ a random variable X_ϕ, such that $X_\phi = 1$ when ϕ satisfy $\mathscr{C}$ and $X_\phi = 0$ otherwise. Then $|S_\mathscr{C}| = \sum_{\phi \in \mathscr{L}} X_\phi$. Considering the expected value of $|S_\mathscr{C}|$, by linearity of the expectation operator we have $E(|S_\mathscr{C}|) = \sum_{\phi \in \mathscr{L}} E(X_\phi)$. Since $E(X_\phi) = 1 \times P(X_\phi = 1) + 0 \times P(X_\phi = 0)$, then $E(|S_\mathscr{C}|) = \sum_{\phi \in \mathscr{L}} P(\phi\ sat.\ \mathscr{C})$. Let S_L be the set of patterns in $\mathscr{L}$ that satisfy $|\phi| = L$. As $P(\phi \; sat. \; \mathscr{C}) = 0$ for all patterns that do not satisfy $|\phi| = L$, we have $E(|S_{\mathscr{C}}|) = \sum_{\phi \in S_L} P(\phi \; sat. \; \mathscr{C})$.

Computing this sum over S_L could be prohibitive, since we want to obtain the values of $E(|S_{\mathscr{C}}|)$ for a large number of points in the parameter space. Thus we estimate $E(|S_{\mathscr{C}}|)$ using only a sample of the patterns in S_L. Let S_{samp} be such a sample, then we use the following value as an estimate of $E(|S_{\mathscr{C}}|)$:

$$\frac{|S_L|}{|S_{samp}|} \times \sum_{\phi \in S_{samp}} P(\phi \; sat. \; \mathscr{C})$$

In practice, many techniques can be used to compute the sample. In our experiments, we use the following process:

- Step 1: build an initial sample S_{samp} of S_L (sampling with replacement) of size 5% of $|S_L|$ and compute the estimate of $E(|S_{\mathscr{C}}|)$.
- Step 2: go on sampling with replacement to add 1000 elements to S_{samp}. Compute the estimate, and if the absolute value of the difference between the new estimate and the previous one is greater than 5% of the previous estimate, then iterate on Step 2.

17.5 An Objective Interestingness Measure

The notion of Twilight Zone (TZ) [11] has been originally proposed to characterize the *subtle motifs*, i.e., motifs that can not be distinguished (no statistically significant difference) from random patterns (patterns due to the random background). In this context, the TZ was defined as the set of values of the scoring function for which we can expect to have some random patterns exhibiting such score values. Let us consider the notion of extraction parameters in a broad sense, including structural properties of the dataset (e.g., number of sequences, length of the sequences) and mining constraints (e.g., selection threshold according to one or several measures, length of the patterns). Then, the TZ can be seen as a region (or a set of regions) in the parameter space, where we are likely to obtain random patterns among the extracted patterns, these random pattern having scores as good (or event better) than the *true* patterns.

We can now define a Twilight Zone Indicator (TZI) to rank the patterns in the case of differential extractions. Let ϕ be a pattern, occurring in $support^+(\phi)$ sequences of the positive dataset, and in $support^-(\phi)$ sequences of the negative dataset. Then, TZI(ϕ) is an estimate of the number of random patterns, having the same length as ϕ, that will be extracted using $\alpha_{min} = support^+(\phi)$ and $\alpha_{max} = support^-(\phi)$, i.e., using the most selective constraints that still permit to obtain ϕ (since for larger α_{min} and/or lower α_{max} threshold values, ϕ will not satisfy the constraints and will not be retained during the extraction). The higher is $TZI(\phi)$, the *deeper* is ϕ in the twilight zone, and thus likely to be retrieved among a larger collection of patterns due to the random background that cannot statistically be distinguished from ϕ. Then, in practice, we will select patterns having a low TZI, to expect to have patterns that are not due to the random background.

At first glance, the number of patterns satisfying $\alpha_{min} = support^+(\phi)$ and $\alpha_{max} = support^-(\phi)$ could be obtained using the sampling based technique presented in Section 17.4. Unfortunately, if this approach can help the user to find estimates of the number of patterns in wide ranges of parameter values, the extracted patterns themselves can represent many much more $(support^+, support^-)$ pairs, than the number of $(\alpha_{min}, \alpha_{max})$ pairs considered during the parameter setting stage. For instance, it can make sense for the expert to explore the α_{min} setting between 20 and 40, while real patterns that are extracted using $\alpha_{min} \in [20, 40]$ could have support larger than 40, and not only in $[20, 40]$. In order to avoid the cost of computing the sampling based estimate for each extracted pattern, we now discuss an alternative way to obtain such an estimate. This second estimate is less accurate, in the sense that it does not take into account the difference among the frequencies of the symbols, but it uses a direct analytical estimate, i.e., without sampling. It can be much more relevant in practice.

We consider that all the sequences have the same length, denoted G. In this context, we want to estimate the number of SMP patterns of length L that will be extracted under the thresholds α_{min}, α_{max} and α_{dist}. Let us notice that estimating the number of EMP is a particular case, where α_{dist} is set to 0. As in [11], we suppose that the data sequences are composed of independent and uniformly distributed symbols, having the same occurrence probability, and that the overlapping of the occurrences of the patterns has a negligible impact on the number of patterns extracted (since $L \ll G$). Additionally, as in the previous section, we suppose that the two datasets are independent.

Occurrences at a given position. The data sequences are gene promoter sequences. On such a given vocabulary, we have 4^L different possible strings of length L. The hypotheses made on the distribution of the symbols imply that the probability that a pattern ϕ of length L has an exact occurrence starting at a given position in a sequence[4] is:

$$P(exact\ occ.\ of\ \phi\ at\ one\ position) = \frac{1}{4^L}.$$

From an exact occurrence of ϕ, one can construct the soft occurrences of ϕ within an Hamming distance α_{dist} by placing k substitutions in $\binom{L}{k}$ possible ways, with $k \in \{0, \ldots \alpha_{dist}\}$. Since we have 4 symbols, then for each position were we have a substitution, we have 3 different possible substitutions. Thus, for a pattern ϕ, there are $\sum_{k=0}^{\alpha_{dist}} \binom{L}{k} \times 3^k$ strings that are soft occurrences of ϕ. Then, the probability that a pattern has a soft occurrence starting at a given position in a sequence is:

$$P(soft\ occ.\ of\ \phi\ at\ one\ position) = \frac{\sum_{k=0}^{\alpha_{dist}} \binom{L}{k} \times 3^k}{4^L}.$$

---

[4] Except the last $L - 1$ positions.

In the following, we also need the probability that a pattern ϕ has a *strict* soft occurrence starting at a given position (a *strict* soft occurrence of ϕ is a soft occurrences of ϕ that is not an exact occurrence). In this case we have simply:

$$P(strict\ soft\ occ.\ of\ \phi\ at\ one\ position) = \frac{\sum_{k=1}^{\alpha_{dist}} \binom{L}{k} \times 3^k}{4^L}.$$

Occurrences in a random sequence. In a sequence, there are $(G-L+1)$ possible positions to place the beginning of an occurrence of ϕ. Since $L \ll G$, for the sake of simplicity, we approximate a number of possible positions by G. Then, considering that the occurrence overlap has a negligible impact, the probability that there is no soft occurrence of ϕ in a random sequence is:

$$P(no\ soft\ occ.\ of\ \phi\ in\ a\ seq.) = (1 - P(soft\ occ.\ of\ \phi\ at\ one\ position))^G.$$

The probability that there is at least one soft occurrence of ϕ in a sequence is:

$$P(exists\ soft\ occ.\ of\ \phi\ in\ a\ seq.) = 1 - (1 - P(soft\ occ.\ of\ \phi\ at\ one\ position))^G.$$

Similarly, the probability that there is at least one strict soft occurrence of ϕ is:

$$P(exists\ strict\ soft\ occ.\ of\ \phi\ in\ a\ seq.) = 1 - (1 - P(strict\ soft\ occ.\ of\ \phi\ at\ one\ position))^G.$$

Finally, the probability that there is at least one exact occurrence is:

$$P(exists\ exact\ occ.\ of\ \phi\ in\ a\ seq.) = 1 - (1 - \tfrac{1}{4^L})^G.$$

Minimum support constraint. To determine $P(\phi\ sat.\ min.\ supp.)$, i.e., the probability of ϕ to satisfy the minimum support constraint, let us define X as the number of sequences, in the positive dataset, that contains at least one exact occurrence of ϕ. The probability $P(\phi\ sat.\ min.\ supp.)$ can be decomposed using the conditional probability of $\phi\ sat.\ min.\ supp.$ given the value of X, as follows:

$$P(\phi\ sat.\ min.\ supp.) = \sum_{i=1}^{N^+} (P(X = i) \times P(\phi\ sat.\ min.\ supp.|X = i)) \quad (17.1)$$

Notice that the sum starts at $i = 1$, and not at $i = 0$, since the pattern must have at least one exact occurrence in the positive dataset (see Section 17.3).

The variable X follows a binomial distribution $B(N^+, P(exists\ exact\ occ.\ of\ \phi\ in\ a\ seq.))$, where N^+ is the number of sequences in the positive dataset. Thus we have:

$$P(X = i) = \binom{N^+}{i} \times P(exists\ exact\ occ.\ of\ \phi\ in\ a\ seq.)^i$$
$$\times (1 - P(exists\ exact\ occ.\ of\ \phi\ in\ a\ seq.))^{N^+ - i}.$$

$P(\phi\ sat.\ min.\ supp.|X = i)$ is the probability that ϕ satisfies the minimum support constraint, given that exactly i sequences contain at least one exact occurrence of ϕ. This also means that $(N^+ - i)$ sequences do not have any exact occurrence of a pattern. Then, according to i, there are two cases:

1. If $i \geq \alpha_{min}$ then $P(\phi\ sat.\ min.\ supp.|X = i)) = 1$ since the constraint is already satisfied by the i sequences that contain each at least one exact occurrence of ϕ.

2. If $i < \alpha_{min}$ then $P(\phi\ sat.\ min.\ supp.|X = i)$ is equal to the probability that at least $(\alpha_{min} - i)$ of the $(N^+ - i)$ remaining sequences contain at least one strict soft occurrence. This number of sequences that contain at least one strict soft occurrence of ϕ also follows a binomial distribution $B(N^+ - i, P(exists\ strict\ soft\ occ.\ of\ \phi\ in\ a\ seq.))$. Then we have:

$$P(\phi\ sat.\ min.\ supp.|X = i)) = \sum_{z=\alpha_{min}-i}^{N^+-i} \left(\binom{N^+-i}{z} \right.$$
$$\times P(exists\ strict\ soft\ occ.\ of\ \phi\ in\ a\ seq.)^z$$
$$\times (1 - P(exists\ strict\ soft\ occ.\ of\ \phi\ in\ a\ seq.))^{N^+-i-z}).$$

It means that we can provide $P(\phi\ sat.\ min.\ supp.)$ by computing the sum in Equation 17.1 and $P(\phi\ sat.\ min.\ supp.|X = i)$ according to the two cases above.

Maximum Support constraint. Let Y be the number of sequences that support ϕ in the negative dataset. A pattern ϕ satisfies the maximum support constraint with threshold α_{max} if $Y \leq \alpha_{max}$. The variable Y follows a binomial distribution $B(N^-, P(exists\ soft\ occ.\ of\ \phi\ in\ a\ seq.))$, where N^- is the number of sequences in the negative dataset. Then the probability that ϕ satisfies the maximum support constraint is:

$$P(\phi\ sat.\ max.\ supp.) = \sum_{z=0}^{\alpha_{max}} \binom{N^-}{z}$$
$$\times P(exists\ soft\ occ.\ of\ \phi\ in\ a\ seq.)^z$$
$$\times (1 - P(exists\ soft\ occ.\ of\ \phi\ in\ a\ seq.))^{N^--z}.$$

Conjunction of Minimum Support and Maximum Support constraints. Given our hypothesis that the positive and negative datasets are independent, the probability that a pattern satisfies a conjunction of minimum support and maximum support constraints is:

$$P(\phi\ sat.\ min.\ and\ max.\ supp.) = P(\phi\ sat.\ min.\ supp.) \times P(\phi\ sat.\ max.\ supp.).$$

Number of expected patterns and Twilight Zone Indicator. Let $ENP(L, \alpha_{min}, \alpha_{max}, \alpha_{dist})$ be the Expected Number of Patterns of length L that will be extracted under the thresholds α_{min}, α_{max} and α_{dist}. Since there are 4^L possible patterns of length L, and given the hypothesis that the overlapping of the occurrences of the patterns has a negligible impact on the number of extracted patterns, we can approximate $ENP(L, \alpha_{min}, \alpha_{max}, \alpha_{dist})$ by $P(\phi\ sat.\ min.\ and\ max.\ supp.) \times 4^L$.

Finally, let ϕ be a pattern, occurring in $support^+(\phi)$ sequences of the positive dataset, and in $support^-(\phi)$ sequences of the negative dataset for a given α_{dist} threshold. Then, $TZI(\phi)$ is defined as $ENP(|\phi|, support^+(\phi), support^-(\phi), \alpha_{dist})$.

17.6 Execution of the Scenario

In this section, we present a typical concrete execution of the whole scenario, in the context of the study of the TFs and TFBSs involved in the activation/repression of genes in reaction to the presence of the v-erbA oncogene, a chemical compound involved in the cell self-renewal process.

17.6.1 Data preparation

Using the SAGE technique [21], we identified two sets of genes: a set R of 29 genes repressed by v-ErbA and a set A of 21 genes activated by v-ErbA. Then, we collected the promoter sequences of all these genes (taking 4000 bases for each promoter). These promoter sequences have been extracted as described in [4]. Finally, we have built two datasets: D^+ (resp. D^-) containing the promoter sequences of the genes of set R (resp. A).

These two datasets represent two biologically opposite situations. As a result, we assume that computing string patterns that have a high support in D^+ and a small support in D^- is a way to identify putative binding sites of transcription factors involved in this activation/repression process induced by v-ErbA.

17.6.2 Parameter tuning

Patterns having slightly degenerated occurrences can be interesting in our context. Therefore, we look for SMP patterns using $\alpha_{dist} = 1$ for the soft support definition. The estimates are computed according to the sampling technique presented in Section 17.4 with respective frequencies of 0.23, 0.26, 0.27, 0.24 for symbols A, C, G and T. Representative graphics depicting portions of the extraction landscape, are presented in Figure 17.2, on the right.

A typical use of such graphics is, for instance, to look for points, in the parameter space, corresponding to a large support on D^+, but a low support on D^-, a large pattern size, and a rather small number of expected patterns. Such a point can be used as an initial guess of the parameters to perform the extractions. For instance, we may consider pattern size = 10, minimal support on D^+ of 15, and maximal support on D^- of 5. The graphic in the middle on the right for Figure 17.2 indicates that, for this setting, only about 1 pattern is expected.

Additionally, in Figure 17.2 on the left, we give the real numbers of extracted patterns. In practice, these graphics are not easily accessible, since in these experiments the running time of a single extraction with *Marguerite* (on a Linux platform with an Intel 2Ghz processor and 1Gb of RAM) ranges from tens of minutes to several hours[5], while for an estimate (graphics on the right) only a few tens of seconds is needed. Even though the global trends correspond to the estimates on the right, there are differences in some portions of the parameter space. For example, for the setting *pattern size* = 10, *minimal support* = 15, and *maximal support* = 5, we have about 100 extracted patterns while we expected only one. Such a difference suggests that these 100 patterns capture an underlying structure of the datasets, and that they are not simply due to the random background.

---

[5] Notice that for experiments using EMP (exact support) on these datasets, with similar parameter values, the running time is only about a few tens of seconds to a few minutes.

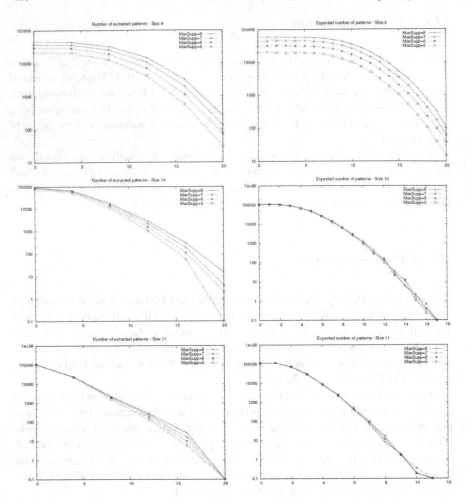

Fig. 17.2 Expected and real numbers of extracted patterns. The minimal support α_{min} corresponds to the horizontal axis, and the number of patterns corresponds to the vertical axis (log scale).

17.6.3 Post-processing and biological pattern discovery

Hierarchical clustering of SMPs. The hierarchical clustering of the SMPs patterns is performed using the *hclust* function of the package *stat* of the *R* environment [18]. The proximity between clusters is computed using the complete linkage method. To improve the quality and efficiency of the clustering, we process the SMPs by groups of patterns having the same length. To construct a distance matrix, we estimate the dissimilarity of each pair of SMPs as follows. For each pair, we compute its optimal pairwise global alignment [17] with the following parameters:

the score for a mismatch is 1, the score for a match is 0, the insertions and deletions inside an alignment are not allowed, the terminal gaps are not penalized, and the length of an alignment (terminal gaps are not included in the alignment length) must be at least a half of the length of the patterns in the pair. Finally, the dissimilarity of a pair of SMPs is simply the score of its best alignment (i.e., alignment having the lowest score).

Finding a consensus pattern within a cluster. To find the consensus pattern of each cluster of SMPs we align the patterns in each cluster using the multiple alignment tool *MultAlin* [5]. We use the following alignment scoring parameters: gap creation and extension penalty is −5, terminal gaps are not penalized, score for a match is 2, and score for any mismatch is 0. Once a consensus SMP is computed we consult *Transfac*® [12] to check whether it is a known TFBS. Figure 17.3 gives an example of a cluster, whose consensus SMP that has been selected because of its rather low TZI value (i.e., not likely to be due to the random background), and that is reported by *Transfac*® as a binding site of the TF c-Myb-isomorf1. In the consensus pattern in this figure, the bases that are highly conserved appear as uppercase letters in the consensus, and the weakly conserved ones appear as lowercase. Positions with no conserved bases are indicated as dots.

```
                                   .CGGCCGTT...       23.94
                                   .GCGCCGTT...        0.68
                                   ...GCCGTTAT.        4.4
                                   ....CCGTTCGT        4.4
                                   ...GCCGTTCG.       23.75
                                   ....CCGTTAGG        0.68
                                   TTGGCCGT....       23.75
                                   ...GCCGTAAC.      107.37
                                   ..TGCCGTAA..        0.58

**Fig. 17.3** A cluster of SMPs        *Consensus*    ...gCCGTt...
and its consensus computed         *Transfac:*    c-Myb-isoform1
by a multiple alignment.           *Mean of TZI*:  21.06
```

Biological interpretation. The application of the scenario therefore allowed us to identify a c-Myb binding site as a signature motif of many newly identified v-ErbA repressed target genes compared with v-ErbA activated target genes. This suggests a potential role for c-Myb in the v-ErbA induced transformation. To determine the role of c-Myb in this transformation process, we used a gene reporter assay to test the ability of v-ErbA to transactivate c-Myb [4]. This experiment demonstrated that v-ErbA can indeed functionally interacts directly or indirectly with the transcriptional activity of endogenous c-Myb in T2ECs, constituting an experimental validation of the *in silico* extracted motif.

17.7 Conclusion

In this chapter, we presented a complete scenario that has been designed and used to support knowledge discovery from promoter sequences. Indeed, it can be applied to suggest putative TFBSs. The description of this application has been made at different levels: the corresponding abstract scenario, its concrete instantiation and a typical execution on a real dataset. To perform the main extraction step, we propose to use a solver developped for inductive querying over the string pattern domain. We also discussed all the additional processing required to use a solver, i.e., a data mining algorithm, in such a realistic context. This includes a parameter tuning tool, a support to pattern ranking and typical post-processing facilities dedicated to this kind of discovery task.

Acknowledgements This work has been partly funded by EU contract IST-FET IQ FP6-516169 (Inductive Queries for Mining Patterns and Models) and by the French contract ANR-07-MDCO-014 Bingo2 (Knowledge Discovery For and By Inductive Queries). We would like to thank Stéphane Schicklin for his contribution to pattern matching against the *Transfac®* database, and the anonymous reviewers for their helpful comments and suggestions.

References

1. Besson, J., Rigotti, C., Mitasiunaité, I., Boulicaut, J.F.: Parameter tuning for differential mining of string patterns. In: Proceedings IEEE Workshop DDDM'08 co-olocated with ICDM'08, pp. 77–86 (2008)
2. Boulicaut, J.F., De Raedt, L., Mannila, H. (eds.): Constraint-Based Mining and Inductive Databases, *LNCS*, vol. 3848. Springer (2005). 400 pages
3. Brazma, A., Jonassen, I., Vilo, J., Ukkonen, E.: Predicting gene regulatory elements in silico on a genomic scale. Genome Res. **8**(11), 1202–1215 (1998)
4. Bresson, C., Keime, C., Faure, C., Letrillard, Y., Barbado, M., Sanfilippo, S., Benhra, N., Gandrillon, O., Gonin-Giraud, S.: Large-scale analysis by SAGE revealed new mechanisms of v-erba oncogene action. BMC Genomics **8**(390) (2007)
5. Corpet, F.: Multiple sequence alignment with hierarchical clustering. Nucl. Acids Res. **16**(22), 10,881–10,890 (1988)
6. Dan Lee, S., De Raedt, L.: An efficient algorithm for mining string databases under constraints. In: Proceedings KDID'04, pp. 108–129. Springer (2004)
7. De Raedt, L.: A perspective on inductive databases. SIGKDD Explorations **4**(2), 69–77 (2003)
8. De Raedt, L., Jaeger, M., Lee, S.D., Mannila, H.: A theory of inductive query answering. In: Proceedings IEEE ICDM'02, pp. 123–130 (2002)
9. Eden, E., Lipson, D., Yogev, S., Yakhini, Z.: Discovering motifs in ranked lists of DNA sequences. PLOS Computational Biology **3**(3), 508–522 (2007)
10. Imielinski, T., Mannila, H.: A database perspective on knowledge discovery. CACM **39**(11), 58–64 (1996)
11. Keich, U., Pevzner, P.A.: Subtle motifs: defining the limits of motif finding algorithms. Bioinformatics **18**(10), 1382–1390 (2002)
12. Matys, V., Fricke, E., Geffers, R., Gössling, E., Haubrock, M., Hehl, R., Hornischer, K., Karas, D., Kel, A.E., Kel-Margoulis, O.V., Kloos, D.U., Land, S., Lewicki-Potapov, B., Michael, H., Münch, R., Reuter, I., Rotert, S., Saxel, H., Scheer, M., Thiele, S., E., Wingender: Transfac : transcriptional regulation, from patterns to profiles. Nucl. Acids Res. **31**(1), 374–378 (2003)

13. Mitasiunaite, I.: Mining string data under similarity and soft-frequency constraints: Application to promoter sequence analysis. Ph.D. thesis, INSA Lyon (2009)
14. Mitasiunaite, I., Boulicaut, J.F.: Looking for monotonicity properties of a similarity constraint on sequences. In: Proceedings of ACM SAC'06 Data Mining, pp. 546–552 (2006)
15. Mitasiunaite, I., Boulicaut, J.F.: Introducing softness into inductive queries on string databases. In: Databases and Information Systems IV, *Frontiers in Artificial Intelligence and Applications*, vol. 155, pp. 117–132. IOS Press (2007)
16. Mitasiunaite, I., Rigotti, C., Schicklin, S., Meyniel, L., j. F. Boulicaut, Gandrillon, O.: Extracting signature motifs from promoter sets of differentially expressed genes. In Silico Biology **8**(43) (2008)
17. Needleman, S., Wunsch, C.: A general method applicable to the search for similarities in the amino acid sequence of two proteins. J. Mol. Biol. **48**(3), 443–453 (1970)
18. The R Project for Statistical Computing: http://www.r-project.org/
19. Tompa, M., Li, N., Bailey, T.L., Church, G.M., Moor, B.D., Eskin, E., Favorov, A.V., Frith, M.C., Fu, Y., Kent, W.J., Makeev, V.J., Mironov, A.A., Noble, W.S., Pavesi, G., Pesole, G., RÃ©gnier, M., Simonis, N., Sinha, S., Thijs, G., van Helden, J., Vandenbogaert, M., Weng, Z., Workman, C., Ye, C., Zhu, Z.: Assessing computational tools for the discovery of transciption factor binding sites. Nat. Biotechnol. **23**(1), 137–144 (2005)
20. Vanet, A., Marsan, L., Sagot, M.F.: Promoter sequences and algorithmical methods for identifying them. Res. Microbiol. **150**(9-10), 779–799 (1999)
21. Velculescu, V.E., Zhang, L., Vogelstein, B., Kinzler, K.: Serial analysis of gene expression. Science **270**(5235), 484–487 (1995)

Chapter 18
Inductive Queries for a Drug Designing Robot Scientist

Ross D. King, Amanda Schierz, Amanda Clare, Jem Rowland, Andrew Sparkes, Siegfried Nijssen, and Jan Ramon

Abstract It is increasingly clear that machine learning algorithms need to be integrated in an iterative scientific discovery loop, in which data is queried repeatedly by means of *inductive queries* and where the computer provides guidance to the experiments that are being performed. In this chapter, we summarise several key challenges in achieving this integration of machine learning and data mining algorithms in methods for the discovery of Quantitative Structure Activity Relationships (QSARs). We introduce the concept of a robot scientist, in which all steps of the discovery process are automated; we discuss the representation of molecular data such that knowledge discovery tools can analyse it, and we discuss the adaptation of machine learning and data mining algorithms to guide QSAR experiments.

18.1 Introduction

The problem of learning Quantitative Structure Activity Relationships (QSARs) is an important inductive learning task. It is central to the rational design of new drugs and therefore critical to improvements in medical care. It is also of economic im-

Ross D. King · Amanda Clare · Jem Rowland · Andrew Sparkes
Department of Computer Science, Llandinam Building,
Aberystwyth University, Aberystwyth, Ceredigion, SY23 3DB, United Kingdom
e-mail: rdk@aber.ac.uk, afc@aber.ac.uk, jjr@aber.ac.uk, nds@aber.ac.uk

Amanda Schierz
2DEC, Poole House, Bournemouth University, Poole, Dorset, BH12 5BB, United Kingdom
e-mail: aschierz@bournemouth.ac.uk

Siegfried Nijssen · Jan Ramon
Departement Computerwetenschappen, Katholieke Universiteit Leuven,
Celestijnenlaan 200A, 3001, Leuven, Belgium
e-mail: siegfried.nijssen@cs.kuleuven.be, jan.ramon@cs.kuleuven.be

portance to the pharmaceutical industry. The QSAR problem is as follows: given a set of molecules with associated pharmacological activities (e.g., killing cancer cells), find a predictive mapping from structure to activity, which enables the design of a new molecule with maximum activity. Due to its importance, the problem has received a lot of attention from academic researchers in data mining and machine learning. In these approaches, a dataset is usually constructed by a chemist by means of experiments in a wet laboratory and machine learners and data miners use the resulting datasets to illustrate the performance of newly developed predictive algorithms. However, such an approach is divorced from the actual practice of drug design, where cycles of QSAR learning and new compound synthesis are typical. Hence, it is necessary that data mining and machine learning algorithms become a more integrated part of the scientific discovery loop. In this loop, algorithms are not only used to find relationships in data, but also provide feedback as to which experiments should be performed and provide scientists with interpretable representations of the hypotheses under consideration.

Ultimately, the most ambitious goal one could hope to achieve is the development of a *robot scientist for drug design,* which integrates the entire iterative scientific loop in an automated machine, i.e., the robot not only performs experiments, but also analyses them and proposes new experiments. Robot Scientists have the potential to change the way drug design is done, and enable the rapid adoption of novel machine-learning/data-mining methodologies for QSAR. They however pose particular types of problems, several of which involve machine learning and data mining. These challenges are introduced further in Section 18.2.

The point of view advocated in this book is that one way to support iterative processes of data analysis, is by turning isolated data mining tools into *inductive querying* systems. In such a system, a run of a data mining algorithm is seen as calculating an answer to a query by a user, whether this user is a human or a computerized system, such as a robot scientist. Compared to traditional data mining algorithms, the distinguishing feature of an inductive querying system is that it provides the user considerably more freedom in formulating alternative mining tasks, often by means of *constraints.* In the context of QSAR, this means that the user is provided with more freedom in how to deal with representations of molecular data, can choose the constraints under which to perform a mining task, and has freedom in how the results of a data mining algorithm are processed.

This chapter summarizes several of the challenges in developing and using inductive querying systems for QSAR. We will discuss in more detail three technical challenges that are particular to iterative drug design: the representation of molecular data, the application of such representations to determine an initial set of compounds for use in experiments, and mechanisms for providing feedback to machines or human scientists performing experiments.

A particular feature of molecular data is that, essentially, a molecule is a structure consisting of atoms connected by bonds. Many well-known machine learning and data mining algorithms assume that data is provided in a tabular (attribute-value) form. To be able to learn from molecular data, we either need strategies for transforming the structural information into a tabular form or we need to develop algorithms that no longer require data in such form. This choice of representation is important both to obtain reasonable predictive accuracy and to make the interpretation of models easier. Furthermore, within an inductive querying context, one may wish to provide users with the flexibility to tweak the representation if needed. These issues of representation will be discussed in more detail in the Section 18.3.

An application of the use of one representation is discussed in Section 18.4, in which we discuss the selection of compound libraries for a robot scientist. In this application, it turns out to be of particular interest to have the flexibility to include background knowledge in the mining process by means of *language bias*. The goal in this application is to determine the library of compounds available to the robot: even though the experiments in a robot scientist are automated, in its initial runs it would not be economical to synthesise compounds from scratch and the use of an existing library is preferable. This selection is, however, important for the quality of the results and hence a careful selection using data mining and machine learning tools is important.

When using the resulting representation in learning algorithms, the next challenge is how to improve the selection of experiments based on the feedback of these algorithms. The algorithms will predict that some molecules are more active than others. One may choose to exploit this result and perform experiments on predicted active molecules to confirm the hypothesis; one may also choose to explore further and test molecules about which the algorithm is unsure. Finding an appropriate balance between exploration and exploitation is the topic of Section 18.5 of this chapter.

18.2 The Robot Scientist Eve

A Robot Scientist is a physically implemented laboratory automation system that exploits techniques from the field of artificial intelligence to execute cycles of scientific experimentation. A Robot Scientist automatically originates hypotheses to explain observations, devises experiments to test these hypotheses, physically runs the experiments using laboratory robotics, interprets the results to change the probability that the hypotheses are correct, and then repeats the cycle (Figure 18.1). We believe that the development of Robot scientists will change the relationship between machine-learning/data-mining and industrial QSAR.

The University of Aberystwyth demonstrated the utility of the Robot Scientist "Adam", which can automate growth experiments in yeast. Adam is the first machine to have autonomously discovered novel scientific knowledge [34]. We have now built a new Robot Scientist for chemical genetics and drug design: Eve. This was physically commissioned in the early part of 2009 (see Figure 18.2). Eve is a prototype system to demonstrate the automation of closed-loop learning in drug-screening and design. Eve's robotic system is capable of moderately high throughput compound screening (greater than 10,000 compounds per day) and is

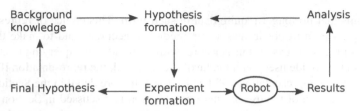

Fig. 18.1 The Robot Scientist hypothesis generation, experimentation, and knowledge formation loop.

designed to be flexible enough such that it can be rapidly re-configured to carry out a number of different biological assays.

One goal with Eve is to integrate an automated QSAR approach into the drug-screening process. Eve will monitor the initial mass screening assay results, generate hypotheses about what it considers would be useful compounds to test next based on the QSAR analysis, test these compounds, learn from the results and iteratively feed back the information to more intelligently home in on the best lead compounds.

Eve will help the rapid adoption of novel machine-learning/data-mining methodologies to QSAR in two ways:

1. It tightly couples the inductive methodology to the testing and design of new compounds, enabling chemists to step back and concentrate on the chemical and pharmacological problems rather than the inductive ones.
2. It enables inductive methodologies to be tested under industrially realistic conditions.

Fig. 18.2 Photographs of Eve, a Robot Scientist for chemical genetics and drug design.

18.2.1 Eve's Robotics

Eve's robotic system contains various instruments including a number of liquid handlers covering a diverse range of volumes, and so has the ability to prepare and execute a broad variety of assays. One of these liquid handlers uses advanced non-contact acoustic transfer, as used by many large pharmaceutical companies. For observation of assays, the system contains two multi-functional microplate readers. There is also a cellular imager that can be used to collect cell morphological information, for example to see how cells change size and shape over time after the addition of specific compounds.

18.2.2 Compound Library and Screening

In drug screening, compounds are selected from a "library" (a set of stored compounds) and applied to an "assay" (a test to determine if the compound is active – a "hit"). This is a form of "Baconian" experimentation – what will happen if I execute this action [45]. In standard drug screening there is no selection in the ordering of compounds to assay: "Start at the beginning, go on until you get to the end: then stop" (Mad Hatter, Lewis Carroll). In contrast, Eve is designed to test an active learning approach to screening.

Eve is initially using an automation-accessible compound library of 14,400 chemical compounds, the Maybridge 'Hit-Finder' library (http://www.maybridge.com). This compound library is cluster-based and was developed specifically to contain a diverse range of compounds. We realise this is not a large compound library – a pharmaceutical company may have many hundreds of thousands or even millions of compounds in its primary screening library. Our aim is to demonstrate the proof-of-principle that incorporating intelligence within the screening process can work better than the current brute-force approach.

18.2.3 QSAR Learning

In the typical drug design process, after screening has found a set of hits, the next task is to learn a QSAR. This is initially formed from the hits, and then new compounds are acquired (possibly synthesised) and used to test the model. This process is repeated until some particular criterion of success is reached, or too many resources are consumed to make it economical to continue the process. If the QSAR learning process has been successful, a "lead" compound is the result which can then go for pharmacological testing. In machine learning terms such QSAR learning is an example of "active learning" - where statistical/machine learning methods select examples they would like to examine next in order to optimise learning [12]. In pharmaceutical drug design the *ad hoc* selection of new compounds to test is done by QSAR experts and medicinal chemists based on their collective experience and intuition – there is a tradition of tension between the modellers and the synthetic chemists about what to do next. Eve aims to automate this QSAR learning. Given a set of "hits" from Baconian screening, Eve will switch to QSAR modelling.

Eve will employ both standard attribute based, graph based, and ILP based QSAR learning methods to model relationships between chemical structure and assay activity (see below). Little previous work has been done on combining active learning and QSARs, although active learning is becoming an important area of machine learning.

18.3 Representations of Molecular Data

Many industrial QSAR methods are based around using tuples of attributes or features to describe molecules [19, 43]. An attribute is a proposition which is either true or false about a molecule, for example, solubility in water, the existence of a benzene ring, etc. A list of such propositions is often determined by hand by an expert, and the attributes are measured or calculated for each molecule before the QSAR analysis starts. This representational approach typically results in a matrix where the examples are rows and the columns are attributes. The procedure of turning molecular structures into tuples of attributes is sometimes called *propositionalization*.

This way of representing molecules has a number of important disadvantages. Chemists think of molecules as structured objects (atom/bond structures, connected molecular groups, 3D structures, etc.). Attribute-value representations no longer express these relationships and hence may be harder to reason about. Furthermore, in most cases some information will be lost in the transformation. How harmful it is to ignore certain information is not always easy to determine in advance.

Another important disadvantage of the attribute-based approach is that is computationally inefficient in terms of space, i.e., to avoid as much loss of information as possible, an exponential number of attributes needs to be created. It is not unusual in chemoinformatics to see molecules described using hundreds if not thousands of attributes.

Within the machine learning and data mining communities, many methods have been proposed to address this problem, which we can categorize along two dimensions. In the first dimension, we can distinguish machine learning and data mining algorithms based on whether they compute *features explicitly,* or operate on the data *directly*, often by having *implicit* feature spaces.

Methods that compute *explicit* feature spaces are similar to the methods traditionally used in chemoinformatics for computing attribute-value representations: given an input dataset, they compute a table with attribute-values, on which traditional attribute-value machine learning algorithms can be applied to obtain classification or regression models. The main difference with traditional methods in chemoinformatics is that the attributes are not fixed in advance by an expert, but instead the data mining algorithm determines from the data which attributes to use. Compared with the traditional methods, this means that the features are chosen much more dynamically; consequently smaller representations can be obtained that still capture the information necessary for effective prediction.The calculation of explicit feature spaces is one of the most common applications of inductive queries, and will hence receive special attention in this chapter.

Methods that compute *implicit* feature spaces or operate directly on the structured data are more radically different: they do not compute a table with attribute-values, and do not propositionalize the data beforehand. Typically, these methods either directly compute a distance between two molecule structures, or greedily learn rules from the molecules. In many such models the absence or presence of a feature in the molecule is still used in order to derive a prediction; the main difference is that both during learning and prediction the presence of these features is only determined when really needed; in this sense, these algorithms operate on an *implicit* feature space, in which all features do not need to be calculated on every example, but only on demand as necessary. Popular examples of measures based on implicit feature spaces are *graph kernels*.

For some methods it can be argued that they operate neither on an implicit nor on an explicit feature space. An example is a largest common substructure distance between molecules. In this case, even though the conceptual feature space consists of substructures, the distance measure is not based on determining the number of common features, but rather on the size of one such feature; this makes it hard to apply most *kernel methods* that assume implicit feature spaces.

The second dimension along which we can categorise methods is the kind of features that are used, whether implicit or explicit:

1. *Traditional features* are typically numerical values computed from each molecule by an apriori fixed procedure, such as *structural keys* or *fingerprints,* or features computed through *comparative field analysis.*
2. *Graph-based features* are features that check the presence or absence of a sub-graph in a molecule; the features are computed implicitly or explicitly through a data mining or machine learning technique; these techniques are typically referred to as *Graph Mining* techniques.
3. *First-order logic features* are features that are represented in a first-order logic formula; the features are computed implicitly or explicitly through a data mining or machine learning technique. These techniques have been studied in the area of *Inductive Logic Programming (ILP).*

We will see in the following sections that these representations can be seen as increasing in complexity; many traditional features are usually easily computed, while applying ILP techniques can demand large computational resources. Graph mining is an attempt to find a middle ground between these two approaches, both from a practical and a theoretical perspective.

18.3.1 Traditional Representations

The input of the analysis is usually a set of molecules stored in SMILES, SDF or InChi notation. In these files, at least the following information about a molecule is described:

1. Types of the atoms (Carbon, Oxygen, Nitrogen);
2. Types of the bonds between the atoms (single, double).

Additionally, these formats support the representation of:

1. Charges of atoms (positively or negatively charged, how much?);
2. Aromaticity of atoms or bond (an atom part of an aromatic ring?);
3. Stereochemistry of bonds (if we have two groups connected by one bond, how can the rotation with respect to each other be categorized?);

Further information is available in some formats, for instance, detailed 3D information of atoms can also be stored in the SDF format. Experimental measurements may also be available, such as the solubility of a molecule in water. The atom-bond information is the minimal set of information available in most databases.

The simplest and oldest approach for propositionalizing the molecular structure is the use of *structural keys*, which means that a finite amount of features are specified beforehand and computed for every molecule in the database. There are many possible structural keys, and it is beyond the scope of this chapter to describe all of these; examples are molecular weight, histograms of atom types, number of heteroatoms, or more complex features, such as the sum of van der Waals volumes. One particular possibility is to provide an *a priori* list of substructures (OH groups, aromatic rings, ...) and either count their occurrences in a molecule, or use binary features that represent the presence or absence of each a priori specified group.

Another example of a widely used attribute-based method is comparative field analysis (CoMFA) [7]. The electrostatic potential or similar distributions are estimated by placing each molecule in a 3D grid and calculating the interaction between a probe atom at each grid point and the molecule. When the molecules are properly aligned in a common reference frame, each point in space becomes comparable and can be assigned an attribute such that attribute-based learning methods can be used. However, CoMFA fails to provide accurate results when the lack of a common skeleton prevents a reasonable alignment. The need for alignment is a result of the attribute-based description of the problem.

It generally depends on the application which features are most appropriate. Particularly in the case of substructures, it may be undesirable to provide an exhaustive list beforehand by hand. *Fingerprints* were developed to alleviate this problem. Common fingerprints are based on the *graph representation* of molecules: a molecule is then seen as a labelled graph (G, V, λ, Σ) with nodes V and edges E; labels, as defined by a function λ from $V \cup E$ to Σ, represent atom types and bond types. A fingerprint is a binary vector of a priori fixed length n, which is computed as follows:

1. All substructures of a certain type occurring in the molecule are enumerated (usually all *paths* up to a certain length);
2. A hashing algorithm is used to transform the string of atom and bond labels on each path into an integer number k between 1 and n;
3. The kth element of the fingerprint is incremented or set to 1.

The advantage of this method is that one can compute a feature table in a single pass through a database. There is a large variety of substructures that can be used, but in practice paths are only considered, as this simplifies the problems of enumerating substructures and choosing hashing algorithms. An essential property of fingerprints is thus that multiple substructures can be represented by a single feature, and that the meaning of a feature is not always transparent. In the extreme case, one can choose n to be the total number of possible paths up to a certain length; in this case, each feature would correspond to a single substructure. Even though theoretically possible, this approach may be undesirable, as one can expect many paths not to occur in a database at all, which leads to useless attributes. Graph mining, as discussed in the next section, proposes a solution to this sparsity problem.

18.3.2 Graph Mining

The starting point of most graph mining algorithms is the representation of molecules as labelled graphs. In most approaches no additional information is assumed – consequently, the nodes and edges in the graphs are often labelled only with bond and atom types. These graphs can be used to derive explicit features, or can be used directly in machine learning algorithms.

18.3.2.1 Explicit Features

Explicit features are usually computed through constraint-based mining (inductive querying) systems, and will hence be given special attention.

The most basic setting of graph mining is the following.

Definition 18.1 (Graph Isomorphism). Graphs $G = (V, E, \lambda, \Sigma)$ and $G' = (V', E', \lambda', \Sigma')$ are called isomorphic if there exists a bijective function f such that: $\forall v \in V : \lambda(v) = \lambda'(f(v))$ and $E = \{\{f(v_1), f(v_2)\} \mid \{v_1, v_2\} \in E'\}$ and $\forall e \in E : \lambda(e) = \lambda'(f(e))$.

Definition 18.2 (Subgraph). Given a graph $G = (V, E, \lambda, \Sigma)$, $G' = (V', E', \lambda', \Sigma')$ is called a subgraph of G iff $V' \subseteq V, E' \subseteq E$, $\forall v \in V' : \lambda'(v) = \lambda(v)$ and $\forall e \in E' : \lambda'(e) = \lambda(e)$.

Definition 18.3 (Subgraph Isomorphism). Given two graphs $G = (V, E, \lambda, \Sigma)$ and $G' = (V', E', \lambda', \Sigma')$, G is called subgraph isomorphic with G', denoted by $G' \succeq G$, iff there is a subgraph G'' of G' to which G is isomorphic.

Definition 18.4 (Frequent Subgraph Mining). Given a dataset of graphs $\mathscr{D}$, and a graph G, the *frequency* of G in $\mathscr{D}$, denoted by $freq(G, \mathscr{D})$, is the cardinality of the set $\{G' \in \mathscr{D} \mid G' \succeq G\}$. A graph G is *frequent* if $freq(G, \mathscr{D}) \geq minsup$, for a predefined threshold *minsup*. The frequent (connected) subgraph mining is the problem of finding a set of frequent (connected) graphs F such that for every possible frequent (connected) graph G there is exactly one graph $G' \in F$ such that G' and G are isomorphic.

We generate as features those subgraphs which are contained in a certain minimum number of examples in the data. In this way, the eventual feature representation of a molecule is dynamically determined depending on the database it occurs in.

There are now many algorithms that address the general frequent subgraph mining problem; examples include AGM [27], FSG [30], gSpan [54], MoFA [1], FFSM [24] and Gaston [47]. Some of the early algorithms imposed restrictions on the types of structures considered [35, 36].

If we set the threshold *minsup* very low, and if the database is large, even if finite, the number of subgraphs can be very large. One can easily find more frequent subgraphs than examples in the database. Consequently, there are two issues with this approach:

1. Computational complexity: considering a large amount of subgraphs could require large computational resources.
2. Usability: if the number of features is too large, it could be hard to interpret a feature vector.

These two issues are discussed below.

Complexity. Given that the number of frequent subgraphs can be exponential for a database, we cannot expect the computation of frequent subgraphs to proceed in polynomial time. For enumeration problems it is therefore common to use alternative definitions of complexity. The most important are:

1. Enumeration with *polynomial delay*. A set of objects is enumerated with polynomial delay if the time spent between listing every pair of objects is bounded by a polynomial in the size of the input (in our case, the dataset).
2. Enumeration with *incremental polynomial time*. Objects are enumerated in incremental polynomial time if the time spent between listing the k and $(k+1)$th object is bounded by a polynomial in the size of the input and the size of the output till the kth object.

Polynomial delay is more desirable than incremental polynomial time. Can frequent subgraph mining be performed in polynomial time?

Subgraph mining requires two essential capabilities:

1. Being able to enumerate a space of graphs such that no two graphs are isomorphic.
2. Being able to evaluate subgraph isomorphism to determine which examples in a database contain an enumerated graph.

Table 18.1 The number of graphs with certain properties in the NCI database

Graph property	Number
All graphs	250251
Graphs without cycles	21963
Outerplanar graphs	236180
Graphs of tree width 0, 1 or 2	243638
Graphs of tree width 0, 1, 2 or 3	250186

The theoretical complexity of subgraph mining derives mainly from the fact that the general subgraph isomorphism problem is a well-known NP complete problem, which in practice means that the best known algorithms have exponential complexity. Another complicating issue is that no polynomial algorithm is known to determine if two arbitrary graphs are isomorphic, even though this problem is not known to be NP complete.

However, in practice it is often feasible to compute the frequent subgraphs in molecular databases, as witnessed by the success of the many graph miners mentioned earlier. The main reason for this is that most molecular graphs have properties that make them both theoretically and practically easier to deal with. Types of graphs that have been studied in the literature include;

1. Planar graphs, which are graphs that can be drawn on a plane without edges crossing each other [14];
2. Outerplanar graphs, which are planar graphs in which there is a Hamilton cycle that walks only around one (outer) face [40];
3. Graphs with bounded degree and bounded tree width, which are tree-like graphs[1] in which the degree of every node is bounded by a constant [44].

Graphs of these kinds are common in molecular databases (see Table 18.1, where we calculated the number of occurrences of certain graph types in the NCI database, a commonly used benchmark for graph mining algorithms).

No polynomial algorithm is however known for (outer)planar subgraph isomorphism, nor for graphs of bounded tree width without bounded degree and bounded size. However, in recent work we have shown that this does *not* necessarily imply that subgraph mining with polynomial delay or in incremental polynomial time is impossible:

1. If subgraph isomorphism can be evaluated in polynomial time for a class of graphs, then we showed that there is an algorithm for solving the frequent subgraph mining algorithm with polynomial delay, hence showing that the graph isomorphism problem can always be solved efficiently in pattern mining [48].
2. Graphs with bounded tree width can be enumerated in incremental polynomial time, even if no bound on degree is assumed [22].

---

[1] A formal definition is beyond the scope of this chapter.

3. For the block-and-bridges subgraph isomorphism relation between outerplanar graphs (see next section), we can solve the frequent subgraph mining problem in incremental polynomial time [23].

These results provide a theoretical foundation for efficient graph mining in molecular databases.

Usability. The second problem is that under a frequency threshold, the number of frequent subgraphs is still very large in practice, which affects interpretability and efficiency, and takes away one of the main arguments for using data mining techniques in QSAR.

One can distinguish at least two approaches to limit the number of subgraphs that is considered:

1. Modify the subgraph isomorphism relation;
2. Apply additional constraints to subgraphs.

We will first look at the reasons for changing the subgraph isomorphism relation.

Changing Isomorphism. Assume we have a molecule containing Pyridine, that is, an aromatic 6-ring in which one atom is a nitrogen. How many subgraphs are contained in this ring only? As it turns out, Pyridine has 2+2+3+3+4+3=17 different subgraphs next to Pyridine itself (ignoring possible edge labels):

```
N   C
C-C   N-C
C-C-C  N-C-C   C-N-C
C-C-C-C   N-C-C-C   C-N-C-C
C-C-C-C-C   N-C-C-C-C   C-N-C-C-C   C-C-N-C-C
N-C-C-C-C-C   C-N-C-C-C-C   C-C-N-C-C-C
```

It is possible that each of these subgraphs has a different support; for example, some of these subgraphs also occur in Pyrazine (an aromatic ring with two nitrogens). The support of each of these subgraphs can be hard to interpret without visually inspecting their occurrences in the data. Given the large number of subgraphs, this can be infeasible.

Some publications have argued that the main source of difficulty is that we allow subgraphs which are not rings to be matched with rings, and there are applications in which it could make more sense to treat rings as basic building blocks. This can be formalized by adding additional conditions to subgraph isomorphism matching:

1. In [20] one identifies all rings up to length 6 in both the subgraph and the database graph; only a ring is allowed to match with a ring.
2. In [23] a *block and bridge preserving* subgraph isomorphism relation is defined, in which bridges in a graph may only be matched with bridges in another graph, and edges in cycles may only be matched with edges in cycles; a *bridge* is an edge that is not part of a cycle.

Comparing both approaches, in [20] only rings up to length 6 or considered; in [23] this limitation is not imposed.

Most subgraph mining algorithms need to be changed significantly to deal with a different definition of subgraph isomorphism. To solve this [20, 23] introduce procedures to deal with ring structures.

We are not aware of an experimental comparison between these approaches.

Additional Constraints. The use of constraints is a general methodology to obtain a smaller set of more meaningful subgraphs [35, 36]. One can distinguish two types of constraints:

1. Structural constraints;
2. Data based constraints.

Minimum frequency is one example of a constraint based on data. Many other subgraph types have been proposed based on data constraints:

1. Maximally frequent subgraphs, which are subgraphs such that every supergraph in a database is infrequent [35, 36, 25];
2. Closed subgraphs, which are subgraphs such that every supergraph has a different frequency [55].
3. Correlated subgraphs, which are subgraphs whose occurrences have a significant correlation with a desired target attribute [4];
4. Infrequent subgraphs [35, 36].

These constraints can be combined. For instance, one can be interested in finding subgraphs that occur frequently in molecules exhibiting a desired property, but not in other molecules.

In practice, these constraints are often not sufficient to obtain small representations. Additional inductive queries can be used to reduce the set of patterns further. A more detailed overview of approaches to obtain smaller sets of patterns is given by Bringmann et al. in Chapter 6 of this volume.

An issue of special interest in QSAR applications is which graph types lead to the best results: even though molecules contain cycles, is it really necessary to find cyclic patterns? Experiments investigating this issue can be found in [46, 4, 53]. The conclusion that may be drawn from these investigations is that in many approaches that use patterns, paths perform equally well as graphs; naïve use of cyclic patterns can even lead to significantly worse results.

18.3.2.2 Implicit Features and Direct Classification

The alternative to graph mining is to learn classifiers directly on the graph data. The most popular approaches are based on the computation of a *distance* between every pair of graphs in the data. Such distance functions can be used in algorithms that require distance functions, such as k-nearest neighbour classification, or support vector machines (SVMs). In SVMs a special type of distance function is needed, the so-called *kernel function*.

One popular type of kernel is the *decomposition kernel*, in which the distance is defined by an implicit feature space. If this implicit feature space is finite, the kernel value between molecules can in principle be computed by first computing two feature vectors for the pair, and then computing a distance from these feature vectors; the advantage of kernels is that in practice only the (weighted) number of substructures that two particular graphs have in common is computed.

The most commonly used graph kernels are based on the idea of *random walks*: given two molecules, we count the number of walks that both molecules have in common. Note that *walks* differ from *paths* as walks are allowed to visit the same node more than once. If a maximum walk length is given, we could represent two molecules by binary feature vectors with one bit for each possible walk. In practice, though, it is more efficient to scan the two molecules in parallel to make sure we search for common walks. This methodology has further possible advantages. For instance, if we give all walks in graphs a weight which (carefully) shrinks with the length of the walk, a kernel can be defined in which we sum the *infinite* number of such common weighted walks. This number is efficiently computable without explicitly enumerating all walks [17]. Many *kernel* methods have the advantage that they deal easily with possibly infinite representations of structures in a feature space. An early overview of graph kernels can be found in [16], while a more recent overview of walk-based kernels can be found in [52].

Another type of distance function is obtained by computing the *largest common subgraph* of two graphs. The assumption is here that the larger the subgraph is that two molecules have in common, the more similar they are. It is easy to see that this problem is at least as hard as computing subgraph isomorphism. However, the problem may become easier for the types of graphs identified in the previous section. In [51] it was shown how to compute the largest common subgraph in polynomial time for outer-planar graphs under the block-and-bridges subgraph relation.

18.3.2.3 Extended Graph Representations

So far we have considered representations in which nodes correspond to atoms and edges to bonds. This limits the types of knowledge that can be used in the classification. It may be desirable to extend the representation: in some cases it is necessary to classify atom types, e.g. *halogen* (F, Cl, Br, I); to say an atom in an aromatic ring but not specify the atom type; to extend the notion of bond from that of a covalent bond to include non-covalent ones, e.g. hydrogen bonds; etc.

To deal with such issues of ambiguity the common solution is to assume given a *hierarchy* of edge and node labels. In this hierarchy more general labels, such as 'halogen' and 'hydrogen donor', are included, as well as the generalization relationships. There are two ways to use these hierarchies:

1. We change the subgraph isomorphism operator, such that more general labels are allowed to match with their specialisations [20, 26];
2. We exploit the fact that in some hierarchies every atom has at most one generalization, by changing the graph representation of the data: we replace the atom

type label with the parent label in the hierarchy, and introduce a new node, which is labeled with the original atom type. Optionally, we add additional nodes labeled with other attributes, such as charges [31].

These approaches have mainly been investigated when computing explicit features. An essential problem is then in both approaches the increased number of patterns. Without additional constraints we could find patterns such as C-Aromatic-C-Aromatic-C in aromatic rings, that is, patterns in which the labels iterate between specific and general labels. The approaches listed above differ in their approach to avoid or limit such patterns.

18.3.3 Inductive Logic Programming

In QSAR applications such as toxicity and mutagenicity prediction, where structure is important, Inductive Logic Programming is among the more powerful approaches, and has found solutions not accessible to standard statistical, neural network, or genetic algorithms [8, 13, 32, 33]. The main distinguishing feature of ILP is that data and models are represented in *first order logic (FOL)*. The classical atom/bond representation in first-order logic is based on the molecular structure hypothesis. Atoms are represented in the form: atom(127,127_1,c,22,0.191), stating that the first atom in compound 127 is a carbon atom of type 22 (aromatic) with a positive charge of 0.191. Similarly, bond(127,127_1,127_6,7) states that there is a type 7 bond (here aromatic) between the first and sixth atom in compound 127. Bonds are represented in a similar fashion.

When only atoms, bonds and their types are represented in FOL facts, the resulting representation is essentially a graph. The main advantage of ILP is the possibility of including additional information, such as charges, and of including background knowledge in the form of computer programs. One example of this is to define a distance measure which enables three-dimensional representations with rules in the form: "A molecule is active if it has a benzene ring and a nitro group separated by a distance of $4 \pm 0.5°A$". The key advantage of this approach to representing three-dimensional structures is that it does not require an explicit alignment of the molecules. It is also straightforward to include more than one conformation of each compound which allows the consideration of conformation flexibility which is often a major drawback by conventional QSAR/SAR methodologies.

Since chemists often study molecules in terms of molecular groups, the atom-/bond representation can be extended with programs that define such high-level chemical concepts. Contrary to propositional algorithms and graph mining, ILP can learn rules which use structural combinations of these multiple types of concepts.

A downside of ILP is the lack of results with respect to efficient theoretical complexity. As shown in the previous section, for many classes of graphs efficient mining algorithms are known. As a result, graph mining is usually efficient, both in theory and in practice. For ILP algorithms no similar theoretical results are available and the algorithms typically require more computational power, both in theory and in practice.

The number of ILP algorithms is very large, and the discussion of this area is beyond the scope of this article. We will limit our discussion here to the relationship between graph mining and ILP algorithms, and approaches that we will need later in this chapter. For a more complete discussion of ILP see [10]. An important aspect of ILP algorithms is the background knowledge used. We will conclude this section with a discussion of the details of a library of background knowledge for SAR applications that we recently developed, and is important in allowing users to formulate alternative inductive queries.

18.3.3.1 Explicit Features

A problem similar to the frequent subgraph mining problem can be formulated in ILP. The data is conceived as a set of definite clauses and facts, for instance:

```
halogen(X,Y)  :- atom(X,Y,f,_,_).
halogen(X,Y)  :- atom(X,Y,cl,_,_).
...
atom(127,127_1,c,22,0.191).
atom(127,127_2,c,22,0.191).
bond(127,127_1,127_2,single).
```

The database is usually represented as a program in Prolog. The clauses can be thought of as *background knowledge*, while the facts describe the original data. Assume now we are given the following clause, which is not part of the database:

```
f1(X)  :- molecule(X),halogen(X,Y),atom(X,Z,c,_,_),
          bond(X,Y,Z,_).
```

Then for a given constant, for instance 127 in our example, we can evaluate using a Prolog engine whether f1(127) is true. If this is the case, we may see f1 as a feature which describes molecule 127. We may call a clause frequent if it evaluates to true for a sufficient number of examples. The problem of finding frequent clauses is the problem that was addressed in the WARMR algorithm [9].

Definition 18.5 (Frequent Clause Mining). Given clause $C = $ h(X) :- b, where b is the body of the clause C, and a Prolog database $\mathscr{D}$ with constants C, the *frequency* of clause C in $\mathscr{D}$, denoted by *freq*$(C,\mathscr{D})$, is the cardinality of the set $\{c \in C \,|\, \mathscr{D} \cup \{C\} \models $ h(c) $\}$; in other words, the number of constants for which we can prove the head of the clause using a Prolog engine, assuming C were added to the data. A clause C is *frequent* if *freq*$(C,\mathscr{D}) \geq minsup$, for a predefined threshold *minsup*. Assume given a language L of clauses. The frequent clause mining is the problem of finding a set of clauses F such that for every possible frequent clause C in L there is exactly one clause $C' \in F$ such that C' and C are equivalent.

It is of interest here to point towards the differences between frequent graph mining and frequent clause mining.

The first practical difference is that most algorithms require an explicit definition of the space of clauses C to be considered. This space is usually defined in a *bias specification language*. In such a bias specification language, it can be specified for instance that only clauses starting with a molecule predicate will be considered, and next to this predicate only atom and bond predicates may be used. Note that such clauses would essentially represent graphs. The bias specification language can be considered a part of the language of an inductive querying system and provides users the possibility to carefully formulate data mining tasks.

The second difference is the use of traditional Prolog engines to evaluate the support of clauses. Prolog engines are based on a technique called *resolution*. There is an important practical difference between resolution and subgraph isomorphism, as typically used in graph mining algorithms. Assume we are given a clause over only atoms and bonds, for instance,

```
h(X)  :-  molecule(X), atom(X,Y,c,_,_),bond(X,Y,Z1,_),
          bond(X,Y,Z2,_)
```

then this clause is equivalent to the following clause:

```
h(X)  :- molecule(X), atom(X,Y,c,_,_),bond(X,Y,Z1,_)
```

The reason is that if constants are found for which the second clause succeeds, we can use the same constants to satisfy the first clause, as there is no requirement that Z1 and Z2 are different constants. On the other hand, when using subgraph isomorphism, two atoms in a subgraph may never be matched to the same atom in a molecule.

The use of resolution has important consequences for the procedure that is used for eliminating equivalent clauses. Whereas in graph mining, it is possible to avoid equivalent subgraphs during the search, it can be proved that there are languages of clauses for which this is impossible; the only solution in such cases is to first generate a highly redundant set of clauses, and eliminate duplicates in a post-processing step.

To address this problem, an alternative to resolution was proposed, in which two different variables are no longer allowed to be resolved to the same constant. This approach is known as theta-subsumption under *Object Identity* [10].

Similar constraints as proposed in graph mining, can also be applied when mining clauses. However, this has not yet been extensively applied in practice.

18.3.3.2 Implicit Features and Direct Classification

The alternative to separate feature construction and learning phases is also in ILP to learn a model directly from the data. Contrary to the case of graphs, however, the use of distance functions has only received limited attention in the ILP literature; see [15] for a kernel on logical representations of data and [11] for a distance based on the *least general generalization* of two sets of literals. The application of these methods on molecular data is yet unexplored; one reason for this is the expected prohibitive complexity of these methods, in particular when one wishes to include background knowledge in the *lgg* based methods.

On the other hand, a very common procedure in ILP is to greedily learn a rule-based or tree-based classifier directly from training data; examples of such algorithms include FOIL, Tilde and Progol [10]. In graph mining such approaches are rare; the main reason for this is that greedy heuristics are expected to be easily misled when the search proceeds in very "small", uninformative steps, as common in graph mining when growing fragments bond by bond.

To illustrate one such greedy algorithm, we will discuss the Tilde algorithm here [2]. Essentially, Tilde starts from a similar database as WARMR, and evaluates the support of a clause in a similar way as WARMR; however, as the algorithm is aware of the class labels, it can compute a score for each clause that evaluates how well it separates examples of one or more two classes from each other. For instance, the clause

```
h(X)  :- molecule(X), benzene(X,Y)
```

may hold for 15 out of 20 constants identifying positive molecules, and only 15 out of 30 negative molecules; from these numbers we may compute a score, such as information gain:

$$(-0.4\log 0.4 - 0.6\log 0.6)$$
$$- 0.3(-0.5\log 0.5 - -0.5\log 0.5) - 0.7(-0.25\log 0.25 - 0.75\log 0.75)$$

Here the first term denotes the information of the original class distribution (20/50 positives, 30/50 negatives), the second term denotes the information of the examples for which the query succeeds, and the third term denotes the information of the examples for which it fails.

Using such a score, we can compare several alternative clauses. In Tilde clauses are grown greedily, i.e. for a given clause, all possible literals are enumerated that can be added to it, and only the extended clause that achieves the best score is chosen for further extension. If the improvement is too small, the molecules are split in two sets based on whether the clause succeeds. For these two sets of examples, the search for clauses recursively continues. The end result of this procedure constitutes a tree in which internal nodes are labeled with clauses; we can label a leaf by the majority class of the examples ending up in the leaf. This tree can be used directly for classification.

The problem of learning accurate decision trees has been studied extensively, and many techniques, such as pruning, can also be applied on relational decision trees [2]. The main downside of algorithms such as Tilde is that the greedy procedure will prevent large carbon-based substructures from being found automatically, as the intermediate steps through which the greedy search would have to go usually do not score exceptionally well on commonly used heuristics. Hence, it is advisable in ILP to specify larger substructures in advance by means of background clauses.

Fig. 18.3 Chemical structure
of 8-nitroquinoline

18.3.3.3 A Library of Chemical Knowledge for Relational QSAR

An important benefit of ILP algorithms is the ability to incorporate background knowledge, for instance, to represent special groups in molecules. The availability of such background knowledge in a data mining system may allow data analysts to query a database from additional perspectives, as will be illustrated in the next section when studying the problem of selecting a library for use in a robot scientist.

To exploit this benefit, it is essential that a comprehensive library of background knowledge is available. We developed a chemical structure background-knowledge-for-learning (Molecular Structure Generator MSG). This consists of a large library of chemical substructures, rings and functional groups, including details of isomers and analogues.

This library consists of three main parts (see Appendix 1): a functional group library, a ring library, and a polycycle library. We encoded the standard functional groups have been pre-coded in the library (Appendix 1, Table 5). The ring library consists of predominantly 3, 4, 5 and 6 length rings. Rings that are identified but do not have specific chemical names are given a *standard* label, e.g., *other_six_ring*. Unnamed rings of up to 15 atoms in length are pre-coded in this way. Appendix 1, Table 4 shows the specific rings that are in the library. Rings with isomers have been defined individually but they will have a corresponding *parent* predicate held in the library, for eg, isomer_parent(1,3-cyclohexadiene, cyclohexadiene); isomer_parent(1,4-cyclohexadiene, cyclohexadiene). This will mean that inductions may be made over either the specific isomer or for the whole family.

The polycycle library consists of predominantly 2 and 3 ring polycycles that have been pre-coded and held in the MSG Prolog library. Polycycles that are not specifically named have been given an *other* label, i.e., *other_carbon*. All polycycles will be identified regardless if specifically named in the library. Appendix 1, Table 6 shows the specific polycyles that are in the library. Structures are built up from substructures, e.g., an anthracene would have facts for 3 benzene rings, 2 fused pair naphthalenes and a polycycle anthracene; an aryl-nitro structure would have facts for a nitro and an aromatic ring. The data have been fully normalised according to Boyce-Codd relational data standards [5]. The example of the representation of the molecule 8-nitroquinoline is shown in Figure 18.3 and Table 18.2.

18.4 Selecting Compounds for a Drug Screening Library

This MSG library will be used to generate ILP representations of the compounds that will be screened by Eve. To test the efficacy of the representation and the method, this library was used to aid the decision-making process for the selection of a compound library to be used with Eve.

The two main criteria for selecting compounds for screening libraries are that they resemble existing approved pharmaceuticals, and that they are structurally diverse. The requirement for a compound in a screening-library to resemble existing pharmaceutically active compounds maximizes the *a priori* probability of an individual compound being active and non-toxic because existing pharmaceutically-active compounds have this property. The requirement for diversity is usually justified by the fact that structurally similar compounds tend to exhibit similar activity - a structurally diverse set of compounds should cover the activity search space and therefore contain fewer redundant compounds [39].

Drug-like properties are usually defined in terms of ADME - Absorption, Distribution, Metabolism, and Excretion - and describe the action of the drug within an organism, such as intestinal absorption or blood-brain-barrier penetration. One of the first methods, and still the most popular, to model the absorption property was the "Rule of 5" [41] which identifies the compounds where the probability of useful oral activity is low. The "rule of 5" states that poor absorption or permeation is more likely when:

1. There are more than 5 Hydrogen-bond donors
2. The Molecular Weight is over 500.
3. The LogP (partition coefficient) is over 5 (or MLogP is over 4.15).
4. There are more than 10 Hydrogen-bond acceptors

The negation of the Lipinski rules are usually used as the main selection criteria for the compounds to include in a screening-library. Though these rules are not

Table 18.2 Ground background knowledge generated for 8-nitroquinoline

ring_length(2,1,6).	ring_atom(2,2,4).	group_atom(2,5,13).
aromatic_ring(2,1).	ring_atom(2,2,5).	r_atom(2,5,9).
carbon_ring(2,1).	ring_atom(2,2,6).	group(2,6,aryl_nitro).
ring(2,1,benzene).	ring_atom(2,2,7).	part_of_group_structure(2,6,1).
ring_atom(2,1,1).	ring_atom(2,2,8).	part_of_group_structure(2,6,5).
ring_atom(2,1,2).	fused_ring_pair(2,3,1).	count_ring(2,benzene,1).
ring_atom(2,1,3).	fused_ring_pair_share_atom(2,3,8).	count_ring(2,pyridine,1).
ring_length(2,2,6).	polycycle(2,4,quinoline).	count_poly(2,quinoline,1).
n_containing(2,2).	hetero_poly(2,4).	count_group(2,nitro,1).
aromatic_ring(2,2).	poly_no_rings(2,4,2).	count_group(2,aryl_nitro,1).
hetero_ring(2,2).	polycycle_pair(2,4,3).	parent(2,6,nitro).
ring(2,2,pyridine).	group(2,5,nitro).	nextto(2,1,2,fused).
ring_atom(2,2,3).	group_atom(2,5,11).	nextto(2,1,5,bonded).

definitive, the properties are simple to calculate, and provide a good guideline for drug-likeness.

We have taken an operational approach to determining the drug-likeness of compounds [50]. The basic idea is to use machine learning techniques to learn a discrimination function to distinguish between pharmaceutically-active compounds and compounds in screening-libraries. If it is possible to discriminate pharmaceutically-active compounds from compounds in a screening-library then the compounds in the library are considered not drug-like; conversely, if they cannot be discriminated then the compounds are drug-like.

Two compound-screening libraries were chosen for analysis – the target-based NatDiverse collection from Analyticon Discovery (Version 070914) and the diversity-based HitFinder (Version 5) collection from Maybridge. The libraries from these companies are publicly available and this was the main reason for their inclusion in this research. The HitFinder collection includes 14,400 compounds representing the drug-like diversity of the Maybridge Screening Collection ($\approx$60,000 compounds). Compounds have generally been selected to be non-reactive and meeting Lipinski's Rule of 5. AnalytiCon Discovery currently offers 13 NatDiverse libraries which are tailor-made synthetic nitrogen-containing compounds. The total number of compounds is 17,402. The approved pharmaceuticals dataset was obtained from the KEGG Drug database and contains 5,294 drugs from the United States and Japan. The data was represented using the Molecular Structure Generator, mentioned above, and the ILP decision tree learner Tilde, was used to learn the discrimination functions between the set of approved pharmaceuticals and the two compound screening-libraries.

Three tests per dataset were carried out – one based on structural information only, another on quantitative attributes only, and the other based on both structural information and the quantitative attributes. The complete datasets were split into a training and validation set and an independent test set. A ten-fold cross-validation was used for Tilde to learn the decision trees. For each of the three scenarios, the ten-fold cross-validation was carried out with identical training and validation sets. For each scenario, the classification tree that provided the best accuracy when applied to the validation set was applied to the independent test set, see Table 3. The independent test results are good and consistent with validation results. They indicate that the inclusion of quantitative attributes resulted in increasing the classification accuracy only slightly. The best accuracy was achieved by the decision trees when the data is represented by both structures and properties. These decision trees were represented as a set of Prolog rules and the most accurate rules were selected to build the smallest decision list that had a minimum accuracy of 85%. A complication is the the problem of uneven class distributions (approximately 3:1, screening-library: approved pharmaceuticals).

The classification system had more difficulty discriminating approved pharmaceuticals from the diversity-based HitFinder library than the target-based NATDiverse library. However, the ILP method had 91% success in classifying compounds in the HitFinder library and 99% in classifying compounds from the NATDiverse collection when applied to an independent test set. These discrimination functions

Table 18.3 Accuracy of the classification trees when applied to the independent test set

Testing Dataset	Accuracy	True Neg- atives	True Pos- itives
HitFinder / App structures only	90%	92%	74%
NAT / App structures only	99%	99%	96%
HitFinder / App properties only	83%	90%	62%
NAT / App properties only	89%	92%	74%
HitFinder / App structures & properties	91%	93%	75%
NAT / App structures & properties	99%	99%	97%

were expressed in easy to understand rules, are relational in nature and provide useful insights into the design of a successful compound screening-library.

Given a set of rules that can discriminate between drugs and non-drug compounds, the question arises how best to use them in the drug design process. The simplest way to use them would be as filters, and to remove from consideration any compound classed as being non-drug-like. This is what is generally done with the original Lipinski rules - any compounds that satisfy the rules are removed from drug libraries. This approach is non-optimal because such rules are *soft,*as they are probabilistic and can be contravened under some circumstances. However, new data mining research such as multi-target learning research [56] has originated better ways of using prior rules than simply using them as filters. We believe that such approaches could be successfully applied to the drug design problem.

18.5 Active learning

In many experiment-driven research areas, it is important to select experiments as optimally as possible in order to reduce the number and the costs of the experiments. This is in particular true for high-throughput screening in the drug discovery process, as thousands of compounds are available for testing. QSAR methods can help to model the results obtained so far. When fit into an active learning strategy, they can be used to predict the expected benefit one can obtain from experiments.

However, in QSAR applications there is an important difference with classical active learning approaches. Usually, one is not interested to get an accurate model for all molecules. It is only important to distinguish the best molecules (and therefore to have an accurate model for the good ones). So instead of active learning where one chooses experiments to improve the global performance of the learned model, in these applications an active optimization approach is desired where one chooses experiments to find the example with the highest target value.

There may be two major reasons why an experiment is interesting. First, one may believe that the molecule being tested has a high probability of being active. In that case, one exploits the available experience to gain more value. Second, the molecule may be dissimilar to the bulk of the molecules tried so far. In that case, the experiment is explorative and one gains new experience from it.

18.5.1 Selection strategies

Different example selection strategies exist. In geostatistics, they are called infill sampling criteria [49].

In active learning, in line with the customary goal of inducing a model with maximal accuracy on future examples, most approaches involve a strategy aiming to greedily improve the quality of the model in regions of the example space where its quality is lowest. One can select new examples for which the predictions of the model are least certain or most ambiguous. Depending on the learning algorithm, this translates to near decision boundary selection, ensemble entropy reduction, version space shrinking, and others. In our model, it translates to *maximum variance* on the predicted value, or $\mathrm{argmax}(\mathrm{var}(t))$.

Likely more appropriate for our optimization problem is to select the example that the current model predicts to have the best target value, or $\mathrm{argmax}(\bar{t})$. We will refer to this as the *maximum predicted* strategy. For continuous domains, it is well known that it is liable to get stuck in local minima.

A less vulnerable strategy is to always choose the example for which the optimistic guess is maximal. In reinforcement learning, one has shown that with this strategy the regret is bounded (Explicit Explore or Exploit, [37]). In that case, the idea is to not (re)sample the example in the database where the expected reward $\bar{t}$ is maximal, but the example where $\bar{t} + b \times \mathrm{var}(t)$ is maximal. The parameter b is the level of optimism. In this paper we do not consider repeated measurements, unlike reinforcement learning, where actions can be reconsidered. This *optimistic* strategy is similar to Cox and John's lower confidence bound criterion [6]. It is obvious that the maximum predicted and maximum variance strategies are special cases of the optimistic strategy, with $b = 0$ and $b = \infty$ respectively. In a continuous domain, this strategy is not guaranteed to find the global optimum because its sampling is not dense [28].

Another strategy is to select the example that has the highest probability of improving the current solution [38]. One can estimate this probability as follows.

Let the current step be N, the value of the set of k best examples be $\|T_N\|_{best-k}$ and the k-th best example be $x_{(k,N)}$ with target value $t_{(k,N)}$. When we query example x_{N+1}, either t_{N+1} is smaller than or equal to $t_{\#(k,N)}$, or t_{N+1} is greater. In the first case, our set of k best examples does not change, and $\|T_{N+1}\|_{best-k} = \|T_N\|_{best-k}$. In the latter case, x_{N+1} will replace the k-th best example in the set and the solution will improve. Therefore, this strategy selects the example x_{N+1} that maximizes $P(t_{N+1} > t_{(k,N)})$. We can evaluate this probability by computing the cumulative Gaussian

$$P(t_{N+1} > t_{(k,N)}) = \int N(\bar{t}, \mathrm{var}(t))dt. \tag{18.1}$$

In agreement with [42], we call this the *most probable improvement* (MPI) strategy.

Yet another variant is the strategy used in the Efficient Global Optimization (EGO) algorithm [29]. EGO selects the example it expects to improve most upon the current best, i.e. the one with highest

$$E[\max(0, t - t_{(k,N)})] = \int (t - t_{(k,N)}) N(\bar{t}, \text{var}(t)) dt. \qquad (18.2)$$

This criterion is called *maximum expected improvement* (MEI).

18.5.2 Effects of properties of experimental equipment

Most approaches assume an alternation between the algorithm proposing one single experiment and the environment performing one experiment producing a definite answer to the proposed question. After a number of iterations, the algorithm converges then to one optimal solution. However, in practice such a procedure is not always acceptable.

First, in some cases, not all parameters are evaluated during the first stage of experimentation. E.g. in the drug discovery process, active compounds may be rejected at a later stage due to other adverse properties such as toxicity, and therefore one prefers to discover in the first stage several dissimilar candidates instead of one optimal one.

Second, in many applications among which high throughput screening, the equipment can perform several experiments at the same time. E.g. several compounds can be tested on a single plate, or the experiments happen in a pipeline such that several experiments are under way before the result of the first one is known. In such cases, the algorithm has to choose several experiments without knowing the result of all earlier experiments. Therefore, apart from exploitation and exploration, the algorithm also needs diversification.

Third, noise is a common factor in real-world experiments. It means that results are not always exact or trustworthy. Depending on the domain, one may want to perform the same experiment several times, or design different experiments to jointly measure a set of related values.

18.6 Conclusions

In this chapter we have first introduced the challenges involved in automating the discovery process of new drugs, of which the development of a robot scientist is the arguably the most ambitious. We have provided a more detailed discussion of several of the challenges particular to iterative drug discovery: the representation of molecular data, the use of active learning and the development of libraries that serve as input for the former two tasks.

Even though we made an attempt to provide a reasonably complete summary of the areas and issues involved, the overview in this chapter is far from complete. An important element which is missing from this chapter is an all-encompassing experimental comparison of the representation methods presented (both ILP and graph mining), as well as detailed recommendations with respect to which algorithms to use for which types of data, under which types of constraints or under which type of language bias. Desirable as this may be, to the best of our knowledge no such comparison is currently available in the literature and most studies have focused on a subset of methods and limited types of data (mostly NCI, see [53, 4] for instance). This type of analysis could be a useful topic for further research, for which we hope that this chapter provides some useful hints.

References

1. C. Borgelt and M.R. Berthold. Mining molecular fragments: Finding relevant substructures of molecules. In *ICDM*, pages 51–58. IEEE Computer Society, 2002.
2. H. Blockeel, L. De Raedt. Top-Down Induction of First-Order Logical Decision Trees. Artif. Intell. 101(1-2): 285-297 (1998).
3. H. Blockeel, S. Dzeroski, B. Kompare, S. Kramer, B. Pfahringer, and W. Van Laer. Experiments in predicting biodegradability. In *Appl. Art. Int. 18*, pages 157–181, 2004.
4. B. Bringmann, A. Zimmermann, L. De Raedt, and S. Nijssen. Don't be afraid of simpler patterns. In J. Fürnkranz, T. Scheffer, and M. Spiliopoulou, editors, *PKDD*, volume 4213 of *Lecture Notes in Computer Science*, pages 55–66. Springer, 2006.
5. E.F. Codd. Recent Investigations into Relational Data Base Systems. *IBM Research Report RJ1385* (April 23rd, 1974). Republished in Proc. 1974 Congress (Stockholm, Sweden, 1974). New York, N.Y.: North-Holland, 1974.
6. Dennis D. Cox and Susan John. SDO: a statistical method for global optimization. In *Multidisciplinary design optimization (Hampton, VA, 1995)*, pages 315–329. SIAM, 1997.
7. R.D. III Cramer, D.E. Patterson, and Bunce J.D. Comparative Field Analysis (CoMFA). The effect of shape on binding of steroids to carrier proteins. *J. Am. Chem. Soc.* 110: 5959–5967, 1988.
8. L. Dehaspe, H. Toivonen, and R.D. King. Finding frequent substructures in chemical compounds. In: *The Fourth International Conference on Knowledge Discovery and Data Mining*. AAAI Press, Menlo Park, Ca. 30-36, 1998.
9. L. Dehaspe, L. De Raedt. Mining Association Rules in Multiple Relations. In: *ILP 1997*: 125-132.
10. L. De Raedt. *Statistical and Relational Learning*. Springer, 2008.
11. L. De Raedt, J. Ramon. Deriving distance metrics from generality relations. *Pattern Recognition Letters* 30(3): 187-191 (2009).
12. R.O.Duda, P.E. Hart, and D.G. Stork. *Pattern Classification*. Wiley, 2001.
13. D. Enot and R.D. King. Application of inductive logic programming to structure-based drug design. Proceedings of the 7th European Conference on Principles and Practice of Knowledge Discovery in Databases (PKDD), 2003.
14. D. Eppstein. Subgraph isomorphism in planar graphs and related problems. In *Symposium on Discrete Algorithms*, pages 632-640, 1995.
15. P. Frasconi, A. Passerini. Learning with Kernels and Logical Representations. *Probabilistic Inductive Logic Programming*, 2008: 56 91.
16. T. Gärtner. A survey of kernels for structured data. *SIGKDD Explorations*, 5(18.1):49–58, 2003.

17. T. Gärtner, Peter A. Flach, and Stefan Wrobel. On graph kernels: Hardness results and efficient alternatives. In B. Schölkopf and M.K. Warmuth, editors, *COLT*, volume 2777 of *Lecture Notes in Computer Science*, pages 129–143. Springer, 2003.

18. J. Gasteiger and T. Engel. *Chemoinformatics: A Textbook*. Wiley-VCH, 2003.

19. C. Hansch, P.P. Malony, T. Fujiya, and R.M. Muir. Correlation of biological activity of phenoxyacetic acids with Hammett substituent constants and partition coefficients. *Nature* 194, 178-180, 1965.

20. H. Hofer, C. Borgelt, and M.R. Berthold. Large scale mining of molecular fragments with wildcards. In M.R. Berthold, H-J. Lenz, E. Bradley, R. Kruse, and C. Borgelt, editors, *IDA*, volume 2810 of *Lecture Notes in Computer Science*, pages 376–385. Springer, 2003.

21. C. Helma, T. Cramer, S. Kramer, and L. De Raedt. Data mining and machine learning techniques for the identification of mutagenicity inducing substructures and structure activity relationships of noncongeneric compounds. In *Journal of Chemical Information and Computer Systems 44*, pages 1402–1411, 2004.

22. T. Horváth and J. Ramon. Efficient frequent connected subgraph mining in graphs of bounded treewidth. In W. Daelemans, B. Goethals, and K. Morik, editors, *ECML/PKDD* (18.1), volume 5211 of *Lecture Notes in Computer Science*, pages 520–535. Springer, 2008.

23. T. Horváth, J. Ramon, and S. Wrobel. Frequent subgraph mining in outerplanar graphs. In *KDD*, pages 197–206. ACM, 2006.

24. J. Huan, W. Wang, and J. Prins. Efficient mining of frequent subgraphs in the presence of isomorphism. In *Proceedings of the Third IEEE International Conference on Data Mining (ICDM)*, pages 549–552. IEEE Press, 2003.

25. Jun Huan, Wei Wang, Jan Prins, and Jiong Yang. Spin: mining maximal frequent subgraphs from graph databases. In Won Kim, Ron Kohavi, Johannes Gehrke, and William DuMouchel, editors, *KDD*, pages 581–586. ACM, 2004.

26. Akihiro Inokuchi. Mining generalized substructures from a set of labeled graphs. In *ICDM*, pages 415–418. IEEE Computer Society, 2004.

27. A. Inokuchi, T. Washio, and H. Motoda. An apriori-based algorithm for mining frequent substructures from graph data. In *Proceedings of the 4th European Conference on Principles and Practice of Knowledge Discovery in Databases (PKDD)*, volume 1910 of *Lecture Notes in Artificial Intelligence*, pages 13–23. Springer-Verlag, 2000.

28. D.R. Jones. A taxonomy of global optimization methods based on response surfaces. *Journal of Global Optimization*, 21:345–383, 2001.

29. D.R. Jones and M. Schonlau. Efficient global optimization of expensive black-box functions. *Journal of Global Optimization*, 13(4):455–492, December 1998.

30. M. Kuramochi and G. Karypis. Frequent subgraph discovery. In *Proceedings of the First IEEE International Conference on Data Mining (ICDM)*, pages 313–320. IEEE Press, 2001.

31. J. Kazius, S. Nijssen, J.N. Kok, T. Bäck, and A. IJzerman. Substructure mining using elaborate chemical representation. In *Journal of Chemical Information and Modeling 46*, 2006.

32. R.D. King, S. Muggleton, R.A Lewis, and M.J.E Sternberg. Drug design by machine learning: The use of inductive logic programming to model the structure-activity relationships of trimethoprim analogues binding to dihydrofolate reductase. *Proc. Nat. Acad. Sci. U.S.A.* 89, 11322-11326, 1992.

33. R.D. King, S. Muggleton, A. Srinivasan, and M.J.E. Sternberg. Structure-activity relationships derived by machine learning: The use of atoms and their bond connectivities to predict mutagenicity by inductive logic programming. *Proc. Nat. Acad. Sci. USA* 93, 438-442, 1996.

34. R.D. King, J. Rowland, S.G. Oliver, M. Young, W. Aubrey, E. Byrne, M. Liakata, M. Markham, P. Pir, L.N. Soldatova, A. Sparkes, K.E. Whelan, A. Clare. The Automation of Science. *Science*. Vol. 324, no. 5923, pp. 85 - 89.

35. S. Kramer and L. De Raedt. Feature construction with version spaces for biochemical applications. In *ICML*, pages 258–265. Morgan Kaufmann, 2001.

36. S. Kramer, L. De Raedt, and C. Helma. Molecular feature mining in hiv data. In *KDD*, pages 136–143, 2001.

37. M. Kearns and S. Singh. Near-optimal reinforcement learning in polynomial time. In *Proc. 15th International Conf. on Machine Learning*, pages 260–268. Morgan Kaufmann, 1998.

38. H.J. Kushner. A new method of locating the maximum point of an arbitrary multipeak curve in the presence of noise. *Journal of Basic Engineering*, pages 97–106, March 1964.
39. A.R. Leach, and V.J. Gillet. *An Introduction to Chemoinformatics*, Kluwer, 2003.
40. A. Lingas. Subgraph isomorphism for biconnected outerplanar graphs in cubic time. *Theoretical Computer Science* 63, 295-302, 1989.
41. C.A. Lipinski, F. Lombardo, B.W. Dominy, and P. J. Feeney. Experimental and computational approaches to estimate solubility and permeability in drug discovery and development settings. *Adv. Drug Delivery Rev.*, 23(1-3), pp. 3-25, 1997.
42. D. Lizotte, T. Wang, M. Bowling, and D. Schuurmans. Automatic gait optimization with gaussian process regression. In *Proceedings of the 20th International Joint Conference on Artificial Intelligence*, pages 944–949, 2007.
43. Y.C. Martin. *Quantitative Drug Design: A Critical Introduction*, Marcel Dekker, 1978.
44. J. Matousek and R. Thomas. On the complexity of finding iso- and other morphisms for partial k−trees. Discrete mathemathics, 108(1-3), 343-364, 1992.
45. P.B. Medewar. *Advice to a Young Scientist*. BasicBooks. 1979.
46. S. Nijssen. Mining interpretable subgraphs. In *Proceedings of the International Workshop on Mining and Learning with Graphs (MLG)*, 2006.
47. S. Nijssen and J.N. Kok. A quickstart in frequent structure mining can make a difference. In *Proceedings of the 2004 International Conference on Knowledge Discovery and Data Mining (KDD)*, pages 647–652. ACM Press, 2004.
48. J. Ramon and S. Nijssen. Polynomial-delay enumeration of monotonic graph classes. *Journal of Machine Learning Research*, 2009.
49. M. J. Sasena. *Flexibility and Efficiency Enhancements for Constrained Global Design Optimization with Kriging Approximations*. PhD thesis, University of Michigan, 2002.
50. A. Schierz, and R.D. King. Drugs and Drug-like compounds: Discriminating Approved Pharmaceuticals from Screening Library Compounds. In *Pattern Recognition in Bioinformatics*, pages 331-343, 2009.
51. L. Schietgat, J. Ramon, M. Bruynooghe, H. Blockeel. An Efficiently Computable Graph-Based Metric for the Classification of Small Molecules. In *Discovery Science* 2008: 197-209.
52. S. V. N. Vishwanathan, N.N. Schraudolph, I.R. Kondor, and K.M. Borgwardt. Graph Kernels. *Journal of Machine Learning Research*, 2009.
53. N. Wale and G. Karypis. Comparison of descriptor spaces for chemical compound retrieval and classification. In *ICDM*, pages 678–689. IEEE Computer Society, 2006.
54. X. Yan and J. Han. gSpan: Graph-based substructure pattern mining. In *Proc. of the Second IEEE International Conference on Data Mining (ICDM)*, pages 721–724. IEEE Press, 2002.
55. X. Yan and J. Han. Closegraph: mining closed frequent graph patterns. In *KDD*, pages 286–295. ACM, 2003.
56. B. Zenko, and S. Dzeroski. Learning Classification Rules for Multiple Target Attributes. In *PAKDD*, pages 454-465, 2008.

Appendix

Table 4 Specific ring structures pre-coded in the MSG library

cyclopropane	2,3-dihydropyrrole	benzene
cyclopropene	2,5-dihydropyrrole	pyridine
aziridine	3,4-dihydropyrrole	1,2-dihydropyridine
diaziridine	pyrrolidine	1,4-dihydropyridine
azirine	furan	tetrahydropyridine
diazirine	1,3-dihydrofuran	piperidine
oxirane	2,5-dihydrofuran	4H-pyran
dioxirane	oxolane	2H-pyran
oxirene	1,2-dioxolane	dihydropyran
thiirane	1,3-dioxolane	aromatic_pyran
dithiirane	dioxole	oxane
thiirene	imidazole	thiane
oxathiirane	imidazolidine	dihydrothiopyran
oxaziridine	dihydroimidazole	pyridazine
thiaziridine	pyrazole	1,2-diazinane
dioxathiirane	pyrazoline	1,3-diazinane
cyclobutane	1,2,3-triazole	tetrahydropyridazine
cyclobutene	1,2,4-triazole	pyrimidine
cyclobutadiene	dihydrotriazole	dihydropyrimidine
azetidine	tetrazole	3H-pyrimidine
2,3-dihydroazete	1,3-oxazole	pyrazine
oxetane	1,2-oxazole	tetrahydropyrazine
1,2-dioxetane	dihydrooxazole	piperazine
1,3-dioxetane	1,3,4-oxadiazole	morpholine
thietane	1,2,5-oxadiazole	1,3-oxazinane
1,2-dithietane	1,2,4-oxadiazole	1,2-oxazinane
1,3-dithietane	thiazole	dihydro-1,2-oxazin
cyclopentane	1,3,4-thiadiazole	dihydro-1,3-oxazin
cyclopentene	1,2,5-thiadiazole	1,3-oxazin
cyclopentadiene	1,2,3-thiadiazole	1,3-thiazinane
thiolane	1,2,4-thiadiazole	thiomorpholine
1,2-dithiolane	dihydrothiazole	1,3-dithiane
1,3-dithiolane	thiazolidine	1,4-dithiane
1,2-dithiole	isothiazole	1,4-dioxane
1,3-dithiole	cyclohexane	1,3-dioxane
thiophene	cyclohexene	1,2-dioxane
2,3-dihydrothiophene	1,3-cyclohexadiene	1,4-dioxene
2,5-dihydrothiophene	1,4-cyclohexadiene	dihydrodioxin
pyrrole		triazine
		cycloheptane

Table 5 Specific functional groups pre-coded in the library

alkyl_halide	aryl-thioether	methoxy
aryl-halide	carboxylic acid	chain ether
carboxylic-acid halide	ester	aryl ether
hydroxyl	amide	imine
alcohol	other carbonyl	nitro
hetero aryl alcohol	0H-amine	aryl nitro
phenols	1H-amine	nitroso
aldehyde	2H-amine	aromatic nitroso
ketone	ammonium	azo
thiol	aromatic amine	aromatic azo
sulfonic acid	hydroxylamine	aliphatic chain length 5
sulfonyl	phosphoric acid	butyl
sulfone	phosphate	propyl
sulfonamide	phosphonate	ethyl
cyclic thioether	phosphinate	norm methyl
chain thioether	cyclic ether	haloalkane methyl
methylene single	haloalkane methylene	
methylene double	heteroatoms single bonded	
methylene valence		
aliphatic halide		

Table 6 Specific polycyclic structures pre-coded in the MSG library

benzocyclobutene	acridine	pyrrolizine
benzofuran	perimidine	pyridopyrimidine
indole	beta_carboline	oxanthrene
isoindole	pteridine	chromene
benzothiophene	phenoxazine	isochromene
benzimidazole	phenothiazine	naphthalene
indazole	phenazine	pentalene
benzoxazole	phenanthroline	indene
benzisoxazole	naphthyridine	as-indacene
benzothiazole	carbazole	s-indacene
purine	phthalazine	biphenylene
quinoline	1H-quinolizine	acenaphthylene
isoquinoline	9H,4H- quinolizine	fluorene
quinoxaline	2H-quinolizine	phenalene
quinazoline	indolizine	phenanthrene
cinnoline	pyrrolopyridine	anthracene

Author index